GENERAL AGRICULTURE

for

I.C.A.R. and other Agri Competitive Exams.

(J.R.F., S.R.F., A.R.S., Ph.D., M.Sc. Agri, IBPS, AFO, AO, IFS, NABARD)

- More than 4250 Solved Questions
- Last 18 years 1300 Solved Questions of JRF, Ph.D. (IARI) & BHU (Pre P.G.) Exams.

INCLUDING LATEST CURRENT AFFAIRS

Written by :-

Muniraj Singh Rathore

2024-25

JAIN BROTHERS (New Delhi)

16/873, East Park Road, Karol Bagh,
Near Dr. Joshi Hospital, NEW DELHI-110 005
Phones : (011) 23618426, 45097398, Mob.: 9871078971
E-mail : jain_bros@hotmail.com
Website : www.thejainbrothers.com

Published by :
Shri Sunil Kumar Jain
for **JAIN BROTHERS (New Delhi)**
16/873, East Park Road, Karol Bagh, New Delhi - 110 005.
Phones: 011-23618426, 45097398, Fax: 23638444

Distributor :

1. **M/s Sunil Book Service**
   Fatima Nagar, Opp. Andhra Bank,
   KAZIPET-506004 (A.P.)
   Ph.: 0870-2459617,
   Mob.: 9390115435
   E-mail: kpkulshrestha@yahoo.com
2. **M/s Jain Brothers**
   Vidya Vihar, Pilani (Rajasthan),
   Phone: 01596-242144

Available at:
Pusa Agricultural Book Service
Shishir Hostel, IARI, PUSA
New Delhi
Ph.: 011-25841048

First Published	: 2004
Second edition	: 2005
Eighth edition	: Oct., 2008
Ninth edition	: Jan., 2009
Tenth edition	: March, 2009
Twentieth edition	: 2014
Twenty first edition	: Oct. 2014
Twenty second edition	: Feb., 2015
Twenty third edition	: Sep., 2015
Twenty fourth edition	: Sep., 2016
Twenty fifth edition	: June., 2017
Twenty sixth edition	: Nov., 2017
Twenty seventh edition	: April, 2018
Twenty eighth edition	: Oct, 2018
Twenty ninth edition	: April, 2019
Reprinted with corrections	: Sep, 2019
Thirty edition	: June, 2020
Thirty first edition	: Jan., 2021
Thirty second edition	: July., 2021
Thirty third edition	: July., 2022
Thirty fourth edition	: June., 2023
Thirty fifth edition	: Oct., 2024

ISBN : 978-81-8360-141-2

Price : ₹ 375.00

Laser Typesetting by :
Niyati *Graphics*
754/23 Khajoor Road, Karol Bagh, New Delhi

Printed by:
Balaji Offset Printers
New Delhi

Preface

There is no text book covering comprehensively the content of all syllabus of general agriculture for M.Sc. & Ph.D. entrance examinations of Agricultural Universities. Students have to depend on the various books, monographs and periodicals to collect the information for various subjects offered in M.Sc. and Ph.D. entrance examinations for general agriculture. The present book is written to satisfy the various subjects syllabus which needs in M.Sc. & Ph.D. entrance examinations in a systematic manner.

I hope that the latest 35th revised edition of this book has covered all the general agricultural questions in a systematic manner according to syllabus to the view of competitive examinations. I would like to express my heartfelt gratitude's to my best friend Amit (Ph.D, Soil science, I.A.R.I), Dr. Ashok (Ph. D. Agronomy, I.A.R.I.), Dr. Deepak (Asst. Proff., Biotech), Dr. Manish Kumar Singh (Associate Prof. Agriculture Economics, Udai Pratap Autonomous College, Varanasi) & my best friend Nimisha Sharma (M.Sc. Agri), without whose whole hearted support and inspiration, this book would have never seen the light of the day.

I am thankful to Miss Niyati Jain for her support in adding questions and checking all the pages to remove errors and improving matter.

Besides this I am also thankful to the students of I.A.R.I. & IABM Bikaner, and Staff of Rajasthan Agriculture University,

Suggestions of students and teachers are always welcome.

Muniraj Singh Rathore
Rajasthan Agriculture University,
Bikaner (Raj.)
E-mail: - royalraj4@rediffmail.com

This Book is Specially
Dedicated to My Father
&
My Loving Son Manku

CONTENTS

(VI) EXTENSION

(VII) STATISTICS AND AGRICULTURAL ECONOMICS

(VIII) ENVIRONMENTAL SCIENCE AND METEOROLOGY

STATISTICAL DATA ABOUT AGRICULTURE PRODUCTION AND AREA

Note- *Area & Production data FY 2022-23 and 23-24 is based on provision and 1st & 2nd advance estimation (source- Dep. Of Agriculture and statistics)*

Q. 1. Total Targeted food grain production of 2024-25

Ans. 340.40 million tones

Q. 2. Total Targeted Kharif food grain production of 2024-25-

Ans. 159.97 million tones

Q.3. Total Targeted Rabi food grain production of 2024-25-

Ans. 164 million tones

Q.4. Total Targeted Zaid (Summer) food grain production of 2024-25

Ans. 16.43 million tones

Q.5. India's (basis on 3rd advance est) total food production in 2023-24

Ans. 3228.52 lac tones

Q.6. India's (basis on 3rd advance est) total Kharif food grain production in 2023-24

Ans. 1566.76 lac tones

Q.7. India's (basis on 3rd advance est) total Rabi food grain production in 2023-24

Ans. 1576.56 lac tones

Q.8. India's (basis on 3rd advance est) total Summer production in 2023-24 of food grains.

Ans. 145.19 lac tones

Q.9. Total Kharif food grain production of 2022-23—(basis on 3rd advance est)

Ans. 155.71 million tones

Q.10. Total Rabi food grain production of 2022-23—(basis on 3rd advance est)

Ans. 157.84 million tones

Q.11. Total food grain production of 2022-23—(basis on 3rd advance est)

Ans. 329.687 million tones

Q.12. Total Cereals production of 2022-23-(basis on 3rd advance est)

Ans. 303.62 million tones

Q.13. Total Rice production of 2022-23-(basis on 3rd advance est.)

Ans. 135.75 million tones

Q.14. Total Wheat production of 2022-23-(basis on 3rd advance est)

Ans. 110.55 million tones

Q.15. Total Maize production of 2022-23-(basis on 3rd advance est)

Ans. 38.08 million tones

Q.16. Total Pulses production of 2022-23-(basis on 3rd advance est)

Ans. 26.05 million tones

Q.17. Total Oil Seeds production of 2022-23-(basis on 3rd advance est)

Ans. 41.35 million tones

Q.18. Total Cotton production of 2022-23-(basis on 3rd advance est)

Ans. 33.66 million Bales

Q.19. Total Jute production of 2022-23 (basis on 3rd advance est)

Ans. 8.98 million Bales

Q.20. Total Sugarcane production of 2022-23-(basis on 3rd advance est)

Ans. 490.53 million tones

Q.21. Total Kharif food grains production in 2021-22

Ans. 155.36 million tones

Q.22. Total Rabi food grains production in 2021-22

Ans. 160.25 million tones

Q.23. Total Food grain production in 2021-22

Ans. 315.62 million tones

Q.24. Growth in total Food grain production from FY 21-22

Ans. 14.07 million tones

Q.25. Total Food grains production in 2020-21

Ans. 310.74 million tones

Q.26. Total Kharif food grains production in 2020-21

Ans. 150.58 million tones

Q.27. Total Rabi food grains production in 2020-21

Ans. 160.17 million tones

Q.28. Target of 12th Five-year plan for food grains production

Ans. 1308.44 million tones

Q.29. Average yield Kg. /ha of Total food grains in 2021-22

Ans. 2425 kg./ha

Q.30. Average yield Kg. /ha of Total food grains in 2022-23 (Provision)

Ans. 2494 kg./ha

Q.31. Productivity increased in Total food grain from the year 2021-22

Ans. +69 kg/ ha.

Q.32. Average yield Kg. /ha of Kharif food grains in 2021-22

Ans. 2131 kg./ha

Q.33. Average yield Kg. /ha of Kharif food grains in 2022-23 (Provision)

Ans. 2203 kg./ha

Q.34. Productivity increased in Kharif food grain from the year 2021-22

Ans. +72 kg/ ha.

Q.35. Average yield Kg. /ha of Rabi food grains in 2021-22

Ans. 2798 kg./ha

Q.36. Average yield Kg. /ha of Rabi food grains in 2022-23(Provision)

Ans. 2828 kg./ha

Q.37. Productivity increased in Rabi food grains from the year 2021-22

Ans. +30 kg/ ha.

Q.38. Average yield Kg. /ha of total Cereals in 2021-22

Ans. 2899 kg./ha

Q.39. Average yield Kg. /ha of total Cereals in 2022-23- (Provision)

Ans. 2939 kg./ha

Q.40. Productivity increased in Cereals grains from the year 2021-22

Ans. +40 kg/ ha.

Q.41. Average yield Kg/ha of Rice in 2021-22

Ans. 2798 Kg./ha.

Q.42. Average yield Kg. /ha of Rice in 2022-23- (Provision)

Ans. 2838 kg./ha

Q.43. Productivity Increased in Rice from the year 2021-22

Ans. +40 kg/ ha.

Q.44. Average yield Kg/ha of Wheat in 2021-22

Ans. 3537 Kg./ha.

Q.45. Average yield Kg. /ha of Wheat in 2022-23- (Provision)

Ans. 3521 kg./ha

Q.46. Productivity decreased in Wheat from the year 2021-22

Ans. -16 kg/ ha.

Q.47. Average yield Kg./ha of Maize in 2021-22

Ans. 3387 kg./ha

Q.48. Average yield Kg. /ha of Maize in 2022-23- (Provision)

Ans. 3545 kg./ha

Q.49. Productivity increased in Maize from the year 2021-22

Ans. +158 kg/ ha.

Q.50. Average yield Kg./ha of Cotton in 2021-22

Ans. 428 Kg/ha

Q.51. Average yield Kg. /ha of Cotton in 2022-23- (Provision)

Ans. 443 Kg/ha

Q.52. Productivity increased in Cotton from the year 2021-22-

Ans. +15 Kg/ ha.

Q.53. Average yield Kg./ha of Sugarcane in 2021-22

Ans. 84906 kg./ha

Q.54. Average yield Kg. /ha of Sugarcane in 2022-23- (Provision)

Ans. 83349 Kg/ha

Q.55. Productivity Decreased in Sugarcane from year 2021-22

Ans. -1557 Kg/ ha.

Q.56. Average yield kg./ha of Total Pulses in 2021-22

Ans. 888 kg./ha

Q.57. Average yield kg. /ha of Pulses in 2022-23- (Provision)

Ans. 902 Kg/ha

Q.58. Productivity increased in Pulses from the year 2021-22

Ans. +14 kg/ ha.

Q.59. Average yield kg./ha of Gram in 2021-22

Ans. 1261 kg./ha

Q.60. Average yield kg. /ha of Gram in 2022-23- (Provision)

Ans. 1172 Kg/ha

Q.61. Productivity Decreased in Gram from the year 2021-22-

Ans. -89 Kg/ ha

Q.62. Average yield kg./ha of Ground nut in 2021-22

Ans. 1777 kg./ha

Q.63. Average yield kg. /ha of Ground nut in 2022-23- (Provision)

Ans. 2075 Kg/ha

Q.64. Productivity increased in Groundnut from the year 2021-22

Ans. +298 kg/ ha.

Q.65. Average yield kg./ha of total Nine Oil seeds in 2021-22

Ans. 1312 kg./ha

Q.66. Average yield kg. /ha of total Nine Oil seeds in 2022-23- (Provision)

Ans. 1368 Kg/ha

Q.67. Productivity increased in total Nine oil seeds from the year 2021-22

Ans. +56 kg/ ha.

Q.68. Gross area under Total Food grains in 2021-22
Ans. 130.17 M. ha.

Q.69. Gross area under Total Food grains in 2022-23 (Provision)
Ans. 132.20 M. ha.

Q.70. Area Increased in Food grains from the year 2021-22
Ans. +2.03 M. ha.

Q.71. Gross area under Total Kharif Food grains in 2021-22
Ans. 72.90 M. ha.

Q.72. Gross area under Total Kharif Food grains in 2022-23 (Provision)
Ans. 70.70 M. ha.

Q.73. Area Decreased in Kharif Food grains from the year 2021-22
Ans. -2.2 M. ha.

Q.74. Gross area under Total Rabi Food grains in 2021-22
Ans. 57.27 M. ha.

Q.75. Gross area under Total Rabi Food grains in 2022-23 (Provision)
Ans. 61.51 M. ha.

Q.76. Area Increased in Rabi Food grains from the year 2021-22
Ans. +4.24 M. ha.

Q.77. Gross area under Rice of India in 2021-22
Ans. 46.27 M. ha.

Q.78. Gross area under Rice of India in 2022-23-(Provision)
Ans. 47.83 M. ha.

Q.79. Gross area under Wheat in 2021-22
Ans. 30.45 M. ha.

Q.80. Gross area under Wheat of India in 2022-23-(Provision)
Ans. 31.40 M. ha.

Q.81. Zone which has the highest area in Wheat in India
Ans. Northwestern Plain Zone

Q.82. Gross area under Maize in 2021-22
Ans. 9.95 M. ha.

Q.83. Gross area under Maize of India in 2022-23-(Provision)
Ans. 10.74 M. ha.

Q.84. Gross area under Total Cereals in 2021-22
Ans. 99.43 M. ha.

Q.85. Gross area under Total Cereals in 2022-23-(Provision)
Ans. 103.30 M. ha

Q.86. Gross area under Total Pulses in 2021-22
Ans. 30.73 M. ha.

Q.87. Gross area under Total Pulses in 2022-23-(Provision)
Ans. 28.90 M. ha.

Q.88. Gross area under Gram in 2021-22
Ans. 10.74 M. ha.

Q.89. Gross area under Gram in 2022-23-(Provision)
Ans. 10.47 M. ha.

Q.90. Gross area under Sorghum (Jowar) in 2022-23-(Provision)
Ans. 3.53 M. ha.

Q.91. Gross area under Pearl Millet (Bajra) in 2022-23(Provision)
Ans. 7.5 M. ha.

Q.92. Gross area under Total Oil Seeds in 2021-22
Ans. 28.94 M. ha.

Q.93. Gross area under Total Oil Seeds in 2022-23-(Provision)
Ans. 30.23 M. ha.

Q.94. Gross area under Ground nut in 2021-22
Ans. 5.70 M. ha.

Q.95. Gross area under Ground nut in 2022-23 (Provision)
Ans. 4.96 M. ha.

Q.96. Gross area under Cotton in 2021-22
Ans. 12.37 M. ha.

Q.97. Gross area under Cotton in 2022-23-(Provision)
Ans. 12.92 M. ha.

Q.98. Gross area under Sugarcane in 2021-22
Ans. 5.17 M. ha.

Q.99. Gross area under Sugarcane in 2022-23-(Provision)
Ans. 5.88 M. ha.

Q.100. The gross area under Jute in 2021-22

Ans. 0.629 M. ha.

Q.101. The gross area under Jute in 2022-23-(Provision)

Ans. 0.619 M. ha.

Q.102. All India Organic certified land area registered under NPOP in 2022-23

Ans. 10.17 M.ha

Q.103. The state having the highest area under certified organic farming

Ans. M.P.

Q.104. The state having the second largest area in certified organic farming

Ans. Maharashtra & 3rd Raj.

Q.105. India's highest organic farm export to the country in 2022-23

Ans. The European Union and 2nd U.S.A.

Q.106. The top two India's exporting organic products are

Ans. Processed Food & Oilseeds

Q.107. Agricultural and Processed Food Products Export Development Authority (APEDA) was established in Year

Ans. Dec 1985

Q.108. The Head Office of APEDA is located at

Ans. New Delhi

Q.109. Cereal Crop which has been included in the Second Schedule of APEDA Act

Ans. Basmati Rice

Q.110. How much percent cover organic farming of total net sown area in 2022-23

Ans. 3.92% approx.

Q.111. Country in terms of largest certified organic area

Ans. Australia

Q.112. Country having highest organic producers

Ans. India

Q.113. Sate having highest production of Bio Fertilizers in (2020-21)

Ans. Tamil Nadu.

Q.114. Which country is World leader in Rice production in 2021-22 (Provision)

Ans. China & 2nd India

Q.115. Which country is World leader in Wheat production in 2021-22 (Provision)

Ans. China & 2nd India

Q.116. Which country is World leader in Total Cereals production in 2021-(Provision)

Ans. China & 2nd USA

Q.117. Which country is World leader in Total Pulses production in 2021-22(Provision)

Ans. India & 2nd China

Q.118. Which country is World leader in Maize production in 2021-(Provision)

Ans. USA & 2nd China

Q.119. Contribution of USA in world Maize production in 2021—(Provision)

Ans. 31.72%

Q.120. Which country is World leader in Sorghum production in 2021- (Provision)

Ans. USA & 2nd Nigeria

Q.121. Which country is World leader in Barley production in 2021-(Provision)

Ans. Russia & 2nd Australia

Q.122. Which country is World leader in Oat production in 2021- (Provision)

Ans. Russia & 2nd Canada

Q.123. Which country is World leader in Ground nut production in 2021-(Provision)

Ans. China & 2nd India

Q.124. Which country is World leader in Rape seed production in 2021-(Provision)

Ans. China & 2nd Canada

Q.125. Which country is World leader in Soyabean production in 2021-(Provision)

Ans. 1st Brazil 2nd USA

Q.126. Country leader in Sunflower Seed production in 2021-(Provision)

Ans. Ukraine & 2nd Russia

Q.127. Which country is World leader in Sugarcane production in 2021-(Provision)

Ans. Brazil & 2nd India

Q.128. Which country is World leader in Jute & Fiber production in 2021-(Provision)

Ans. India & 2nd Bangladesh

Q.129. Which country is World leader in Seed Cotton production in 2021—(Provision)

Ans. China & 2nd India

Q.130. Which country is World leader in Tobacco leaves production in 2021—(Provision)

Ans. China & 2nd India

Q.131. Which country is World leader in Vegetable production in 2021-(Provision)

Ans. China & 2nd India

Q.132. Country leader in Ginger production in 2021 (Provision)

Ans. India

Q.133. Country leader in Okra production in 2021-(Provision)

Ans. India

Q.134. Country leader in Tomato production in 2021-(Provision)

Ans. China & 2nd India

Q.135. Country leader in Green Peas production in 2021—(Provision)

Ans. China & 2nd India

Q.136. Country leader in Cauliflower & Broccoli production in 2021—(Provision)

Ans. China & 2nd India

Q.137. Country leader in Cabbage production in 2021—(Provision)

Ans. China & 2nd India

Q.138. Which country is World leader in Potato production in 2021—(Provision)

Ans. China & 2nd India

Q.139. Which country is World leader in Onion (Dry) production in 2021-(Provision)

Ans. India & 2nd China

Q.140. Country leader in Pumpkin, Squash & Gourd production in 2021-(Provision)

Ans. China & 2nd India

Q.141. Country leader in Eggplant production in 2021-(Provision)

Ans. China & 2nd India

Q.142. Which country is World leader in Fruit production in 2021—(Provision)

Ans. China

Q.143. Country leader in Banana production in 2021—(Provision)

Ans. India & 2nd China

Q.144. Country leader in Mango production in 2021—(Provision)

Ans. India & 2nd China

Q.145. Country leader in Papaya production in 2021—(Provision)

Ans. India & Brazil

Q.146. Country leader in Lemon production in 2021—(Provision)

Ans. India & 2nd Mexico

Q.147. Country leader in Guava production in 2021-(Provision)

Ans. India & 2nd Indonesia

Q.148. Country leader in Oranges production in 2021—(Provision)-

Ans. Brazil & 2nd India

Q.149. Country leader in Coconut production in 2021—(Provision)

Ans. Indonesia & 2nd Philippines

Q.150. Country leader in Cashew nut production in 2021—(Provision)

Ans. Ivory Coast & 2nd India

Q.151. Which country is World leader in Tea production in 2021-(Provision)

Ans. China

Q.152. Which country is World leader in Coffee (Green) production in 2021-(Provision)

Ans. Brazil & 2nd Vietnam

Q.153. Which country is World leader in Natural Rubber production in 2021-(Provision)

Ans. Thailand & 2nd Indonesia

Q.154. Country leader in Black pepper production in 2021-(Provision)

Ans. Vietnam & 2nd Brazil

Q.155. Country leader in Chili pepper production in 2021-(Provision)-

Ans. China & 2nd Turkey

Q.156. Largest producer of Olive in world in 2021-(Provision)

Ans. Spain and 2nd Italy

Q.157. Country leader in Egg production in 2021-(Provision)

Ans. China & 2nd India

Q.158. Country leader in Meat production in 2021- -(Provision)

Ans. China & 2nd USA

Q.159. Country leader in Cow Milk production in 2021-(Provision)

Ans. India & 2nd USA

Q.160. Country leader in total Milk production in 2021-(Provision)

Ans. India & 2nd USA

Q.161. Country leader in Buffalo Milk production

Ans. India & 2nd Pakistan

Q.162. Country leader in Goat Milk production

Ans. India & 2nd Bangladesh

Q.163. India's Rank in world in Total Cereals production in 2021-(Provision)

Ans. Third

Q.164. India's Rank in world in Rape seed production in 2021—(Provision)

Ans. Third

Q.165. India's Rank in world in Sugarcane production in 2021—(Provision)

Ans. Second

Q.166. India's Rank in world in Tobacco leaves production in 2021—(Provision)

Ans. Second

Q.167. India's rank in lentil production in 2021—(Provision)

Ans. Second Rank

Q.168. India's Rank in world in Vegetable's production in 2021—(Provision)

Ans. Second

Q.169. India's Rank in world in Fruit production in 2021—(Provision)

Ans. Second

Q.170. India's Rank in world in Coconut production in 2021—(Provision)

Ans. Third

Q.171. India's Rank in world in Coffee (Green) production in 2021—(Provision)

Ans. Ninth

Q.172. India's Rank in the world in Tea (Leaves) production in 2021—(Provision)

Ans. Second (First in China)

Q.173. India's Rank in world in Species production in 2021—(Provision)

Ans. First

Q.174. Which country highest tractor density in the world

Ans. Slovenia

Q.175. The country which is first in transgenic plants in 2024

Ans. USA & 2nd Brazil

Q.176. India's rank in transgenic plants in 2024

Ans. Fifth

Q.177. The largest consumer of Cashew

Ans. India (40%)

Q.178. The largest consumer of Tea

Ans. India

Q.179. Country First in Beef Cattle population

Ans. India & 2nd Brazil

Q.180. Global percent contribution of cattle in total milk production

Ans. 83%

Q.181. Global percent contribution of buffaloes in total milk production

Ans. 14%

Q.182. Global percent contribution of Goats in total milk production

Ans. 2%

Q.183. Global percent contribution of Sheep in total milk production

Ans. 1%

Q.184. World milk production in 2023

Ans. 965.57 million tones

Q.185. Percent contribution of India's in world milk production 2023-24

Ans. 24.64%

Q.186. Country first in Buffalo population

Ans. India (56.7%)

Q.187. India's rank in Egg production in 2022-23

Ans. Second

Q.188. India's rank in total Meat production-2022-23
Ans. Fifth

Q.189. India's rank in Goats population in world
Ans. Second (1st China)

Q.190. India's rank in Sheep population in world
Ans. Third (1st China)

Q.191. India's rank in Camels population in world
Ans. Sixteenth (1st Chad)

Q.192. India's rank in Chicken population in world
Ans. Sixth (1st China)

Q.193. India's rank in arable land in world
Ans. Second (1st USA)

Q.194. India's rank in rural population in world
Ans. First

Q.195. Country having highest productivity in Cereals in 2021 (P)
Ans. New Zealand (8728 Kg/ Ha)

Q.196. Country having highest productivity in Pulses in 2021-(P)
Ans. Belgium & 2nd Denmark

Q.197. Country having highest productivity in Paddy in 2021-(P)
Ans. Egypt & 2nd Australia

Q.198. Country having highest productivity in Wheat in 2021-(P)
Ans. New Zealand

Q.199. Country having highest productivity in Maize in 2021 –(P)
Ans. Israel

Q.200. Country having highest productivity in Soya bean in 2021-(P)
Ans. Turkey & 2nd U.S.A

Q.201. Country having highest GDP/ Capita 2021 (Current Price in USA Dollars)
Ans. USA & 2nd China

Q.202. Country having highest Biofuel production in 2021
Ans. U.S.A. & 2nd Brazil

Q.203. World Rice production in 2021-(P)
Ans. 787.29 million tones

Q.204. World Wheat production in 2021-(P)
Ans. 770.87 million tones

Q.205. World Maize production in 2021-(P)
Ans. 1210.23 million tones

Q.206. World Cereals production in 2021-(P)
Ans. 3070.64 million tones

Q.207. World Pulses production in 2021-(P)
Ans. 88.96 million tones

Q.208. World Fruit Production in 2021- (P)-
Ans. 909.64 million tones

Q.209. World Vegetables Production in 2021-(P)
Ans. 1154.60 million tones

Q.210. World Meat production in 2022(P)
Ans. 360.61 million tones

Q.211. India's Meat production in 2022-23(P)
Ans. 9.77 million tones

Q.212. Highest contribution in Indian meat industries is by
Ans. Poultry (36%)

Q.213. PMART is related to
Ans. To promote hygienic meat production

Q.214. Rashtriya Gokul Mission is related to
Ans. Development and conservation of indigenous bovine breeds

Q.215. Rashtriya Gokul Mission has been initiated in-
Ans. December 2014

Q.216. The Animal Welfare Board of India (AWBI) was set up in
Ans. 1962

Q.217. The Founder & Chairperson of the Animal Welfare Board of India (AWBI) was
Ans. Smt. Rukmini Devi Arundale

Q.218. "Nationwide AHDF KCC Campaign duration was
Ans. From 15th November 2021 to 15th February 2022
Compaign for 2023-24 (Started on 1st May, 2023, 30th March, 2024)

Q.219. "The National Programme for Dairy Development (NPDD) started across the country
Ans. Since Feb-2014

Q.220. One of the highest National Awards in the field of livestock and dairy sector is

Ans. Gopal Ratna Award (2021)

Q.221. Ministry of Fisheries, Animal Husbandry and Dairying a newly formed ministry in India was formed in

Ans. May 2019

Q.222. Union Minister of Ministry of Fisheries, Animal Husbandry and Dairying is-

Ans. Sh. Rajiv Ranan Singh known as Shri Lalan Singh

Q.223. Total Regional Fodder Stations in India are

Ans. Eight Stations

Q.224. Indian Veterinary Council Act was established in

Ans. 1984

Q.225. World Dairy production in 2022 (Estimated)

Ans. 930 million tones

Q.226. Dairy contribution in national economy

Ans. 5% Approx

Q.227. World Fish production in 2023(Est)

Ans. 186.6 million MT

Q.228. Country leader in Fish production 2023-24-

Ans. China & 2nd Indonesia

Q.229. India's rank in fish production 2023-24

Ans. Third

Q.230. Country leader in Poultry production-

Ans. U.S.A.

Q.231. The country having highest consumption of kilo Cal/ Capita/ Day in 2023

Ans. Bahrain & 2nd U.S.A.

Q.232. Country having highest consumption of fruit and vegetables per capita

Ans. China

Q.233. India's consumption of kilo Cal/Capita/ Day in 2018

Ans. 2533 K. Cal/Capita

Q.234. World consumption of kilo Cal/Capita/ Day in 2018

Ans. 2861 K. Cal/ Capita

Q.235. Country having lowest density people per Sq. km-

Ans. Australia (3 person/sq/km)

Q.236. Country having highest people density per Sq./ km-

Ans. Bangladesh (1188 person/sq/km)

Q.237. Country having highest percentage of employment in Agriculture in 2021

Ans. Uganda (62.9%)

Q.238. Country having lowest percentage of employment in Agriculture in 2021

Ans. Israel & Belgium (0.9%)

Q.239. India's percent share of employment in Agriculture in 2021

Ans. 44%

Q.240. World average percent share of employment in Agriculture in 2021

Ans. 26.4%

Q.241. India's population density of per/sq/km. in 2021

Ans. 473.4 Person/sq/km

Q.242. Country having the highest population density per/sq/ km. in 2021

Ans. Bangladesh (1301)

Q.243. Country having the Lowest population density of per/ sq/km. in 2021

Ans. Australia (3.3)

Q.244. World population density of per/sq/km. in 2021

Ans. 60.8 (Taking ino account land area)

Q.245. Country leader in Rock phosphate and Sulphur production

Ans. China

Q.246. World nutrient (N+P+K) Production in 2021

Ans. 213.46 Million Tones

Q.247. World nutrient (N+P+K) Consumption in 2021.

Ans. 195.36 Million Tones

Q.248. Country leader in total nutrient (N+P+K) production in 2021

Ans. China & 2nd India

Q.249. Country leader in total nutrient (N+P+K) consumption in 2021

Ans. China

Q.250. Fertilizer major nutrient for which India is solely dependent on import

Ans. K_2O

Q.251. Country which is highest importer of N+ P +K in 2021

Ans. Brazil

Q.252. Country which is highest exporter of N+ P +K in 2021

Ans. Russian Federation

Q.253. Country leader in Urea production

Ans. China

Q.254. India's rank in Urea production

Ans. 2nd rank

Q.255. Country leader in DAP production

Ans. China

Q.256. India's rank in DAP production

Ans. 2nd rank

Q.257. Country leader in Nitrogen & Phosphate Consumption

Ans. China

Q.258. Country leader in Potash Consumption

Ans. China

Q.259. India's rank in Nitrogen Consumption

Ans. 2nd rank

Q.260. India's rank in Phosphate Consumption

Ans. 2nd rank

Q.261. India's rank in Potash Consumption

Ans. 4th rank

Q.262. India's rank in N + P +K consumption

Ans. 2nd rank

Q.263. India's rank in (N + P) Production

Ans. 2nd rank

Q.264. Food grain crop having the highest Area and Production in 2022-23 in India

Ans. Rice & 2nd Wheat

Q.265. State having highest Area under total Food Grains production in 2021-22 (P)

Ans. U.P. & 2nd M.P.

Q.266. State having highest Area under Rice production in 2021-22 (P)

Ans. U.P. & 2nd W. Bengal

Q.267. State having highest Area under Wheat production in 2021-22 (P)

Ans. U.P. & 2nd M.P.

Q.268. State having highest Area under Maize production in 2021-22 (P)

Ans. Karnataka 2nd M.P.

Q.269. State having highest Area under Rabi Maize production in 2021-22 (P)

Ans. Bihar & 2nd Maharashtra.

Q.270. State having highest Area under Barley production in 2021-22 (P)

Ans. Rajasthan

Q.271. State having highest Area under Pearl millet production in 2021-22 (P)

Ans. Rajasthan

Q.272. State having highest Area under Sorghum production in 2021-22 (P)

Ans. Maharashtra

Q.273. State having highest Area under Small millet production in 2021-22 (P)

Ans. M.P. & 2nd Maharashtra.

Q.274. State having highest Area under total Cereal production in 2021-22 (P)

Ans. U.P. and 2nd M.P.

Q.275. State having highest Area under Gram production in 2021-22-(P)

Ans. Maharashtra & 2nd Raj.

Q.276. State having highest Area under Pigeon Pea production in 2021-22(P)

Ans. Karnataka & 2nd Maharashtra

Q.277. State having highest Area under total Pulses production in 2021-22-(P)

Ans. Rajasthan

Q.278. State having highest Area under Ground nut production in 2021-22-(P)

Ans. Gujarat

Q.279. State having highest Area under Soya bean production in 2021-22-(P)

Ans. M.P. & 2nd Maharashtra

Q.280. State having highest Area under Sugarcane production in 2021-22-(P)

Ans. U.P. & 2nd Maharashtra

Q.281. State having highest Area under Cotton production in 2021-22 (P)

Ans. Maharashtra& 2nd Gujarat

Q.282. State having highest Area under Jute production in 2021-22 (P)

Ans. West Bengal

Q.283. State having highest Area under Mustard production in 2021-22 (P)

Ans. Rajasthan

Q.284. State having highest Area under total Oil seeds production in 2021-22 (P)

Ans. M.P. & 2nd Raj.

Q.285. Which State was No. 1 for Total Food grains production in 2021-22- (P)

Ans. U.P.

Q.286. Which State was No. 1 for Rice production in 2021-22- (P)

Ans. West Bengal

Q.287. Which State was No. 1 for Wheat production in 2021-22 (P)

Ans. U.P.

Q.288. Which State was No. 1 for Maize production in 2021-22 (P)

Ans. Karnataka

Q.289. State No. 1 in Rabi maize production in 2021-22 (P)-

Ans. Bihar

Q.290. Which State was No. 1 for Pearl millet production in 2021-22 (P)

Ans. Rajasthan

Q.291. Which State was No. 1 for total cereals crop production in 2021-22 (P)

Ans. U.P.

Q.292. Which State was No. 1 for Gram production in 2021-22 (P)

Ans. Maharashtra

Q.293. Which State was No. 1 for Pigeon Pea production in 2021-22 (P)

Ans. Maharashtra

Q.294. Which State was No. 1 for total Pulses production in 2021-22 (P)

Ans. M.P.

Q.295. Which State was No. 1 for Ground nut production in 2021-22-(P)

Ans. Gujarat

Q.296. Which State was No. 1 for Soya bean production in 2021-22-(P)

Ans. Maharashtra

Q.296a.Stae No. 1 in Soyabean in 2022-23

Ans. M.P. surpassing Maharashra.

Q.297. Which State was No. 1 for total Oil seed production in 2021-22 (P)

Ans. Rajasthan

Q.298. Which State was No. 1 for Sugarcane production in 2021-22-(P)

Ans. U.P.

Q.299. Which State was No. 1 for Cotton production in 2021-22-(P)

Ans. Maharashtra

Q.300. Which state was No.1 for Mustard production in 2021-22-(P)

Ans. Rajasthan

Q.301. Which state was No.1 for Sunflower production in 2021-22-(P)

Ans. Karnataka

Q.302. Which state was No.1 for Sorghum production in 2021-22-(P)

Ans. Maharashtra

Q.303. Which state was No.1 for Barley production in 2021-22-(P)

Ans. Rajasthan

Q.304. Which state was No.1 for Ragi production in 2021-22-(P)

Ans. Karnataka

Q.305. Which state was No.1 for small millets production in 2021-22-(P)

Ans. M.P.

Q.306. Which state was No.1 for Jute production in 2021-22-(P)

Ans. West Bengal

Q.307. Which state was No.1 for Tobacco production in 2021-22 (P)

Ans. Gujarat

Q.308. Which state was No.2nd for total Food grain production in 2021-22

Ans. M.P.

Q.309. Which state was No.2nd for Rice production in 2021-22- (P)

Ans. U.P.

Q.310. Which state was No.2nd for Wheat production in 2021-22- (P)

Ans. M.P.

Q.311. Which state was No.2nd for Gram production in 2021-22 (P)

Ans. M.P. & 3rd Rajasthan

Q.312. Which State was No. 2nd for Pigeon Pea production in 2021-22 (P)

Ans. Karnataka

Q.313. Which state was No.2nd for total Pulses production in 2021-22-(P)

Ans. Maharashtra

Q.314. Which state was No.2nd for Soya bean production in 2021-22(P)

Ans. M.P.

Q.315. Which state was No.2nd for Ground nut production in 2021-22-(P)

Ans. Rajasthan

Q.316. Which State was No. 2nd for total Oil seed production in 2021-22 (P)

Ans. M.P.

Q.317. Which state was No.2nd for Mustard production in 2021-22-(P)

Ans. M.P.

Q.318. Which state was No.2nd for Sunflower production in 2021-22-(P)

Ans. Telangana

Q.319. Which state was No.2nd for Sorghum production in 2021-22-(P)

Ans. Karnataka

Q.320. Which state was No.2nd for Barley production in 2021-22-(P)

Ans. U.P.

Q.321. Which state was No.2nd for Sugarcane production in 2021-22-(P)

Ans. Maharashtra

Q.322. Which state was No.2nd for Cotton production in 2021-22 (P)

Ans. Gujarat

Q.323. Which State was No. 2nd **for Maize production in 2021-22 (P)**

Ans. M.P.

Q.324. Which State was No. 2nd for Pearl millet production in 2021-22 (P)

Ans. U.P.

Q.325. Which State was No. 2nd for total Cereals crop production in 2021-22 (P)

Ans. M.P.

Q.326. Which state was No.2nd for Jute production in 2021-22-(P)

Ans. Assam

Q.327. Which state was No.2nd for Tobacco production in 2021-22(P)

Ans. Andhra Pradesh

Q.328. Which state has highest average Yield Kg/ha. of Food grains in 2021-22

Ans. Punjab

Q.329. Which state has highest average Yield Kg/ha. of Rice in 2021-22

Ans. Punjab & 2nd Haryana

Q.330. Which state has highest average Yield Kg/ha. of Wheat in 2021-22

Ans. Haryana & 2nd Punjab

Q.331. Which state has highest average Yield Kg/ha. of Maize in 2021-22

Ans. Tamil Nadu

Q.332. Which state has highest average Yield Kg/ha. of Total Pulses in 2021-22

Ans. Himachal Pradesh

Q.333. Which state has highest average Yield Kg/ha. of Gram in 2021-22

Ans. Gujarat

Q.334. Which state has highest average Yield Kg/ha of total Oil Seeds in 2021-22

Ans. Tamil Nadu

Q.335. Which state has highest average Yield Kg/ha of Ground Nut in 2021-22

Ans. Tamil Nadu

Q.336. Which state has highest average Yield Kg /ha of Soyabean in 2021-22

Ans. Meghalaya

Q.337. Which state has highest average Yield Kg/ha of Sugarcane in 2021-22

Ans. Tamil Nadu

Q.338. Which state has highest Average Yield Kg/ha of Cotton in 2021-22

Ans. Kerala

Q.339. Total Imports of India in 2023-24 (Est)

Ans. 854.80 US$ Billion

Q.340. Total Exports of India in 2023-24-(Est)

Ans. 776.68 US$ billion

Q.341. Trade balance of 2023-24-(Est)

Ans. -78.12 US$ Billion

Q.342. India's agricultural sector estimated growth rate in 2022-23

Ans. 3.5%

Q.343. The first advance estimate of GVA in the agriculture sector for 2023-24 is expected to be

Ans. 5.5%

Q.344. Country largest Importer of Indian products

Ans. U.S.A.

Q.345. Total Agricultural Imports in 2021-22(P)

Ans. 231850.27 Crore

Q.346. Which product contributed the highest value in Agri imports in 2021-2

Ans. Vegetable Oils & 2nd Fresh fruits

Q.347. Total Agri exports of India in 2021-22 (P)

Ans. 375662.47 crore

Q.348. Which product contribute the highest value in Agri Exports in 2022 -23

Ans. Rice

Q.349. Cereal Crop which India has the highest percent share of total exports

Ans. Rice

Q.350. The state having the highest Cold storage capacity in India in 2022

Ans. U.P.

Q.351. Per capita availability of total Food grains in India in 2021-22

Ans. 514.6 gm/Day

Q.352. Per capita net availability of Cereals in India in 2021-22

Ans. 460.8 gm/Day

Q.353. Per capita net availability of Rice in India in 2021-22

Ans. 90.5 gm/Day

Q.354. Per capita net availability of Wheat in India in 2021-22

Ans. 188.6 gm/Day

Q.355. Per capita net availability of Pulses in India in 2021-22

Ans. 53.8 gm/Day

Q.356. India's Protein intake of per capita/gm/ day

Ans. 60.3 gm/ Day

Q.357. Availability of animal Protein in Indian diet is

Ans. 10 gm./person/Day

Q.358. Per capita availability of Milk in India in 2022-23

Ans. 459 gm/Person/Day (Prov)

Q.359. Per capita availability of Milk in the World in 2022

Ans. 322 gm/Person/Day (Prov)

Q.360. Per capita availability of Fruits in India in 2017-18

Ans. 207.9 gm/Person/Day

Q.361. Per capita availability of Vegetables in India in 2017-18

Ans. 393.76 gm/Person/Day

Q.362. Per capita availability of Egg in India in 2022-23(P)-
Ans. 101/Person/Year

Q.363. Per capita availability of Meat in India 2022-23(P)-
Ans. 7.1 Kg/Person/Year

Q.364. Per capita availability of Edible Oil in India in 2021
Ans. 19.7 Kg/Person/Year (P)

Q.365. Per capita availability of Cloth in India in 2021
Ans. 53.3 Meter/Person/Year (Prov)

Q.366. Per capita availability of Tea in India in 2021-
Ans. 863 gm/Person/Year(P)

Q.367. ICMR recommendation for Milk
Ans. 280gm/Person/day

Q.368. ICMR recommendation for Egg
Ans. 182 Egg/Person/year

Q.369. ICMR recommendation for Meat -
Ans. 11 kg/Person/year

Q.370. The state having the highest per capita availability of Milk in India in 2022-23
Ans. Punjab, 1283 gm/day

Q.371. State having 2nd in per capita availability of Milk in India in 2022-23
Ans. Rajasthan, 1138 gm/day

Q.372. State having lowest in per capita availability of Milk in India in 2022-23
Ans. Mizoram, 55 gm/day

Q.373. State which is first in Milk production -
Ans. U.P.

Q.374. State which is second in Milk production-
Ans. Rajasthan

Q.375. Total milk production of India in 2022-23-
Ans. 230.6 million tones

Q.376. In total milk production which animal having highest contribution
Ans. Buffalo (45%)

Q.377. Country first in livestock population

Ans. India

Q.378. India's contribution in world livestock population-

Ans. 18%

Q.379. Egg production of India in 2022-23 (Est.)

Ans. 138.38 Billion egg

Q.380. Total wool production of India In 2022-23 (Est.)-

Ans. 33.61 Million kg.

Q.381. Total Fish Production of India in 2022-23-

Ans. 17.5 million Tones

Q.382. Total livestock population in India in 2019-

Ans. 535.8 million

Q.383. Total poultry population in India in 2019-

Ans. 851.81 million

Q.384. Among total livestock species which has highest population -

Ans. Cattle (192.49 million)

Q.385. Among total livestock species which has 2nd highest population

Ans. Goat (148.88. Million)

Q.386. Livestock increased from 2012 in 2019

Ans. 4.6%

Q.387. Among total livestock which species population has highest decreased

Ans. 1st- Donkeys & 2nd Mule

Q.388. Among total livestock which species population has highest increased

Ans. 1st- Mithun & 2nd Sheep

Q.389. Which animal is lowest in population amount livestock in India

Ans. Yaks (60,000)

Q.390. State No.1 in livestock population

Ans. U.P. & 2nd Rajasthan

Q.391. State having highest population of Cows is
Ans. 1st West Bengal & 2nd U.P.
Q.392. State having highest population of Buffalo is
Ans. 1st U.P. & 2nd Rajasthan
Q.393. Sate No.1 in Sheep population
Ans. Telangana & 2nd A.P.
Q.394. Sate No.1 in Goat population
Ans. Rajasthan & 2nd West Bengal
Q.395. Sate No.1 in Poultry population
Ans. Tamil Nadu & 2nd A.P.
Q.396. Sate No.1 in Pig population-
Ans. Assam & 2nd Jharkhand
Q.397. Sate No.1 in Horses population
Ans. U.P. & 2nd J & K
Q.398. Percent share of livestock sector to total GVA
Ans. 5.73%
Q.399. Percent share of livestock sector to Agri & allied sector GVA
Ans. 30.19%
Q.400. State No 1 in Wool production
Ans. Rajasthan & 2nd J & K
Q.401. State No.1 in Egg production-
Ans. Andhra Pradesh
Q.402. State No.1 in Meat production
Ans. Maharashtra & 2nd U.P.
Q.403. State No.1 in Fish production
Ans. A. P. & 2nd West Bengal
Q.404. Import of Urea in 2022-23
Ans. 7.58 million tone's
Q.405. Import of DAP in 2022-23
Ans. 6.58 million tone's
Q.406. Import of MOP in 2022-23
Ans. 1.86 million tone's

Q.407. One Nation One Fertilizer was introduced in

Ans. 24th Aug 2022

Q.408. At present total fertilizers plant which are in operation in India

Ans. 170 Plants

Q.409. Number of Plants for Urea in India

Ans. 36 Plants

Q.410. Number of DAP & Complex Fertilizers Plants

Ans. 20 Plants

Q.411. Number of SSP plants

Ans. 102 plants

Q.412. State having highest No. of Nitrogenous fertilizers plants

Ans. U.P. & Gujarat

Q.413. State having highest No. of DAP & other complex fertilizers plants

Ans. Gujarat (5 Plants)

Q.414. State having highest No. of SSP fertilizers plants

Ans. Rajasthan (23 Plants)

Q.415. Total No. of sales point of fertilizer in 2022-23-(P)

Ans. 3.13 Lack points.

Q.416. India's consumption of total nutrients(N+P+K) in 2022-23

Ans. 29.83 million tone's

Q.417. India's total consumption of Nitrogen in 2022-23

Ans. 20.20 million tone's

Q.418. India's total consumption of Phosphorous in 2022-23

Ans. 7.92 million tone's

Q.419. India's total consumption of Potash in 2022-23

Ans. 1.71 million tone's

Q.420. India's total consumption of Urea in 2022-23

Ans. 35.72 million tone's

Q.421. India's total consumption of DAP in 2022-23

Ans. 10.41 million tone's

Q.422. India's total consumption of MOP in 2022-23

Ans. 1.63 million tone's

Q.423. India's total consumption of SSP in 2022-23

Ans. 5.01 million tones

Q.424. State having highest consumption of NPK/Kg/ha in 2022-23

Ans. Punjab & Andhra Pradesh

Q.425. State having highest total consumption of NPK in 2022-23

U.P. & 2nd Maharashtra.

Q.426. Most widely used fertilizers after Urea & DAP in India in 2021-22

Ans. S.S.P.

Q.427. State having least total consumption of NPK in 2022-23

Ans. H.P.

Q.428. District having highest NPK consumption kg/ ha

Ans. Raichur (Karnataka)

Q.429. Crop which has highest consumption of (N+ P + K)

Ans. Sugar Cane

Q.430. Vegetable crop which has highest consumption of (N+ P + K)

Ans. Chilly

Q.431. India's N P K use ratio during 2022-23

Ans. (N) 11.8: (P) 4.6 :(K) 1

Q.432. India's Fertilizer consumption (N+P+K)/Kg/ha in 2023

Ans. 138.1 Kg/ha.

Q.433. MRP fixed for Urea in 2022-23

Ans. Rs. 5360/tone

Q.434. Subsidy amount on all fertilizers in 2023-24

Ans. 175103.17 Crore

Q.435. NBS subsidy approved for (P & K) fertilizers by Cabinet for Kharif 2024

Ans. Rs. 24420 crores

Q.436. Nutrient based subsidy (NBS) was started in year

Ans. 1st May 2010

Q.437. Which fertilizer was first under NBS scheme

Ans. SSP

Q.438. Duration of New Urea Policy

Ans. 1st June 2015 to 31st March 2019

Q.439. Revised weight of Urea bag is

Ans. 45 Kg

Q.440. Urea Gold Fertilizer is developed by

Ans. RCF

Q.441. Content of Urea Gold

Ans. N-37% & S-17%

Q.442. Weight of Gold Urea Bag is-

Ans. 40 Kg.

Q.443. Projection plan for 12th Five-year plan for N, P, K

Ans. 33516 MT

Q.444. What is Rabi Nutrient Base Subsidy (NBS) rate for Nitrogen in 2023-24

Ans. Rs.47.02/Kg.

Q.445. What is Rabi Nutrient Base Subsidy (NBS) rate for Phosphorous in 2023-24

Ans. Rs.20.82/Kg.

Q.446. What is Rabi Nutrient Base Subsidy (NBS) rate for Potash in 2023-24

Ans. Rs.2.38/Kg.

Q.447. What is Rabi Nutrient Base Subsidy (NBS) rate for Sulphur in 2023-24

Ans. Rs. 1.89/Kg.

Q.448. How many nutrients are included under nutrient-based subsidy

Ans. 4 (N, P K, S)

Q.449. Newly fertilizer grade included in the NBS scheme is

Ans. 8-21-21 & 9-24-24

Q.450. No of Fertilizers notified by FCO in India

Ans. 114 fertilizers

Q.451. No of Fertilizers Quality Control Labs are in India 2022-23

Ans. 83 Labs

Q.452. No of Soil testing labs are in India in 2022-23

Ans. 1597 Labs

Q.453. State having the highest Fertilizers Quality Control Labs in India

Ans. Tamil Nadu (14)

Q.454. Soil Health Card (SHC) contains information about

Ans. 12 Parameters

Q.455. State having highest production of Biofertilizer (Dry)-2022-23(Prov)

Ans. Tamil Nadu

Q.456. The state having the highest production of Biofertilizer (Liquid) 2022-23(Prov)

Ans. Tamil Nadu

Q.457. The state declared the first fully organic state of India in 2016

Ans. Sikkim

Q.458. Zinc deficiency percentage soils in India

Ans. 43.0%

Q.459. Iron deficiency percentage soils in India

Ans. 12.1%

Q.460. Which state soils have the highest percentage of Zinc deficiency in India

Ans. Tamil Nadu

Q.461. Which state soils have the highest percentage of Iron deficiency in India

Ans. Rajasthan

Q.462. Soil group having the highest area of total geographical area in India

Ans. Red Sandy

Q.463. Minimum support price for Common paddy in 2023-24

Ans. 2183/ Qtl

Q.464. Minimum support price for Common paddy in 2024-25
Ans. 2300/Qtl

Q.465. MSP rupees increased per /Qtl for Common Paddy from last year
Ans. Rs. 117/Qtl

Q.466. Minimum support price for grade A paddy in 2023-24
Ans. 2203/Qtl

Q.467. Minimum support price for grade A paddy in 2024-25
Ans. 2320/Qtl

Q.468. MSP rupees increased per /Qtl for grade A paddy from last year
Ans. Rs. 117/Qtl

Q.469. The minimum support price for Jowar (Hybrid Sorghum) in 2023-24
Ans. 3180/Qtl

Q.470. The minimum support price for Jowar (Hybrid) in 2024-25
Ans. 3371/Qtl

Q.471. Minimum support price for Pearl millet in 2023-24
Ans. 2500/Qtl

Q.472. Minimum support price for Pearl millet in 2024-25
Ans. 2625/Qtl

Q.473. Minimum support price for Maize in 2023-24
Ans. 2090/Qtl

Q.474. Minimum support price for Maize in 2024-25
Ans. 2225/Qtl

Q.475. Minimum support price for Ground Nut in 2023-24
Ans. 6377/Qtl

Q.476. Minimum support price for Ground Nut in 2024-25
Ans. 6783/Qtl

Q.477. Minimum support price for Soya bean in 2023-24
Ans. 4600/Qtl

Q.478. Minimum support price for Soya bean in 2024-25
Ans. 4892/Qtl

Q.479. Minimum support price for Sesame in 2023-24
Ans. 8635/Qtl

Q.480. Minimum support price for Sesame in 2024-25
Ans. 9267/Qtl

Q.481. Minimum support price for Cotton long staple in 2023-24
Ans. 7020/Qtl

Q.482. Minimum support price for Cotton long staple in 2024-25
Ans. 7521/Qtl

Q.483. Minimum support price for Cotton Medium staple in 2023-24
Ans. 6620/Qtl

Q.484. Minimum support price for Cotton Medium staple in 2024-25
Ans. 7121/Qtl

Q.485. Minimum support price for Arhar (Pigeon Pea) in 2023-24
Ans. 7000/Qtl

Q.486. Minimum support price for Arhar in 2024-25
Ans. 7550/Qtl

Q.487. Minimum support price for Moong in 2023-24
Ans. 8558/Qtl

Q.488. Minimum support price for Moong in 2024-25
Ans. 8682/Qtl

Q.489. Minimum support price for Urad in 2024-25
Ans. 7400/Qtl

Q.490. Minimum support price for Sunflower seed in 2024-25
Ans. 7280/Qtl

Q.491. Minimum support price for Niger seeds in 2024-25
Ans. 8717/Qtl

Q.492. Minimum support price for Ragi in 2024-25
Ans. 4290/Qtl

Q.493. Minimum support price for Jute in 2024-25
Ans. 5335/Qtl

Q.494. Minimum support price for Copra (milling) -2023-24
Ans. 11160/Qtl

Q.495. Highest MSP increased in Kharif Crops in 2024-25
Ans. 1st Niger seed, 2nd Sesamum, 3rd Arhar

Q.496. Minimum support price for Wheat in 2022-23

Ans. 2125/Qtl

Q.497. Minimum support price for Wheat in 2023-24

Ans. 2275/Qtl

Q.498. MSP rupees increased per /Qtl for Wheat from last year

Ans. Rs. 150/Qtl

Q.499. Minimum support price for Barley in 2022-23

Ans. 1735/Qtl

Q.500. Minimum support price for Barley in-2023-24-

Ans. 1850/Qtl (Increase of Rs. 115000 per Qtl.)

Q.501. FRP price for Sugar cane in 2024-25

Ans. 340/Qtl

Q.502. Minimum support price for Rapeseed & Mustard in 2022-23

Ans. 5450/Qtl

Q.503. Minimum support price for Rapeseed & Mustard in 2023-24

Ans. 5650/Qtl (Increase of Rs. 200 per Qtl.)

Q.504. Minimum support price for Gram in 2022-23

Ans. 5335/Qtl

Q.505. Minimum support price for Gram in 2023-24

Ans. 5440/Qtl (Increase of Rs.105 per Qtl.)

Q.506. Minimum support price for Lentil in 2023-24

Ans. 6425/Qtl

Q.507. Minimum support price for Safflower seed in 2023-24

Ans. 5800/Qtl

Q.508. Highest MSP increased in Rs. for the Rabi 2023-24

Ans. Lentil (Rs. 425)

Q.509. Crop which has been removed from MSP category list-

Ans. Sugarcane

Q.510. In which year MSP was announced by Govt

Ans. 1970

Q.511. Which committee approve the MSP-

Ans. Cabinet Committee on Economic Affairs (CCEA)

Q.512. Which committee recommend the MSP

Ans. CACP (Commission for Agricultural Costs and Prices)

Q.513. Chairman of Cabinet committee on Economic Affairs

Ans. Prime Minister

Q.514. No. of Cereals covered under MSP

Ans. Seven

Q.515. No. of Pulses covered under MSP-

Ans. Five

Q.516. No. of Oil seeds covered under MSP-

Ans. Eight

Q.517. No. of Fiber crops covered under MSP-

Ans. Two

Q.518. No of Commercial crops under MSP-

Ans. 3 Crops

Q.519. Total No. of crops are in MSP regime-

Ans. 22 Crops

Q.520. No of Kharif crop include in MSP regime

Ans. 14 Crops

Q.521. No of Rabi crop include in MSP regime-

Ans. 8 Crops

Q.522. Minor crop which is considered under MSP-

Ans. Ragi

Q.523. National year for Millets-

Ans. 2018

Q.524. Foxtail is a type of millet-

Ans. Minor millet

Q.525. State which has highest cost of cultivation in paddy-

Ans. A.P.

Q.526. State which has highest cost of cultivation in Wheat-

Ans. Rajasthan

Q.527. State which has highest cost of cultivation in Cotton-

Ans. Andhra Pradesh

Q.528. State which has highest cost of cultivation in Sugar cane

Ans. Tamil Nadu

Q.529. State which has highest cost of cultivation in Ground Nut-

Ans. Gujarat

Q.530. State which has highest cost of cultivation in Maize-

Ans. Andhra Pradesh

Q.531. No of Farmers covered under PMFBY Scheme till 2022-23

Ans. 45.12 Crore

Q.532. State having highest farmers registered under crop insurance scheme in 2022-23

Ans. Rajasthan

Q.533. Total No. of Operative Kisan Credit cards in 2021-22-(Prov)

Ans. 7.37 crore

Q.534. State having highest operative Kisan Credit card in 2021-22 (Prov)

Ans. U.P. & 2nd Maharashtra

Q.535. All India production of Certified seeds in 2022-23

Ans. 51.42 Lac Tones

Q.536. Targeted Production of Breeder seed in 2022-23-(Prov)

Ans. 7603 tones

Q.537. Production of Foundation seed in 2022-23-

Ans. 2.38 Lac tone

Q.538. Tractor sales in India in 2022-23-

Ans. 8.34 Lac

Q.539. Sate having highest tractors in 2022-23-

Ans. U.P.

Q.540. Power tillers sales in India in 2022-23-

Ans. 34471

Q.541. Percent share of total consumption of electricity for agricultural purposes in 2021-22

Ans. 20.11% (Prov)

Q.542. The state has the highest electricity consumption for agri purposes in 2021-22-(Prov)

Ans. Rajasthan

Q.543. Total GVA of India based on constant price in 2022-23
Ans. 14764840 Crore (Provisional Est)

Q.544. Total Agriculture GVA based on constant price in 2022-23
Ans. 2234269 Crore (Provisional Est)

Q.545. Percent growth in GVA in total economic activity at the constant price in 2022-23
Ans. 7% (Prov. Est.)

Q.546. Percent growth in GVA in Agriculture activity at the constant price in 2022-23
Ans. 4% (Prov. Est.)

Q.547. Total GVA of India based on the current price in 2022-23
Ans. 24742871 Crore (Provisional Est)

Q.548. Total Agriculture GVA based on the current price in 2022-23
Ans. 4557599 Crore (Provisional Est)

Q.549. Percent contribution of agriculture in GVA (current price) in 2022-23
Ans. 18.4% (Provisional Est)

Q.550. Percent growth from last year in GVA by Agriculture (Constant Price) in FY 2022-23
Ans. 0.5% (Prov)

Q.551. Per capita net national income (Current Price) in 2022-23
Ans. Rs.172276 (Provisional)

Q.552. Per capita net national income (Constant Price) in 2022-23
Ans. Rs.98374 (Provisional)

Q.553. Total Districts in India (2023-24)-
Ans. 788 Districts

Q.554. Largest district of India by territorial area
Ans. Kachchh (Gujarat)

Q.555. Total villages in India (Senses 2011)
Ans. 640867 village

Q.556. State having highest no. of villages-
Ans. U.P.

Q.557. Percent increase in population over the last year in 2021-22

Ans. 0.9 %

Q.558. State having highest population in India

Ans. U.P.

Q.559. State has the lowest population in India

Ans. Sikkim

Q.560. State having the highest population below the poverty line is

Ans. Odisha

Q.561. The density of Indian population Per/ Sq./ km is (as per 2011 Census)

Ans. 382

Q.562. Percentage of population below poverty line in India (As per 2011-12)

Ans. 21.9%

Q.563. Percent of Agriculture Workers of total workers

Ans. 54.9% (P)

Q.564. Percent of rural population of total population

Ans. 68.8% (P)

Q.565. Total layout plan (in value) for 11th Five-year Plan (2007-2012)

Ans. 3644718 Crore

Q.566. layout plan (in value) for 11th Five-year Plan for Agri. (2007-2012)

Ans. 136381 Crore

Q.567. Total layout plan (in value) for 12th Five-year Plan (2012-2017)

Ans. 7,669,807 Crore

Q.568. layout plan (in value) for 12th Five-year Plan for Agri. (2012-2017)

Ans. 363,273 Crore

Q.569. Top 3 sectors which has given max outlay in 12th Five Year Plan

Ans. 1. Social Service, 2. Energy, 3. Transport

Q.570. Percent share of Agri. & allied sectors in 12th layout plan-

Ans. 4.7%

Q.571. Percent share of Agri. & allied sectors in 11th layout plan

Ans. 3.7%

Q.572. Budget for Agriculture sector in FY 2024-25

Ans. **Rs 1.27 lac Crore**

Q.573. Budget for PM Kisan Samman for FY 2024-25

Ans. Rs. 60.000 Crore

Q.574. Central Govt expenditure budget for Agriculture Research & Education in 2024-25

Ans. 9941.09 Crore

Q.575. Budget allocation for Department of Agriculture and Farmers Welfare in 2024-25

Ans. 117528.79 Crore

Q.576. Budget allocation for Department of Fertilizer in 2024-25

Ans. 164150.81 Crore

Q.577. Budget allocation for Department of Animal Husbandry and Dairying in 2024-25

Ans. 4521.24 Crore

Q.578. Budget allocation for Department of Fisheries in 2024-25

Ans. 2584.50 Crore

Q.579. Average size of agricultural land holding in India is

Ans. 1.1 Ha

Q.580. India's percent share contribution in world production of total Cereals in 2021

Ans. 11.6%(P)

Q.581. India's percent share contribution in world production of Wheat in 2021

Ans. 14.2%(P)

Q.582. India's percent share contribution in world production of Rice in 2021

Ans. 24.8%(P)

Q.583. India's percent share contribution in world production of total Pulses in 2021

Ans. 28.9%(P)

Q.584. India's percent share contribution in world production of Ground Nut in 2021

Ans. 19.1% (P)

Q.585. India's percent share contribution in world production of Rape seed in 2021

Ans. 14.3%(P)

Q.586. India's percent share contribution in world production of Vegetables primary in 2021

Ans. 12 %(P)

Q.587. India's percent share contribution in world production of Fruit primary in 2021

Ans. 11.9%(P)

Q.588. India's percent share contribution in world production of Potato's in 2021

Ans. 14.4%(P)

Q.589. India's percent share contribution in world production of Onion dry in 2021

Ans. 25 %(P)

Q.590. India's percent share contribution in world production of Sugar cane in 2021

Ans. 21.8%(P)

Q.591. India's percent share contribution in world production of Tea in 2021

Ans. 19.4%(P)

Q.592. India's percent share contribution in world production of Coffee in 2021

Ans. 3.3%(P)

Q.593. India's percent share contribution in world production of Jute in 2021

Ans. 49.7%(P)

Q.594. India's percent share contribution in world production of Seed cotton in 2021

Ans. 23.3%(P)

Q.595. India's percent share contribution in world production of Tobacco in 2021

Ans. 12.9%(P)

Q.596. Country having highest average size of agricultural land holding is

Ans. Australia & 2nd Argentina.

Q.597. Country having lowest average size of agricultural land holding is

Ans. Nepal

Q.598. State having highest area & production under Tea

Ans. Assam

Q.599. All India Area under Tea production in 2022-23

Ans. 6.24 lac ha.

Q.600. All India Tea production of India in 2021-22

Ans. 13.66 lac Kg.

Q.601. Average yield/ha. of Tea in India in 2021-22

Ans. 2205 Kg/ha

Q.602. State having highest Area & Production for Coffee

Ans. Karnataka

Q.603. All India Coffee production in 2022-23

Ans. 360.5 thousand tones (Prov)

Q.604. State having highest production and area of Rubber

Ans. Kerala

Q.605. All India Rubber production in 2022-23

Ans. 839 thousand tones. (Prov)

Q.606. All India Rubber Area in 2022-23

Ans. 850 thousand Ha. (Prov)

Q.607. Total fruit and vegetable production in India in 2022-23

Ans. 3212.5 Lack tones(P)

Q.608. Total area under Fruits in India in 2022-23-

Ans. 7,009 Thousand ha.(P)

Q.609. Total production of Fruit in 2022-23

Ans. 1083.4 Lack tones(P)

Q.610. Productivity (tone/ha) of fruits in India in 2022-23

Ans. 15.5 tone/ha

Q.611. The fruit which covers the highest area in India

Ans. Mango & 2nd Citrus

Q.612. The fruit that has the highest production in India is

Ans. Banana

Q.613. Fruit which is second in production in India

Ans. Mango

Q.614. Fruit having highest yield tone/ha

Ans. Banana & 2nd Papaya

Q.615. Fruit which is 3rd in production in India

Ans. Citrus

Q.616. The state has the highest area under Fruits in 2022-23

Ans. Maharashtra & 2nd A.P.

Q.617. The state has the highest production of Fruits in 2022-23

Ans. A.P. & 2nd Maharashtra.

Q.618. The state has the highest productivity tones/ha in fruits in 2022-23

Ans. A.P.

Q.619. Total area under Vegetables in India in 2022-23

Ans. 11,358 thousand ha.

Q.620. Total production of vegetables in India in 2022-23

Ans. 212,908 thousand tones

Q.621. Productivity (tone/ha) of vegetables in India in 2022-23

Ans. 18.7 tone/ha. (Prov)

Q.622. Vegetables having highest area in India in 2022-23

Ans. Potato & 2nd Onion

Q.623. Vegetable which is highest in production in India in 2022-23

Ans. Potato

Q.624. State having highest area under vegetables in 2022-23

Ans. West Bengal & 2nd U.P.

Q.625. State having highest production of vegetables in 2022-23

Ans. U.P. & 2nd West Bengal

Q.626. State having highest productivity tones/ha in vegetables in 2022-23

Ans. A.P. & 2nd Kerala

Q.627. Vegetable which is 2nd in production in India

Ans. Onion

Q.628. Vegetable which is 3rd in production in India

Ans. Tomato

Q.629. Area under Mango cultivation of India in 2022-23

Ans. 2332 thousand ha. (P)

Q.630. Production of Mango in India year 2022-23

Ans. 20,958 thousand tones. (P)

Q.631. Mango productivity tone/ha of India in 2022-23-

Ans. 9.0 tone/ha(P)

Q.632. Area under Banana cultivation of India in 2022-23

Ans. 938 Thousand ha.(P)

Q.633. Production of Banana in India year 2022-23

Ans. 35,365 Thousand tones(P)

Q.634. Banana productivity tone/ha. of India year 2022-23

Ans. 37.7 tone/ha (P)

Q.635. Area under Citrus cultivation of India in 2022-23

Ans. 1096 Thousand ha.(P)

Q.636. Production of Citrus of India in 2022-23

Ans. 14,250 thousand tones (P)

Q.637. Citrus productivity tone/ha of India in 2022-23

Ans. 13.0 tone/ha. (P)

Q.638. Tomato area of India in 2022-23

Ans. 849 thousand ha. (P)

Q.639. Tomato production of India in 2022-23

Ans. 20,402 thousand tones. (P)

Q.640. Potato area of India in 2022-23

Ans. 2352 thousand Ha. (P)

Q.641. Potato production of India in 2022-23

Ans. 60,540 thousand tones. (P)

Q.642. Potato productivity tone/ha of India in 2022-23

Ans. 25.7 tone/ha(P)

Q.643. The total area of major spices crops of India in 2022-23

Ans. 4300 thousand ha. (P)

Q.644. Total production of major spices crops of India in 2022-23

Ans. 10,986 thousand tones(P)

Q.645. Spices productivity tone/ha of India in 2022-23

Ans. 2.6 tone/ha(P)

Q.646. Spice crop which covers the highest area in spices of India in 2022-23

Ans. Chilly & 2nd Cumin

Q.647. The spice which has the highest production in India in 2022-23

Ans. Garlic & 2nd Ginger

Q.648. Spices having the highest yield/ha. of India in 2022-23

Ans. Ginger & 2nd Garlic

Q.649. Total Gross area sown in the country

Ans. 216.10 M. ha (Prov)

Q.650. Net area sown of the country

Ans. 141.44 M. ha. (Prov)

Q.651. The zone which has the highest net sown area

Ans. West Zone

Q.652. The state having the highest net sown area

Ans. Rajasthan

Q.653. The state having 2nd highest net sown area

Ans. Maharashtra

Q.654. Gross area irrigated in India

Ans. 118.93 M. ha. (Prov)

Q.655. Net irrigated area of India -

Ans. 77.72 M. ha. (Provisional)

Q.656. The state having the highest total cropped area -

Ans. M.P & 2nd Rajasthan

Q.657. India's Share of gross irrigated to the gross sown area-

Ans. 55.0% (Prov)

Q.658. India's share of net irrigated area to net cultivated area-

Ans. 53.03%

Q.659. The state having the highest irrigation percentage to net cultivated area

Ans. Punjab (100%)

Q.660. The state having 2nd highest irrigation percent to net cultivated area

Ans. Haryana (99.1%)

Q.661. Country having the highest irrigated area

Ans. India

Q.662. Crop having max area under irrigation-

Ans. Wheat

Q.663. Crop having 2nd in max area under irrigation-

Ans. Rice

Q.664. Net area irrigated by canals in India-

Ans. 18.6- M. ha. (Prov)

Q.665. State which has highest area irrigated by canals-

Ans. U.P.

Q.666. Net area irrigated by tube wells in India -

Ans. 37.20 M. ha.

Q.667. The state which has the highest irrigated area by tube wells-

Ans. U.P.

Q.668. Crop having the highest percent of the irrigated area -

Ans. Wheat & 2nd Rice

Q.669. The total area under Drip irrigation in 2022-

Ans. 66.79 Lac ha.

Q.670. Total area under Sprinkler irrigation in 2022

Ans. 78.12 Lac ha.

Q.671. The total area under Micro irrigation in 2022—

Ans. 144.91 Lac ha.

Q.672. The state having the highest area under Drip irrigation in 2022-

Ans. Maharashtra & 2nd A.P.

Q.673. State having a highest area under Sprinkler irrigation in 2022-

Ans. Rajasthan & 2nd Karnataka

Q.674. State having highest area under Micro irrigation in 2022-

Ans. Karnataka & 2nd Rajasthan

Q.675. Application efficiency of Drip irrigation is-

Ans. 90%

Q.676. Highest source of irrigation in India in 2021-

Ans. Tube Wells (47.85%)

Q.677. Percent share of Irrigation area through Canal in 2021

Ans. 23.92%

Q.678. Current cropping intensity of India in 2020-21-

Ans. 152.7% (P)

Q.679. Geographical area of India in 2020-21-

Ans. 328.726 M. ha.

Q.680. Reporting area for land utilization statistics of India in 2020-21

Ans. 306.98 M. ha.

Q.681. Highest geographical area for the state-

Ans. Rajasthan

Q.682. Highest net cultivated area for the state-

Ans. Rajasthan

Q.683. Highest gross cultivated area for the state-

Ans. M.P.

Q.684. Highest cropping intensity for the state in 2020-21

Ans. West Bengal (193.2%)

Q.685. Second highest cropping intensity for the state in 2020-21-

Ans. 2nd Tripura & 3rd Punjab

Q.686. The state having the lowest cropping intensity in 2020-21

Ans. Manipur (100.0%)

Q.687. Highest net irrigated area for the state-

Ans. U.P. & 2nd M.P.

Q.688. State highest net irrigated area which is irrigated through canals & tube wells

Ans. U.P.

Q.689. Total forest cover area of India in 2021

Ans. 7,13789 Sq. K.M

Q.690. India's rank in forest and tree cover in world-

Ans. 10th rank

Q.691. Total forest and tree cover area of India in 2021—

Ans. 80.9 M. ha

Q.692. National Forest Policy of India was established in year-

Ans. 1988

Q.693. NYM (Nagar Van Yojana) was implemented since-

Ans. 2020

Q.694. As per National Forest Policy goal of achieving of forest and tree cover area percentage of total Geographical area is-

Ans. 33% (Forest+ Tree Cover)

Q.695. The current per capita forest area in India is-

Ans. 0.07 ha

Q.696. Percent area of forest of total geographical area in India in 2021

Ans. 21.71%

Q.697. Percent of total forest and tree cover area of total geographical area in 2021

Ans. 24.62%

Q.698. Percent area for very dense forest in India in 2021

Ans. 3.04%

Q.699. State having highest dense forest area in 2021-

Ans. Arunachal Pradesh.

Q.700. National Forest commission was set up in year-

Ans. 2002

Q.701. Highest area under forest in state in 2021-

Ans. M.P. (8.6 M.ha.)

Q.702. State which is 2nd in forests area in 2021-

Ans. Chhattisgarh

Q.703. State which has lowest forest cover area -

Ans. Haryana

Q.704. State having highest percentage of forest cover area of total geographical area is

Ans. Mizoram (84.53%)

Q.705. Highest forest area increased for the state in 2021-

Ans. A.P.

Q.706. Highest forest area Decreased for the state in 2021-

Ans. Arunachal Pradesh

Q.707. Percent contribution of forest area of India in world in 2020

Ans. 2 %

Q.708. Country having the highest area of forest of the world in 2020-

Ans. Russian Federation (20%)

Q.709. The area under the barren land of the country

Ans. 16.68 M. ha. (Prov)

Q.710. The area under permanent pasture & grazing lands-

Ans. 10.32 M. ha. (Prov)

Q.711. The area under non -agricultural uses in India-

Ans. 27.72 M. ha. (Prov)

Q.712. Percent share of total cropped area for total food crops-

Ans. 73.7%(P)

Q.713. Percent share of total cropped area for total food grain crops

Ans. 63.1%(P)

Q.714. Percent share of total cropped area for cereal crops-

Ans. 50.09%(P)

Q.715. Cereal which having the highest percent share of the total cropped area

Ans. Paddy

Q.716. Percent share of total cropped area for Pulses

Ans. 12.2%(Prov)

Q.717. Pulse having the highest percent share of total cropped area

Ans. Gram

Q.718. Percent share of total cropped area for Oil seed crops

Ans. 14.3%(Prov)

Q.719. Oil seed having highest percent share of total cropped area

Ans. Soyabean

Q.720. Percent share of total cropped area for fiber crops-

Ans. 6.5%(Prov)

Q.721. Which Fiber having highest percent share of total cropped area

Ans. Cotton

Q.722. Percent share of total cropped area for fodder crops

Ans. 3.5%(Prov)

Q.723. Percent share of total cropped area for fruits & vegetables-

Ans. 5.7%(Prov)

Q.724. Percent share of total cropped area for Fruits -

Ans. 2.3% (Prov)

Q.725. Fruit having highest percent share of total cropped area

Ans. Mango

Q.726. Percent share of total cropped area for Dry fruits

Ans. 0.3%(Prov)

Q.727. Percent share of total cropped area for Vegetables-

Ans. 3.4%(Prov)

Q.728. Vegetable having highest percent share of total cropped area

Ans. Potato

Q.729. Percent share of total cropped area for Spices & condiments-

Ans. 2.0%(Prov)

Q.730. Spice having highest percent share of total cropped area

Ans. Chilly

Q.731. Percent share of total cropped area for Drug, Narcotics & Plantation crops

Ans. 1.4%(Prov)

Q.732. Percent share of total cropped area for Rice

Ans. 23.3%(Prov)

Q.733. Percent share of total cropped area for Wheat-

Ans. 16.4%(Prov)

Q.734. Percent share of total cropped area for Sugarcane-

Ans. 2.7%(Prov)

Q.735. Percent share of total cropped area for Cotton

Ans. 6.1%(Prov)

Q.736. Percent share of total cropped area for Gram-

Ans. 4.3%(Prov)

Q.737. Percent share of total cropped area for Potato-

Ans. 0.8%(Prov)

Q.738. Percent share of total cropped area for Tobacco-

Ans. 0.2%(Prov)

Q.739. Percent share of total cropped area for Mango-

Ans. 0.7%(Prov)

Q.740. Percent share of total cropped area for Banana-

Ans. 0.4%(Prov)

Q.741. Percent share of total cropped area for Tea-

Ans. 0.3%(Prov)

Q.742. Percent share of total cropped area for Coffee-

Ans. 0.2%(Prov)

Q.743. Average size of operational holdings by marginal farmers-

Ans. 0.38 ha.

Q.744. Average size of operational holdings by small farmers-

Ans. 1.40 ha.

Q.745. Average size of operational holdings by medium farmers

Ans. 5.72 ha.

Q.746. Average size of operational holdings by large farmers-

Ans. 17.07 ha.

Q.747. Average size of operational holdings by all group's farmers

Ans. 1.08 ha.

Q.748. Highest average size land holding in state-

Ans. Nagaland (4.87ha)

Q.749. Lowest average size land holding in state-

Ans. Kerala (0.18 ha)

Q.750. State which having highest land holding farmers >10 ha-

Ans. Rajasthan 2nd Maharashtra

Q.751. Rainfall amount for Dry season

Ans. 0-750 mm

Q.752. Approximate percentage of area receiving rainfall in dry season

Ans. 30%

Q.753. Rainfall amount for medium season

Ans. 750-2000 mm

Q.754. Rainfall amount for assured season-

Ans. >2000 mm

Q.755. Approximate percentage of area receiving rainfall in assured season

Ans. 8%

Q.756. Highest percent annual rainfall is received from

Ans. S.W. monsoon (June to Sep.)

Q.757. Duration of Pre monsoon in India-

Ans. March- May

Q.758. Approx percentage of annual rainfall by pre-monsoon-

Ans. 10.40%

Q.759. Total percent amount received from Southwest monsoon

Ans. 73.7%

Q.760. Duration of post monsoon

Ans. Oct- Dec.

Q.761. Approx percentage of annual rainfall received by post monsoon

Ans. 13.30%

Q.762. Duration of North-east (Winter) monsoon

Ans. Jan- March

Q.763. Approx percentage of annual rainfall received by Winter monsoon

Ans. 2.60%

Q.764. Average annual rainfall of India is (2012-2021)

Ans. 1236 mm

Q.765. Total no. of reservoirs in 2023-

Ans. 150 Reservoirs

Q.766. Category defined of drought for 2009

Ans. Severe drought

Q.767. Average lowest rainfall occurs in

Ans. Jaisalmer (Rajasthan)

Q.768. Extremely severe cyclonic storm appears in 2019-

Ans. Fani

Q.769. The generated rain fall volume is-

Ans. 400 m ha-m

Q.770. The generated rain off volume is-

Ans. 185 m ha-m

Q.771. Maximum average annual runoff occurs in-

Ans. Brahmaputra Basin

Q.772. Annual soil loss due to erosion-

Ans. 53334 million tonnes

Q.773. The concentration of nitrogen in atmosphere up to 50 km from the ground surface is

Ans. 78%

Q.774. Per capita water availability in India in 2021

Ans. 1486 Cubic. M/Person/ Annum (Prov) (For 2031 esimated to be 1367 cum/person/year)

Q.775. Government of India has launched AMRUT 2.0 on-

Ans. 1st October 2021

Q.776. The Pradhan Mantri Krishi Sinchayee Yojana (PMKSY) was launched on

Ans. July 1, 2015

Q.777. Pradhan Mantri Krishi Sinchayee Yojana (PMKSY) motto was

Ans. "Har Khet Ko Paani".

Q.778. Percentage of fresh water on earth-

Ans. 2.8%

Q.779. Conversation ratio of seed lac from stick lac-

Ans. 66% of Stick Lac.

Q.780. Conversation ratio of Cashew nuts from cashew kernels-

Ans. 25% of Cashew nuts

Q.781. Conversation of Riçe (Cleaned) of paddy-

Ans. 2/3 of Paddy production

Q.782. Beushening practices is common in-

Ans. Rice

Q.783. Conversion of Cotton lint from Kapas production-

Ans. 1/3 of Kapas production

Q.784. Conversion of Ground Nut kernel to nuts in shell-

Ans. 70%

Q.785. One candy of cotton is equal to

Ans. 356 kg. or 782.42 lbs (Ginned Cotton)

Q.786. One tone copra (Coconut) is equal to-

Ans. 6773 nuts

Q.787. Bale size of Jute

Ans. 180 Kg.

Q.788. Spiked millet is English name of-

Ans. Bajra (Pearl Millet)

Q.789. Cholam (great millet) is English name of-

Ans. Jowar (Sorghum)

Q.790. Mundua is a Hindi name of-

Ans. Ragi

Q.791. One hectare is equal to acres-

Ans. 2.471 Acres

Q.792. One acre is equal to hectare-

Ans. 0.40468 ha.

Q.793. Conversation of Butter from mix milk is-

Ans. 6.9 %

Q.794. Conversion of Ghee from mixed milk is-

Ans. 5.5%

Q.795. Conversation of Meal to soya bean seed crushed-

Ans. 73%

Q.796. Conversion ratio of Gur from cane crushed-

Ans. 10%

Q.797. Conversion ratio of Khandsari sugar from Gur refined-

Ans. 37.5%

Q.798. Volume of Oxygen in atmosphere is-

Ans. 21%

Q.799. Longest river of India is

Ans. Ganga

Q.800. Major group of vegetable sown in Jayad-

Ans. Cucurbits

Q.801. Hanumantha Rao Committee is associated with-

Ans. Fertilizers

Q.802. History of Indian Agriculture was written by-

Ans. M.S. Randhawa

Q.803. Fort of soya bean known as-

Ans. M.P.

Q.804. Garden city is known as

Ans. Bangalore

Q.805. The quickest technique for lawn making is -

Ans. Turfing

Q.806. Botanical garden "Lal Bagh "is situated at-

Ans. Bangalore (Founded by King Hyder Ali in 1760)

Q.807. Rashtrapati Bhawan Garden, New Delhi laid out on the pattern of

Ans. Mughal Garden

Q.808. What is the new name of Mughal Garden of Rashtrapati Bhavan, New Delhi

Ans. Amrit Udyan

Q.809. Which was the first hybrid Tea rose in world-

Ans. La France

Q.810. Square watermelon was discovered by the country

Ans. Japan

Q.811. Lowest forest area in India is-

Ans. Western Rajasthan

Q.812. Arable land per head highest in world-

Ans. Australia

Q.813. Non reporting area maximum is in-

Ans. Jammu & Kashmir

Q.814. Highest water erosion in state-

Ans. West Bengal

Q.815. Highest wind erosion in state is-

Ans. Rajasthan

Q.816. State which is 2nd in geographical area is

Ans. M.P.

Q.817. Highest alluvial soils found in state-

Ans. U.P.

Q.818. State having India's first agricultural University-

Ans. Uttarakhand (Pant Nagar)

Q.819. First Agriculture University

Ans. G.B. Pant Agri. University, Pant Nagar

Q.820. Agri. University which was named on 1st India's President

Ans. Rajendra Agri. University, Bihar

Q.821. State which has highest saline soils-

Ans. Gujarat

Q.822. State which has highest acidic soils

Ans. West Bengal

Q.823. Consumption of chemical Pesticide (Technical) In 2022-23

Ans. 52466 MT

Q.824. State highest consumption of Pesticide (Technical) In 2022-23

Ans. U.P. & 2nd Maharashtra

Q.825. States having Zero use of pesticide (Organic State)-

Ans. Sikkim & Meghalaya

Q.826. State which has highest use of pesticides per /ha

Ans. Punjab

Q.827. Consumption of Bio Pesticide (Technical) In 2022-23-

Ans. 7248 MT

Q.828. State which is No. 1 in Bio pesticide consumption in 2022-23

Ans. West Bengal % 2nd Tamil Nadu

Q.829. State which is No. 1 in total pesticide consumption

Ans. Maharashtra (13175 MT Tech Grade)

Q.830. Total Pesticide testing (SPTL) laboratories in India are

Ans. 71 Labs

Q.831. Central Government has established Central Insecticides Laboratory (CIL) at Feridabad on

Ans. 28th February 1981

Q.832. Draft Pesticide Management bill was passed in Year

Ans. 2017

Q.833. Per ha Herbicide consumption of India is

Ans. 40 gm/ha

Q.834. Per ha Pesticide consumption of India is

Ans. 0.315 kg/ha

Q.835. Crop which accounts max share of pesticide in India-

Ans. Paddy (29%) & 2nd Cotton

Q.836. Country having highest consumption of pesticides in world

Ans. China

Q.837. Total dry land area of India is-

Ans. 34.5 M. ha.

Q.838. Percent arid area of total area of India is

Ans. 17%

Q.839. Percent of Humid area of total area of India is

Ans. 1.10 %

Q.840. Environmental Protection Act (EPA) was passed by Govt of India in year

Ans. May, 1986

Q.841. The Web of digital dissemination system of Indian Agricultural Research is

Ans. AGROWEB

Q.842. The Indian Journal of Agriculture Sciences published by-

Ans. ICAR

Q.843. Indian Farming Journal published by-

Ans. ICAR

Q.844. King Baudouin International Development Prize is given for-

Ans. Green revolution

Q.845. Agro ecological zones classified by NBSS & LUP-

Ans. 20 zones

Q.846. The Bay of Bengal Initiative for MultiSectoral Technical and Economic Cooperation (BIMSTEC) was started

Ans. Since 2007

Q.847. Present Director, ICAR-NBSS & LUP, Nagpur in 2023-

Ans. Dr. N.G. Patil

Q.848. Present President, ICAR Society

Ans. Miniser of Agriculture Shri Shivraj Singh Chouhan

Q.849. Present Vice President, ICAR Society

Ans. Shri Kailash Choudhary

Q.850. All-India Soil and Land Use Survey Organization (AIS&LUS) was established in year

Ans. 1969

Q.851. Agro climatic zones classified by planning commission are

Ans. 15 zones.

Q.852. Agro climatic zones classified by NARP of ICAR are

Ans. 131

Q.853. Total State Agriculture Universities under ICAR in 2022-23

Ans. 63

Q.854. Total Agricultural Universities under ICAR are-

Ans. 74

Q.855. Highest No. of Agricultural Universities is in state-

Ans. Rajasthan (6) & Karnataka (6)

Q.856. Sam Higginbottom University of Agriculture, technology and Sciences is situated at

Ans. Allahabad

Q.857. National Organic Farming Research Institute (NOFRI) will be set up at

Ans. Sikkim

Q.858. National Biotic Stress Management Institute (NIBSM) is situated at

Ans. Raipur (Chhattisgarh)

Q.859. National Abiotic Stress Management Institute (NIASM) is situated at

Ans. Malegaon (Maharashtra)

Q.860. No. of Deemed Universities are in 2023-24-

Ans. 4

Q.861. No. of Central Agriculture University under ICAR in 2023-24

Ans. 3

Q.862. No. of National institutes under ICAR in 2023-24-

Ans. 66

Q.863. No. of Directorate under ICAR in 2023-24-

Ans. 13

Q.864. Agricultural Technology Application Research Institutes under (ATARI) ICAR are

Ans. 11

Q.865. No. of AICRP under ICAR in 2024-

Ans. 60

Q.866. Net work projects under ICAR are-
Ans. 19

Q.867. Other Projects under ICAR are-
Ans. 10

Q.868. Total No of National Bureaus are in 2023-24-
Ans. 6 Bureaus.

Q.869. Number of NRC centers under ICAR in 2023-24-
Ans. 15 NRC

Q.870. First NRC was started for which crop-
Ans. Ground Nut (Junagarh, Guj.)

Q.871. For which crop Dr. G.S.Khush awarded by wolf prize -
Ans. Rice in the year 2000

Q.872. National Bureau of Agriculturally Important Microorganism is situated at
Ans. Mau (U.P.)

Q.873. National Bureau of Fish Genetic Resources is situated at
Ans. Lucknow (U. P.)

Q.874. No. of KVK's in India in 2023-24
Ans. 731 KVK

Q.875. No. of KVK's under ICAR in 2023-24-
Ans. 66 KVK

Q.876. KVK under Agricultural Universities in 2023-24
Ans. 487 KVK

Q.877. KVK under Central Agricultural Universities-2023-24-
Ans. 22 KVK

Q.878. KVK Under NGO- in 2023-24
Ans. 101 KVK

Q.879. KVK under Public sector undertaking-2023-24
Ans. 2 KVK

Q.880. KVK Under state government 2023-24-
Ans. 38 KVK

Q.881. KVK under deemed university-2023-24-
Ans. 7 KVK

Q.882. State having highest numbers of KVK's in 2023-24
Ans. U.P. (89) 2nd M.P.

Q.883. Krishi Kalyan Abhiyan was commenced on
Ans. From 1 June 2018 to 31st July 2018

Q.884. NBAIR (National Bureau of Agricultural Insect Resources) is situated at
Ans. Bangalore

Q.885. NASF (National Agricultural Science Fund) is situated at
Ans. PUSA, New Delhi

Q.886. Present Chairman of NASF is
Ans. Dr. Panjab Singh

Q.887. Present ADG of NASF
Ans. Dr. Jitendra Kumar

Q.888. First chairman of NASF was
Ans. Dr. C.N.R. Rao

Q.889. National Gene Bank was established at-
Ans. New Delhi, 1996

Q.890. Network Project on Agricultural Bioinformatics and Computational Biology is situated at-
Ans. New Delhi

Q.891. Latest NRC was established for
Ans. NRCIF for Integrated Farming (Mothihari, Bihar)

Q.892. AICRP on Sesame and Niger is located at-
Ans. Jabalpur (M.P.)

Q.893. AICRP Vegetables including NSP vegetable is situated at
Ans. Varanasi (U.P.)

Q.894. National Academy of Agricultural Research Management is located at-
Ans. Hyderabad

Q.895. No. of Agricultural Technology Information Centers (ATIC) established under ICAR institutes and State Agricultural Universities.
Ans. 44 centers

Q.896. India's first all India Co- ordinate research project was for the crop-
Ans. Maize

Q.897. First Agriculture minister of independent India was-
Ans. Rafi Ahmed Kidwai

Q.898. Present Director General of ICAR
Ans. Dr. Himanshu Pathak

Q.899. Present Chairman of ASRB
Ans. Dr. Sanjay Kumar

Q.900. Foundation Day of ASRB is celebrated on
Ans. 1st November.

Q.901. Agricultural Education Day is celebrated on
Ans. 3rd December.

Q.902. Slogan of ICAR
Ans. Agri Search with Human touch

Q.903. ICAR day is celebrated on
Ans. 16th July

Q.904. No of Deputy Director general under ICAR-
Ans. 8

Q.905. No. of Assistant Director general under ICAR-
Ans. 24

Q.906. Present Secretary, Department of Agricultural Research and Education (DARE)-
Ans. Dr. Himanshu Pathak

Q.907. Present Additional Secretary (DARE) and Secretary (ICAR)
Ans. Shri Sanjay Garg (From 25 August 2021)

Q.908. Additional Secretary and Financial Adviser DARE/ ICAR
Ans. Ms Alka Nangia Arora

Q.909. The Department of Agricultural Research and Education (DARE) was established in the Ministry of Agriculture in -
Ans. Dec. 1973

Q.910. First Chairman of National Commission for Farmers
Ans. Dr. M.S. Swaminathan

Q.911. National commission of farmers (NCF) was constituted on-

Ans. Nov 18, 2004

Q.912. First Deputy chairman of planning commission was-

Ans. Gulzari Lal Nanda

Q.913. Present Chairman & M.D. of F.C. I. -

Ans. Ashok K Meena

Q.914. Director of Food and Agriculture Organization (FAO)-

Ans. QU Dongyu

Q.915. Who was the first Director of FAO-

Ans. Sir John Boyd Orr

Q.916. Foundation day of FAO

Ans. Oct 16, 1945

Q.917. How many countries are covered by FAO -

Ans. 130 Countries

Q.918. Present Chairman of Tea Board of India

Ans. Amandeep Singh Bhatia, IAS

Q.919. Set up of Indian Tea Board in year-

Ans. 1st April 1954

Q.920. Head Quarter of Indian Tea Board is situated at-

Ans. Kolkata

Q.921. Present Director General of International Rice Research Institute (IRRI) is

Ans. Dr. Yvonne Pinto

Q.922. IRRI South Asia Regional Centre (IRRI SARC) is located at-

Ans. Varanasi, Uttar Pradesh,

Q.923. Present Director Of IARI-

Ans. Dr. Tilak Raj Sharma

Q.924. First Director of IARI-(Pre-Independence)

Ans. Bernard Coventry (1904-1916)

Q.925. Present Director of IARI belongs to which department-

Ans. Crop Science & Biotech

Q.926. Head office of NABARD is situated at-

Ans. Mumbai

Q.927. Directorate of Knowledge Management in Agriculture (DKMA) is situated at

Ans. New Delhi

Q.928. Directorate of Floricultural Research-

Ans. Pusa, New Delhi

Q.929. Black revolution is related to

Ans. Biofuel /Jatropha production.

Q.930. Set up of National Fisheries Development Board (NFDB)

Ans. Sep 2006

Q.931. Year known as International Rice year-

Ans. 2004

Q.932. Year known as International Forest year-

Ans. 2011

Q.933. Year known as International Horticulture year-

Ans. 2012

Q.934. International Year for Fruits and Vegetable is -

Ans. 2021

Q.935. Year known as International Pulse year-

Ans. 2016

Q.936. Year known as International Soils Year-

Ans. 2015

Q.937. Year known as International Potato year-

Ans. 2008

Q.938. World food day is celebrated on -

Ans. 16th Oct.

Q.939. World Pulses day is celebrated on-

Ans. 10th Feb

Q.940. International Carrot day is celebrated on-

Ans. 4th April

Q.941. World environment day is celebrated on -

Ans. 5th June.

Q.942. World water day celebrated on-

Ans. 22 March.

Q.943. World forest day is celebrated on -

Ans. 21st March.

Q.944. World Metrological day is celebrated on-

Ans. 23rd March.

Q.945. ISOPOM scheme launched in -

Ans. 2004 (In 14 Major States)

Q.946. Block Delivery Scheme (BDS)is related to-

Ans. Fertilizer (1980-81)

Q.947. HPC committee is related to-

Ans. Fertilizer consumer price.

Q.948. National mission on Micro Irrigation (NMMI) stared on-

Ans. 2010

Q.949. All India Bt cotton area of total cotton area-

Ans. 86% of cotton (8.4 M ha.)

Q.950. Red soils are dominated in state-

Ans. Tamil Nadu

Q.951. First soil testing lab was starting in year

Ans. 1955-56 (IARI)

Q.952. Agri scientist who awarded for Noble Prize for peace in 1970-

Ans. Dr. N.E. Borlaug

Q.953. Cash prize given in ICAR's N. Borlaug Award-

Ans. 10 Lac

Q.954. NBS policy for SSP was effective from the year-

Ans. 1st May-2010

Q.955. National committee on the use of Plastics in Agriculture (NCPA) was constituted in -

Ans. March, 1981

Q.956. First RBI Governor of India

Ans. Osborne Smith (April 1935)

Q.957. Present RBI Governor of India

Ans. Shri Shaktikanta Das (25th Governor)

Q.958. Founder Chairman of NABARD-

Ans. M. Ramakrishnayy

Q.959. Present Chairman of NABARD-

Ans. Shri Shaji K V

Q.960. Largest cooperative society of India-

Ans. IFFCO

Q.961. Present Chairman of IFFCO

Ans. Mr. Dileep Sanghani

Q.962. Present Chairman of KRIBHCO-

Ans. Dr. Chandra Pal Singh

Q.963. Present Director of NDRI (Karnal) is-

Ans. Dr. Dheer Singh (Acting Director)

Q.964. First Director of NDRI (Karnal) was-

Ans. William Smith (1920-1931)

Q.965. National Institute of Nutrition is based at-

Ans. Hyderabad

Q.966. GCHERA World Agriculture Prize for the year 2021 given to

Ans. Dr. Marc Van Montagu & Dr. Mitiku Haile

Q.967. World Food Prize 2023 has given to

Ans. Heidi Kuhn

Q.968. World Food Prize for 2024 won by

Ans. Geoffrey Hawtin and Cary Fowler

Q.969. Founder of World Food Prize

Ans. Dr. Norman Borlaug

Q.970. Max cash amount given for ICAR Challenge award prize is

Ans. Rs.10 Lac

Q.971. Who was awarded by IFA Norman Borlaug Award in 2023-

Ans. Dr. Jian Feng Ma

Q.972. Who was winner for Norman Borlaug Field award 2023-

Ans. Dr. Swati Nayak

Q.973. Internationally recognized system for reducing the risk of safety hazard in crop is

Ans. HACCP (Hazard Analysis and Critical Control Point)

Q.974. First Chairman of National Dairy Development Board was

Ans. Dr. V. Kurien

Q.975. Present Chairman of (NDDB) National Dairy Development Board is

Ans. Meenesh C Shah

Q.976. National Livestock Mission (NLM) was launched in-

Ans. 2014-15

Q.977. Present president of Indian Sugar Mills Association (ISMA) in 2024

Ans. Mr. Mandava Prabhakar Rao

Q.978. Present Director General of ICRISAT in 2024

Ans. Dr. Jacqueline d'Arros Hughes

Q.979. ICRISAT was established in year

Ans. 1972

Q.980. Present D. G. of MANAGE in 2024

Ans. Dr. P. Chandra Shekara

Q.981. Chairman of NITI Aayog

Ans. Prime Minister

Q.982. Present Vice Chairman of NITI Aayog-2024-

Ans. Shri Suman K Bery

Q.983. Present President of World bank in 2024-

Ans. Ajay Banga

Q.984. Present Director General of WHO - 2024-

Ans. Dr. Tedros Adhanom Ghebreyesus

Q.985. Present Chairman & MD of National Seed Corporation (NSC)

Ans. Dr. Maninder Kaur Dwivedi (IAS)

Q.986. National Agro forestry Policy (NAP) was launched in Year

Ans. 2014

Q.987. National Seed research and training Centre (NSRTC) is located at

Ans. Varanasi (U.P.)

Q.988. International Livestock Research Institute (ILRI) is located at

Ans. Nairobi, Kenya

Q.989. Duration of Agriculture year is-

Ans. 1st June to 31st May

Q.990. Average rainy days in country is-

Ans. 130 days/ year

Q.991. South India state which receives most of rainfall during N.E. Monsoon

Ans. Tamil Nadu

Q.992. (NC-MRWF) National center for Medium Range weather forecasting is based at

Ans. New Delhi

Q.993. Crop known as vegetable meat

Ans. Cow pea

Q.994. Crop known as Queen of cereals

Ans. Maize

Q.995. Country having highest irrigation efficiency-

Ans. Israel

Q.996. Country first in sugar production and consumption-

Ans. India

Q.997. State which is the largest producer of cut flowers

Ans. West Bengal

Q.998. Zabo Indian farming system is mainly practiced in-

Ans. Nagaland

Q.999. City known as Foliage capital of World-

Ans. Apopka, Florida

Q.1000. First transgenic crop approved for cultivation in USA during year 1994 was

Ans. Tomato

Q.1001. Crop contributes highest revenue through excise among Agri commodities

Ans. Tobacco

Q.1002. Crop known as king of coarse grains-

Ans. Sorghum (Camel Crop)

Q.1003. Variety of wheat which had major contribution in green revolution

Ans. HD2329

Q.1004. Highest award presented to agricultural scientist in India is-

Ans. Rafi ahmad kidwai award

Q.1005. Netaji Subhas-ICAR International Fellowships is given to promote for

Ans. Globalization of agricultural education

Q.1006. Father of Golden revolution-

Ans. Dr. K.L. Chadda

Q.1007. Pradhan Mantri Fasal Bima Yojana was launched in

Ans. 18 Feb , 2016

Q.1008. Soil Health Card Scheme (SHC) was launched on-

Ans. 19th Feb 2015

Q.1009. Project "Mera Gaon Mera Gaurav" was launched by-

Ans. I C A R

Q.1010. PM Kisan Scheme was launched in

Ans. 24th Feb 2019 (Gorakhpur)

Q.1011. Amount given to farmer under PM Kisan Scheme-

Ans. Rs.6000 / Year

Q.1012. Premium rate for Kharif crops under PMFBY-

Ans. 2%

Q.1013. Premium rate for Rabi crops under PMFBY-

Ans. 1.5%

Q.1014. Premium rate for commercial/ Horticultural crops under PMFBY-

Ans. 0.5%

Q.1015. National Gender Resource Centre in Agriculture (NGRCA) is located at

Ans. Pusa , New Delhi

Q.1016. Institute of Himalayan Bioresource Technology is located at-

Ans. Palampur (H.P.)

Q.1017. Commission on Agricultural Cost & Price (CACP) was set up in

Ans. Jan, 1965

Q.1018. Chairman of CACP is

Ans. Prof. Vijay Paul Sharma

Q.1019. KCC scheme was launched in year -

Ans. 1998-1999

Q.1020. KCC Scheme was prepared by

Ans. NABARD Bank

Q.1021. KCC Scheme recommendation was given by-

Ans. R.V. Gupta Committee

Q.1022. Present Chairman of NAFED in 2024-

Ans. Sh. Jethabhai Ahir

Q.1023. NAFED was established in-

Ans. 2 Oct, 1958

Q.1024. CFCC was created in year

Ans. 1998

Q.1025. CFCC is responsible for

Ans. Periodic crop forecasting of major crops

Q.1026. Full form of NSSO

Ans. National Sample Survey Organization

Q.1027. Central Sheep Breeding Farm is located at-

Ans. Hisar (Haryana) 1968-70

Q.1028. Department of Revenue and Agriculture and Commerce was set up in year-

Ans. June 1871

Q.1029. National Food Security Mission' (NFSM) was launched in -

Ans. October 2007.

Q.1030. NFSM was launched to increase the production of especially for-

Ans. Rice, Wheat & Pulses

Q.1031. Food grain storage duration max for rural go down-

Ans. 10 Months

Q.1032. Programme which was launched for unemployed Agri graduates by extension Dep

Ans. ACABC (Agri-Clinic and Agri-Business Centers by Agriculture Graduates)

Q.1033. Kisan Call Center agents are known as -

Ans. Farm Tele Advisors (FTAs)

Q.1034. Full form of STRY scheme-

Ans. Skill Training of Rural Youth

Q.1035. Full form of FCAC programme

Ans. Farmers Capacity Assessment & Certification

Q.1036. Directorate of Areca nut and spices development was established on

Ans. 1st April, 1966 (Calicut, Kerala)

Q.1037. National Horticulture Board (NHB) was set up by Government of India in

Ans. April 1984

Q.1038. Present Managing Director of National Horticulture Board-

Ans. Priya Ranjan

Q.1039. The National Horticulture Board (NHB) was established in the year

Ans. 1984

Q.1040. The Indian Society for Horticultural Research and Development (ISHRD), established

Ans. Dehradun, Uttarakhand, 1969

Q.1041. IIHR was established in

Ans. 5 Sep, 1967

Q.1042. Headquarters of Coconut Development Board -

Ans. Kochi in Kerala

Q.1043. E- NAM was launched in year-

Ans. 14th April 2016

Q.1044. Mandis registered under E – NAM-till 2023-

Ans. 1389 Mandi

Q.1045. Agency which is enveloped in enhancing export of processed agricultural products

Ans. APEDA

Q.1046. National Agricultural Women Resource Centre established in

Ans. 2004-05 (Pusa, New Delhi)

Q.1047. Technology which is included in PMFBY

Ans. Smart Sampling Technique and Crop cutting

Q.1048. Key factor which plays major role for recommendation of MSP

Ans. Cost of Cultivation

Q.1049. Soil Health Card Day was celebrated in India

Ans. Feb 19, 2020

Q.1050. National Milk Day Celebrated in India on

Ans. 26th Nov

Q.1051. World Milk Day celebrated on

Ans. 1st June

Q.1052. World Soil Day is celebrated on

Ans. 5th Dec

Q.1053. National Artificial Insemination Programme (NAIP) was launched in-

Ans. 11th Sep 2019

Q.1054. PM Man Dhaan Yojna was approved on-

Ans. 31st May, 2019

Q.1055. State got highest no of Krishi Karman award -

Ans. Tamil Nadu

Q.1056. National Women farmer's Day (Rashtriya Mahila Kisan Diwas) is Celebrated on

Ans. 15th Oct

Q.1057. Present Secretary of Agriculture Corporation & Farmers Welfare is -

Ans. Mr. Manoj Ahuja

Q.1058. Growth rate in 12th Five Year Plan for Agriculture and allied sector-

Ans. 2.40%

Q.1059. Two multiple-stress tolerant rice varieties namely Swarna Samridhi Dhan and Swarna Sukha Dhan have been released by

Ans. ICAR (Year 2020 and 2021)

Q.1060. First Five Year plan implanted in world by-

Ans. Joseph Stalin (1920, Soviet Union)

Q.1061. First Five-Year Plan was based on-

Ans. **Harrod-Domar Model.**

Q.1062. Funder Chairman of NECC (National Egg Co-Coordination Committee)-

Ans. Padmasri Dr.B. V. Rao

Q.1063. Budget given for National Beekeeping and Honey Mission (NBHM)

Ans. 500 Crore (2020-21 to 22-23)

Q.1064. The Farmers' (Empowerment and Protection) Agreement on Price Assurance and Farm Services Act was started in Year

Ans. 2020

Q.1065. Paramparagat Krishi Vikas Yojna (PKVY) is related to-

Ans. Organic farming

Q.1066. In which farming there is use of information and technology-based system

Ans. Precision Farming

Q.1067. Advisor of Directorate of Plant Protection Quarantine & Storage (DPPQ&S)

Ans. Dr. J. P. Singh

Q.1068. Total plant quarantine stations at airports, Seaports and land frontiers are in India

Ans. 73

Q.1069. Full Form of ARYA Project is

Ans. Attracting and Retaining Youth in Agriculture

Q.1070. How many KVK's are operating ARYA project in India-

Ans. 100 KVK's.

Q.1071. Under the ICAR-CIWA-IRRI collaborative project a Women Farmer Producer Company was established, which named was-

Ans. Chitri Dora

Q.1072. To increase the penetration of agricultural technologies and two-way effective communication with farmers, multi-media based multilingual ICT based communication system was launched which named was

Ans. Kisan Sarathi

Q.1073. Rajbhasha Aalok is being published regularly by

Ans. ICAR

1074. Indian Patent Office Head Quarter is located at-

Ans. Mumbai

Q.1075. Foundation day of Indian Patent Office

Ans. 1st Jan, 1912

Q.1076. Directorate of Plant Protection Quarantine & Storage was established in the year

Ans. 1946

Q.1077. Directorate of Plant Protection Quarantine & Storage was established on the recommendation of-

Ans. Woodhead Commission

Q.1078. The United Nations (UN) General Assembly declaring 2023 as the International Year of

Ans. Millets

Q.1079. Country proposed UN for declaring the year 2023 as international year of millets-

Ans. India

Q.1080. Government of India has launched AMRUT 2.0 on-

Ans. 1st October 2021

Q.1081. Which Five-year plan is known as Gadgil Yojna-

Ans. Third Five Year Plan

Q.1082. The National Programme for Organic Production (NPOP) was started in

Ans. 2001

Q.1083. NPOP is being managed by -

Ans. APEDA

Q.1084. Prampragat Krishi Vikas Yojana (PKVY) was started in-

Ans. 2015-16

Q.1085. PKVY is a part of

Ans. National Mission on Sustainable Agriculture (NMSA).

Q.1086. India's rank in terms of the World Organic Agricultural land -

Ans. 2nd Rank.

Q.1087. World Intellectual Property Day is celebrated on-

Ans. 26 April

Q.1088. Area registered under National Programme for Organic Production in India in 2023-24

Ans. 7.3 mha

Q.1089. India produced certified organic products in 2023-24

Ans. 3.6 Million MT

Q.1090. Government constituted an Inter-ministerial Committee in April, 2016 to examine issues relating to

Ans. Doubling of Farmers Income

Q.1091. Central Sector Scheme of financing facility under Agri Infrastructure Fund is operational from the year

Ans. 2020-21 to 2032-33.

Q.1092. "Sweet Revolution" is related to

Ans. Mission Beekeeping & Honey

Q.1093. Funding amount and period for "National Beekeeping & Honey Mission (NBHM)

Ans. 500 Cr, 3 years (FY 2020-21-2022-23)

Q.1094. Full form of (PM-AASHA)

Ans. Pradhan Mantri Annadata Aay Sanrakshan Abhiyan

Q.1095. Full form of (HMNEH)

Ans. Horticulture Mission for North East and Himalayan States

Q.1096. Basmati Export Development Foundation (BEDF) is located at

Ans. Meerut (U.P.)

Q.1097. Chairperson of International Grains Council

Ans. Ms. Anita Katial

Q.1098. Head office of International Grains Council-

Ans. London (UK)

Q.1099. Cabinet approves ambitious programme for Horticulture

Ans. Clean Plant Centres (CPCs) established across India in the ICAR Institutes.

Q.1100. No. of crop varieties of Field and Horticulture Crops dedicated to farmers by PM (Sh. Narendra Modi)

Ans. 109 (Field Crops 69, Horticultural Crops, 40)

BASIC PRINCIPLES OF CROP PRODUCTION

Q. 703. Term Agronomy is derived from
Ans. Greek language

Q. 703a. Father of Agronomy.
Ans. Peter De-cresenzi

Q. 704. Who wrote Horse Hoeing Husbandry.
Ans. Jethro Tull

Q. 705. Father of weed science.
Ans. Jethro Tull

Q. 706. Jhum cultivation is mostly found in which part of India?
Ans. Eastern part

Q. 707. Demerit of shifting cultivation is.
Ans. Soil loss

Q. 708. Objective of sustainable Agriculture is.
Ans. Ecological balance

Q. 709. Which pattern of planting has maximum plant population.
Ans. Cubodial pattern

Q. 710. Farming which includes crop production and live-stock.
Ans. Mixed farming (First done by La-Flitze in the year 1928)

Q. 711. A system of growing the same crop on the same land year after year is known as.
Ans. Mono cropping

Q. 712. Cultivation of such crops which have different natural habit and zero competition is known as.
Ans. Parallel cropping *e.g.*, Green gram + Maize

Q. 713. The cropping system in which the yields of both crops are higher than of their pure crops on unit area basis is called as.
Ans. Synergetic cropping

Q. 714. Growing hard and thorny crops. (Mesta/Safflower) around the main crop.
Ans. Guard crops

Q.715. Growing of different crops in the same area in sequential seasons
Ans. Rotation cropping

Q. 716. Cultivation of two or more than two crops of different heights simultaneously on a certain piece of land in any certain period is called as.

Ans. Multistoried cropping

Q. 717. Cropping intensity will be always 100% in.

Ans. Monocropping

Q. 718. Peira cropping is mostly adopted in.

Ans. Bihar & West Bengal

Q. 719. Kind of cropping in which a line is left unsown in the regular row series.

Ans. Skip cropping

Q. 720. Herbicide resides for longest period in soil

Ans. Linuron

Q. 721. The farming system used to develop the mass (planet) according to the Earth's environment is called.

Ans. Terra farming

Q. 722. Formula of Harvest Index is.

Ans. $\text{H.I.} = \frac{\text{Economic yield}}{\text{Biological yield}} \times 100$

Q. 723. Cropping Intensity is given by

Ans. $\text{C.I.} = \frac{\text{Total cropped area}}{\text{Total sown area}} \times 100$

(Menegay proposed cropping intensity index in 1978)

Q. 724. Agro climatic Regional planning in India was initiated in which five year plan.

Ans. 7th Five year plan (1988)

Q. 725. Total agro climatic zones in India are.

Ans. 15 Zones

Q. 726. Cultivation of crops in areas where annual rainfall is less than 750m.m. is called as.

Ans. Dry Farming

Q. 727. Cultivation of crops in areas where annual rainfall is more than 750m.m. but less than 1150 mm is called as.

Ans. Dry land Farming

Q. 727a. Iron share invented by
Ans. Robert Ransome (1789)

Q. 727b. Iron plough invented by
Ans. James Small (Scottish Inventor) (1789-80)

Q. 727c. First Agricultural Eperimentation station started by
Ans. JB Bossinggault in 1836 at Pechelbronn in Alsace) France

Q. 728. Means of cultivation of crops in regions where annual rainfall is more than 1150m.m. is called as.
Ans. Rain fed Farming

Q. 729. When annual rainfall is less than 75% of normal the situation will be called as.
Ans. Drought

Q. 730. If deficiency of rainfall is above 50% of the normal the situation will be called as.
Ans. Severe drought (1935)

Q. 731. Supplemental irrigation is known as.
Ans. Life saving irrigation

Q. 732. P.M.A. phenyl mercuric acetate is a type of anti transpirant.
Ans. Stomata closing type

Q. 733. Kaolin is a type of antitranspirant.
Ans. Reflecting type

Q. 734. Cycocel is a.
Ans. Growth retardant

Q. 735. The direction from which the winds are coming are called as.
Ans. Wind ward side

Q. 736. For the regular cultivation the land capability classes used are.
Ans. I, II & III Classes

Q. 737. Agronomic measures are used to reduce erosion where slope is
Ans. Less than 2%

Q. 738. Bench terracing usually practiced on slopes ranging from.
Ans. 16-33%

Q. 739. Family of tobacco.
Ans. Solanaceae

Q. 739a. Mould Board Plough was invented by
Ans. Thomas Jefferson (1793 AD)

Q.739a Agricultural Experimental Station at Rothamsted , England was started by
Ans. Lawes and Sir Gilbert
Q. 740. Family of jute.
Ans. Tiliaceae
Q. 741. Family of sesame.
Ans. Pedoliaceae
Q.742. Family of castor.
Ans. Euphorbiaceae
Q. 743. Which family of crops is more exhaustive.
Ans. Graminae
Q. 744. Tobacco is a.
Ans. Rabi crop
Q. 745. Origin of maize
Ans. Mexico
Q. 746. Origin of soyabean.
Ans. China
Q. 747. Origin of potato.
Ans. Peru (S. America)
Q. 748. Origin of tobacco
Ans. Mexico & Central America
Q. 749. Highest area of summer maize.
Ans. Bihar
Q. 750. Highest consumption of K fertilizer is in.
Ans. Maharashtra
Q. 751. Highest area of cereals in India.
Ans. 1. Rice 2. Wheat
Q. 752. Highest production of Cereal in world.
Ans. Wheat
Q. 753. De suckering is a process of a crop.
Ans. Tobacco.
Q. 754. Harvest Index is low in.
Ans. Pulses
Q. 755. Weight of 100 seeds is known as.
Ans. Seed Index
Q. 756. Weight of 1000 seeds is known as.
Ans. Test weight
Q. 757. Dockage is the.
Ans. Impurity percentage

Q. 758. Formula of Real value is.

Ans. $RV = \frac{\text{Purity \%} \times \text{Germination\%}}{100}$

Q. 760. Term Legume taken from

Ans. Latin word *legera* mean to *collect* or *gather*.

Q. 761. The weed which was first biologically controlled.

Ans. Lantana Camaera

Q. 762. Trade name of Alachlor.

Ans. Lasso

Q. 763. Trade name of Butachlor

Ans. Machete

Q. 764. Trade name of Fluchloralin.

Ans. Basalin

Q. 765. Trade name of Propanil.

Ans. Stam F-34

Q. 766. Trade name of Pendamethalin.

Ans. Stomp

Q. 767. Total root parasite.

Ans. Orobanche

Q. 768. Total stem parasite.

Ans. Cuscuta

Q. 769. Semi root parasite.

Ans. Striga

Q. 770. Semi stem parasite.

Ans. Loranthus

Q. 771. Orobanche is also known as.

Ans. Broom rape

Q. 772. Striga is also known as.

Ans. Witch weed (Parasitic plant) occur naturally

Q. 773. Host crops of orobanche.

Ans. Tobacco, chili, brinjal, potato

Q. 774. Host plants of striga.

Ans. Sorghum, maize, sugarcane, sunflower

Q. 775. Cuscuta (Parasitic weed) is serious problem with.

Ans. Lucerne crop

Q.776. Horizontal flow of water in channel is called as.

Ans. Seepage (*JRF 2015*)

Q. 777. Vertical movement of water in soil known as.

Ans. Percolation

Q. 778. PF value was first time introduced by.

Ans. Schofield (1935)

Q. 779. Soil moisture tension was directly measured by.

Ans. Tensiometer

Q. 780. Tensiometer was first designed by

Ans. Richards and Gardner (1936)

Q. 781. Lysimeter is used for measurement of.

Ans. Evapo - transpiration

Q. 782. Parshall flume is used for measurement of.

Ans. Water flow

Q. 783. Flooding irrigation method commonly used for.

Ans. Rice

Q. 784. Fertilizer application through irrigation is known as.

Ans. Fertigation

Q. 785. Important cultural practice in rice field.

Ans. Puddling

Q. 786. Power tiller is most suitable for the cultivation of.

Ans. Paddy

Q. 787. Post harvest losses for cereals accounts for.

Ans. 10%

Q. 788. Saltation is a process of.

Ans. Wind erosion

Q. 789. Highly salt tolerant cereal crop is.

Ans. Barley

Q. 790. Irrigation efficiency of loam soil is.

Ans. 70%

Q. 791. In saline soil the method for irrigation used is

Ans. Flood method

Q. 792. Crops between the main crop are called as.

Ans. Inter crop

Q. 793. Sowing pattern used in dry land.

Ans. Broadcasting

Q. 794. Crop lodging was first adopted in which crop.

Ans. Sugarcane

Q. 795. Most efficient method of irrigation is.

Ans. Drip irrigation

Q. 796. Extensively grown pulse crop in India.

Ans. Chick pea

Q. 797. Relative humidity is measured by.

Ans. Psychrometer

Q. 798. Crops grown to conserve soil moisture are known as.

Ans. Mulch crops

Q. 799. Mat type nursery is related to crop

Ans. Paddy

Q. 800. Fertilizer (most concentrated) used for nutrient supply to crops is

Ans. Anhydrous Ammonia

Q. 801. Weed Amaranth is also known as

Ans. Pigweed

Q. 802. Annidation

Ans. Complementry interaction between inter crops.

Q. 803. Competitive interaction between the crop species in intercropping is

Ans. Allelopathy

Q.804. The practice of taking the second crop from previous one is known as

Ans. Ratooning

Q. 805. Competition Index was proposed by
Ans. Donald (1968)

Q. 806. Curing is associated with which crop
Ans. Tobacco & Tea.

Q. 807. Trashing is associated with which crop
Ans. Sugarcane

Q. 808. Stripping is associated with which crop
Ans. Jute

Q. 809. Wrapping is associated with which crop
Ans. Sugarcane

Q. 810. Wind circulation in cyclone in Northern hemisphere is called as
Ans. Anticlockwise

Q. 811. Tender annuals with semi-aquatic habit
Ans. Weland weeds

Q. 812. The important turfgrass nutrient which is needed in larger amoutns and supplied more often is
Ans. Nitrogen

Q. 813. Secondary cultivation is a done during which season
Ans. Spring, Summer or Autumn.

Q. 814. LER stand for
Ans. Land Equivalent Ratio

Q. 815. Rotavator is used for
Ans. Primary & secondary cultivation

Q. 816. Chemical contained by Roundup herbicide
Ans. Glyphosate

Q. 817. Roundup is registered product of which company
Ans. Monsanto company

Q.818. Collective farming system has originated from
Ans. Soviet Union (1927-1933)

Q.819. The "miracle plant" which is useful for fuel, fodder fire wood and manure is
Ans. *Leucaena leucocephala* (River tamarind or Pearl wattle)

Q.820. Oldest tractor manufacturing company

Ans. New Holland (1895)

Q.821. Crop which is used to make car fuel in Brazil

Ans. Sugarcane

Q.822. First chemically manufactured fertilizer is

Ans. ammonium sulphate

Q.823. Optimum temperature for crop production

Ans. 15-20°C *(MPPSC, 2011)*

Q.824. Water use efficiency highest under which method

Ans. Drip Irrigation (90-95%) (*JRF 2013*)

Q. 825. What is Inter cropping

Ans. Growing two or more crops simultaneously in alternate rows

Q. 826. What is Utera cropping

Ans. Sowing of next seeds before harvesting of standing crop in order to utilize moisture efficiency under rainfed agro-ecosystems. Utera cropping is adopted in Rabi seasons only (Mostly in M.P.).

Q.827. The country who introduced Zero tillage first

Ans. U.S.A

Q.828. Tetrazoloin test is used to determine

Ans. Seed viability

Q.829. Collego (C. gloeosphoroides) is type of herbicide

Ans. Myco herbicide

Q.830. The largest producer state for Moth bean is

Ans. Rajasthan

Q.831. The method of irrigation employed where surface is undulating is

Ans. Sprinkler irrigation (Efficiency 75%)

Q.832. Bund former is used for making

Ans. Ridges/ Bunds

Q.833. Sunflower act as an indicator plant to measure the deficiency of

Ans. Boron

Q.834. Topping in safflower plant is done

Ans. To promote branching and flowering

Q.835. Growing of coconut, black pepper and ginger simultaneously in the same field is called as

Ans. Multistoried cropping

Q.836. Stevenson screen is related to

Ans. Agro meteorology. (It is a shelter or enclosure to shield meteorological instruments used to record humidity and air temperatures). It is kept at a height of 1·25 m/4·1 ft above the ground.

Q.837. The term "Evergreen Revolution" has been given by

Ans. Dr. M. S. Swaminathan (JRF 2019)

Q.838. The most important potential contaminant of food produced on sewage sludge amended soils is

Ans. Cadmium

Q.839. PMA (phenyl mercuric acetate) chemical is used in agri plants to

Ans. Reduce Transpiration

Q.840. The concept of Q/I relationship of labile soil was developed by

Ans. Beckett 1964 (Used to predict K availability to plants in which K^+ in the soil solution (an intensity factor I) is related to changes in exchangebled K^+ in the soil (a quantity factor Q)

Q.841. At field capacity the value of pF is

Ans. 2.5

Q.842. At Permanent wilting point (PWP) the pF value is

Ans. 4.2

Q.843. LER is also known as

Ans. Relative Yield Toal (RYT)

Q.844. "A" value concept was given by

Ans. Fried and Dean

Q.845. Application of "K" increase

Ans. Disease resistance in plants

Q.846. Crops grown along irrigation and drainage channel are

Ans. Riparian crops

Q.847. Growing of field crops in between two widely spaced rows of fast growing trees is known as

Ans. Alley cropping

Q.848. Growing of four crops on the same land in a year in such a manner that the following crop is sown before the harvest of preceding crop is known as

Ans. Relay Cropping

Q.849. A short duration crop in between two main seasonal crops is known as

Ans. Catch crop

Q.850. Rainfall intensity is generally higher at

Ans. Near the equator

Q.851. Form of "N" which is absorbed by paddy in water logged condition

Ans. NH_4 ion

Q.852. Nutrient which if taken in excess is known as luxury consumption

Ans. Potassium

Q.853. Seed work board is required for

Ans. Blending of seed

Q.854. LEISA (Low External Input sustainable Agriculture) is related to

Ans. Organic farming and Environmental conservation

Q.855. All the grain legumes have a high photo respiration because of

Ans. C_3 mechanism

Q.856. Hydraulic conductivity of soils varies directly with

Ans. Total Porosity

Q.857. Organic carbon is a measure of

Ans. Available of N in soil

Q.858. In a good quality manure C/N ratio should not be more then
Ans. 10

Q.859. 'Tarai' soils are deficient in
Ans. Zn

Q.860. Permanent soil property is
Ans. Soil texture

Q.861. Blind tillage refers to
Ans. Hoeing before germination

Q.862. N fixation in soil by Rhizobium is increased by
Ans. Phosphorus

Q.863. Crops whose main root goes deep into the soil example Cotton, Tur
Ans. Tap roots

Q.863. Crop whose roots are fibrous, shallow and spreading like cereal crops
Ans. Adventitious roots

Q.864. First Dry farming research station, Manjri (Pune) was started in India
Ans. 1923

Q.865. Head quarter of Rainfed Authority of India is located at
Ans. New Delhi

Q.866. The study of the relationship of Agri crops and environment is known as
Ans. Agro Ecology

Q.867. Primitive method of breeding
Ans. Mass selection

Q.867. Winds are mainly caused due to
Ans. Pressure gradients.

Q. 868. LAI (Leaf Area Index) concept was given by
Ans. Watson, 1948

Q. 869. In which soils group drainage facility especially required
Ans. Black Soils

Q. 870. Anti transpirants also reduces-
Ans. Plant Photosynthesis

Q. 871. Which gas is a effective anti transpirant-
Ans. CO_2

Q. 872. Growing of Agriculture crops + Forest Trees known as-
Ans. Agri-Silviculture

Q. 873. Silviculture + Pasture management known as-
Ans. Silvo-Pastoral System

Q. 874. Agriculture crops + Forest crops + Fish production known as-
Ans. Agri-Silvi-Aquaculture System

Q. 875. Agriculture + Fruit Trees + Silviculture is known as-
Ans. Agri- Horti- Silvicultural system

Q.875a Trees + Crops + Pasture for Animals
Ans. Agri Silve Pasture

Q. 876. Which soil conservation measures are adopted if land slope is > 2% -
Ans. Mechanical measures

Q. 877. Contour bunding is mostly used in-
Ans. Arid and Semi arid areas where slope is up to 6%

Q. 878. Graded bunding is recommended for those area where annual rainfall is-
Ans. > 800 mm

Q. 879. BBF (*Broad Bed & Furrow system*) is mostly used for which type of soils-
Ans. Deep Black Soils

Q. 880. BBF method widely practiced in which crop-
Ans. Grount Nut

Q. 881. Plant most suitable for soil conservation as a vegetative barriers-
Ans. Khus (*Vetteria Zelanica*)

Q. 882. Soil conservation method mostly used in coffee gardens-
Ans. Vertical mulching

Q. 883. Lines joining equal elevations called as-
Ans. Contour

Q. 884. Lines connecting points of equal sunshine hours called as-
Ans. Isohels

Q. 885. Lines connecting points of equal temperature called as-
Ans. Isotherm

Q. 886. Instrument used to measure height of a location above sea level.
Ans. Altimeter

Q. 886a. Imaginary lines connecting points of equal amount of rainfall called as
Ans. Isohyets

Q. 886b. Imaginary lines connecting points of equal atmospheric pressure is called as
Ans. Isobar

Q. 886c. Imaginary lines connecting equal depth of rainfall is called as
Ans. Isopluvial

Q. 886d. Imaginary lines connecting equal points of equal wind velocities called as
Ans. Isotech

Q. 887. Instrument used to measure sunshine-
Ans. Campbell Stokes recorder

Q. 888. Instrument used to measure specific gravity of soil-
Ans. Pyconmeter

Q. 889. Irrigation project having 2000- 10000 CCA(ha) will be categorized as-
Ans. Medium Project

Q. 889a. Minimum amount of rainfall necessary for taking up sowing
Ans. 20 mm

Q. 890. > 2.5 mm rains during 24 hours called as-
Ans. Rainy day

Q. 891. Average speed of S. W. monsoon is-
Ans. 30 km / day

Q. 892. Contribution of N. E. Monsoon in Indian rainfall amount-
Ans. 15%
Q. 893. Which monsoon also known as Retreating monsoon
Ans. North East monsoon
Q. 894. Chemical used for warm clouds
Ans. NaCl (Sodium chloride)
Q. 895. Short range of weather forecasting is-
Ans. < 3 Days or up to 72 hrs
Q. 896. Long range of weather forecasting is-
Ans. > 10 Days
Q. 897. Latent heat of vaporization(water) is-
Ans. 540 Cal/g at 0°C
Q. 898. Highest specific heat of water for any substance (except Liq ammonia) is-
Ans. 1 Cal/g
Q. 899. One Cusec water is equal to-
Ans. 28.3 liters
Q. 900. Situation where all macro and micro pores are filled with water temporarily/ permanently called as-
Ans. Water logging
Q. 901. Water oxygen diffusion rate always reduces-
Ans. ODR and EH (Oxygen diffusion rate and redox potential)
Q. 902. Amount of water transpired by a crop to produce unit amount of dry matter is known as-
Ans. Transpiration ratio approach
Q. 903. Highest IW/CPE (Irrigation water /comulative pan evaporation) ratio for the cereal crop-
Ans. Rice (1.2)
Q. 904. Lowest IW/ CPE ratio for the crop-
Ans. Safflower (0.4)
Q. 905. According to CRIDA shallow ploughing depth is-
Ans. 5-6 cm
Q. 906. According to CRIDA Deep ploughing depth is-
Ans. 25-30 cm
Q. 907. Area of Macro watershed-
Ans. >50,000ha

Q. 908. Area of Micro watershed-
Ans. 100-1000ha

Q. 909. Area of Mini watershed-
Ans. 10-100ha

Q. 910. Most common irrigation method in India is-
Ans. Surface Irrigation

Q. 911. Irrigation method which is most common among surface method-
Ans. Check Basin Method

Q. 912. Suitable irrigation method for fruits is-
Ans. Ring basin method

Q. 913. Irrigation method which is suitable with a slope of 0.5-1% but not for sandy soils-
Ans. Border strip method

Q. 914. Sprinkler system is not suitable for-
Ans. Rice & Jute crops and heavy clay soils

Q. 915. Water discharge rate of sprinkler irrigation-
Ans. >1000 lit/hr

Q. 916. Sprinkler pressure required for sprinkler system-
Ans. 2.5 – 4.5 kg/cm^2

Q. 917. Emitters are important part of-
Ans. Drip irrigation

Q. 918. Irrigation method suitable for salt affected soils-
Ans. Drip irrigation

Q. 919. Size of BBF (Broad bed and Furrow method)
Ans. Recommended 90 cm width of bed and 45 cm wide furrow, sowing of crop 30 cm spacing

Q. 920. Mole drainage is suitable for-
Ans. Clay soils

Q. 921. The energy by which water is held by soils is known as-
Ans. Capillary potential

Q. 922. Water loss form irrigation channel/canal is mostly due to-
Ans. Seepage

Q. 923. Entry of water from the upper layer of soils is called-
Ans. Infiltration (Mostly in unsaturated soils)

Q. 924. Downward movement of nutrients and salts from the root zone with water known as-

Ans. Leaching

Q. 925. When all the soils pores are filled with water at a tension of zero bars it known as-

Ans. Saturation capacity

Q. 925a. Optimum depth of most field crops range between

Ans. 3-5 cm.

Q. 926. Permanent wilting point (PWP) was proposed by-

Ans. Briggs & Shantz in 1912

Q. 927. Which is upper and lower limit of water availability in plants-

Ans. Upper-Field Capacity & Lower- PWP

Q. 928. Soil moisture tension at PWP is-

Ans. 15 Bars or 15 ATM

Q. 929. At field capacity, the soil water potential is-

Ans. – 0.1 to – 0.3 bars.

Q. 930. The percentage of moisture in root zone at the PWP of plants called-

Ans. Wilting coefficient

Q. 931. Portion of capillary water held between field capacity to PWP known as-

Ans. Available water

Q. 932. pH term was given by-

Ans. SPL Sorenson (1909)

Q. 933. Range of pH-

Ans. 0 to 14

Q. 934. Tensiometer is not sutable for-

Ans. Clay oils

Q. 935. Best method for quantities estimation of changes in soil moisture-

Ans. Neutron moisture meter

Q. 936. Formula of water use efficiency-

Ans. WUE (Field)= Y (yield)/ ET (Evapo- transpiration)

Q. 937. Evapo- transpiration constitutes how much percentage of the total water uptake-
Ans. 99%

Q. 938. Tensiometer are sensitive up to-
Ans. 0.85 bars of soil moisture

Q. 939. Largest cultivated crops family is-
Ans. Graminae (Poaceae)

Q. 940. Concept of Ideotype was given by-
Ans. Donald (1968)

Q. 941. Growing of plants under soil less condition called as-
Ans. Hydroponics

Q. 942. Plants which are growing in salt water known as-
Ans. Halophytes

Q. 943. Father of Zero tillage-
Ans. Gluber B. Triplets (1950)

Q. 944. Mycorrhiza is symbiotic association between-
Ans. Fungus and higher plants roots

Q. 945. VAM (Vasicular Arbuscular Mycorrhiza) is used as a-
Ans. "P" solubilizer in plant roots

Q. 946. Crop canopy temperature is measured by-
Ans. Infra red thermometer

Q. 947. Vertical mulch is mostly used in-
Ans. Black cotton soil

Q. 948. Spike tooth harrow is a type of implement-
Ans. Secondary tillage

Q. 949. Dropsy disease caused by which weed in human-
Ans. *Argemone Mexicana* (Satyanasi)

Q. 950. Aminophillic plants are-
Ans. Paddy and Sugarcane

Q. 951. Mostly cultivated crops have in nature
Ans. Acidic nature

Q. 952. Practice of grazing of animals on large farms is called as-
Ans. Ranching

Q. 953. Size of small gully in soil water erosion-
Ans. 3m deep & 18m wide

Q. 954. How many times a ratoon crop of sugarcane is advisable-

Ans. One Time

Q. 955. Father of field plot experiment-

Ans. J.B. Boussingault

Q. 956. Rancidity in sunflower oil is caused by-

Ans. Oxidation

Q. 957. Crops which causes soil sickness are-

Ans. Sorghum & Linseed

Q. 958. Which Vitamin is synthesized during sprouting of moong beans -

Ans. Vitamin C

Q. 759. The Seed which is known as mother Seed-

Ans. Foundation Seed

Q. 960. How much percent nitrogen is saved by legumes crop if recommended level of nitrogen application to the associated cereals when grown as intercrop-

Ans. 25%

Q. 961. Exception in cereal crops which crop is not affected by sorghum effect-

Ans. Ragi

Q. 962. The part of plant where photosynthates accumulate called as-

Sink

Q. 963. Jalshakti chemical is associated with-

Ans. Improvement of Soil aeration, infiltration and WHC

Q. 964. Crop which is known as wonder crop as well as meat for poor-

Ans. Soya bean

Q. 965. Maize was first time cultivated in the year

Ans. 4400 BC

Q. 967a. Nucleus seed

Ans. Genetically most pure

Q. 968. Crop in which weed control is done through only herbicides -

Ans. Tea & Coffee

Q. 969. 2-4 –D is a type of Herbicide-
Ans. Selective, Post emergence & Systematic type
In India 2-4-D first time use in the year 1946.

Q. 970. For environmentally sound country ,ideal forest % should be-
Ans. 33%

Q. 971. Formula of wilting coefficient is-
Ans. WC = Hygroscopic coeff/0.68

Q. 972. Best method of tea plucking is-
Ans. Two leaves and a bud

Q. 973. Best example of perennial deep rooted absolute weed is-
Ans. *Cyperus rotundus*

Q. 974. Notorious weed which was imported with food grain from USA-
Ans. Congress Grass

Q. 975. Weed which was introduced in India as ornamental plant form Australia-
Ans. *Lantana Camara*

Q. 976. Weeds which are grown from the fallen seeds of previous or preceding crop in field known as-
Ans. Volunteer weeds

Q. 977. Pesticides consumption order in world-
Ans. Herbicides > Insecticides > Fungicides

Q. 978. Pesticides consumption order in India-
Ans. Insecticides > Fungicides > Herbicides

Q. 979. Largest pesticides consuming crop in India-
Ans. Rice > Cotton > Vegetables

Q. 980. Largest herbicides consuming crops in India-
Ans. Wheat > Rice & Tea

Q. 981. Largest production and consumption of herbicide in India is-
Ans. Isoproturon

Q. 982. Leader is a trade name of-
Ans. Sulfosulfuron

Q. 983. Dual is a trade name of-
Ans. Metalochlor

Q. 984. Sencor is a trade name of-

Ans. Metribuzin (Triazine herbicide)

Q. 985. Goal is a trade name of-

Ans. Oxyflurofen

Q. 986. Tok E-25 is a trade name of-

Ans. Nitrofen

Q. 987. Ronak is trade name of-

Ans. Isoproturon

Q. 988. Most common and widely used herbicide in soya bean is-

Ans. Imazethapyr

Q. 989. Process by which dispersed clay particles are migrated from upper to lower horizons resulting a textural horizon known as-

Ans. Argillation

Q. 990. Blind cultivation is mostly practiced in crops-

Ans. Potato and Sugarcane. (Cultivating before a seedling or planted crop emerges).

Q. 991. First irrigation of crops is known as-

Ans. Core Watering

Q. 992. The ratio of maximum dose tolerated by the crop to the minimum dose required to kill weeds known as-

Ans. Selectivity ratio

Q. 993. Base temp for wheat and Rice are-

Ans. Wheat 4°C and Rice 10 °C

Q. 994. Cupping occurs due to deficiency of-

Ans. Mo Deficiency

Q. 994a. Major function of inter-row tillage

Ans. Moisture conservation through weed destruction.

Q. 994b. Weed that spreads Asthama

Ans. *Parthenium*

CULTIVATION OF SOME IMPORTANT CROPS

1. Rice

Q. 995. Scientific name of rice.

Ans. *Oryza sativa* 2n = 24

Q. 996. Family of rice.

Ans. Graminae

Q. 996. Origin of rice.

Ans. South East Asia

Q. 998. Sowing time of Aus/Autumn rice.

Ans. March-April

Q. 999. Sowing time of Aman rice.

Ans. May-June

Q. 1000. Sowing time of Boro rice.

Ans. December-January

Q. 1001. Aman rice also known as.

Ans. Kharif/winter rice

Q. 1002. Optimum temperature for cultivation

Ans. 30-35ºC

Q. 1003. Boro. rice also known as.

Ans. Summer rice

Q. 1004. Lodging doesn't occur in.

Ans. Japonica rice

Q. 1005. Rice inflorescence is known as.

Ans. Panicle (*RAS 2008*)

Q. 1006. Stem of rice is called

Ans. Hull

Q. 1007. Hull is combination of.

Ans. Lemma + Palea together

Q. 1008. SRI stands for

Ans. System of Rice Intensification.

Q. 1009. SRI concept

Ans. Increasing productivity by changing management of plants, soil, water and nutrients.

Q. 1010. SRI was first synthesized by

Ans. Fr. Henri de Laulanie under drought conditions at Madagascar in1983.

Q. 1011. Yield increased by adopting SRI is

Ans. 50 to 90%

Q. 1012. Highest rice producing state in India.

Ans. West Bengal

Q. 1013. Test weight of basmati rice grain is.

Ans. 21 gm.

Q. 1014. Rice is a type of plant.

Ans. Self pollinated & Short day plant

Q. 1015. Cardinal temperature of rice is.

Ans. 30-32ºC

Q. 1016. Hulling % of rice is.

Ans. 70-75%

Q. 1017. First developed rice variety.

Ans. T.N.-1

Q. 1018. Jagannath is the mutant variety of.

Ans. T.N.-141

Q. 1019. The World's first high yielding variety (dwarf) of rice that has been developed by IARI.

Ans. Pusa Basmati-1

Q. 1020. Hybrid rice which is released by private organization.

Ans. PHB-71

Q. 1021. Most critical stage for water.

Ans. Booting stage

Q. 1022. Seedling ready for transplanting in Depog method.

Ans. 12th day

Q. 1023. Best Biofertilizer for rice.

Ans. Azolla

Q. 1024. Area required for rice nursery.
Ans. 700-800 m^2 (.05 ha)

Q. 1025. Fruit of rice is known as.
Ans. Caryopsis

Q. 1026. Widely used nitrogenous fertilizer in rice.
Ans. Ammonium Sulphate

Q. 1027. Most dominant weed sp. in rice.
Ans. *Echinochloa* sp.

Q. 1028. White eye of rice is due to.
Ans. Fe deficiency

Q. 1029. Killer diseases of rice are:
Ans. Tunger virus

Q. 1029a. Rice Foolish seedling disease is known as
Ans. Bakanae Disease

Q. 1029b. Who identified first Bakanae disease symptoms
Ans. E. Kurosawa (1926)

Q. 1030a. Foolish Rice seedling disease led to the discovery of
Ans. Gibberellic acid

Q. 1030. The gas emitted from rice field is.
Ans. Methane (*MPPSC 2011*)

Q. 1031. Akiochi disease is due to.
Ans. H_2S

Q. 1032. Highest N losses in rice field is by.
Ans. Denitrification

Q. 1033. Paira & Utera cropping system is closely related to
Ans. Rice

Q. 1034. Puddler & puddling is related to.
Ans. Rice

Q.1035. Hydrothermal treatment of rice grain before milling.
Ans. Paraboiling

Q. 1036. Moisture requirement for rice growing
Ans. 1200 mm

Q. 1037. Parboiling of rice conserve the vitamin.
Ans. Vitamin - B_{12}

Q. 1038. Seed rate in depog method for rice.
Ans. 3–4 kg/m^2

Q. 1039. Area required for seedling preparation in Depog method.
Ans. 25–30 m^2

Q. 1040. Miracle rice (first dwarf variety) of India is.
Ans. Jaya

Q. 1040a. Jaya variety was developed in India by
Ans. Dr. S V S Shastry

Q. 1040b. Jaya is a cross of
Ans. T(N)1 x T-141

Q. 1041. Gene responsible for dwarfness in rice.
Ans. Dee-Gee-Woo-gene

Q. 1042. Space for sowing of seedlings of rice.
Ans. 20 × 10 cm

Q. 1043. Khaira diseases of rice is due to
Ans. Zinc deficiency (AAO 2009)

Q. 1044. Optimum pH of soil required for growing rice is
Ans. 4-6

Q. 1045. Seed rate of hybrid rice is
Ans. 6-10 Kg/ acre

Q. 1046. First Super aromatic Basmati rice is
Ans. Pusa RH – 10

Q.1047. Saline tolerant rice variety is
Ans. SR 26 B

Q.1048. Vikramarya rice variety is resistant for
Ans. Tungro virus

Q.1049. Ajaya rice variety is resistant for
Ans. Bacterial Blight

Q. 1050. Phalguna rice variety is resistant for
Ans. Gall midge

Q. 1051. State known as "*Bowl of Rice*" is
Ans. Chhattisgarh or Andhra Pradesh (Krishna & Godavari Delta region)

Q. 1052. Vitamin rich in golden yellow rice is
Ans. Vitamin A

Q.1053. Total districts of India in paddy growing
Ans. 564 districts

Q.1054. The Bio herbicide used in rice cultivation is
Ans. Collego

Q. 1055. Bronzing disease of rice is associated with high concentration of

Ans. Ferrous (*JRF 2013*)

Q. 1056. Value of Albedo for rice crop is about

Ans. 12% (*JRF 2013*)

Q. 1057. Harvest Index of Rice

Ans. 0·40

Q. 1058. What percentage of world's dietary energy source is provided by rice?

Ans. 20 percent

Q. 1059. From where does black rice obtain its color?

Ans. It obtains it from anthocyanins and tocols which contain antioxidants useful for human health.

Q. 1060. Where are anthocyanins and gamma tocols located in black rice?

Ans. They are located in the inner portion of black rice

Q.1061. What do red and black rice provide?

Ans. The consumption of red and black rice reduces or retards the progression atherosclerotic plaque development, induced by dietary cholesterol, in mammals.

Q.1062. What does white rice provide?

Ans. It is a good source of magnesium, phosphorus, manganese, selenium, iron, folic acid, thiamine and niacin

Q.1063. Golden rice is genetically modified crop plant where the incorporated genes are meant for the biosynthesis of?

Ans. Beta Carotene

Q.1064. Which country produces the largest quantity of rice in the world?

Ans. China

Q.1065. What are the different varieties of rice?

Ans. Brown rice, black/purple rice, red rice, wehani rice and white rice.

Q.1066. Green revolution in India has so far been the most successful in the case of?

Ans. Rice and Wheat

Q.1067. Inter-racial hybridization programme between japonicas and indicas varieties of which among the following crops was initiated during 1950-54?

Ans. Rice

Q.1068. Why using cooked rice for the next day can be toxic?

Ans. Cooked rice contains *Bacillus Cereus* spores which produce toxicity at a temperature of 4-60 degree Celsius.

Q.1069. The de hulled rice grain is known as

Ans. Brown Rice

Q.1070. Climate suitable for Rice crop is

Ans. Hot & humid

Q.1071. Optimum temperature range for ripening of rice is

Ans. 20-25 °C

Q.1072. The BGA bio fertilizer is most useful for the crop

Ans. Rice

Q.1073. IR-8 variety of rice is introduced in India from the country

Ans. Philippines (1966)

Q.1074. In wet bed method of nursery the seedling would be ready for transplanting in stage of

Ans. 20-25 Days

Q.1075. Idea of super Rice was given by

Ans. G. S. Khush

Q.1076. Nitrogen losses in rice can be reduced by placing NH_4 fertilizer in

Ans. Reduced Zone

Q.1077. Among the cereal crop which consumed highest water for the production of 1 kg dry matter

Ans. Rice

Q.1078. Main protein of Rice is

Ans. Oryzenin

Q.1079. Rice grain is relatively rich in

Ans. lysine

Q.1080. Most widely used post emergence herbicide in Rice is

Ans. Propanil ($C_9H_9Cl_2NO$)

Q.1081. Highest area of rice in world

Ans. India

Q.1082. New Paddy variety useful for salt affected areas

Ans. CSR-43 (Released by U.P.)

Q.1083. Rice variety suitable for direct seeding is

Ans. Bala

Q.1084. BLB tolerant variety is

Ans. TKM-6

Q.1085. Variety suitable for water logged condition is

Ans. Jalmagna

Q.1086. First hybrid of basmati in world

Ans. PRH-10 (*IARI 2007*)

Q.1087. First hybrid rice in India

Ans. CORH-1 (1994)

Q.1088. Rice seed rate for Drilling

Ans. 60 kg/ha

Q.1089. Rice seed rate for broadcasting

Ans. 100kg/ha

Q.1090. Optimum population of rice hill per m^2 is

Ans. 50 hills per m^2

Q.1091. Nitrogen use efficiency of rice is

Ans. 30-40%

Q.1092. Micro nutrient which is more essential for rice

Ans. Silicon

Q.1093. Effect of pudling on soil bulk density

Ans. It increase the bulk density

Q.1094. Two critical stage of water in rice

Ans. Booting and Tellering stage

Q.1095. Country which developed first hybrid rice

Ans. China (By. Sh. Yuan Longping (known as father of hybrid rice) This variety can produce 13.9 tons of rice per hectare).

Q.1096. During milling process rice loose the vitamin

Ans. Vitamin B

Q.1097. Spacing of rice sowing of seedling is

Ans. 20 × 10 cm

Q.1098. Length of extra long rice is

Ans. 7 mm or more

Q.1099. Long grain rice are higher in amylose and medium and short grain rice are high in

Ans. amylopectin (starch molecule)

Q.1100. What is plastic rice?

Ans. Plastic rice is made mixing potato, sweet potatoes and resin (plastic) together, shaping into grain

Q.1101. Hybrid Rice given by

Ans. Yuan Longping (Known as father of Hybrid Rice). This variety produces 13.9 tonnes per hectare) 1974

Q.1102. Golden rice was developed by

Ans. Ingo Potrykus and Peter Bayer (Year 1999)

Q.1102a. Jaya first rice variety of India given by

Ans. Dr. Shastry, 1968

2. Wheat

Q. 1103. Botanical name of wheat is.

Ans. *Triticum aestivum* (Chromosome no. 2n = 42)

Q. 1104. Origin of wheat.

Ans. South West Asia (Turkey)

Q. 1105. The state which has highest production and highest cultivated area

Ans. Uttar Pradesh (*JRF 2006*)

Q. 1106. The state which has highest productivity.

Ans. Punjab

Q. 1107. Marconi wheat is known as.

Ans. Triticum durum 2n = 48

Q. 1108. Emmer wheat is known as.

Ans. *T. dicoccum*

Q. 1109. Suitable late sown variety is.

Ans. Sonalika

Q. 1110. Sowing time of wheat is.

Ans. 15 Nov. - 20 Nov.

Q. 1111. Spacing of row to row.

Ans. 22.5cm.

Q. 1112. Most critical stage for irrigation.

Ans. C.R.I. Stage (*ARS 2009*) (*Ph.D 2009*)

Q. 1113. C.R.I. stage comes after days.

Ans. 21 D.A.S.

Q. 1114. Gene responsible for dwarfness in wheat.

Ans. Norin 10

Q. 1115. Important mimicry weed of wheat is.

Ans. Phalaris minor

Q. 1116. Common bread wheat is known as.

Ans. Triticum aestivum

Q. 1117. Flowering portion of wheat is called.

Ans. Ear/Head/Spike

Q. 1118. Test weight of wheat is.

Ans. 40 gm.

Q. 1119. Shelling percentage of wheat is.

Ans. 60%

Q. 1120. Triticum. Spherococcum also known as.

Ans. Indian dwarf wheat/club wheat

Q.1121. Central zig zag axis of wheat grain is called.

Ans. Rachis

Q. 1122. Triple gene dwarf varieties were released in the year.

Ans. 1970

Q. 1123. Optimum depth of sowing

Ans. 5 cm

Q. 1124. If there are only three irrigation it will be applied-

Ans. 1. C.R.I. stage (Crown Root Initiation)

2. Late jointing (hoot) stage

3. Milking stage

Q.1125. Seminal roots are.

Ans. Temporary roots

Q. 1126. Super Wheat given by

Ans. S. Nagarajan

Q. 1127. Fruit type of wheat is (Grain).

Ans. Caryopsis

Q. 1128. First man made cereal is.

Ans. Triticale (By Rhimpus 1890)

Q. 1129. Triticale is a cross of.

Ans. Wheat × Rye (*PhD 1994*)

Q. 1130. Temperature for germination of wheat seed is.

Ans. 20–25°C

Q.1131. Protein content in wheat.

Ans. 8–11%

Q. 1132. Triticum aestivum is a type of wheat.

Ans. Hexaploid

Q.1133. When seed is dropped by hand in furrow it's known as.

Ans. Kera method

Q.1134. Word staple food grain.

Ans. Wheat

Q.1135. Seed rate of wheat is.

Ans. 100 kg./ha.

Q.1136. Most of the present days India's wheat varieties contains the gene.

Ans. Rht_1, & Rht_2 (Rht : Reduced height)

Q.1137. Biofertilizer useful for wheat is

Ans. Azatobacter

Q.1138. Heera, Moti, Arjun and Lal Bahadur are the type of varieties.

Ans. 3 gene dwarf varieties

Q.1139. The protein which is essential for good bread quality and chapati making is

Ans. Gluten

Q.1140. If sowing is done by the dibbler the seed rate will be.

Ans. 25–30 kg./ha.

Q.1141. Starch in wheat grain is about.

Ans. 60–68%

Q.1142. Moisture content at harvesting stage in wheat is.

Ans. 25–30%

Q.1143. National average yield of wheat is .

Ans. 22 kg/ha.

Q.1144. Triticum aestivium was first introduced in India by

Ans. N. E. Borlaug of Mexico

Q.1145. Flowering portion of wheat is called

Ans. Spike

Q.1146. Species which can be grown with minimum irrigation

Ans. Triticum durum

Q.1147. The variety which contributed more production during Green revolution

Ans. HD2329 (Early maturity 126-134 days)

Q.1148. FIRB

Ans. Furrow Irrigated Raised Bed method in wheat for conserving irrigation water

Q.1149. Highest cultivated area for the crop in world

Ans. Wheat

Q.1150. *Phalaris minor* belongs to which family

Ans. Gramineae

Q.1151. Herbicide which is best for control of *Phalaris minor* is

Ans. Isoproturon

Q.1152. Test weight of *Phalaris minor is*

Ans. 2 gm

Q.1152a. Major shortcoming seen during green revolution

Ans. Pulse crops were neglected

Q.1153. Directorate of wheat research (DWR) of wheat is located at
Ans. Karnal (Haryana)

Q. 1153a. Most prominent variety of wheat
Ans. DBW-327 high yield.

Q.1154. King of cereals is known as
Ans. Wheat

Q.1155. Perling index in wheat is used to determine
Ans. Hardness of grain

Q.1156. Most common crop usually intercropped with wheat is
Ans. Mustard

Q.1157. Recommended ratio for Wheat + Mustard intercropping is
Ans. 9:1

Q.1158. Wheat is type of a plant
Ans. C_3, Long day plant

Q.1159. Climate suitable during vegetative growth
Ans. Cool and moist

Q.1160. Climate suitable during grain formation
Ans. Warm & Dry

Q.1161. Variety which is suitable for making good quality Chapatti
Ans. WH-147

Q.1162. Variety suitable for bread making
Ans. DBW-17

Q.1163. Variety suitable for dry land area
Ans. Sujata

Q.1164. Dwarfing agent for wheat crop is
Ans. Cycocel

Q.1165. Objectionable weed of wheat is
Ans. *Convolves arvensis*

Q.1166. Absolute weed of wheat is
Ans. *Chenopodium album*

Q.1167. Export quality of wheat is influenced by
Ans. karnal bunt

Q.1167a. Biofortiled variety of wheat developed by IIWR (Karnal)
Ans. WB-02 (High in Zinc content 42 ppm and Iron 40 ppm)

Q.1168. Most common hybridization method in wheat is

Ans. Goojo method

3. Maize (Queen of cereals)

Q.1169. Botanical name of Maize (C_4 plant) (Origin: Maxico)

Ans. *Zea mays L. Cromosome number* ($2n = 20$)

Q.1170. Yellow colour in maize due to

Ans. Crytoxanthin

Q.1171. Critical stage of water application

Ans. Tasseling to silking

Q.1172. Suitable temperature for crop growing

Ans. 32°C

Q.1173. Common Herbicide used in Maize is

Ans. Simazine

Q.1174. First hybrid varieties in Maize in 1961

Ans. Ganga 1, Ganga 101, Ranjit and Deccan

Q.1175. Double cross techniques of production of Maize introduced by

Ans. D. F. Jones (1920)

Q.1176. On Maize, plants first appearance is of

Ans. Tassels (Male Inflorescence)

Q.1177. Female Inflorescence is called

Ans. Cob or ear

Q.1178. Maximum root zone depth for Maize

Ans. 400–600 mm

Q.1179. Water requirement for Maize crop is

Ans. 500–800 mm

Q.1180. Optimum harvesting moisture content for maize is

Ans. 20–25% (wb)

Q.1181. Protein content of Maize

Ans. 10% (Zein). It is deficient in Tryptophan and Lysine

Q.1182. Maize variety used for snack food

Ans. *Zea mays amylacea*

Q.1183a Germination percentage in maize

Ans. 90% (Highest in field crops).

Q.1183. Minimum temperature for germination of maize

Ans. 6-7°C

Q. 183a. Composite cultivar of Maize

Ans. Jawahar, Ambar, Vijay, Sona, Vikram, etc.

Q.1183b. Indian scientist shared world food prize

Ans. Dr. Surinder K. Vasal (2000) with Evangelina Villegas (CIMMYT) (For QPM)

Q.1184. Seed rate

Ans. 18–20 kg/ha

Q.1185. Serious pest for Maize (Pop corn)

Ans. Maize borer (*Chilo partellus*)

Q.1186. Waxy Maize (Originated in China)

Ans. *Zea mays ceratina*

Q.1187. Hybrid maize was first reported by

Ans. E. M. East and G. H. Hull in year 1910

Q.1188. Most suitable temperature for germination and growth

Ans. 21-30°C, 30-32 °C

Q.1189. Type of maize plant is

Ans. Non tiller, Cross pollinated plant

Q.1190. In maize yellow color is due to

Ans. Cryptoxanthin

Q.1191. Maize is sensitive for

Ans. Water logging

Q.1192. Best suitable direction of maize sowing is

Ans. East-West direction

Q.1193. A prson wholly depended on maize lead to a disease called as

Ans. Pellagra

Q.1194. Single cross technique was given by

Ans. EM East & G H Shull

Q.1195. In India mostly used technique is

Ans. Double cross technique (Developed by D.F. Jones, 1920)

Q.1196. Dent corn (*Zea mays Indentata*) is mostly grown in

Ans. U.S.A.

Q.1197. *Zea mays amylacea* is also known as

Ans. Soft corn (Flour Corn)

Q.1198. Maize which is used for pop corn is
Ans. *Zea mays everta*
Q.1199. First (1961) 4 double cross hybrids released in India
Ans. Ganga-1, Ganga-101, Deccan & Ranjeet
Q.1200. Synthetic variety of maize
Ans. Amber
Q.1201. Optimum plant population for kharif maize is
Ans. 65000 to 70000 plants
Q.1202. Disease occurs due to deficiency of Zinc in maize
Ans. White bud
Q.1203. Optimum range of soil moisture for effective flowering
Ans. 25-50%
Q.1203a. One inch of surface soil over one hectare land weight about
Ans. 375 tonnes
Q. 1203b. Zea Maize indurate
Ans. Flint corn

4. Sorghum

Q.1204. Botanical name of Sorghum/Jowar
Ans. *Sorghum bicolour* 2n = 20
Q. 1205. Harvesting of sorghum is done at grain moisture range
Ans. 21–24% by weight
Q.1206. Jowar is native of
Ans. Africa
Q.1207. Sorghum hybrid is
Ans. CSH-1 (released in the year 1964)
Q.1208. Protein and Calcium, content of Sorghum /Jowar
Ans. 10–12% and 25%
Q.1209. The best high yield varieties of Rabi Jowar is
Ans. M. 35-1
Q.1210. Seed rate
Ans. 8–12 kg/ha
Q.1211. In India maximum producing state.
Ans. Maharastra
Q. 1212. Inflorescence of Sorghum
Ans. Panicle (head)
Q.1213. Minimum temperature for germination
Ans. 7-8 ºC

Q.1214. Rainfall requirement
Ans. 40-60 cm

Q.1215. Drought and salinity tolerant variety of sorghum is
Ans. CSH-9

Q.1216. Due to resistance to drought sorghum also known as
Ans. Camel crop

Q.1217. Most important male sterile variety is
Ans. Combine (CK 60-kafir)

Q.1218. Major weed of sorghum
Ans. Striga Spp. (Witch weed)

Q.1219. Seed rate of Jowar for fodder purpose
Ans. 40-50 kg./ ha.

Q.1220. Sorghum is called
Ans. Water sipping crop due to drought tolerance

Q.1221. Chemical which is produced by sorghum at flowering stage which is poisonous for animals
Ans. HCN Acid

Q.1222. Cholam (Great millet) is known as
Ans. Jowar (Sorghum)

5. Barley

Q. 1223. Scientific name of Barley
Ans. *Hordeum vulgare* ($2n = 14$)

Q. 1224. Top growing state
Ans. Rajasthan (30% of total area)

Q. 1225. Inflorescene of barley
Ans. Spike (Head of ear)

Q.1226. Growin g temperature
Ans. 12-15ºC

Q.1227. Protein content
Ans. 10% (Protein lacks essential amino acid lysine)

Q.1228. pH range
Ansl. 7-8

Q.1229. Growth stages
Ans. 1. germination and initial seedling establishment 2. tillering, 3. jointing and booting 4. heading 5. ripening, 6. maturity and drying.

Q. 1230. Grain maturity temperature
Ans. 30ºC

Q. 1231. Seed rate of Barley
Ans. 75-80 kg/ha

Q.1232. Main characteristics of Barley
Ans. Malting quality

Q.1233. Water requirement for good crop growth is
Ans. 35-50 cm

Q.1234. Type of barley which is most growing in India is
Ans. Six rowed barley

Q.1235. Crop which is more tolerant for saline and alkali soils among cereals
Ans. Barley

Q.1236. Barley is mostly used in
Ans. Malt production (husk less barley is not suitable for malting)

Q.1237. Content which is not found in barley flour is
Ans. Gluten

Q.1238. Best varieties of Barley
Ans. Ratna, Jyoti, RS-6, Rajkiran, RD 2052

Q.1239. Dwarfing gene in barley
Ans. Uzu (discovered by Japan)

Q. 1239a. Important cultivar of Barley
Ans. Clipper released in 1968

6. Pearl Millet

Q.1240. Botanical name of Pearl Millet (2n=2x=14)
Ans. *Pennisetum glaucum L.* Family: Poaceae (It is a short day plant)

Q.1240a. Transplanting in Pearl millet was given by
Ans. Dr. R.C. Gautam

Q.1241. Native country
Ans. Sudan (Africa)

Q.1242. Also known as
Ans. Cattail and bulrush millet and Bajra

Q.1243. Rainfall requirement
Ans. 25-35 cm
Q.1244. Seed rate
Ans. 5 kg/hectare
Q.1244a. Transplanting in Pearl millet was given by
Ans. Dr. R.C. Gautam
Q.1245. Largest producing state of Pearl Millet in India
Ans. Rajasthan followed by Maharastra
Q. 1246. Largest producer of Pearl Millet in the world
Ans. India
Q.1247. Temp. range for maximum output
Ans. 30–35 °C
Q.1248. Hybrid varieties
Ans. HB 1, 2, 3, 4 & 5, Pusa 23
Q.1249. Downey mildew is due to
Ans. *Sclerospora graminicola*
Q.1250. Ergot an important disease in India is due to
Ans. *Claviceps fusiformis* (*AAO, 2010*)
Q.1251. Protein and fat content
Ans. 11.6% and 3.5%
Q.1252. The country first developed hybrid pearl millet
Ans. India
Q.1253. Cereals which have highest minerals
Ans. Bajra
Q.1254. Crop which require lowest water requirement among all cereals
Ans. Bajra
Q.1255. First hybrid of Bajra
Ans. HB-1 (1965, PAU, Ludhiana) Tift 23A × BIL 3B
Q.1256. Spiked millet is known as
Ans. Pearl millet (Bajra)
Q.1256a. Parasitic (Purple witch) weed can reduce productivity
Ans. *Striga Hermonthica* (Purple Witch weed)
Q. 1256b. Bajra is sensitive to
Ans. Water logging and acidic soils.

> Pearl Millet is highest drought tolerant crop among Cereals and Millets.

7. Chick Pea (Bengal Gram)

Q.1257. Botanical name of Gram (Chick pea).
Ans. *Cicer arietinum* (Browngram), *Cicer Kabulianum* (Kabuli white gram)

Q. 1258. Origin of gram
Ans. South West Asia or India

Q. 1259. Root system of gram is.
Ans. Tap root system

Q.1260. Type of soil best for chick pea is
Ans. Light Alluvial soil

Q.1261. Pollination type of gram is.
Ans. Self pollinated

Q.1262. Sour taste of gram leaf is due to.
Ans. Maleic & Oxalic acid

Q.1263. Deep sowing of gram is protection of.
Ans. Wilt disease

Q.1264. Best variety for dry land is.
Ans. C-235

Q.1265. Early maturing variety having medium sized grain is.
Ans. Chaffa (1948)

Q.1266. Optimum sowing time of gram is.
Ans. 15 Oct.–20 Oct.

Q.1267. Seed rate of gram is.
Ans. 80–100 kg/ha.

Q.1268. Important operation which is done in gram crop is.
Ans. Nipping (Process of plucking the apical buds)

Q.1269. Time of Nipping operation is.
Ans. 30-40 Days after sowing

Q.1270. Seed bed for gram should be.
Ans. Rough

Q.1271. Gram fruit is known as.
Ans. Pod

Q.1272. Spacing for gram is.
Ans. 30 × 10 cm.

Q.1273. Protein content in gram.
Ans. 21%

Q.1274. Gram is a.
Ans. Rabi crop

Q.1275. Variety developed by I.A.R.I.
Ans. Pusa 209, BG 203

Q.1276. Critical stage for irrigation.
Ans. Pre-flowering, Pod developing stage

Q.1277. Largest producer of Gram
Ans. India

Q.1278. Pulse having largest area and production among pulses
Ans. Gram

Q.1279. Test weight of Desi gram (*C. aretinum*)
Ans. 140-200 gm

Q.1280. Test weight of Kabuli/white gram
Ans. 340-375 gm

Q.1281. Inflorescence of Chickpea
Ans. Rosette

Q.1282. Plant type of gram
Ans. C_3, Long day plant

Q.1283. Wilt resistant variety of gram
Ans. Avarodhi and Vikash

Q.1284. Ascochyta blight resistant variety of gram
Ans. Gaurav

Q.1285. Mutant variety of Gram
Ans. RS-11

Q.1286. Short duration variety of gram
Ans. ICCC-2

Q.1287. Gram variety which grains remain green colour is
Ans. Aparna

Q.1288. Effect of late sowing of gram
Ans. Increase the infestation of Pod borer

Q.1289. Effect of shallow and early sowing of gram
Ans. Increase the wilt disease attack

Q.1290. Yield of gram (Maturity when leaves get dry and turn reddish brown and start shedding)

Ans. 20-25 Q/ha

Q.1291. Chemical used for gram nipping is

Ans. TIBA @ 75 PPM

8. Pigeon Pea/Red Gram/Arhar

(Highly drought resistant crop)

Q.1292. Botanical name of arhar

Ans. *Cajanus cajan.* Family; *Fabaceae* (it is a perennial legume)

Q.1293. Origin of arhar is

Ans. South Africa (Domestication in Indian sub-continent)

Q.1294. Temperature for germination is

Ans. 30–35°C

Q.1295. Temperature for growth is

Ans. 20–25°C

Q.1296. Sowing time of arhar is

Ans. 15 June

Q.1297. Spacing for arhar is

Ans. 60 × 15cm.

Q.1298. Seed rate for arhar is

Ans. 12–15 kg/ha.

Q.1299. Proportion of seeds to pods

Ans. 50-60%

Q.1300. Highest production in the state.

Ans. Uttar Pradesh (about 20% of total production)

Q.1301. Highest productivity in the state of.

Ans. Bihar (1115 kg/ha)

Q.1302. Early maturing variety.

Ans. Prabhat, UPAS 120 days

Q.1303. Harvest index of arhar is.

Ans. 19% (very low)

Q.1304. Hybrid variety released by ICRISAT.

Ans. ICPH-8 (yield 20 q/ha). It recorded yield potential of 2057 kg/hectare in 1991)

Q.1305. Protein content of arhar is.
Ans. 25%

Q.1306. Earliest variety
Ans. UPAS-120

Q.1307. Pulse which is highest drought tolerant among pulses
Ans. Pigeon Pea

Q.1308. Type of Pigeon pea plant
Ans. C_3, Short day, Self pollinated plant

Q.1309. Wilt resistant variety of Arhar
Ans. Mukta

Q.1310. New variety of Arhar from IARI PUSA
Ans. PUSA Arhar-16

Q. 1310a. Shelling percentage of Pea is
Ans. 49%.

9. Sugarcane

Q.1311. Botanical name of Sugarcane
Ans. *Sccharum officinarum (Tropical cane)*
Sccharum barberi (Indian cane)

Q.1312. Inflorescence of sugarcane is called as.
Ans. Arrow or open panicle (*JRF, 1994*)

Q.1313. Arrowing (Flowering of sugarcane plant) stage arrives on.
Ans. 300-350 days after planting

Q.1314. Sugar mills are highest in India with highest sugar production
Ans. Uttar Pradesh (105 mills) (*JRF. 1995*)

Q.1315. In sugarcane upper 1/3 part is used for sowing due to.
Ans. High nitrogenous substance & glucose for better germination.

Q.1316. Highest producer of sugar from unit area is
Ans. Maharashtra

Q.1317. Which country is World leader in Sugarcane production-
Ans. Brazil & 2nd India

Q.1318. Higher dose of nitrogen decrease the
Ans. Sucrose content

Q.1319. Most critical stage for irrigation is
Ans. Formative stage

Q.1320. Germination phase comes after
Ans. 0–60 DAP

Q.1321. Maturity phase comes after
Ans. 250–365 DAP

Q.1322. Formative stage comes after
Ans. 60-130 days after planting

Q.1323. Most popular planting method in north India.
Ans. Flat bed planting

Q.1324. Brix reading should be for proper maturity
Ans. 16-18%

Q.1325. Jaggery extracted from juice is
Ans. 9-10%

Q.1326. Sucrose content in cane.
Ans. 13-14%

Q.1327. Sugar from juice is
Ans. 6-10%

Q.1328. Soils which are unsuitable for sugarcane are
Ans. Saline soils

Q.1329. Noble cane is
Ans. S. officinarium

Q.1330. Adsali sugarcane is planted in
Ans. June-July

Q.1331. Duration takes adsali sugarcane
Ans. 18 months

Q.1332. Brix meter is used for
Ans. Measuring maturity of sugarcane

Q.1333. Seed rate for 3 budded sets
Ans. 25000-30,000 sets

Q.1334. Seed rate for 2 budded sets
Ans. 45,000-50,000 sets

Q.1335. Seed rate for single budded sets
Ans. 1,25000 sets

Q.1336. Origin of sugarcane
Ans. India

Q.1337. Sets roots are
Ans. Temporary roots

Q.1338. Temperature required for grand phase in sugarcane.
Ans. 70 °F

Q.1339. Earthing should be done in the month of
Ans. June-July

Q.1340. Tying should be done in the month of
Ans. August and Sept. when plant of about 5 feet height. Tying is done to prevent lodging economize water use efficiency, control weeds and draining excess water.

Q.1341. Sowing time of Eksali crop
Ans. Jan.-Feb. (South India)

Q.1342. Wonder cane is
Ans. COC-671 (High sugar%)

Q.1343. Sugarcane ripener chemical.
Ans. Glyphosate

Q.1343a. Detrashing and Popping in Sugarcane
Ans. Detrashing refers to removal of unwanted bottom dry and green leaves at regular intervals. *Propping:* The operation of tying the leaves together using dry, green leaves. Popping can be either done for each row or two rows can be brought together and tied. It is done to check lodging of cane. **Lodging** reduces sucrose accumulation of sugarcane in wet and dry tropics.

Q.1343b. Hot water treatment of Sugarcane setts is done
Ans. To control seed borne diseases

Q.1344. Spacing of row to row
Ans. 90cm.

Q.1345. Name of wild type cane.
Ans. *S. spontaneum*

Q.1346. Family of sugarcane.
Ans. Gramineae

Q.1346a. Align plnting method of sugarcane developed by
Ans. Allahabad Agriculture Institute

Q.1347. Burning of canes is done to
Ans. Improve sucrose & juice quality

Q.1348. Indian cane indigenous to North and Eastern India
Ans. *Saccharum barberi*

Q.1349. Major disease of sugarcane is
Ans. Red rot

Q.1350. Noble cane was developed by
Ans. T.S. Vankataraman

Q.1351. Type of sugarcane plant is
Ans. C_4 *Intermediate short day* plant

Q.1352. Which S. cane species having very low sugar and high fiber content
Ans. *S. Spontaneum*

Q.1353. Material used for S. cane sowing is known as
Ans. Sets

Q.1354. Element necessary for sugar translocation
Ans. Pottassium (K)

Q.1355. Most common irrigation method used in S. cane is
Ans. Furrow method to reduce lodging

Q.1356. Ratooning is related to
Ans. Sugarcane

Q.1357. For brix reading juice should be taken from which part of cane
Ans. Middle part

Q.1358. Fehling test is done for
Ans. Maturation of cane

Q.1359. In Fehling test the reading should be less then
Ans. 0.5% glucose

Q.1360. Who developed the term crop logging in sugarcane
Ans. H. F. Clements (1960)

Q.1361. By products of cane
Ans. Bagasses (for fuel & paper), Molasses (for Alcohol), Press mud

Q.1362. Press mud is used for
Ans. Reclamation of acidic soils

Q.1363. Energy product which is produced by cane is
Ans. Gasohol

Q.1364. At present total Sugar mills in India
Ans. 515 operational (Junuary, 2023) with Maharashtra maximum, 198 mills

Q.1364a. Variety developed by Sugarcane Breeding Institute Regional Centre, Karnal.
Ans. Karnal-6 (C00239)

10. Ground Nut

Q.1365. Botanical name of Ground nut.

Ans. *Arachis hypogea.* (*Arachis* means legume and *hypogea* mean below ground). It is geocarpic crop ripening fruit under ground. Native to South America and originated in Eastern slopes of he Andes.

Q.1366. Ground Nut is Legume crop grown mainly for

Ans. edible seeds

Q.1367. State which is largest producer.

Ans. Gujarat

Q.1368. Shelling percentage.

Ans. 70%

Q.1369. Oil percentage in Ground nut.

Ans. 40-45%

Q.1370. Nitrogen % in Ground nut cake.

Ans. 7-8%

Q.1371. Best soil for Ground nut cultivation.

Ans. Sandy

Q.1372. Seed rate of Ground nut.

Ans. 100-120 kg/ha.

Q.1373. Spacing for Ground nut and pH requirement

Ans. 30 × 10 cm. sandy loam soil with pH of 5.9-7

Q.1374. High yielding type of Ground nut.

Ans. Spreading type (Late maturity)

Q.1375. Pegging stage comes after.

Ans. 55 D.A.S.

Q.1376. Important variety.

Ans. Jyoti (Bunch type), Chandra (spreading type)

Q.1377. Chemical used for floral initiation.

Ans. NAA@40 PPM

Q.1378. Fruit of Ground nut is known as.

Ans. Nut (Lomentum)

Q.1379. Botanical name of bunchy type Ground nut.

Ans. *Archis hypogaea* sub spp. fastigata

Q.1380. Botanical name of spreading type Ground nut.

Ans. *Arachis hypogea* sub spp. procumbens

Q.1381. Bitterness of kernel due to.
Ans. Aflatoxin (*Aspergillus flavus*)

Q.1382. Sowing time of Ground nut.
Ans. 20 June to 31 July

Q.1383. Protein content in Ground nut.
Ans. 26%

Q.1384. Protein content in shell.
Ans. 7%

Q.1385. Gynophore of Ground nut is known as.
Ans. Peg

Q.1386. Rostee disease is due to.
Ans. Virus

Q.1387. Vector of virus.
Ans. Aphid

Q.1388. Major and serious pest of Ground nut.
Ans. White grub

Q.1389. Tikka disease (an important diseases of ground nut) is due to.
Ans. Cercospora personata & C. arachidicola

Q.1390. Best and suitable methods of irrigation
Ans. Check Basin irrigation

Q.1391. Critical stages for irrigation
Ans. Flowering stage-Pegging stage-Pod formation stage. (Water requirements 400-700 mm)

Q.1392. Ground nut is also known as
Ans. Peanut/ Earthnut/ Monkey nut

Q.1393. Which G. nut species is non dormant type
Ans. Bunch type

Q.1394. G. nut is type of plant
Ans. **C_3, *short day* plant, self pollinated**

Q.1395. Popping in G. Nut caused by
Ans. Due to Ca deficiency

Q.1396. Herbicide used in G. Nut is
Ans. Tok E-25

Q.1397. Yellowing of G. nut leaves due to
Ans. Iron deficiency
Q.1398. Bud necrosis in G. Nut due to
Ans. Thrips
Q.1398a. Synthetic textile fiber Ardil is produced from
Ans. Groundnut Protein

11. Sunflower

Q.1399. Botanical name is
Ans. *Helianthus annuus* (Family: Asteraceae 2n = 34) (Origin: South America and Mexico)
Q.1400. Sunflower is known as
Ans. C_3 Day neutral crop (short duration crop 90-100 days)
Q.1401. Head of sun flower is called
Ans. Capitulum
Q.1402. Important disease of sunflower
Ans. Downey mildew, root rot, stem rot and Alternaria blight and headrot
Q.1402a. Important pests of sunflower
Ans. Cutworms, Head borer whitefly and termites
Q.1403. Critical stages
Ans. 1. Seedling (10–12 DAS) 2. Bud initiation (130–35 DAS)
3. Flowering (50–55 DAS)
4. Seed Development stage (70–90 DAS)
Q.1404. Important varieties
Ans. Modern (80 days duration), Jawalamukhi, MSFH8,17, JS-1 etc.
Q.1405. Yield
Ans. 20-30 Qt/ha.
Q.1405a. Rancidity in sunflower oil is due to
Ans. oxidation
Q.1405b. Antinutritional factors
Ans. Chlorogenic acid and phenolic compounds.
Q.1405c. Oil content
Ans. 30-43% (Rabi sunflower has higher oil content)
Q.1405d. Protein content
Ans. 14-19%

12. Soyabean

Q.1406. Botanical name. Family: Fabaceae
Ans. *Glycine max* also (known as Gold crop of America)
Q.1407. The crop is known as
Ans. Wonder crop and yellow jewel

Q.1407a. First introduced in India
Ans. Year 1880

Q.1407b. Hybrid varieties
Ans. NRC-86, JS 97-52, RVS-2002-4, JS-2069, JS-2029, JS-2098

Q.1408. Protein content
Ans. 40–42%

Q.1409. Oil content
Ans. 20–22%

Q.1410. Beary taste in soybean is due to
Ans. Sulphur compound

Q.1411. Critical stages
Ans. 1. Sprouting 2. Flowering
3. Pod filling 4. Grain development.

Q.1412. Optimum temperature of germination
Ans. 30°C (min 5 - max. 40°C)

Q.1413. Crop which is first in source of vegetable oil
Ans. Soya bean

Q.1414. Origin of Soyabean
Ans. China

Q.1415. Fort of soya bean known as
Ans. M. P. (Highest producing state)

Q.1416. Optimum temperature for germination
Ans. 30°C

Q.1417. Plant type of soyabean
Ans. C_3 *short day* plant

Q.1418. USA imported variety
Ans. Bragg, Lee, Clark 63, Kent, Black heart

Q.1419. Indian varieties
Ans. Shilajeet, Alankar, Ankur, Punjab-1, Durga and Gaurav

Q.1420. Sowing time of soyabean
Ans. 20 June to 15 July

Q.1421. Spacing in soyabean
Ans. Kharif crop (row-row) 45 cm
inter row - 4-5 cm, spring crop inter row - 30 cm

Q.1422. Depth of sowing for soyabean
Ans. 3-5 cm optimum

Q.1423. Seed rate of Soyabean
Ans. 75-80 kg.

Q.1424. Plant population for good yield
Ans. 0·45 million plants/ha

Q.1425. **Major pest of soya bean**
Ans. Heliothis and Girdle bettle

Q.1426. **SOPA (Soyabean Processors Association of India) is situated at**
Ans. M.P. (INDORE)

Q.1427. **Inflorescence of soya bean**
Ans. Raceme

13. Linseed (Alsi)

Q.1428. **Botanical Name**
Ans. *Linum Usitatissium.* Family: Linaceae (Center of origin: India)

Q.1429. **Fibre of lineseed called**
Ans. Flax

Q.1420. **Percent oil content**
Ans. 40–42%

Q.1421. **Percent protein.**
Ans. 20-30%

Q.1421a. **Inflorescence**
Ans. Carymbose

Q.1421b. **pH requirement of soil**
Ans. 5-7 (Best 6%)

Q.1421c. **Best soil for growth**
Ans. Silty loam clay loam and silty clays.

Q.1422. **Percent Linolinic acid**
Ans. 50–60%

Q.1423. **Optimum seed rate**
Ans. 25–30 kg/ha

Q.1424. **Critical stages**
Ans. 1. Leaf stage, 2. Tillering stage, 3. Flower stage, 4. Pod filling.

Q.1425. **Varieties**
Ans. Mukta, Sweta, Jawahar-7, Jeevan (DPL-21), Gourav, Shital, Himalini, Garima, Hira, Sheela, Shikha, Padmani, NL-17, LC-2063, LC-2023.

Q.1426. **Fruit of Linseed is**
Ans. Seedball

Q.1426a. **Sowing Temperature / Seed rate**
Ans. 20-21°C / 25-30 kg/ha under rainfed condition

14. Rapeseed Mustard

Q.1427. **Botanical name:**
Ans. *Brassica spp.* Family: Brassicaceae, White Mustard: *Brassia alba*

Q.1428. **Centre of origin:**
Ans. Asia Minor and China

Q.1429. Pungency in mustard oil due to
Ans. Glucosinolates or Allyl-isothiocyanate

Q.1430. Oil content:
Ans. 37 to 49% (High in yellow sarson 42·1%, toria 41%)

Q.1431. Highest producing state:
Ans. Rajasthan

Q.1431a. First tolerant variety
Ans. Swarna jyothi

Q.1432. Highest productivity
Ans. Haryana

Q.1433. Fruit of mustard
Ans. Siliqua (*JRF 2013*)

Q.1434. Protein content
Ans. 24-30%

Q.1435. Seed rate
Ans. 3·5-10 kg/ha

Q.1436. Yield rate
Ans. 500-800 kg/ha

Q.1437. Varieties
Ans. Sangam, Type-9, Type-36, Brown BSH-1, Pusa Kalyani, Suphala Yellow, Jhumka

Q.1438. Disease caused by regular use of mustard cakes
Ans. Goiter

Q.1439. Upper limit fixed by Govt. for Glucosinolates in mustard oil is
Ans. 0.5%

Q.1440. Types of Rapseed
Ans. Rapseed Mustard is important economic crop comprising eight different species like, Indian sarson, Toria, Yellow sarson, Brown sarson, Gobhi sarson, Karan rai, Black mustard and Taramira

Q.1441. Botanical name of Indian mustard
Ans. *Brassica juncea*

Q.1442. Indian mustard also known as
Ans. Rai

Q.1443. Varieties of Indian mustard
Ans. Kranti, Varuna (T-59), Rohini, Pusa bold

Q.1444. New released varieties of mustard
Ans. Laxmi, Vasundhra, Ashirwad, Basanti, Pusa Jai Kisaan, NRC-HB-506, DMH-1, RH-0408

Q.1444a. First Indian variety of Mustard with low euric acid is
Ans. Pusa Karishma

Q.1444b. First double zero mustard variety in the country released for cultivation

Ans. PDZM-31 (Pusa double zero mustard)

Q.1445. Mustard Variety which has first developed through biotechnology

Ans. Pusa Jai Kisaan (Bio-902)

Q.1440a. Hybrid variety developed at National Research Center on Rapseed mustard, Bharatpur

Ans. NRCHB-506

Q.1445b. Mustard type which having less than 2% erucic acid is called-

Ans. CANOLA Type Mustard

Q.1446. Variety suitable for late sowing irrigated area

Ans. NRCHB-101
(Variety of Indian Mustard by Pusa Mustard-26, Pusa Mustard-27, Pusa Mustard-28 (2012)

Q.1447. Nutrient which is essential for mustard

Ans. Sulphur

Q.1448. Major pest of mustard

Ans. Mustard Aphid & Mustard saw fly

Q.1449. National research centre for mustard is situated at

Ans. Bharatpur (Raj)

Q.1450. Various varieties of Rapeseed mustard

Ans. Brown sarson - Pusa Kalyani, KOS-1, KBS-1
Yellow sarson - Jhumka
Mustard - Varun, Pusa Bold, Bandan, Pusa kranti

Q.1450a. Important diseases

Ans. Alternaria blight white rust downy mildew and stem rot.

Q.1365. Adulteration in mustard oil

Ans. Argemone mexicana oil

15. Potato

Introduced in India by Portuguese

Q.1451. Botanical name is

Ans. *Solanum tuberosum* (Solanum tuberosum)

Q.1452. Potato introduced in India

Ans. From Portuguese in 1965

Q.1453. Important Nutrient for potato crop

Ans. Nitrogen

Q.1454. Protein content of potato

Ans. 1.6%

Q.1455. Maximum soil area under potato production in India

Ans. Alluvial soils

Q.1456. Dehaulming in potato is done by

Ans. $CuSO_4$ (Copper sulphate) to get quality seed tuber

Q.1457. Solanin content
Ans. 5 mg/100 gm of potato

Q.1457a. Tuber represents
Ans. Enlarged underground stem

Q.1458. Most critical stage for irrigation
Ans. 25% tuber formation stage

Q.1459. Excess Mg produce type of toxicity in potato
Ans. stem streak necrosis

Q.1460. Best suitable soil for potato
Ans. Sandy loam soil

Q.1461. Protein rich Prototo variety by
Ans. Asish Data of NIPGR (New Delhi)

Q. 1462. Crop known as King of Vegetable is
Ans. Potato

Q. 1463. Cash crop which has high market value per unit area
Ans. Potato

Q.1464. Origin of Potato
Ans. Peru

Q.1465. True potato seeds (TPS) developed by
Ans. Ramanujan

Q.1466. Seed rate for potato
Ans. 100-150gm/ha in nursery

Q.1467. Crop which having highest area and production among all vegetable crop
Ans. Potato

Q.1468. Potato propagation is done by
Ans. Tubers

Q.1469. Plant type and fruit type of potato
Ans. C_3 Plant and fruit type Berry

Q.1470. Temp & pH for best for tuber growth is
Ans. 17-19° C temp and pH 5.0-6.5

Q.1471. Variety which is extra early
Ans. Kufri Ashoka

Q.1472. Frost resistant variety is
Ans. Kufri sheetman

Q. 1473. Variety which are suitable for processing
Ans. Kufri chipsona-1 & Kufri chipsona-2

Q. 1474. Early maturity (75 days)varieties
Ans. Kufri chnadramukhi, K. alankar, K. lavkar, K. navtal

Q. 1475. Late maturity varieties are

Ans. K. Deva, K. Sunduri

Q.1476. Seed rate

Ans. 25 gm cut of tuber with 3-5 buds @ 15-20 Qtl/ha

Q.1477. Spacing for potato

Ans. 60 × 20 cm

Q.1478. A technique which is used for multiplication of disease free seed in northern plains of country known as

Ans. Seed Plot technique (SPT)

Q.1479. SPT was developed by

Ans. Pushkarnath (1967)

Q.1480. The disorder which is related to lack of 0_2 in stored condition

Ans. Black heart disorder

Q.1481. Due to excessive use of N-fertilizer an irregular cavity in the centre of tubers develops, it is known as

Ans. Hollow heart

Q.1482. Which potassium fertilizer is not suitable for Potato

Ans. Potassium Chloride

Q.1483. Most Critical stages of irrigation is

Ans. Stolen formation (20-50 DAP) and Tuber elongation (50-80 DAP)

Q.1484. Potato leaf roll virus is transmitted by

Ans. Aphids (*Myzus persicae*)

Q.1485. Domestic quarantine is done mainly to protect from

Ans. Wart Disease

Q.1486. Wart disease resistant Varity

Ans. Kufri kanchan & Kufri Jyoti

Q.1487. Most common disease of Potato

Ans. Early Blight

Q.1488. Late blight resistant variety

Ans. K. Jawahar, K. chipsona 1 & 2, K. Sutlaj

Q.1489. Important inter culture practice of potato

Ans. Earthing

Q.1490. Insect which damage standing potato as well as stored potato

Ans. Potato tuber moth

Q.1491. Yield of Potato

Ans. 30-40 tones/ha

Greening in Potato: Due to accumulation of solanine and chaconine alkaloids on tuber and exposure of sun rays

Q.1491a. International Potato Center (CIP) was founded in

Ans. 1971, Headquarters Lima, Peru

16. Cotton

Q.1492. Botanical name of cotton

Ans. *Gossypium sp.* Family: *Malvaceae*

Q.1493. Origin of cotton

Ans. India

Q.1494. Cotton crop also known as

Ans. White Gold & King of appraisal fiber

Q.1495. Cotton species which covered highest area in India is

Ans. G. *hirsutum*

Q.1496. Recently developed terminator technology has been used in crop

Ans. Cotton

Q.1497. Root system of cotton

Ans. Tap root system

Q.1498. In cotton monopodial are

Ans. Vegetative branches

Q.1499. In cotton sympodial are

Ans. Reproductive branches

Q.1500. How many species are grown in world out of 20 species of cotton

Ans. 4 species (*G. arboreum, G. herbaceum, G. barbadense* and *G. hirsutum*)

Q.1501. G. *arborium* & G.*herbaceum* are
Ans. 26, diploid (Desi cotton)

Q.1502. G. *hirsutum* & G. *barbadence* (Sea island cotton) are
Ans. 52, tetraploids

Q.1503. Highest fiber length cotton is
Ans. G. *barbadence*

Q.1504. American cotton is
Ans. G. *hirsutum*

Q.1505. Country where all 4 species of cotton grown
Ans. India

Q.1506. First spinning variety of Egyptian cotton released in India is
Ans. Sujata

Q.1507. First commercially hybrid variety of cotton
Ans. H-4 (1970) By C.T. Patel

Q.1508. H4 is cross of
Ans. G-67 × American Nectorless

Q.1509. First inter specific hybrid is
Ans. Varalaxmi (Evolved at Surat)

Q.1510. Varalaxmi is cross of
Ans. G. *hirsutum* & G. *barbadence*

Q.1511. Cotton extra long variety is
Ans. MCU-5

Q.1512. Chemical used in cotton delinting
Ans. Sulphuric Acid (1:10 ratio) (H_2SO_4)

Q.1513. Seed rate for desi cotton
Ans. 10-18 kg at 67.5 × 30 cm spacing

Q.1514. Seed rate for America cotton
Ans. 15-25 kg at 67.5 × 30 cm spacing

Q.1515. Seed rate for Hybrid cotton
Ans. 2-3 kg at 100 × 60 cm spacing

Q.1516. Optimum population of plants
Ans. 50 000 to 80000 (Average 66000)

Q.1517. Sowing time of cotton in North India
Ans. 15th May

Q.1518. Sowing time in central India
Ans. 20 June to 10 July

Q.1519. Sowing time in south India
Ans. Sep. to Oct.

Q.1520. Ultra Narrow row (UNR) was developed by
Ans. USA

Q.1521. Spacing and Plant population in UNR
Ans. 19×19 cm^2, & PP 277000/ha

Q.1522. Appearance of flower bud in cotton called as
Ans. Square formation

Q.1523. Removal of terminal growing point of each plant called as
Ans. Topping

Q.1524. Height and days for topping in cotton
Ans. Height 1-1.2 m and 80-90 DAS

Q.1525. Separation of fiber from the seed cotton is known is
Ans. Ginning

Q.1526. Ginning % in cotton
Ans. 30-35 %

Q.1527. Common weedicide used in cotton
Ans. Diuron

Q.1528. Cotton is highly sensitive for the weedicide
Ans. 2,4-D

Q.1529. 2,4-D resistant cotton gene
Ans. tfd-A

Q.1530. Little leaf in cotton caused by
Ans. Zn deficiency

Q.1531. Crinkle leaf in cotton caused by
Ans. Mn toxicity

Q.1532. An elongation/outgrowth of an epidermal cell of seed coat is known as
Ans. Cotton fiber

Q.1533. The long outgrowth forms
Ans. Staple/Lint

Q.1533a. India position in Cotton export (globally)
Ans. 3rd

Q.1534. The shorter outgrowth forms
Ans. Fuzz

Q.1535. No of fibers in one seed of cotton
Ans. 60000-80000 fibers

Q.1536. Bad opening of bolls is known as
Ans. Tirak

Q.1537. Tirak in cotton occurs due to
Ans. Salinity in soil and decrease in N and increase in tannin

Q.1538. Genetic traits of cotton for fiber quality
Ans. Fiber length and fiber fineness

Q.1539. A measure of the air permeability of compressed cotton fibre
Ans. Micronarie

Q.1540. If fiber average weight is below 3.0 microgram then it is
Ans. Very fine fiber

Q.1541. Fiber is very strong when breaking point is more than
Ans. 95 kg/sq.cm

Q.1542. Maturity of fiber is judged by
Ans. Arealometer

Q.1543. Formula of Ginning %
Ans. $= \frac{\text{Weight of lint}}{\text{Weight of seed cotton taken for ginning}} \times 100$

Q.1544. Formula of Lint Index
Ans. $= \frac{\text{Weight of 100 seeds} \times \text{Ginning\%}}{100 - \text{Ginning\%}} \times 100$

Q.1545. Lint percent in cotton is
Ans. 33%

Q.1546. Nippiness is related to
Ans. Fiber thickness measurement

Q.1547. Spinning performance is measured by
Ans. Number of counts

Q.1547a. Boll shedding in cotton is due to
Ans. Inhibitory effect of ABA (Abscisic Acid)

Q.1548. Cotton corporation of India established in
Ans. 1970 for market intervention

Q.1549. Oil content in cotton seed is
Ans. 14.5% to 25.6 %

Q.1550. Toxic pigment/phenol compound present in cotton seed is
Ans. Gossypol

Q.1551. No of counts in Indian cotton
Ans. 22 counts

Q.1552. No of counts required for best quality cotton
Ans. 80-400 counts

Q.1553. Seed + lint called as
Ans. Seed cotton

Q.1554. Seed after removing lint called as
Ans. Cotton seed

Q.1555. World's earliest maturity variety released is India
Ans. By CICR Scientist in 2017 (YUGANK) 120 maturity days

Q.1556. Bt Cotton in India given by
Ans. CD Mayee (Also known as Father of Bt cotton in India)

Q.1556a. Best variety of cotton
Ans. Sea Island cotton, Extra long staple length upto 34 mm

17. Tobacco

Q.1557. Botanical name of tobacco
Ans. *Nicotiana Spp*

Q.1558. How many spp of Indian tobacco
Ans. 2, *Nicotiana tabacum & Nicotiana rustica*

Q.1559. Key feature of Nicotiana tabacum
Ans. 1. Plant height 150-250cm,
2. Large and narrow leaf
3. Nicotine content is 0.5-5.5%,
4. Used for smoking and chewing purpose,
5. Grown in light soils

Q.1560. Key feature of Nicotiana rustica
Ans. 1. Plant smaller than tabacum,
2. Nicotine content is 3.5-8.0%,
3. Used for hookah, chewing and snuff purpose,
4. Mostly grown in heavy soils

Q.1561. Best soils for Tobacco
Ans. Mild acidc (pH 5-6)

Q.1562. Sodic soils are unfit for tobacco because of
Ans. Due to Chloride ions it results poor burning quality of leaves

Q1563. Nicotine mainly produced in
Ans. Roots and stored in Leaves (Economic part)

Q.1564. Best tobacco for Cigarette
Ans. Flue Cured Virginia (FCV)

Q.1565. Tobacco best for Bidi & hookah
Ans. Non Virginia type

Q.1566. Tobacco type which having max area and production is
Ans. Non Virginia type

Q.1567. Max exportable tobacco type is
Ans. FCV

Q.1568. First hybrid variety of tobacco in India is
Ans. GTH-1

Q.1569. Central Tobacco Research Institute situated at
Ans. Rajahmundri, A.P. (1947)

Q.1570. State having highest production of non Virginia tobacco
Ans. Gujarat

Q.1571. Seed rate of tobacco
Ans. N-tabacum 3kg/ha N. Rustica- 6 kg/ha

Q.1572. Seed rate for nursery of tobacco
Ans. 20-30 gm per $100m^2$

Q.1573. No of seeds in one gm of tobacco
Ans. Seed weight of *N-tabacum* (0.08-0.09 mg) 11,000-12,000 seeds per gram *N. rustica* is three times heavy.

Q.1574. Time of nursery sowing
Ans. 2nd fortnight of Aug

Q.1575. Removal of flower heads to improve size and quality of leaves is known as
Ans. Topping

Q.1576. Removal of lateral suckers/auxiliary buds is called as
Ans. De- Suckering

Q.1577. Process in which matured leaves are removed from plant called as
Ans. Priming

Q.1578. How many priming required in entire harvest
Ans. 5-6 priming

Q.1579. When top leaves develop red rusty spots it is called as
Ans. Spangles

Q.1580. Leaf priming harvesting is done mainly for
Ans. Cigarette & wrapper tobacco **(JRF Agronomy, 2019)**

Q.1581. Stalk cut harvesting is done mainly for
Ans. Bidi, Hookah, Cigar, Cheroot

Q.1582. Drying process of tobacco called as
Ans. Curing

Q.1583. Flue curing mainly used for
Ans. Cigarette

Q.1584. Principle of curing
Ans. Yellowing, fixing color, leaf drying and mid-rub drying.

Q.1585. Which pottasic fertilizer is best for tobacco
Ans. Potassium sulphate

Q.1586. Most widely distributed viral disease of tobacco
Ans. Tobacco Mosaic

Q.1587. Most important pest of tobacco nursery
Ans. Tobacco Caterpillar

18. Jute

Q.1588. Botanical name of Jute
Ans. *Corchorus Spp.* n = 7

Q.1589. India's rank in area and production
Ans. First Rank followed by Bangladesh and China

Q.1588a. Bitter gycoside present in leaves
Ans. Corchorin

Q.1590. State which is first in area and production
Ans. W. Bengal (7592.8 thousand Bales in 2020-21)

Q.1591. Seed rate of Jute
Ans. 6-8 Kg/ha

Q.1592. Harvesting time to obtain good quality fiber
Ans. 135-140 DAS
Q.1593. Ribboning process is related to
Ans. Jute
Q.1594. Retting is a
Ans. Microbial process
Q.1595. Optimum temp and days for retting
Ans. 34°C and 8-30 days.
Q.1596. Botanical name of white jute
Ans. *Corchorus Capsularis (most grown)*
Q.1597. National Institute for Research on Jute and Allied Fiber Technology is situated at
Ans. Kolkata
Q.1598. From which plant part the fiber is obtained
Ans. Phloem
Q.1599. Beat- Break-jerk method used for
Ans. Fiber extraction
Q.1600. Yield of jute
Ans. 20-27 Q/ha and 3-5 Q seed/ha
Q.1600a. Jute Packaging Material Act to protect the interest of Jute Industry was brought in the year
Ans. 1987
Q.1548. A fast-growing plant of the hibiscus, or mallow, family (Malvaceae) and its fibre, one of the bast fibre group. It is used mainly as a jute substitute.
Ans. Kenaf (*Hibiscus cannabinus*) or (Deccan hemp, Mesta)

CULTIVATION OF VEGETABLES

(Science concerned with vegetable growing - Olericulture)

Tomato

Q.1601. Botanical name of tomato is.
Ans. *Solanum lycopersicum* (old name is Lycopersicon esculentium) Common Name: Wolf apple, Common man's Orange
Q.1602. Family of tomato is known as.
Ans. *Solanaceae genus Lycopericon esculentum* (cultivated) and *pimpinellifolium* (small fruited) or currant tomato wild species.

Q.1603. Origin of tomato was in.
Ans. Peru

Q.1604. Fruit type of tomato is.
Ans. Berry

Q.1605. Popular variety of tomato.
Ans. Pusa ruby, Arka vikas, Sawrab, Ashish

Q.1606. Seed rate of tomato.
Ans. 400–500 gm/ha.

Q.1607. BER (Blossom end rot) disease is due to.
Ans. Ca deficiency

Q.1608. Major pest of tomato.
Ans. Fruit borer (Helicoverpo spp.)

Q.1609. Chemical used for tomato sauce preservation.
Ans. Sodium benzoate

Q.1610. Tomato is susceptible for.
Ans. Frost

Q.1611. Red colour of tomato is due to
Ans. Lycopene

Q.1612. Tomato is also known as.
Ans. Wolf apple

Q.1613. Most serious virus disease of Tomato
Ans. Leaf curl virus

Q.1614. Best suitable soil for tomato
Ans. Rich loam soil

Q.1615. Curly top disease is caused by
Ans. Beet leaf hopper

Q.1616. Edible Part of Tomato
Ans. Pericap and Placenta

Q.1617. Tomato fruits generally regarded as
Ans. Protective Food

Q.1617a. Blossom end rot in tomato due to
Ans. Calcium deficiency

Q.1617b. Maximum acidity found at
Ans. Pink stage

Q.1618. Determinate variety of Tomato
Ans. Punjab chhuhara
Q.1619. A period of drought followed by sudden heavy irrigation during fruiting may cause
Ans. Cracking of Fruits
Q.1620. Plant type of tomato
Ans. Day neutral and self pollinated plant
Q.1621. Seed rate of hybrid tomato
Ans. 100-150 gm/ha
Q.1622. Nematode resistant varieties are
Ans. Pusa-120, Hissar lalit, SL-120
Q.1623. Variety suitable for winter season
Ans. Pusa sheetal
Q.1624. Variety suitable for transport and processing
Ans. Roma and Punjab Chhuhara
Q.1625. Variety resistant for fruit borer
Ans. Parker
Q.1626. Tomato leaf curl virus is transmitted by
Ans. White fly
Q.1627. Fruit cracking occurs due to deficiency of
Ans. Boron deficiency
Q.1628. Sun scald occurs due to
Ans. Expose of fruits in high temperature > 40ºC
Q.1629. Puffiness occurs due to
Ans. Poor pollinating and temp fluctuation and heavy rains
Q.1630. Most popular variety of tomato
Ans. Pusa ruby

17. Cole Crops (Includes Cabbage, Cauliflower and Broccoli)

CABBAGE

Q.1631. All cole crops belongs to family
Ans. *Curciferae*
Q.1632. Botanical name of cabbage
Ans. *Brassica oleracea var capitata*

Q.1633. Type of Infloresecence
Ans. Catkin

Q.1634. Anticancer properties due to
Ans. Indole-3-carbinol

Q.1635. Hybrid varieties
Ans. Pusa Synthetic, Pusa Mukta and Pusa Drumhead, Pride of India, Green express.

Q.1636. Tipburn disease is due to
Ans. Calcium deficiency

Q.1637. The incidence of club root of cabbage is 100% at pH
Ans. pH 5.7

Q.1638. Seed rate of cabbage
Ans. 500-750 gm.

Q.1639. Improved Varieties of cabbage
Ans. Golden Acre, Copenhagen Market, Pride of India (Early & Round Variety)

Q.1640. Major pest
Ans. DBM (*Plutella xylostella*)

Q.1641. Fruit of cabbage (Edible part) is known as
Ans. Head

Q.1641a. Highly sensitive crop to irrigation
Ans. Cabbage

Q.1641b. Flavour in cabbage leaves
Ans. Due to glucoside 'singrin'

Q.1641c. Optimum pH range
Ans. 5.5-6.5

18. Cauliflower

Q.1642. Botanical name of Cauliflower
Ans. *Brassica oleracea var botrytis*

Q.1643. Introduced in india by
Ans. Dr. Jemson in 1822

Q.1644. Native country
Ans. Cyprus

Q.1645. Hybrid variety
Ans. Pusa Aghani, Snowball, Pusa hybrid2

Q.1646. Variety which can be grown April to July at hills.
Ans. Pusa Himjyoti
Q.1647. Edible part of cauliflower
Ans. Curd
Q.1648. Blanching is a important process of
Ans. cauliflower
Q.1649. Whiptail in cauliflower due to
Ans. Mo deficiency *(JRF 2004)*
Q.1650. pH of the soil for maximum production should be
Ans. 5.5-6.6
Q.1651. Browing in cauliflower is due to
Ans. Boron deficiency
Q.1652. Blanching is done to protect from
Ans. Attaining yellow colour of curd
Q.1653. All Cole crops are
Ans. Protogyny
Q.1654. India rank in Clauliflower production in the world
Ans. 2nd
Q.1655. Process of removal of central portion of curd for earlier initiation of flower stalk in cauliflower known as
Ans. Scooping
Q.1656. Variety which having self blanched habit
Ans. Pusa himjyoti & Pusa snowball
Q.1657. Seed rate of Cauliflower
Ans. 450-700 gm.
Q.1658. Blindness disorder occurs due to
Ans. Low temp and frost
Q.1659. Buttoning occurs due to
Ans. Nitrogen deficiency
Q.1660. Hollow stem in cauliflower occurs due to-
Ans. Excess Nitrogen and Boron deficiency
Q.1661. Chlorisis occurs due to
Ans. Mg deficiency

Q.1662. Riceyness occurs due to

Ans. Fluctuation of temp, Heavy application of N & high humidity

Q.1663. The characteristic flavor or aroma of cole crops is due to presence of

Ans. *Dimethyl trisulfide*

19. Other Vegetables

Q.1664. Richest source of riboflavin

Ans. Radish leaves

Q.1665. Richest source of thiamine

Ans. Giant chillies

Q.1666. Richest source of fat and vitamin C

Ans. Drumstick

Q.1667. Richest source of carbohydrate

Ans. Tapioca

Q.1668. Largest producing state in India for Brinjal

Ans. West Bengal

Q.1669. Main pigment in brinjal crop

Ans. Anthocyanin

Q.1670. Use of white coloured brinjal

Ans. For diabetic patients

Q.1671. Serious pest of brinjal vegetable

Ans. Shoot and Fruit borer

Q.1672. Largest producer of chillies in world

Ans. India

Q.1673. Pungency in Radish is due to

Ans. Isothiocyanates

Q.1674. Pungency in chilli is due to

Ans. Capsaicin

Q.1675. Bitterness in cucumber is due to

Ans. Cucurbitacin (*JRF 2010*)

Q.1676. Brinjal is native to

Ans. India

Q.1677. Country leading in okra production (2021-22)
Ans. India (54.45 MT)

Q.1678. Fruits of bitter gourd are rich in
Ans. Iron

Q.1679. Vegetable counted as riped fruit
Ans. Brinjal

Q.1680. Brinjal is a good source of
Ans. Vitamin B

Q.1681. Varieties of Brinjal
Ans. Pusa purple long, pusa Samrat, Azad Kranati, Black beauty, Pusa Bhairav

Q.1682. Little leaf of brinjal due to
Ans. Mycoplasma (transmitted by Leaf hoppers)

Q.1683. Red colour of chili is due to
Ans. *Capcyanthin*

Q.1684. Major pest of Chili is
Ans. Thrips

Q.1685. State having max chili production
Ans. A.P.

Q.1686. Seed rate of chili
Ans. 1.0 to 1.5 kg/ha

Q.1687. Seed rate for capsicum annum
Ans. 250gm/ha

Q.1688. Varieties of chili
Ans. California wonder, Pusa jwala, Jwalamukhi, Punjab lal, Mathania long, Bharat, Indra, Bhagyaluxmi, Bhasker, Arka Mohini

Q.1689. Responsible for transmit chili mosaic virus
Ans. White fly

Q.1690. Yellow vein mosaic of okra is transmitted by
Ans. White fly

Q.1691. Varieties of okra
Ans. Pusa sawani , Pusa makhmali, Punjab padmini, Varsha uphar, Prabhani kranti, Hissar unnat, Arka Anamica

Q.1692. Onion is a rich source of
Ans. Vitamin B

Q.1693. Pungency in onion is due to
Ans. Allyl propyl di sulphide **(IBPS, AFO, 2021)**

Q.1694. Bolting in onion occurs due to
Ans. Low temperature

Q.1695. The colour of onion outer skin is due to
Ans. Quecertin

Q.1696. Onion sowing is done by
Ans. Seeds (10-12 kg) & bulbs (750 kg/ha)

Q.1697. Varieties of onion
Ans. Pusa red, Nasik red, Arka Pragati,Pusa ratnar, Arka Bindu, Arka Kalyan

Q.1698. Antibacterial substance of garlic is
Ans. Allicin (diallylthiosulfinate)

Q.1699. Roots of radish prepared for seed production is known as
Ans. Stackling

Q.1700. Brown heart of radish occurs due to
Ans. Deficiency of Boron

Q.1701. Multiple disease resistant variety of radish
Ans. Arka Nishant

Q.1702. All season variety of radish
Ans. Pusa Himani

Q.1703. Forking of carrot occurs due to
Ans. Hard Pan of soil

Q.1704. Variety of carrot having highest Vitamin A
Ans. Pusa Meghali

Q.1705. Carrot variety which is suitable for canning and storage
Ans. Chantenay

Q.1706. A Kanji beverage is prepared from
Ans. Black carrot

Q.1707. Powdery mildew in cucurbits due to
Ans. *Erysiphe cichoraaceearum*

Q.1708. Downey mildew in cucurbits due to

Ans. *Pseudoperanospora cubensis*

Q.1709. Pusa viswas is the variety of

Ans. Pumpkin

Q.1710. Pusa Chikni and Arka chandan is variety of

Ans. Sponge guard (Turai)

Q.1711. Key pest of cucurbits

Ans. Red pumpkin beetle (Both adult & grub)

Q.1712. Pillow disease of cucumber is due to

Ans. Ca Deficiency

Q.1713. Vector of cucumber mosaic virus is

Ans. Aphid

Q.1714. Variety of cucumber

Ans. Pusa Sanyog, (Gynoecious var.), Japanese long green

Q.1715. Water content in watermelon is

Ans. 92-95%

Q.1716. Heavy dull sound at the time of harvesting of W. melon shows

Ans. Maturity

Q.1717. Bud necrosis in watermelon is due to

Ans. Aphids (*Aphis Craccivora*)

Q.1718. Sweetest variety of Water melon

Ans. Sugar baby (11-13% TSS)

Q.1719. Seedless hybrid variety of water melon

Ans. Pusa bedana

Q.1720. Yellow fleshed variety of watermelon

Ans. Durgapura Kesar

Q.1721. High temperature at the time of fruit maturity of musk melon will results to

Ans. Increased sweetness

Q.1722. Musk melon is harvested at the stage of

Ans. Full slip stage

Q.1723. First hybrid Onion

Ans. By IARI in 2002 with a potential of increasing yield by 20%

Q.1724. Musk melon variety which is not harvested at full slip stage

Ans. Hara Madhu (Late Var.)

Q.1725. Seed rate of musk melon

Ans. 1.5- 2.0 kg/ha

Q.1726. Early variety of Musk melon

Ans. Durgapura Madhu

Q.1727. Variety good for transportation of musk melon

Ans. Arka Rajhans

Q.1728. First hybrid of bottle guard

Ans. Pusa Maghdut

Q.1729. Highest yielding variety of bottle guard

Ans. Samarat

Q.1730. Seed rate of bottle guard

Ans. 3-6 kg/ha

Q.1731. Pea is rich source of

Ans. Protein (25%)

Q.1732. State having highest production of pea

Ans. U.P.

Q.1733. Powdery mildew of pea caused by

Ans. *Erysiphe polygoni*

Q.1734. Pea variety which are P. mildew resistant

Ans. Arka Ajit & Rachna

Q.1735. Extra early variety of Pea

Ans. Harbhajan

Q.1736. Pea variety suitable for dehydration

Ans. Bonneville

Q.1737. First dwarf variety of pea

Ans. Aparna/ HFP-4

Q.1738. Vegetable which is used as vegetable as well as fodder crop also

Ans. Cow pea

Q.1739. Dwarf variety of cow pea

Ans. Pusa Phalguni

Q.1740. Improved variety of cow pea
Ans. Pusa barsati, P. Dofasli, P.rituraj, P.komal, P.gouri

Q.1741. Vegetable crop which is used as vegetable, fodder and industrial use
Ans. Cluster bean

Q.1742. State first in production of cluster bean
Ans. Rajasthan

Q.1743. Varieties of cluster bean
Ans. Pusa sadabahar, P. mousami, P.navbahar, Dugapura safed

Q.1744. Varieties of Palak
Ans. Punjab green, All green, Pusa harit, P. Jyoti, Pusa bharti

Q.1745. Mutant variety of Palak
Ans. Jobner green

Q.1746. Intensive method of vegetable cultivation
Ans. Market garden

Q.1747. Extensive method of vegetable cultivation
Ans. Truck garden

Q.1748. Green leafy vegetables are rich source of
Ans. Folic acid

Q.1749. Vegetable which is non climacteric vegetable fruit
Ans. Cucumber

Q.1750. Flower colour of Chilli, point gourd and bottle guard
Ans. White

Q.1751. Flower colour of cruciferous is
Ans. Yellow

Q.1752. Sweet potato, ginger & Garlic is example of
Ans. Perennial vegetables

Q.1753. Daily requirement of vegetables
Ans. 285gm/day/capita

Q.1754. Vegetable can be processed at temperature of
Ans. 115-121°C

Q.1755. Production of off season vegetable is known as
Ans. Vegetable forcing

Kitchen Gardening: Growing of vegetable crops and small fruit plants in the backyard of house to meet the requirement of family year round.

FRUITS

1. Mango (King of Fruits)

Q.1756. Botanical name of mango is
Ans. *Mangifera Indica*

Q.1757. Fruit type of mango is
Ans. Drupe

Q.1758. Origin of mango was
Ans. Indo-Burma

Q.1759. India's rank in mango production
Ans. First, (>50% share of world)

Q.1760. Cagging technique of breeding first used by
Ans. Dr. R.N. Singh

Q.1761. Edible part of mango is known as
Ans. Mesocarp

Q.1762. Good mango varieties have a TSS of
Ans. 20%

Q.1763. Dwarf variety of mango
Ans. Amrapali

Q.1764. Seedless variety of mango is
Ans. Sindhu

Q.1765. Sweetest variety of mango is
Ans. Chausa

Q.1766. Suitable variety for processing in mango
Ans. Alphanso

Q.1767. Propagation of mango is done by the method.
Ans. Veneer grafting

Q.1768. Pollination in mango carried by
Ans. House fly

Q.1769. Major pest of mango
Ans. Mango hoppers

Q.1770. Major disease of mango
Ans. Powdery mildew

Q.1771. Mango malformation is due to.
Ans. Low temperature

Q.1772. In mango black tip disease mostly occurs due to the following.
Ans. SO_2, CO, C_2H_2

Q.1773. Inflorescence of mango
Ans. Panicle

Q.1774. Control of Black tip by.
Ans. Borax spray

Q.1775. Spongy tissue is due to.
Ans. Convection heats (Highly susceptible Alphonso)

Q.1776. Internal fruit necrosis is due to.
Ans. Boron deficiency

Q.1777. Deblossoming is done for.
Ans. Control of malformation.

Q.1778. Paclobutrazol & Kuttar chemical is used in mango to.
Ans. Avoid the alternate bearing

Q.1779. Fruit known as national fruit/ King of fruit is
Ans. Mango

Q.1780. State first in mango production
Ans. U.P.

Q.1781. Mango is richest source of
Ans. Vitamin A

Q.1782. Fruit having highest area under fruit cultivation
Ans. Mango

Q.1783. Clustering/ Jhumka in mango occurs due to
Ans. Low temp/improper pollination and fertilization

Q.1784. Mango is a type of fruit
Ans. Climacteric and tropical fruit

Q.1785. Percent of flowers develops fruits to maturity
Ans. 0.1%

Q.1786. Vapor Heat treatment (VHT) is done for
Ans. Disinfection of mango fruit flies and Stone weevil

Q.1787. Variety which is most commonly grown in Maharashtra

Ans. Alphanso

Q.1788. Hybrid variety which is regular bearing and free from malformation is

Ans. Sai sugandha

Q.1789. Recent method of grafting

Ans. Epicotyl stone grafting (MH)

Q.1790. Famous north Indian variety which has characteristic turpentine flavor

Ans. Langra

Q.1791. Off season variety is

Ans. Niranjan & Madhulica

Q.1792. Variety which is resistant from powdery mildew

Ans. Lal sindhuri

Q.1793. Variety resistant for anthracnose disease

Ans. Edward

Q.1794. Amrapali (Dwarf Var.) is a cross of

Ans. Dasheri × Neelam

Q.1795. Ratna is a cross of

Ans. Neelam × Alphanso

Q.1796. Arka aruna is a cross of

Ans. Banganpali × Alphanso

Q.1797. Arka anmol is a cross of

Ans. Alphanso × Janardan pasand

Q.1798. Arka puneet is a cross of

Ans. Alphanso × Banganpali

Q.1799. Sindhu is a cross of

Ans. Ratna × Alphanso

Q.1800. Malika is a cross of

Ans. Neelam × Dashari

Q.1801. Sai sugandha is cross of

Ans. Totapuri × Kesar

Q.1802. Neelam and Totapuri are

Ans. Regular bearing varieties

Q.1803. Langra and Dashari are

Ans. Alternate bearing varieties

Q.1804. Quarantine importance pest of mango is

Ans. Stone weevil

Q.1805. Leaf scorching occurs due to

Ans. Potassium deficiency

2. Banana

Q.1806. Scientific name of Banana is

Ans. *Musa acuminate*

Q.1807. Ripe Banana contains sugar

Ans. 26%

Q.1808. Banana is native to

Ans. South East Asia

Q.1809. Largest producer in the world

Ans. India is the largest producer

Q.1810. Leading banana producing state in India

Ans. Tamilnadu

Q.1811. Ideal temperature range to grow banana is

Ans. 20–40 °C

Q.1812. Serious fungal disease of Banana

Ans. Sigatoka leaf spot

Q.1813. Spadix is a

Ans. Inflorescence of banana

Q.1814. Seedlessness of banana is due to

Ans. Vegetative parthenocarpy

Q.1815. Most serious disease of Banana

Ans. Bunchy top

Q.1816. Crop which has highest production among fruits

Ans. Banana

Q.1817. Banana is propagated by

Ans. Suckers/ Corms

Q.1818. Most of cultivated edible banana is

Ans. Triploid in nature

Q.1819. Weight of sword suckers generally used

Ans. 500-750gm.

Q.1820. Cut rhizomes used for propagation is known as

Ans. Peepers

Q.1821. Varieties of Banana

Ans. Dwarf cavandish, Robusta, Rasthali, Poovan, Honda, Champa, Nendran (Nendran is best variety for making chips).

Q.1822. Diploid banana variety is

Ans. Lady finger

Q.1823. Banana disease which are transmitted by aphids

Ans. Bunchy top & Kokkan disease

Q.1824. Temperature and humidity for banana storage

Ans. Temp 13°C and 85-90% humidity

Q.1825. Panama wilt is associated with

Ans. Banana

Q.1826. Panama disease is prevented by

Ans. Application of lime in soil

Q.1827. Infectious chlorosis in banana is caused by

Ans. Virus

Q.1828. Vegetative parthenocarpy is found in

Ans. Banana

Q.1829. The stem above the ground in banana is known as

Ans. Pseudostem

Q.1830. How many times banana crop gives fruit in life

Ans. Once time

Q.1831. Vector of bunchy top virus

Ans. Aphids

3. Apple

Q.1832. Scientific name of Apple
Ans. *Malus* × *domestica*

Q.1833. Apple is propagated by
Ans. Shield budding and grafting

Q.1834. Apple bowl of India
Ans. Himachal Pradesh

Q.1835. Origin or Native to
Ans. Asia minor

Q.1836. Redness in Apple is due to
Ans. Anthocyanin (*ARS 2010*)

Q.1837. Country leading in Apple production
Ans. China

Q.1838. Aroma in ripe apple in also due to
Ans. Isopentanol

4. Guava

Q.1839. Scientific name of Guava
Ans. *Psidium guajava.* Family: *Myrtaceae*
Chromosome no. 2n = 22

Q.1840. Origin of Guava fruit
Ans. West Indies to Peru

Q.1841. State with highest production in India
Ans. Uttar Pradesh

Q.1842. Hybrid Varieties
Ans. Kohir safed and safed jam

Q.1843. Sardar is a variety of
Ans. Guava

Q.1844. Ideal time for pruning
Ans. Feb.-March

Q.1845. Dual purpose variety of guava is
Ans. Lalit

Q.1846. Guava most popular variety
Ans. Allahabad Safeda

Q.1847. The most common and cheapest method of Guava propagation

Ans. Stooling/Air layering

Q.1848. Guava is susceptible for the soils

Ans. Acidic soils

Q.1849. Major insect pest of Guava

Ans. Fruit fly, Bark caterpillar

Q.1850. Serious disease of Guava

Ans. Wilt

Q.1851. Red fleshed variety of Guava

Ans. Hafsi

Q.1852. Seedless Guava variety is

Ans. Behat coconut

Q.1853. Hybrid 45 is a cross of

Ans. Allahabad safeda× Sardar (Lucknow -49)

Q.1854. Wilt resistant variety of Guava

Ans. Allahabad safeda

Q.1855. Name of Guava tea mosquito bug

Ans. *heliopeltis antonil*

Q.1856. Bronzing in Guava occurs due to

Ans. Zn deficiency

Q.1857. Soft seeded dwarf variety of Guava

Ans. Arka mirdula

5. Papaya

Q.1858. Scientific name of Papaya

Ans. *Carica papaya*

Q.1859. Native to country

Ans. Tropical America

Q.1860. Yellow pigment in papaya is due to

Ans. Caricaxanthin

Q.1861. Richness of papaya in Vitamin A

Ans. Second after Mango

Q.1862. Papaya is most susceptible to
Ans. Water logging
Q.1863. Papin is extracted from which part
Ans. Unmatured fruit
Q.1864. Plant type of papaya
Ans. Polygamous plant and tropical fruit
Q.1865. Gynodioecious varieties of papaya
Ans. Pusa delicious, Pusa majesty, Coorg honey dew, Taiwan, Surya, Sunrise solo
Q.1866. Nematode resistant varieties of papaya
Ans. Pusa majesty
Q.1867. Milk obtained from unripe fruit of papaya known as
Ans. Papain
Q.1868. Percent of male plants required in dioecious varieties
Ans. 10%
Q.1869. Dioecious varieties of papaya
Ans. Pusa giant, Pusa dwarf, CO-5, Pant CO-1, Betty, Hatras gold, Pusa Nanha
Q.1870. Climate factor which is most limiting factor for papaya
Ans. Frost (Frost resistant variety is carica-candomarcensis and c-pentagona)
Q.1871. Variety suitable for canning
Ans. Pusa giant
Q.1872. Variety suitable for papain production
Ans. CO-5
Q.1873. Enzyme present in dried latex of papaya
Ans. Pepsin
Q.1874. Papaya is commercially propagated by
Ans. Seed
Q.1875. Seed rate of papaya
Ans. 200-300gm/ha
Q.1876. Most serious disease of papaya seedling is
Ans. Damping off
Q.1877. Papaya leaf curl virus is transmitted by
Ans. White fly
Q.1877a. Gynodioecious variety
Ans. Taiwan-786, Pusa delicious, CO3, CO7

6. Grapes

Q.1878. Scientific name of grape
Ans. *Vitis vinifera*, (15-25% sucrose in present)

Q.1878a. Science of grapes
Ans. Viticulture

Q.1878b. Milky latex secreted by unripe sapota fruits is known as
Ans. Chuckle (It forms the base for making chicklets and chewing gums)

Q.1879. Grape is propagated by
Ans. Hard wood cuttings

Q.1880. Training system of grapes which is most used in India
Ans. Bower system

Q.1881. Planting time of unrooted cutting of grapes is
Ans. Oct

Q.1882. Planting time of rooted cutting of grapes is
Ans. Jan-Feb

Q.1883. Most cultivated famous variety of grapes
Ans. Thompson seedless

Q.1884. Grape is a
Ans. Deciduous crop

Q.1885. Element which is universally deficient in grapes
Ans. Mg

Q.1886. Major disease of grapes
Ans. Downey Mildew

Q.1887. Causal organism of Downey mildew
Ans. *Plasmophora viticola*

Q.1888. Acid present in grapes fruits
Ans. Tartaric acid

Q.1889. Best variety for Raisin is
Ans. Kismis Bely

Q.1890. Pruning time of grapes in North India
Ans. Dec- Jan

Q.1891. Chemical used to reduce post harvest fruit drop in grape is
Ans. NAA (50 ppm)

Q.1892. Powdery mildew of grapes is caused by
Ans. *Erysiphe viti*

Q.1893. Varieties suitable for wine
Ans. Arka trishna and Arka Soma

Q.1894. Variety which is cultivated for white wine
Ans. Arka Hans

Q.1895. Coloured seeded variety
Ans. Mascot

Q.1896. Hen & chicken disease occurs due to
Ans. Bo deficiency

Q.1897. Calyx end rot occurs due to
Ans. Ca deficiency

Q.1898. Due to improper pollination and fertilization in grapes it caused
Ans. Berry/ Blossom drop

Q.1899. Pink berry formation occurs due to
Ans. High temperature

Q.1900. Grapes are generally dried in
Ans. Sun light

7. Other Fruits

Q.1901. Cider is prepared by
Ans. Apple & Pear

Q.1902. Origin of Date palm
Ans. Iraq

Q.1903. Highly salt tolerant fruit tree is
Ans. Date palm

Q.1904. Fruit having twice energy value of banana
Ans. Avocado (Butter fruit)

Q.1905. National Research centre for Date palm is situated at
Ans. Bikaner

Q.1906. Common chemical which is used for colourless fruits preservation
Ans. KMS

Q.1907. State leader in litchi production
Ans. Bihar

Q.1908. Country leading in Litchi and Peaches production
Ans. China

Q.1909. Hardness of citrus fruit is due to
Ans. Boron deficiency

Q.1910. Polyembryony present in fruits like
Ans. Mango, Citrus, Jamun

Q.1911. Cricket ball and Murrabba is famous variety of
Ans. Sapota

Q.1911a. Milky latex secreted by unripe sapota fruits known as
Ans. Chuckle (forms the base for making chicklet and chewing gum).

Q.1912. Rayan as rootstock is commercial method of
Ans. Sapota

Q.1913. Acid content is a very important consideration for the preparation of
Ans. Jelly

Q.1914. Fruit type of Grape, Phalsa, Banana, Date palm & Guava is
Ans. Berries type

Q.1915. Element which improves quality of fruits
Ans. Boron

Q.1916. Vitamin lost during processing is
Ans. Vitamin C

Q.1917. Central Research Institute for Chikoo is situated at
Ans. Muzaffarpur (U.P.)

Q.1918. Central Research Institute for Banana is situated at
Ans. Tiruchirappalli (TN)

Q.1919. National Horticulture Board is situated at
Ans. Gurugram (Haryana)

Q.1920. Fruits which don't ripe after picking called
Ans. Non climacteric fruits like Citrus, Grape,

Q.1921. Daily requirement of fruits
Ans. 120gm/day/capita

Q.1922. Fruits having max vitamin C
Ans. 1st Barbadas Cherry (2000mg/100gm & 2nd Aonla (600mg/100gm)

Q.1923. Edible part of Papaya & Mango
Ans. Mesocarp

Q.1924. Edible part of Litchi
Ans. Fleshy aril
Q.1925. Edible part of Ber
Ans. Pericarp
Q.1926. Edible part of Apple & strawberry
Ans. Thalamus
Q.1927. Edible part of Coconut
Ans. Endosperm
Q.1928. Edible part of Pineapple
Ans. Bracts
Q.1929. Edible part of Pomegranate
Ans. Juicy seed coat (April)
Q.1930. Edible part of Citrus
Ans. Juicy placental hairs
Q.1931. Date palm is rich in
Ans. Iron
Q.1932. Queen of fruits
Ans. Litchi
Q.1933. King of arid fruit and poor man fruit/ Apple
Ans. Ber
Q.1934. Century plant is known as
Ans. Date palm
Q.1935. Citrus is commercially propagated by
Ans. Seeds/ Air layering
Q.1936. Pomegranate is commercially propagated by
Ans. Cutting/ Air layering
Q.1937. Guava is commercially propagated by
Ans. Stooling
Q.1938. Date palm is commercially propagated by
Ans. Offshoot/ suckers
Q.1939. Sapota is commercially propagated by
Ans. Inarching
Q.1930. Anola is commercially propagated by
Ans. T-budding/Patch

Q.1931. Ber is commercially propagated by
Ans. Ring & T- budding

Q.1932. Family of cashew
Ans. Anacardiaceae

Q.1933. Leathery rind skin found in fruits
Ans. Citrus

Q.1934. Fruit type of Citrus
Ans. Hesperidum

Q.1935. Kinnow is cross between
Ans. King sweet × Willow Leaf mendarins

Q.1936. Laddu is a variety of
Ans. Mandarin (Citrus fruit)

Q.1937. A spiny, disease resistant variety produces tall fruits (most delicious variety)
Ans. Abacaxi

Q.1938. Seed less variety of acid lime (Kagzi Lime)
Ans. Chakradhar

Q.1939. Little leaf and Die back of citrus due to
Ans. Cu deficiency

Q.1940. Yellow leaf of citrus due to
Ans. Mo Deficiency

Q.1941. Little leaf of Litchi
Ans. Due to zinc deficiency

Q.1942. Crop highly drought tolerant after date palm and Ber
Ans. Pomegranate

Q.1943. Red colour of pomegranate is due to
Ans. Anthocyanin

Q.1944. Country first in area and production of pomegranate
Ans. India

Q.1945. State first in pomegranate production
Ans. Maharashtra

Q.1946. Daru is a
Ans. Wild type pomegranate

Q.1947. Popular varieties of Pomegranate
Ans. Ambe bahar, Bhagawa, Hasta, Ganesh, Dolka, Wonderful, Jalore seedless, Kandhari, Ruby

Q.1948. Pomegranate variety Amlidana is across of
Ans. Ganesh × Nanha
Q.1949. Best varieties of Ber
Ans. Dodhia, Gola, Kaithali, Seb, Mundia, Jogia, Meharun, Banarasi, Umran, Illaichi, Katha, Pathani
Q.1950. Fruit known as Indian gooseberry
Ans. Aonla
Q.1951. Best varieties of Aonla
Ans. Banarasi, Kanchan, Chakia, Hathijhool
Q.1952. Fruits can be processed at a temperature of
Ans. 100°C
Q.1953. Concentration of salt preservation
Ans. 10-25%
Q.1954. Concentration of sugar used for preservation is
Ans. 60-70%
Q.1955. Fruit ripening hormone is
Ans. Ethylene
Q.1956. Pajaro is a variety of
Ans. Straw berry
Q.1957. Development of fruits without fertilization called as
Ans. Parthenocarpy
Q.1958. Development of embryo without fertilization is called
Ans. Apomixis
Q.1959. All fruits by nature are type of
Ans. Acidic type
Q.1960. Date palm variety which used to made for Chhuhara is
Ans. Halawi
Q.1961. Multistoried cropping is popular in
Ans. Coconut
(First hybrid variety of coconut by CPCRI Kerala in 1932)
Q.1962. Coconut is propagated by
Ans. Seeds
Q.1962a. Makhana also known as
Ans. Fox nuts
Q.1962b. Pineapple's largest producers in the world
Ans. Costa Rica, Brazil and Phillipines (Pineapple contains, Vit. C, Manganese and Dietary fiber)

FLORICULTURE

Q.1963. Leading flower producing country
Ans. Netherlands

Q.1964. Flower capital of world
Ans. California (U.S.A.)

Q.1965. State having max production under floriculture is
Ans. Tamilnadu

Q.1966. State having max area under floriculture is
Ans. Karnataka

Q.1967. State which produce max cut flowers
Ans. W. Bengal

Q.1968. Country leading in cut flowers market
Ans. Germany

Q.1969. Largest importer of floriculture products from India is
Ans. U.S.A.

Q.1970. Flower crop having max area in India
Ans. Jasmine

Q.1971. Japanese flower arrangement is known as
Ans. Ikebana

Q.1972. Largest producer of loose flowers in world
Ans. India

Q.1973. Mughal garden & Pinjore garden is type of garden
Ans. Formal style garden

Q.1974. Japanese & English gardens style are
Ans. Informal style

Q.1975. Rose garden, Chandigarh is a example of which type garden
Ans. Free style garden

Q.1976. Book *Beautiful garden* is written by
Ans. M.S.Randhawa

Q.1977. Queen of flowers is
Ans. Rose

Q.1978. Example of Japanese garden in India
Ans. Buddha Jayanti Park, New Delhi

Q.1979. Important cultivars of Rose
Ans. Arjun, Raktgandha and Sindhur

Q.1980. Rastrapati garden is a example of
Ans. English garden

Q.1981. Baradari & Terminal building are key element of
Ans. Mughal garden

Q.1981. Garden Lanterns & dry landscape are key element of
Ans. Japanese garden

Q.1982. Biggest formal garden in India is
Ans. Brindavan garden (Mysore)

Q.1983. Mughal garden is situated at
Ans. Pinjore (Haryana)

Q.1984. National Botanical garden is situated at
Ans. Lukhnow

Q.1985. Rockery & Lawn are the key elements of
Ans. English garden

Q.1986. Best and fast method of lawn development
Ans. Turf method

Q.1987. Key difference of formal and Informal garden is
Ans. Symmetrical and Asymmetrical

Q.1988. Origin place of rose
Ans. India

Q.1989. Symbol of beauty is
Ans. Rose

Q.1990. Commercial method of propagation of rose is
Ans. T- Budding

Q.1991. Best time of budding
Ans. Nov. to Feb.

Q.1992. Best time of planting of rose
Ans. Sep. to Oct.

Q.1993. Rose is a national flower of
Ans. England

Q.1994. Fruit of rose called (IBPS, AFO, 2021)
Ans. Rose hips

Q.1995. Thorn less variety of rose
Ans. Suchitra & Pusa Mohit

Q.1996. Scientist who is associated with rose breeding
Ans. Dr. B.P. Pal

Q.1997. Tushar is a variety of
Ans. White rose

Q.1998. Crimson glory is a variety of
Ans. Red rose

Q.1999. Gulkand is prepared by
Ans. Rose Petal & Sugar (1:1 ratio)

Q.2000. Native of marigold
Ans. Mexico

Q.2001. Family of Gladiolus
Ans. Iridaceae

Q.2002. Gladiolus is commercially propagated by
Ans. Corms

Q.2003. Varieties of gladiolus
Ans. Priyadrashani, Friendship

Q.2004. Botanical name of Chrysanthemum
Ans. *Dendranthema gradiflora*

Q.2005. Family of Chrysanthemum
Ans. Compopsiteae

Q.2006. Common name of Chrysanthemum
Ans. Guldaudi

Q.2007. Origin of Chrysanthemum
Ans. China

Q.2008. Propagation method of chrysanthemum
Ans. Root suckers

Q.2009. Varieties of chrysanthemum
Ans. Diana, Indra, Red gold, Birbal sahani, Basanti

Q.2010. Shrub planted grown on boundary for fencing is known as
Ans. Hedge

Q.2011. The low growing perennial plants grown on the border of plot known as
Ans. Edge plants

Q.2012. Which plant produces the largest flower
Ans. Rafflesia amoldii

SOIL SCIENCE

Q.2013. Granite & Basalt are type of rocks.
Ans. Igneous rocks

Q.2014. Lime stone, Dolomite, Sandstone are type of rocks.
Ans. Sedimentary rocks

Q.2015. Gneiss & Marble are type of rocks.
Ans. Metamorphic rocks

Q.2016. Chief constituent of Sandy fraction.
Ans. Quartz

Q.2017. Most dominant mineral on Earth crust.
Ans. Feldspars (48%)

Q.2018. Major source of Mg.
Ans. Dolomite

Q.2019. Major source of P.
Ans. Apatite

Q.2020. Major source of B.
Ans. Tourmaline

Q.2021. Major source of Mo.
Ans. Olivine

Q.2022. Major source of K.
Ans. Orthoclase

Q.2023. Most resistant to weathering.
Ans. Quartz

Q.2024. Least resistant to weathering.
Ans. Calcite

Q.2025. Kankar nodules are found mostly in.
Ans. Red soils

Q.2026. Process of moving out of sesquioxide is known as.
Ans. Podzolization

Q.2027. First Indian President of the ISSS
Ans. Dr. J.S. Kanwar (1978-1982)

Q.2028. Process of mixing of soils is known as.
Ans. Pedoturbation

Q.2029. Soils having at least 20% organic matter are known as
Ans. Organic soils

Q.2030. Law of minimum was proposed by
Ans. Von Liebig (1840)

Q. 2031. 'O' horizon is absent in
Ans. Arable soils

Q.2032. Top most mineral horizon is
Ans. 'A' horizon

Q.2033. Maximum Eluviation horizon is
Ans. E horizon

Q.2034. Alluvial horizon is
Ans. B horizon

Q.2035. Physical property which can't be changed
Ans. Soil texture

Q.2036. Mechanical analysis of soils separation is done by.
Ans. Hydrometric method

Q.2037. Particle more than 250 mm in diameter is known as
Ans. Stone

Q.2038. Diameter of coarse sand particle is
Ans. 0.2-2 mm

Q.2039. Bulk density formula for soils is
Ans. $$B.D. = \frac{\text{Weight of oven dry soil}}{\text{Volume of soil (Solid + Pores)}}$$

Q.2040. Diameter of fine sand particle is
Ans. 0.02-0.2 mm

Q.2041. Diameter of silt particle is
Ans. 0.02-0.002 mm

Q.2042. Diameter of clay particle is
Ans. Less than 0.002 mm

Q.2043. The soils which are most suitable for most of the crops are

Ans. Sandy loams

Q.2044. Soil structure which is best for cultivation is

Ans. Crumby structure

Q.2045. Particle Density of most of the soils is

Ans. 2.65 g/cm^3

Q.2046. Bulk Density of general soils is

Ans. 1.33 mg/cm^3

Q.2047. Total pore space are highest in.

Ans. Clay soils

Q.2047a. Water holding capacity of various soils in descending order

Ans. Clay > Silt > Sand **(AFO, 2021)**

Q.2048. Portion of capillary water lying between field capacity (1/3 atm) and wilting coefficient (15 atm) is known as.

Ans. Available water

Q.2049. Water held up to the tension about 31 atm is known as

Ans. Hygroscopic coefficient

Q.2050. Major source of Mn

Ans. Pirolusice.

Q.2051. Kaolinite is a type of mineral.

Ans. 1 : 1 type

Q.2052. Montmorillonite is a type of mineral.

Ans. 2 : 1 type expanding type mineral

Q.2053. Vermiculate is a type of mineral.

Ans. Limited expanding 2 : 1 type mineral.

Q.2054. Micas are type of mineral.

Ans. Non Expanding type mineral

Q.2055. Chlorites are type of mineral.

Ans. 2 : 1 : 1 type mineral

Q.2056. Recently formed soil order is

Ans. Entisols

Q.2057. Black soils found in

Ans. Maharashtra

Q.2058. Red soils are dominant in.
Ans. TamilNadu

Q.2059. Laterite soils are dominant in.
Ans. Karnataka & Kerala

Q.2060. Weight of soil furrow slice.
Ans. 2×10^6 kg/ha.

Q.2061. Humic acid is soluble in.
Ans. Alkali solution

Q.2062. C : N ratio of humus.
Ans. 10 : 1

Q.2063. CEC of humus.
Ans. 150–300 C mol (p^+)/kg soil

Q.2064. C : N ratio of Legumes.
Ans. 20 : 1 to 30 : 1

Q.2065. C : N ratio of F.Y.M.
Ans. 100 : 1

Q.2066. C : N ratio of micro organism.
Ans. 4 : 1 to 9 : 1

Q.2067. Wood is mainly decomposed by.
Ans. Actinomycetes

Q.2068. Organic matter content in Indian soils is generally.
Ans. <0.5%

Q.2069. Optimum temperature & pH for Nitrifying bacteria is.
Ans. Temp. 30–35°C, pH 6.5–7.5

Q.2070. Loss of N_2 in the form of NH_3 in alkaline medium known as.
Ans. Volatilization

Q.2071. C.E.C. of vermiculate.
Ans. 80–150 C mol(P^+) /kg soil

Q.2072. CEC of montmorillonite.
Ans. 80-100 C mol(P^+) /kg soil

Q.2073. N, P, K % in F.Y.M.

Ans. 0.5% N, 0.2% P_2O_5, 0.5% N_2O

Q.2074. Most outstanding green manure crop.

Ans. Sun hemp

Q.2075. Fastest N fixing plant.

Ans. Sesbania rostrata

Q.2076. N% in cotton cake.

Ans. 6.5% N

Q.2077. N% in safflower cake.

Ans. 7.8% N

Q.2078. Substances added to soils for the improvement of their condition known as.

Ans. Amendments

Q.2079. The fertilizer having less than 25% of the primary nutrients known as.

Ans. Low analysis fertilizers.

Q.2080. Equivalent acidity of NH_4Cl.

Ans. 128

Q.2081. Equivalent acidity of $(NH_4)_2$ SO_4.

Ans. 110

Q.2082. Equivalent acidity of urea.

Ans. 80-85

Q.2083. Equivalent acidity of DAP.

Ans. 77

Q.2084. Equivalent acidity of Ammonium Nitrate.

Ans. 60

Q.2085. Equivalent basicity of $NaNO_3$.

Ans. 29

Q.2086. Equivalent basicity of $Ca(NO_3)_2$.

Ans. 21

Q.2087. The crops which absorb the ammonical form directly.

Ans. Paddy & Potato

Q.2088. N% in sodium nitrate.

Ans. 16% N

Q.2089. N% in ammonium sulphate.

Ans. 20.6% N

Q.2090. Sulphur % in Ammonium sulphate.

Ans. 24% S

Q.2091. N% in Ammonium Nitrate.

Ans. 33-35% N

Q.2092. N% in Ammonium Sulphate Nitrate.

Ans. 26% N

Q.2093. Sulphur % in Ammonium sulphate Nitrate.

Ans. 15% S

Q.2094. N% in kisan khad (CAN).

Ans. 25-28% N

Q.2095. N% in Ammonium Chloride.

Ans. 26% N

Q.2096. N% in urea.

Ans. 46% N

Q.2097. Biuret should be according to fertilizer control order.

Ans. < 1.5%

Q.2098. N% in calcium cyanamide.

Ans. 20.6%

Q.2099. Highest N% in fertilizer.

Ans. Aqueous Amonia (80% N)

Q.2100. Water soluble phosphatic fertilizer.

Ans. 1. S.S.P., D.S.P., T.S.P., D.A.P.

Q.2101. Citrate soluble P fertilizer,

Ans.
1. Di calcium phosphate
2. Basic slag
3. Rhemania Phosphate

Q.2102. Citrate & water Insoluble P fertilizer.

Ans. Rock phosphate, Raw bone meal, steamed bone meal

Q.2103. P_2O_5% in single super phosphate.
Ans. 16-20% P_2O_5

Q.2104. P_2O_5% in Double super phosphate (D.S.P.).
Ans. 32% P_2O_5

Q.2105. N & P_2O_5 % in DAP
Ans. 16% N, 48% P_2O_5

Q.2106. P_2O_5 % in Di calcium phosphate (D.C.P.).
Ans. 33-40% P_2O_5

Q.2107. P_2O_5 % in Basic slag.
Ans. 14-18% P_2O_5

Q.2108. P_2O_5 % in Rock phosphate.
Ans. 20-30% P_2O_5

Q.2109. P_2O_5 % in Raw bone meal.
Ans. 20-25% P_2O_5

Q.2110. P_2O_5 % in steamed bone meal.
Ans. 20-30% P_2O_5

Q.2111. Sulphur % in S.S.P.
Ans. 12% S

Q.2112. Ca% in S.S.P.
Ans. 18-21% Ca

Q.2113. Ca% in Gypsum ($CaSO_4 . 2H_2O$)
Ans. 29.2% Ca

Q.2114. S% in Gypsum.
Ans. 18.6% S

Q.2115. S% in potassium sulphate.
Ans. 17.5% S

Q.2116. K_2O% in Murate of potash (KCl).
Ans. 60%/20

Q.2117. K_2O% in potassium sulphate.
Ans. 48–52% (20)

Q.2118. Indian soils are deficient in.
Ans. Zinc

Q.2119. N% in Thio urea.
Ans. 36.8%

Q.2120. AM and N-serve are
Ans. Nitrification inhibitors
AM (2-Amino-4-chloro-6-methylpyrimidine)
N-Serve (2-chloro-6-trichloromethyl Pyridine)

Q.2121. Rhizobium melilots is used for the crops
Ans. Medicago (Alfalfa), Trigonella (Fenugreek)

Q.2122. R. trifoli is used for
Ans. Trifoliem (Clover)

Q.2123. R. leguminosarum is used for
Ans. Pisum (Pea), Lens (Lentil)

Q.2124. R. Phaseoli is used for
Ans. Phaseolus (beans)

Q.2125. R. Japonicum is used for.
Ans. Glycine (Soyabean), Vigna (Cow pea), Arachis (Ground nut), Crotolaria

Q.2126. Azotobacter & Beijernckia are
Ans. Aerobic bacteria

Q.2127. Anaerobic bacteria
Ans. Clostridum

Q.2128. Azotobacter is used for the crops like.
Ans. Rice, Cotton and Sugarcane

Q.2129. Azospirillum is used for.
Ans. Sorghum

Q.2130. Low status of N in soil when it is less than.
Ans. 250 kg/ha.

Q.2131. Analyzing process for determination of available N.
Ans. Alkaline permanganate method

Q.2132. Phosphorus is extracted by
Ans. Olsen's method & Bray No. 1 method

Q.2133. K^+ & Na^+ is determined by
Ans. Flame photometer

Q.2134. Organic carbon is determined by
Ans. 1. Walkey & Black method 2. Morgan's method

Q.2135. Application of clay to sandy soils is known as
Ans. Marling

Q.2136. The form of N is preferable for saline soils.
Ans. Nitrate form

Q.2137. Amonical fertilizers should be applied in.
Ans. Reduced zone

Q.2138. Nitrate fertilizers should be applied in.
Ans. Oxidized zone

Q.2139. Criteria of essentiality was given by.
Ans. Arnon & Stout (1939)

Q.2140. Functional nutrients are in number.
Ans. 20

Q.2141. Source of N for plants for absorption.
Ans. NO_3 (Mostly)

Q.2142. Source of P for plants for absorption.
Ans. $H_2PO_4^-$, HPO_4^{2-}

Q.2143. Source of B for plants for absorption.
Ans. $B_4O_7^{2-}$ $H_2BO_3^-$

Q.2144. Source of Mo for plants for absorption.
Ans. MoO_4^{2-}

Q.2145. Elements provide basic structure.
Ans. C,H,O

Q.2146. Energy Exchange elements.
Ans. H_2O

Q.2147. Highly mobile nutrients in plant.
Ans. N, P, & K

Q.2148. Immobile nutrients in plant.
Ans. Ca & B

Q.2149. Secondary nutrients are.
Ans. Ca, Mg, S

Q.2150. Zn, Fe, are.
Ans. Micro nutrients

Q.2151. Co is the structural component of.
Ans. Vit. B_{12}

Q.2152. S is essential for.
Ans. Oil seed crops

Q.2153. Na is essential for.
Ans. Sugarbeets

Q.2154. 'V' shaped yellowing at the tip of the lower leaves shows.
Ans. N deficiency

Q.2155. Chlorosis in between the veins and veins remain green shows.
Ans. Deficiency of Mg

Q.2156. Tip burn, margin scorching shows the deficiency of
Ans. K^+

Q.2157. Interveinal chlorosis occurs due to.
Ans. Fe deficiency

Q.2158. Complete interveinal chlorosis occurs due to.
Ans. Mn deficiency

Q.2159. Grey speck of oat, Phala blight of sugarcane caused by.
Ans. Mn deficiency

Q.2160. Rosetting & excess gumming occurs due to.
Ans. Cu deficiency

Q.2161. Top sickness of tobacco occurs due to.
Ans. B deficiency

Q.2162. Beneficial elements are.
Ans. Co, Si, Na, Ni, V

Q.2163. Ni & Co is most useful for the crops.
Ans. Legumes

Q.2164. Silicon is essential for.
Ans. Rice, Maize

Q.2165. Excess vegetative growth is due to.
Ans. High supply of N

Q.2166. Major constituent of chlorophyll is.
Ans. Mg

Q.2167. White tip & whitebud of maize occurs due to
Ans. Zn deficiency

Q.2168. Sickle leaf disease occurs due to
Ans. P deficiency

Q.2169. Whiptail of cauliflower occurs due to
Ans. Mo deficiency

Q.2170. Tea yellow disease occurs due to
Ans. S deficiency

Q.2171. Failure of terminal bud & root tip due to
Ans. Ca deficiency

Q.2172. Die back of citrus occurs due to.
Ans. Cu deficiency

Q.2172. Brittle leaf occurs due to.
Ans. Ca deficiency

Q.2173. Little leaf of cotton occurs due to
Ans. Zn deficiency

Q.2174. High lime requirement crops are
Ans. Soybean & Sugarbeet

Q.2175. Problems of soils are highest in
Ans. Uttar 'Pradesh

Q.2176. Alluvial soils are dominant in.
Ans. Uttar Pradesh

Q.2177. If pH < 8.5, EC > 4 (ds/m). ESP < 15% the soil will be.
Ans. Saline

Q.2178. If pH > 8.5, EC < 4 (ds/m), ESP > 15% the soil will be.
Ans. Alkali

Q.2179. If pH < 8.5, EC > 4 (ds/m). ESP > 15% the soil will be.
Ans. Saline alkali

Q.2180. Leaching is used for the treatment of
Ans. Saline soils

Q.2181. Gypsum is used for the reclamation of.
Ans. Alkali soils

Q.2182. Lime stone is used for the reclamation of.
Ans. Acidic soils

Q.2183. Urea is type of fertilizer.
Ans. Organic fertilizer

Q.2184. Highly salt tolerant crops are.
Ans. Barley, Sugarbeet

Q.2185. Acid tolerant crop is.
Ans. Rice

Q.2186. The soils which have pH < 4.0 are known as.
Ans. Cat soils

Q.2187. The soils which have organic matter + Na are known as.
Ans. Black alkali soils

Q.2188. Potato scab disease favours.
Ans. Alkaline soils

Q.2189. According to USDA, Soils have been divided into
Ans. 10 orders
(Entisol, Vertisol, Inceptisol, Acidisol, Mollisol, Spodosol, Alfisol, Ultisol, Oxisol and Histosol)

Q.2190. No. of soil orders according to seventh approximation soil classification are
Ans. 11 orders

Q.2191. Major source of Mn
Ans. Pyrolusite

Q.2192. Electrical Conductivity is used to express
Ans. Salinity of the soil

Q.2193. First country to introduce zero tillage
Ans. USA.

Q.2194. Special kinds of alkaline soils.
Ans. Takyr and Lunettes

Q.2195. Total fertilizers plant in India in 2016
Ans. 165 Plants

Q.2196. Total Urea Production plant in operation in India in 2016

Ans. 30 plants

Q.2197. No of Soil testing labs in India in 2016

Ans. 1424 Lab

Q.2198. No of Mobile testing labs are in India in 2016

Ans. 273 Labs

Q.2199. Fertilizer Control Act (FCO) was introduced in year

Ans. 1957

Q.2200. Central fertilizer pool was started in year

Ans. 1944

Q.2201. Sivaraman Committee was constituted for

Ans. Fertilizer (1964)

Q.2202. Fertilizer prices committee was constituted in

Ans. Jan, 1976

Q.2203. Block delivery scheme (BDS) for fertilizer was launched in

Ans. 1980-81

Q.2204. Highest SSP producer plants are in state

Ans. Rajasthan

Q.2205. Highest No. of Nitrogenous fertilizer plants are in state

Ans. U. P. & Gujarat

Q.2206. Highest production of FYM in state

Ans. Karnataka

Q.2207. Soil type vertisol is related to

Ans. Black soil

Q.2208. Indian Institute of Soil Science (IISS) is situated at

Ans. Bhopal (M.P.)

Q.2209. Excess of Ca in soil then availability of phosphorus

Ans. Decreases

Q.2210. C:N ratio for normal soils are-

Ans. C:N ratio 10:1 (*JRF 2012, 2005*)

Q.2211. Study of soils in relation to higher plants is called-
Ans. Edaphology *(JRF 2012, 2006)*

Q.2212. Dimeter of clay particles is-
Ans. < 0.002 mm *(JRF 2012)*

Q.2213. Nutrient essential for energy transformation-
Ans. P *(JRF 2012, 2003, BHU 2009)*

Q.2214. Soil have highest cation exchange capacity
Ans. Black clayey soil *(BHU, 2011)*

Q.2215. Soils having pH value < 7·0
Ans. Acid soil

Q.2216. The word soil is originated from
Ans. Latin word "Solum"

Q.2217. Father of Pedology
Ans. V. V. Dokuchaev

Q.2218. Horizon "A + B" combination known as a
Ans. Solum/ True soil

Q.2219. Physical condition of soil in relation to plant growth known as
Ans. Soil tilth

Q.2220. "O" horizon mostly found in
Ans. Forest area

Q.2221. Highest acidic soils found in
Ans. West Bengal

Q.2222. Nutrient which is essential constituent of protein
Ans. Nitrogen

Q.2223. Most essential function of K^+ is
Ans. Stomata regulation

Q.2224. Sodium (Na^+) is essential for
Ans. Osmo-regulation

Q.2225. Nutrient which is required for biosynthesis of hormones
Ans. Zinc

Q.2226. Nutrient which is known as animal protein factor
Ans. Cobalt

Q.2227. Silicon is absorbed in plants in the form of
Ans. Monosilicic acid

Q.2228. Pollen grain sterility of wheat occurs due to deficiency of
Ans. Boron

Q.2229. The susceptibility nature of soil to erosion is known as
Ans. Soil eradibility

Q.2230. Size of soil colloids
Ans. Less than 1 μm

Q.2231. The smallest volume of soil is called
Ans. Pedon

Q.2232. Order of strength of adsorption/flocculate soil colloids
Ans. $Al^{3+} > H^{+} > Ca^{2+} > Mg^{2+} > K^{2+} > Na^{+}$

Q.2233. Mineral soil has usually percent of organic matter
Ans. Less the 20%

Q.2234. Permissible soil loss by water is
Ans. 12 tone/ha.

Q.2235. First manufactured fertilizer in India is
Ans. SSP (Single super phosphate)

Q.2236. Pulse crop which doesn't fix Nitrogen from atmosphere
Ans. Rajma

Q.2237. Most of exchange of gases in soils is due to
Ans. Diffusion

Q.2238. Ballast elements are
Ans. Al and Si

Q.2239. Gene responsible for N fixation
Ans. Nif gene

Q.2240. Desired soil sample size of soil is
Ans. 0.5 kg.

Q.2240a. Good source of phosphorus in acid soil
Ans. Bone meal

Q.2241. Fine texture soils are most sensitive for
Ans. Water erosion

Q.2242. Most serious form of water erosion
Ans. Sheet erosion

Q.2243. Which form of erosion can't be removed by normal tillage operations
Ans. Gully erosion

Q.2244. Most soil erosion caused by
Ans. Saltation

Q.2245. When erosion is caused by excessive grazing & deforestation it is called
Ans. Anthropogenic erosion

Q.2246. Insoluble fraction of soil organic matter is
Ans. Humin

Q.2247. Soils having highly decomposed organic matter known as
Ans. Muck soil

Q.2248. Lowering bulk density of a soil indicates
Ans. Better Physical property of soil

Q.2249. On the number basis population of microbes in soils
Ans. Bacteria > Actinomycetes > Fungi > Algae

Q.2250. Which nutrient availability increase in rice field due to water logging
Ans. Mn^+

Q.2251. Crop which is highly sensitive to Boron toxicity
Ans. Cabbage

Q.2252. Ideal NPK ration for cereals and pulses
Ans. Cereals 4:2:1 & Pulses 1:2:1

Q.2253. Fertilizer prepared with Rhizobium culture known as
Ans. Nitrogen

Q.2254. Geometry refers to the shape of space available for individual plants known as

Ans. Crop geometry

Q.2255. Geometry refers to plant shape known as

Ans. Plant geometry

Q.2236. Nutrient which required by plant in concentrations > 1 ppm called as

Ans. Major / Micro nutrient

Q.2257. Liebig Law of Minimum was restated as

Ans. Barrel concept

Q.2258. According to Arnon & Stout number of essential nutrients are

Ans. 16 Nutrient

Q.2259. Latest 17th Essential nutrient was added in the list of essential nutrients is

Ans. Nickel **(JRF Agronomy, 2019)**

Q.2260. At present total essential nutrients are

Ans. 17 Nutrients

Q.2261. Functional nutrient was proposed by

Ans. D. J. Nicholas (1963)

Q.2262. Fertilizer which supply only one major plant nutrient called as

Ans. Straight fertilizer (Urea)

Q.2263. Fertilizer which supply two or more major plant nutrient called as

Ans. Complex fertilizer (DAP)

Q.2264. Fertilizer which having all three major plant nutrient called as

Ans. Complete fertilizer (NPK)

Q.2265. Fertilizer having > 25% of the total primary nutrient called as

Ans. High analysis fertilizer (Urea)

Q. 2265a. Balance sheet of nutrients was given by

Ans. Boussingault

Q.2266. For quality crop production which Pottassic fertilizer is mostly used

Ans. Potassium sulphate

Q.2267. Example of free living aerobic bacteria

Ans. *Azatobactor chroccum*

Q.2268. Example of Associative symbiotic micro aerobic bacteria

Ans. *Azospirillum*

Q.2269. Example of symbiotic nodule forming aerobic & heterotrophic bacteria

Ans. *Rhizobium*

Q.2270. Bacteria used for Sorghum and Pearl millet

Ans. *Azospirillum*

Q.2271. Best pH range for bacteria growth

Ans. 6.5 to 8.0

Q.2272. Best pH range for fungi growth

Ans. 4.5 to 6.5

Q.2273. Soil condition prefers for actionmycetes growth

Ans. Slightly alkaline condition

Q.2274. Indicator plant for Nitrogen deficiency

Ans. Cauliflower, cabbage

Q.2275. Indicator plant for Phosphorus deficiency

Ans. Mustard

Q.2276. Indicator plant for Potassium deficiency

Ans. Potato

Q.2277. Indicator plant for Calcium deficiency

Ans. Cauliflower, Cabbage

Q.2278. Indicator plant for Sodium deficiency

Ans. Sugar beet & Turnip

Q.2279. Indicator plant for Boron deficiency

Ans. Sunflower

Q.2279a. Indicator plant for Ferrous deficiency

Ans. Sorghum

STRUCTURE & FUNCTION OF CELL ORGANELLES CELLS MITOSIS & MEIOSIS & MANDELLIAN GENETICS

Q.2280. Power house of cell.
Ans. Mitochondria *(JRF 2006)*
(Term Mitochondria was coined by Carl Brenda in 1898)

Q.2280a. Who first observed mitochondria in plant cells
Ans. Friedrich Meves (1904)

Q.2281. Ploidy level in seeds.
Ans. Embryo-2n, Endosperm-3n, Testa-2n, Aleuren-2n

Q.2282. Rediscovery of mendelian principles in the year.
Ans. 1900

Q.2283. Chromosomal theory of inheritance (1903) by.
Ans. Sutton & Bovery (It is a fundamental unifying theory of genetics which identifies chromosome as the Carriers of genetic material)

Q.2284. Term genetics was given by.
Ans. Bateson (1905)

Q.2285. Chromosome named by.
Ans. Waldayer (1888)
(Chromosome: A short length of DNA inside a cell's nucleus)

Q.2285a. Chromosome shape is observed during
Ans. Anaphase

Q.2286. First use of X-rays as mutation.
Ans. Muller

Q.2287. The lines are homozygous and homogenous in nature called as.
Ans. Pure line

Q.2288. Allo Hexaploidy found in.
Ans. Wheat

Q.2289. Autopolly ploidy found in.
Ans. Sugarcane, cotton, Brassica

Q.2290. Autotetra ploidy found in.
Ans. Potato, coffee

Q.2290a. Kitchen of the cell (MPPSC 2018)
Ans. Chloroplast

Q.2290b. Crossing over term first used by
Ans. Morgan and Eleth Cattell (1909)

Q.2290c. Mendel paper entitled "Experiments on Plant Hybridization" was published
Ans. 1866

Q.2291. Autotriploidy found in.
Ans. Banana

Q.2292. Two lines different for a single locus called.
Ans. Isogenic line

Q.2293. If Embryo originates from unfertilized egg process is called.
Ans. Parthogenesis

Q.2294. The development of fruit without fertilization the process is called.
Ans. Parthenocarpy

Q.2295. Change in the genome with reference to individual chromosomes called as.
Ans. Aneuloidy

Q.2296. Repeated crossing of hybrid progeny back to one of its parents called as.
Ans. Back cross

Q.2297. If a single gene governing multiple traits it is called as.
Ans. Pleiotrophy

Q.2298. Embryo development without fertilization will be called.
Ans. Apomixis

Q.2299. First interspecific cross was made by.
Ans. Thomas Fairchild

Q.2300. N.E. Borlaug was awarded Noble Prize in.
Ans. 1970

Q.2301. Father of Hybrid cotton.
Ans. C.T. Patel

Q.2302. First hybrid of rice was developed by.
Ans. Y.L. Ping (China)

Q.2303. First transgenic plant was developed by.
Ans. Fraley (1983) Tobacco

Q.2304. Laws of heredity were first discovered by.

Ans. Mandel

Q.2305. Tift 60 is an important source of male sterility in.

Ans. Sorghum.

Q.2306. The term Germplasm was first used by.

Ans. Weismann (1834–1914)

Q.2307. Centers of origin were first given by.

Ans. Vavilov

Vavilov recognized eight main centres of origin

1. China; 2. Hindustan;
3. Central Asia; 4. Asia Minor;
5. Mediterranean; 6. Abyssinya;
7. Central America; 8. South America

Q.2307a. Who is recognised as Indian Vavilov

Ans. Dr. Harbhajan Singh

Q.2308. Origin of cultivated plants

Ans. Alphonse de can dolle (1863) Swiss botanist

Q.2309. NBPGR established in.

Ans. 1976 (New Delhi)

Q.2310. The term genomics was coined by.

Ans. Tom Roderick (1986)

Q.2311. The term parthogenesis (reproduction without fertilization) was coined by.

Ans. Richard Owen (1849)

Q.2312. Development of seed by self pollination refers to.

Ans. Autogamy.

Q.2313. Often cross pollinated crops are.

Ans. Cotton, Sorghum, P. Pea

Q.2314. Concept of pure line theory was developed by.

Ans. Jensen (1952)

Q.2315. Term heterosis was coined by.

Ans. Shull (1914)

Q.2316. Jagannath is a mutant variety of.

Ans. Rice

Q.2317. A nullisomic individual is represented by.
Ans. 2n - 2

Q.2318. Chromosome was discovered by.
Ans Strasburger (1875) (Chromosomes are made up of nucleic acids and proteins)

Q.2319. Microtubles were discovered by
Ans. DeRobertis & Franchi in 1953.

Q.2320. DNA was first synthesized by
Ans. A Kornberg in 1953.

Q.2321. RNA was first synthesized by
Ans. S.Ochoa in 1969.

Q.2322. Largest Organelle in Eukaryotic cell
Ans. Nucleus

Q.2323. Term Protoplast was coined by
Ans. Purkinje in the year 1840

Q.2324. Longest Phase of Mitosis is
Ans. Prophase (ARS 2005)

Q.2325. Golgi complex was described by Camillo Golgi in the year
Ans. 1822

Q.2326. Which company have patent of Bt gene for cotton crop
Ans. Monsanto Ltd

Q.2327. The vegetable crop which is under approval for Bt technology in India is
Ans. Brinjal

Q.2328. Type of Bt gene inserted in cotton
Ans. Cry 1 Ac

Q.2328a. Golgi releases special enzymes
Ans. Lysosomes

Q. 2228b. Flexibility in plants is due to
Ans. Collenchyma

Biological Clock; It is an internal (endogenous) timing system that continues without any external cues and controls activities of plants or animals.

It enables plants to prepare for periodic environmental extremes (winter, draught etc.). It enables plant to flower at the same time / when pollinators are available, increasing the chance of pollination. Biological clock mechanism is genetic in origin.

ELEMENTARY KNOWLEDGE OF PHOTOSYNTHESIS; RESPIRATION AND TRANSPIRATION

Q.2329. Glycolysis occurs in the part of cell.
Ans. Cytoplasm

Q.2329a. Photosynthesis utilizes the portion of the electromagnetic spectrum (solar radiation) known as
Ans. Visible light

Q.2329b. Photosynthesis occurs in
Ans. Chloroplasts

Q.2330. Krebs cycle & ETC (Electron Transport Chain) occurs in
Ans. Mitochondria

Q.2331. Glycolysis is a type of reaction.
Ans. Anaerobic

Q.2332. Total ATP synthesis from one 2 × 2 molecule of glucose in respiration.
Ans. 36 ATP (Net gain), Gross - 38 ATP

Q.2333. Total ATP synthesis in glycolysis is.
Ans. 2 ATP + 2NADP → 8 ATP (Net gain - 2ATP)

Q.2334. Krebs cycle is also called.
Ans. Citric acid or T.C.A. cycle (It occur in Mitochondria)

Q.2335. Final product of glycolysis is.
Ans. Pyruvate

Q.2336. Kerbs cycle starts with.
Ans. Acetyl COA & Oxaloacetate (Acetyle CoA made up of two carbon acetyle group attached to co-enzyme A)

Q.2337. Energy content of molecule of glucose during aerobic respiration ***(JRF, 2019)***
Ans. 686 Kcal
$(C_6H_{12}O_6+6O_2 \rightarrow 6CO_2 + 6H_2O + 686\ Kcal)$

Q.2338. Soaking up of water by dry substances due to hydrophillic colloids is
Ans. Imbibition

Q.2339. Respiration liberates
Ans. Metabolic energy as ATP (Adenosine Triphosphate)

Q.2340. Anaerobic respiration product
Ans. Ethanol and Lactic acid

Q.2341. Shrinkage of protoplasm due to outward flow of water in a Hypertonic solution (strong)

Ans. Plasmolysis.

Q.2342. Chlorophyll mostly occur in

Ans. Grana (Term Grana was given by Warburg 1913)

Q.2343. Photochemical or light or hill reaction takes place in.

Ans. Grana of chloroplast in presence of ferricyanide

Q.2344. Biochemical or Dark reaction or Calvin cycle takes place in.

Ans. Stroma of chloroplast (Dark reaction is pathway by which CO_2 is reduced to sugar)

Q.2345. C_3 pathway (Calvin cycle) found in.

Ans. Rice, Wheat, Pea, Gram, Arhar, Soybean

Q.2346. C_4 (Hatch Slack pathway) found in.

Ans. Sorghum, Maize, Bajra, Sugarcane

Q.2347. CAM (Crassulacean acid metabolism) pathway found in.

Ans. Pineapple, Opuntia, Agave, Cactus and desert plants.

Q.2348. Highest water use efficiency.

Ans. CAM > C_4 > C_3

Q.2349. Kranz type leaf anatomy found in.

Ans. C_4 plants

Q.2349a. Stacks of Thylakoids are known as

Ans. Grana

Kranz Anatomy: It is present in leaves of the C_4 plants (*e.g.* maize) where the tissue equivalent to the spongy mosophyll cells is clustered in a ring around to the leaf veins, outside the bundle-sheath cells. Each pair of veins is thus separated by two (bs) and two (M) cell in a V-bs-M-M- bs-V pattern.

Q.2350. Light reaction of photosynthesis is known as

Ans. Hill reaction

Q.2351. Calvin cycle & Hatch - Slack pathway occurs in.

Ans. Stroma of Chloroplast

Calvin cycle complete in four major phases: (*i*) Caboxylation (Carbon fixation), (*ii*) Reductive phase, (*iii*) Regeneration phase (*iv*) Glycolytic phase

Q. 2351a. First product of Hatch and Slack cycle

Ans. 4COAA (oxaloacetic acid)

Q.2352. Magnesium is a central component of

Ans. Chlorophyll materials

Q.2353. Photosynthetic rate is highest in.
Ans. C_4 plants

Q.2353a. Part of chlorophyll molecule which absorb the sun light
Ans. Porphyrin head

Q.2354. Calvin cycle require
Ans. 6 CO_2, 18 ATP and 12 $NADPH_2$

Q.2355. Citric acid cycle is
Ans. Catabolic while calvin cycle is anabolic in nature

Q.2356. Harvest Index in cereals.
Ans. 0.4–0.5

Q.2357. Harvest Index in pulses.
Ans. 0.2–0.3

Q.2358. First product of C_3 cycle
Ans. 3 PGA (Phosphoglycerate)

Q.2359. Two processes involved in the mechanism of respiration
Ans. Glycolysis and Kreb cycle.

Q.2360. Exchange of gases in plants occurs through
Ans. Stomata and lenticels

Q.2361. Name two phytohormones that promote growth
Ans. Auxins Cytokinins & Gibberellins

Q.2362. Name two phytohormones that inhibit growth
Ans. Abscisic and Ethylene.

Q.2363. Optimum temperature for photosynthesis is
Ans. 20–35° C

Q.2364. Translocation of nutrient and water through xylem is due to
Ans. Transpiration Pull (*JRF 2011, Raj. PG2010, BHU PG 2010*)

Q.2365. Sum of all the chemical transformations taking place in cell or organism
Ans. Metabolism

Q.2366. Name the process by which carbon dioxide is converted to simple organic compounds in green plants?
Ans. Carbon dioxide assimilation or fixation

Q.2366a ATP (Adenosine triphosphate) molecule contain
Ans. 24% C, 14% N and 18% Phosphorus (ATP is energy currency of the cell)

Q.2367. Stem and root nodules are found in.
Ans. Sesbania rostrata

Q.2367a. Photoperiodism term was coined by
Ans. TD Lysenko (1920)

Q.2368. Cellulose is found particularly in the following ?
Ans. Stalks, Stems, Trunks and the woody portions of the plant body.

Q.2369. What do you understand by ECM?
Ans. Extracellular matrix which provides the porous pathway for the diffusion of nutrients and oxygen.
ECM is an intricate network composed of an array of multidomain macromolecules organized in a cell/tissue-specific manner. Components of the ECM link together to form a structurally stable composite, contributing to the mechanical properties of tissues.

Q.2370. What are ECM composed of ?
Ans. Interlocking meshwork of Hetereopolysaccrides and fibrious proteins.

Q.2371. The nitrogen cycle and the citric acid cycle together constitute which cycle?
Ans. Krebs Bicycle.

Q.2372. Where does the Glyoxylate cycle occur?
Ans. Glyoxysome.

Q.2373. Term ascent of SAP (upward movement of the water from the roots) given by
Ans. Priestley

Q.2374. In injured plants, respiration
Ans. Increases

Q.2375. Root nodules of legumes contain the pigment
Ans. Leghaemoglobin.

Q.2376. EMP (Embden-Meyehof –Parnas) pathways are also known as
Ans. Glycolysis

Q.2377. Maximum rate of photosynthesis is possible under the light
Ans. red light

Q.2378. Solar (sun energy) is convereted to ATP in the
Ans. Chloroplast

Q.2379. In Kerb's cycle how many ATP are produced
Ans. 30 ATP

Q.2379a. Term Photorespiration coined by
Ans. Dicker and Tio (1959)

Q.2380. Photorespiration is high in the plant
Ans. Rice plant (Respiration that is initiated in chloroplasts under light conditions is called photorespiration)

Q.2381. Term stomata was given by
Ans. Malpighi (Stomata are main avenue for transpiration)

Q.2382. Nutrient required in hill reaction for splitting of water
Ans. Chlorine (Cl^-)

STRUCTURE AND FUNCTIONS OF CARBOHYDRATES, PROTEINS, NUCLEIC ACIDS, ENZYMES AND VITAMINS

Q.2383. Most abundant protein in the world.

Ans. RUBISCO (Ribulose bisphosphate carboxylase oxygenase)

Q.2383a. Largest protein in plants

Ans. Volvox carteri

Q.2384. First enzyme in CO_2 fixation in C_3 plant.

Ans. Rubisco

Q.2385. Disturbance in digestive system is due to the deficiency of

Ans. Vitamin B_1

Q.2386. Prolong blood clotting time occurs due to the deficiency of

Ans. Vitamin K

Q.2387. Phylloquinone is known as vitamin

Ans. Vitamin K

Q.2388. Cresco graph is used for measuring

Ans. Plant Growth

Q.2389. First enzyme in CO_2 fixation in C_4 plants.

Ans. PEP Carboxylase

Q.2390. Glucose is which type of sugar.

Ans. Monosaccharides

Q.2391. Sugar which is sweetest among all sugars.

Ans. Fructose

Q.2392. Disaccharides are.

Ans. Maltose, Lactose, Sucrose, Cellobiose

Q.2393. Non reducing sugar is.

Ans. Sucrose

Q.2394. Disaccharides have molecular formula

Ans. $C_{12}H_{22}O_{11}$

Q.2395. Which sugar is largely found is germinating seeds?
Ans. Maltose

Q.2396. Lactose is a combination of.
Ans. Glucose + Galactose

Q.2397. Glycogen is present only in.
Ans. Animal cells

Q.2398. Pectin is normally present in.
Ans. Cell wall

Q.2399. Name 'Protein' was suggested by.
Ans. Berzelius (1837)

Q.2400. Proteins are the polymer of.
Ans. Amino acid

Q.2401. Collagen is found in.
Ans. Muscle protein

Q.2402. Keratin is found in.
Ans. Hairs, Wool and Nails

Q.2403. Fibroin is found in.
Ans. Silk

Q.2404. Elastin is found in.
Ans. Insect wings

Q.2405. Regulatory proteins are?
Ans. Enzymes

Q.2406. Term protein was first used by
Ans. Moulder in 1840.

Q.2407. Transport protein.
Ans. Myoglobin, Hemoglobin's

Q.2408. First enzyme was found initially from.
Ans. Yeast

Q.2409. Enzymatic activity was first discovered by.
Ans. Buchner

Q.2409a. Maltose is composed of
Ans. Two glucose molecules.

Q.2410. It is a very longed chained glucose polymer important in structural support of plants

Ans. Cellulose.

Q.2411. Cellulose contains

Ans. More than 50% of plants total carbons.

Q.2411a. Plant cells store extra carbohydrates as energy

Ans. Complex sugars and starch

Q.2412. The term Enzyme was coined by.

Ans. W. Kuhne (1877)

Q.2413. Lock & Key model was proposed by.

Ans. Emil Fischer

Q.2414. Term vitamin was introduced by.

Ans. Casimir Funk (1884-1967)

Q.2415. Water soluble vitamins.

Ans. Vit. B complex & vit. C, & B_6

Q.2416. Fat soluble vitamins.

Ans. Vit. A, D, E, K

Q.2417. Xerophthalmia and night blindness is due to.

Ans. Vit. A (retinal) deficiency

Q.2418. Ariboflavinosis (cracks on skin) is due to.

Ans. B_2 (riboflavin) deficiency

Q.2419. Anemia is due to the deficiency of.

Ans. Vit. B_{12} (cyanocobalomin)

Q.2420. Bond that joins amino acids

Ans. Carboxyl

Q.2421. Pellagra (Black tongue) is due to the deficiency of.

Ans. B_6 (Niacin)

Q.2422. A group of enzymes highly associated with DNA to form nucleosomes

Ans. Histones

Q.2423. Scurvy is due to.

Ans. Vit. C (Ascorbic acid) deficiency

Q.2424. Word Fat was coined by
Ans. Bioor

Q.2425. Rickets is due to
Ans. Vit. D (Caleiferol) deficiency

Q.2426. Sterility is due to
Ans. Vit. E (α Tocopherol) deficiency

Q.2427. Useful enzyme to remove oxidative stress
Ans. SOD in plants

Q.2428. Beri Beri is due to
Ans. Thiamine (Vit. B_1) deficiency

Q.2429. Nucleoside + Phosphate group are called as
Ans. Nucleotide
Nucleotide are constituent of Nucleic acids (DNA, RNA) which store and transmit genetic information

Q.2430. Bacteriophages are.
Ans. Single stranded DNA

Q.2431. Nucleic Acids are
Ans. Family of molecules that include DNA and RNA molecules

Q.2432. The most abundant form of RNA is known as
Ans. t-RNA (80%)

Q.2433. The scientist who used the word cell.
Ans. Robert Hook

Q.2434. The scientist who used the word nucleus
Ans. Robert Brown

Q.2435. Cell theory was given by
Ans. M. Schleiden & Schwann (1839)

Q.2436. Free living N fixing bacteria is
Ans. Azotobacter

Q.2437. What is used for the regulation of Calvin Cycle Enzymes ?
Ans. Thioredoxin

Q.2438. Crossing over can be seen in the stage.
Ans. Pachytene

Q.2439. An organic compound which build protein molecules is
Ans. Amino acid

Q.2440. Proteome

Ans. Sum of all the protein functioning in cell.

Q.2441. What is the energy storage in plants ?

Ans. Starch

Q.2442. Vitamin A is a derivative of?

Ans. Retinoic Acid.

Q.2443. Name the products which are rich in Vitamin A?

Ans. Eggs, whole milk, fish liver oil and butter along with other dietary products.

Q.2444. From what products do we obtain Vitamin E?

Ans. Eggs and vegetable oils and wheat gram.

Q.2445. Name the scientists who discovered Vitamin K?

Ans. Edward A Doisy and Henrik Dam

Q.2446. Where is Vitamin K found?

Ans. Green plant leaves i.e. (Spinach leaves, mustard green, broccoli)

Q.2447. What is the role of Vitamin K?

Ans. It helps in the formation of active prothrombin, a blood plasma protein essential in blood clotting.

Q.2448. What is the plant form of vitamin -K

Ans. Phylloquinone

Q.2449. Deficiency of Vitamin K can lead to?

Ans. It can lead to uncontrolled bleeding. It is not commonly seen in adults but can be seen in infants.

Q.2450. Name the products that are rich in Vitamin C?

Ans. Bell peppers, dark leafy greens, kiwis, broccoli, berries, citrus fruits, tomatoes, peas, and papayas.

Human genome contains 24000 to 25,000 genes and human body is composed of more than 200 different kinds of specialized cell types.

PLANT BIOTECHNOLOGY AND TISSUE CULTURE

Q.2450. Biotechnology is

Ans. Applied and controlled use of molecular biology and recombinant DNA Technology

Q.2450a. Father of Plant Tissue culture

Ans. Haber landt (1884 to 1945)

Q.2450b. Term tissue culture coined by

Ans. American Pathologist- Montrose Thomas Burrows

Q.2451. Term Biotechnology was coined by

Ans. Karl Ereky (1919) (*JRF 2016*)

Q.2452. Totipotency is

Ans. Capability of an isolated single cell to multiply and differentiate into multicellular organism.

Q.2453. RNA (ribonucleic acid) is a

Ans. Molecule similar to DNA, which helps in process of decoding genetic information carried by DNA. It contains four bases that are AGCUT.

A–Adenine, G–Guanine, C–Cytosine, T–Tymine, U–Uracil

Q.2454. Methods used for determination of gene sequences

Ans. 1. Maxm & Gilbert method

2. Sanger dideoxy nucleotide synthetic method

Q.2455. Extrachromosmal genetic element found within bacterial cells and replicates independently of chromosomal DNA.

Ans. Plasmids

Q.2456. DNA (deoxyribonucleic acid)

Ans. A molecule which encodes genetic information. It contains four bases AGCT.

Q.2457. Double haploids are

Ans. Individuals derived from haploid gametes and carry two identical chromosomes

Q.2458. Micropropgation is

Ans. In vitro multiplication of plants from a small tissue explant

Q.2459. What is Electroporation

Ans. A technique whcih uses electric discharge to produce pores on cell membrane for intake of recombinant DNA

Q.2460. In vitro is a

Ans. Biochemical process or reaction taking place in a test tube (in lab).

Q.2461. In vivo is a

Ans. Biological process or reaction taking place in a living cell or organism.

Q.2462. E-coli is a

Ans. Bacterium used in genetic engg. for its small size.

Q.2463. *Agrobacterium rhizogenes*

Ans. Soil bacteria often possesses R_i plasmids and cause the hairy root system in plants

Q.2464. *Agrobacterium tumefaciens*

Ans. Soil bacteria often possesses T_i plasmid

Q.2465. Allele is

Ans. A variant form of a gene

Q.2466. Anticodon

Ans. Triplet of nucleotides in transfer of RNA

Q.2467. Embryogenesis is a

Ans. Process of formation of somatic embryo's from callus

Q.2468. Double haploids are

Ans. Individuals derived from haploid gametes and carry two identical chromosomes

Q.2469. Gene cloning is

Ans. Isolating a gene and the process of producing identical copies.

Q.2470. Crossing of plants through fusion of somatic cell is

Ans. Somatic hybridization

Q.2471. Chromomere is a

Ans. Serially aligned beads or granules of eukaryotic chromosome from local coiling of continuous DNA thread and is visible on a chromosome during prophase of meiosis and metosis.

Q.2472a. First embryo culture in crucifers

Ans. Hanning (1904)

Q.2472. DNA ligase is

Ans. An enzyme that closes nicks or discontinuties in one strand of double stranded DNA by creating a bond.

Q.2473. DNA first synthesised by

Ans. Arthur Kornberg (1956)

Q.2474. Father of genetic engineering

Ans. Paul Berg

Q.2475. Gene bank is a

Ans. Group of genes or cloned DNA fragments

Q.2476. Marker is

Ans. A sequence of bases at a unique physical location in genome

Q.2477. Molecular markers are

Ans. DNA fragments used as finger print in identification of any organism or species.

Q.2478. RFLP

Ans. Restriction Fragment Length Polymorphism

Q.2479. RAPD

Ans. Random Amplified Polymorphic DNA

Q.2480. STMS

Ans. Sequence Tagged Microsattelite Sites

Q.2481. SNP

Ans. Single or Simple Nucleotide Polymorphism

Q.2482. PCR

Ans. Polymerase Chain Reactions

Q.2483. PCR was developed by

Ans. Dr. Karl Mullis in year 1980 and won Nobel Prize in 1993

Q.2484. PCR technique uses

Ans. Bacterial enzymes for in vitro amplification of DNA

Q.2485. DNA cloning

Ans. The ability to construct recombinant DNA molecules and maintain them in cells.

Q.2486. PCR techniques provides

Ans. Several copies of specific DNA from a single copy of DNA

Q.2487. Genetic map is a rounds of DNA replication in vitro

Ans. Map of genome showing relative positions of genes and or markers on chromosomes

Q.2488. Gene pyramiding is

Ans. Technique of combining 2 or more major genes

Q.2489. Genomics is

Ans. Science of identifying the sequence of DNA in species

Q.2490. First sequenced plant is

Ans. *Arabidopasis thaliana* (a weed)

Q.2491. GUS (β-glucaronidase)

Ans. Gene from Eschericha. coli. It is used in transformation as a reporter gene.

Q.2492. Germplasm was discovered by-

Ans. August Weismann

Q.2492a. What is Germplasm

Ans. It is living tissue from which new plants can be grown. It can be a seed or another plant part – a leaf, a piece of stem, pollen or even just a few cells that can be turned into a whole plant. Germplasm contains the information for a species' genetic makeup, a valuable natural resource of plant diversity.

Q.2493. Transgenics plants are

Ans. Genetically modified organisms

Q2494. Transgenics or GMO are

Ans. Organisms with gene or genetic construct introduced by molecular or remcobinant DNA.

Q.2495. Transformation is

Ans. A process by which genetic material carried by a species cell is altered by incorporation of exogenous DNA in genome

Q.2496. Transcription is the

Ans. Process during which the information in a sequence of DNA is used to construct m-RNA

Q.2497. Golden rice was developed by

Ans. Ingo Potrykus and Peter Bayer

Q.2497a. Molecular signal that binds to the receptor is called

Ans. Ligand

Q.2498. Golden rice provides
Ans. Higher β-carotene over normal rice

Q.2499. β-Carotene is
Ans. Precursor of Vitamin-A

Q.2500. β-carotene-indica rice was developed by
Ans. Indian scientists

Q.2501. Project on Indian Mustard oil with higher β-carotene is initiated by
Ans. TERI (Tata Energy Research Institute)

Q.2502. Biotechnology related to Agriculture is also called
Ans. Green biotechnology

Q.2503. Plant tissue culture broadly refer to
Ans. The in vitro cultivation of plants seeds, plant parts etc. on nutrient media under aseptic conditions.

Q.2504. Virus that live in bacteria is
Ans. Bacteriophage

Q.2505. Bacteriophage first observed by
Ans. Twort (1915) and d'Herelle (1917)

Q.2506. Bacteriophage were first described by
Ans. Helmut (1940) Luria and Anderson (1942)

Q.2507. m-RNA is primarily used for
Ans. Transcription of DNA.

Q.2508. Thread like bodies that carry gene
Ans. Chromosomes

Q.2509. Transplanting a cell, tissue or organ from one nutrient medium to another
Ans. Subculture

Q.2510. Stages in micropropagation
Ans. Four stages
1. Initiation of aseptic culture 2. Shoots proliferation stage
3. Root development stage 4. Hardening stage

Q.2511. Hybridization is
Ans. Crossing of two genotypically different plants. Genes from both plants are brought together

Q.2512. m-RNA (Messenger RNA)

Ans. Made up of single stranded molecule, life span is 2 min. at 37°C in E-coli

Q.2513. The insoluble RNA that constitutes the largest part of cellular RNA (About 80%).

Ans. r-RNA

Q.2514. Genome is

Ans. Complete set of chromosomes found in the gamete of true diploid (True diploid are individual with two set of chromosomes)

Q.2515. Explant is

Ans. An excised fragment of a tissue or an organ used to start tissue culture

Q.2516. A group of adjacent genes that are identical or related.

Ans. Gene cluster

Q.2517. Process by which a cell monolayer or a plant explant is transferred without subdivision into a fresh medium

Ans. Reculture

Q.2518. A pheromone is

Ans. A small molecule secreted by one mating type of an organisms in order to interact with a member of the opposite mating type.

Q.2519. DNAs consist of repetitions of extremely short units. (typically < 10 bp).

Ans. Microsatellite

Q.2520. The distance between two genes that recombine with a frequency of 1%.

Ans. Map unit

Q.2521. Enzymes that attack bonds in DNA.

Ans. DNAases

Q.2522. Technique of genetic engineering by which a gene sequence with many identical copies is replicated

Ans. Gene cloning

Q.2523. DNA-synthesizing enzyme required specifically for replication.

Ans. DNA replicase

Q.2524. Molecular scissors used in genetic engg. is

Ans. Restriction endonuclease

Q.2525. Enzyme used in polymerase chain reaction is

Ans. Taq-polymerase

Q.2526. What does abbreviation (*Bt*) means

Ans. *Bacillus thuringiensis*

Q.2507. Bt was discovered by (1901)

Ans. Shigetane Ishiwata

Q.2528. Which Bt crop approval was revoked in 2001

Ans. Potato

Q.2529. Major Bt crop grown in world

Ans. Cotton and Maize

Q.2530. First DNA synthesis in the test tube done by

Ans. Arthur Kornberg (1956)

Q.2531. For which insect group the Bt cotton is resistant

Ans. Lepidoptera Specially (Heliothis & Spodoptera)

Q.2532. A segment of DNA that codes for a specific character is called.

Ans. Gene

Q.2534. Term bolting refers to

Ans. Internodal elongation

Q.2536. Lysosome was discovered by

Ans. Christian de Duve (1955)

Q.2537. Term DNA Fingerprinting was coined by

Ans. Prof A. Jeffrey

Q.2538. Genetic Engineering Appraisal Committee Constituted under the

Ans. Environmental Protection act 1986 (GEAC functions under Ministry of Environment and Climate Change).

Q.2538a. First transgenic plant product touched the market

Ans. Flavr Savr tomato

MAJOR PESTS & DISEASES OF RICE, WHEAT, COTTON, CHICKPEA, SUGARCANE, AND THEIR MANAGEMENT

Q.2539. Most damaging insect of Lepidoptera group is
Ans. Heliothis armigera

Q.2539b. National pest is
Ans. White grubs

Q.2539c. International pest is
Ans. *Schistocerca gregaria* (Desert locust)

Q.2540. Serious Pest of Bt cotton
Ans. Aphid & Jassid

Q.2541. Phorate 10 G is marked by which color triangular sign for poison
Ans. Red color

Q.2542. DBM (Diamond back moth) is a serious pest of
Ans. Cabbage

Q.2543. First chemical pesticide used in India was
Ans. DDT

Q.2544. DDT was banned due to
Ans. Prolong persistence in soil (Banned in India in 1985)

Q.2545. Rice stem borer (scirpophaga incertulas) is a
Ans. Monophagous (Family: Pyrausdtidae) described by Walker in the year 1863.

Q.2546. Trichogramna is a
Ans. Egg parasitoid

Q.2547. Silver shoot or onion leaf is caused by
Ans. Gall midge

Q.2548. Vector of Rice tungro is
Ans. Green Leaf hopper (*Nephotettix spp.*)

Q.2549. Vector of Grassy stunt disease.
Ans. Brown plant hopper (Nilaparvata lugens)

Q.2550. Chaffy grains with black spot is due to.
Ans. Gundhi bug (*Leptocorisa acuta*)

Q.2551. Family of Gundhi bug is.
Ans. Alydidae

Q.2552. Gundhi bug caused at the stage.
Ans. Early Milking stage

Q.2552a. International pest in known as
Ans. Desert Locust (*Schistocerca gregaria*)

Q.2553. White grub is a.
Ans. Beetle, Polyphages

Q.2554. Serious pest of wheat.
Ans. White grub & termite

Q.2555. Ear cockle is due to
Ans. *Anguina tritici* (nematode)

Q.2556. Tundu/yellow ear rot disease is due to.
Ans. *Anguina tritici* + *Corynebacterium tritici*

Q.2557. Control of tundu disease.
Ans. Hot water treatment 50°C for 2 hrs. use of seed free from nematode galls or certified seeds.

Q.2558. Wheat stem borer sesamia inferns attack in.
Ans. Night

Q.2559. Name of rice yellow stem borer.
Ans. *Scirpophaga incertulas* (Described by Francis Walker in 1863)

Q.2560. Highest consumption of pesticide in crop.
Ans. Cotton (54%)

Q.2561. Hopper burn in cotton is due to.
Ans. *Amrasca biguttula*

Q.2562. Family of cotton white fly.
Ans. Aleyrodidae

Q.2563. Vector of cotton leaf curl virus.
Ans. Bemisia tabaci

Q.2564. Flaring of squares in cotton is due to.
Ans. Spotted bollworm (*Earias vitella*)

Q.2565. Rosetting of flowers due to.
Ans. Pink bollworm (*pectinophora gossypiella*)

Q.2566. Double seed formation is due to.
Ans. Pink bollworm

Q.2567. Large circular bore holes with faecal pellets is the symptoms of.
Ans. American bollworm (Helicoverpa armigera)

Q.2568. Dysdercus cingulatus is known as.
Ans. Red cotton bug

Q.2569. Bt formulation is used for.
Ans. Early instars of bollworms

Q.2570. Helicoverpa and Agrotisypsilon is the serious pest of.
Ans. Chickpea

Q.2571. Greasy cut worm attack in.
Ans. Night

Q.2572. Scientific name of sugarcane shoot borer.
Ans. *Chilo infuscatellus*

Q.2573. Bunchy top appearance in sugarcane is due to.
Ans. Top borer (Scirpophaga excerptalis)

Q.2574. Family of Top borer.
Ans. Pryalidae

Q.2575. Family of shoot borer.
Ans. Crambidae

Q.2576. Pyrilla perpusilla (Fam. Lophopidae) is a.
Ans. Leaf hopper

Q.2577. Biological control of borers.
Ans. Trichograma Japonicun

Q.2578. Destructive Insect Pest Act (DIP) passed in year
Ans. 1914

Q.2579. Insecticide act was passed in year
Ans. 1968

Q.2580. Pest that occur most frequently on cultivated crops.
Ans. Regular pest

Q.2581. Pest that occur in a few isolated localities.
Ans. Sporadic pest

Q.2582. Pest that occurs in same area year after year.
Ans. Endemic pest

Q.2583. Pest that occurs in an area in severe form.
Ans. Epidemic pest

Q.2584. Central Plant Protection Training Institute.
Ans. Hyderabad

Q.2585. Safest insecticide for honeybee.
Ans. Endosulfan

Q.2586. Pest population should be kept below.
Ans. Economic threshold level

Q.2587. Serious pest of Rice.
Ans. Yellow stem borer

Q.2588. What is the sequence of coating of seed by insecticide, fungicide & Rhizobium.
Ans. Fungicide + Insecticide + Rhizobium

Q.2589. Insect growth regulators in pest management are:
Ans. (*i*) Ecdysoids, (*ii*) Juvensoids,
(*iii*) Anti JH, (*iv*) Chitin

Q.2590. Term integrated pest management was propounded by
Ans. Geiger & Clark (1961)

Q.2591. Wiggelworth
Ans. Father of Insect Physiology

Q.2592. H.S. Pruthi was the
Ans. First plant protection adviser to the Govt. of India.

Q.2593. Eco-friendly insectcide is
Ans. Nimbicidine

Q.2594. Insectcide can be safely used on vegetables
Ans. Malathion

Q.2595. Widely accepted classification of insects was given by
Ans. A. D. Imms

Q.2596. Out of 29 insects order how many are of agriculture importance
Ans. 8 orders

Q.2597. Termites belongs to order

Ans. Isoptera

Q.2598. Honey bee wasps and ants belongs to order

Ans. Hymenoptera

Q.2599. Cockroach belongs to the order

Ans. Dictyoptera

Q.2600. How many pairs of leg in spiders

Ans. 4 pairs

Q.2601. Insects having Biting and chewing type mouth parts

Ans. Grasshoppers, Beetles, Larva of lepidoptera

Q.2602. Insects having Piercing and sucking type mouth parts

Ans. Mosquito, Aphids, Bugs, Leaf hoppers

Q.2603. Insects having Sponging type mouth parts

Ans. House fly

Q.2604. Insects having Siphoning type mouth parts

Ans. Butterfly, Moths

Q.2605. Insects having chewing and lapping type mouth parts

Ans. Honey bee

Q.2606. Insects having Rasping and sucking type mouth parts

Ans. Thrips

Q.2607. The stage of insects damage where loss can be recognized called

Ans. Damage Boundary

Q.2608. ETL (Economic threshold level) is also known as

Ans. Action threshold level

Q.2609. EIL (Economic injury level) is also known as

Ans. DTL (Damage threshold level)

Q.2610. The average population density of an insect population over a long period of time unaffected by the temporary interventions of pest control known as

Ans. GEP (General Equilibrium Position)

Q.2611. Father of host plant resistance

Ans. R.H. Painter

Q.2612. (NC-IPM) National Centre for Integrated Pest Management is situated at

Ans. New Delhi

Q.2613. Head Quarter of Directorate of plant protection, Quarantine & Storage at

Ans. Faridabad

Q.2614. International Organization of biological control is at

Ans. Zurich, Switerzerland (Since 1962)

Q.2615. NPV (Nuclear Polyhedrosis Virus) is mostly used to control

Ans. Lepidopteron insects (Moths and Butterflies)

Q.2616. Bud necrosis in G. Nut is transmitted by

Ans. Thrips

Q.2617. Scientific name of Maize stem borer

Ans. *Chilo partellus*

Q.2618. Physical method of termite control

Ans. Irrigation

Q.2619. Juvenile hormone is secreted by

Ans. *Corpora allata*

Q.2620. Tree banding is useful for control of

Ans. Mango mealy bug

Q.2621. Main features of insect is

Ans. Three pairs of leg

Q.2622. Best suitable species of honey bee for bee culture

Ans. *Aphis mellifera*

Q.2623. Largest size of honey bee is
Ans. *Aphis dorsata*

Q.2624. Most serious pest of apiculture
Ans. Wax moth (*Galleria Melonella*)

Q.2625. Disease spread by house fly is
Ans. Cholera

Q.2626. A boll guard is
Ans. Transgenic cotton

Q.2627. Most suitable fumigant for quarantine purpose is
Ans. *Methyle bromide*

Q.2628. Insects which survive at higher temperature and low humidity is
Ans. Khapara beetle

Q.2629. Toxin of Bt interferes with insect's
Ans. Digestive system

Q.2630. Effective parasitoid against sugarcane pyrilla is
Ans. *Epiricania melanoleuca*

Q.2631. Absolute monophagus pest of Rice
Ans. Rice stem borer

Q.2632. Rice gundhi bug cause at the stage
Ans. Milky stage

Q.2633. Pest which attacks all parts of plants is
Ans. Termite

Q.2634. In termite the king is developed from
Ans. Unfertilized egg

Q.2635. The production of light by living organism is called
Ans. Bioluminescence

Q.2636. Common insecticide used to control termite
Ans. Chlorpyriphos

Q.2637. Suitable soils for white grubs
Ans. Sandy loam

Q.2638. White grub is kept in the category of
Ans. National Pest

Q.2639. Mostly storage pests belongs to the order of
Ans. Coleoptera

Q.2640. The total heat required for the completion of physiological process in the life history of a species is considered as a
Ans. Thermal Constant

Q.2641. Control of weed by biological agent known as
Ans. Para biological control

Q.2642. Bordeaux mixture was developed by
Ans. P.M.A: Millardet (1885)

Q.2643. The optimum safe concentration of the left over pesticides or its degradation products called as
Ans. MRL (Maximum Residues Limit)

Q.2644. Size for mist spray
Ans. 50-150 micron

Q.2645. Imidacloprid is used to control for mainly
Ans. Sucking pest

Q.2646. Pest known as "International pest" is
Ans. Desert locust (*Schistacera gregarea*)

Aphids are small sap sucking insects and members of super family-Aphidoidea. Aphids are largest groups of insect vectors of about 192 species and transmitting about 275 types of virus.

Bacillus thuringiensis (Bt), soil-dwelling bacterium that naturally produces a toxin that is fatal to certain herbivorous insects. The toxin produced by Bacillus thuringiensis (Bt) has been used as an insecticide spray since the 1920s and is commonly used in organic farming. Bt is also the source of the genes used to genetically modify a number of food crops so that they produce the toxin on their own to deter various insect pests. The toxin is lethal to several orders of insects, including Lepidoptera (butterflies, moths, and skippers), Diptera (flies), and Coleoptera (beetles), though a number of Bt strains are available to make its use more target-specific.

PLANT PATHOLOGY

Q.2647. Father of Pathology

Ans. Anton de Berry

Q.2648. Father of microbiology

Ans. Louis Pasteur

Q.2649. Father of Mycology

Ans. P. A. Miccheli

Q.2650. Father of Indian pathology

Ans. E. J. Butler

Q.2651. Father of modern plant pathology

Ans. J. F. Dastur (11876-1971)

Q.2652. Any organism which desire the material of their needs of growth from living plants called

Ans. Parasite

Q.2653. Exanthema disease caused by

Ans. Bacteria

Q.2654. ELISA (Enzyme-linked immunosorbent assay) test is done to detect for

Ans. Virus disease

Q.2655. Ooze test is done to detect for

Ans. Bacteria

Q.2656. Name of first bacterial disease which was discovered

Ans. Fire blight of Pear

Q.2657. White rust of cruciferous caused by

Ans. *Albugo candida*

Q.2658. Cauliflower mosaic virus contain

Ans. DNA

Q.2659. Bacteria used for bio control

Ans. *Pseudomonas fluorescens*

Q.2660. Most of plant virus are

Ans. single-stranded RNA

Q.2661. Vitavax (Carboxin) is a effective fungicide for

Ans. Smut disease

Q.2662. Hot water treatment was developed by

Ans. J. L. Jensen (1887) (Used for loose smut of wheat)

Q.2663. Bavistin is a type of fungicide

Ans. Systemic fungicide

Q.2664. The downy mildew of pearl millet is suppressed by

Ans. Chlorine

Q.2665. Incidence of blast rice can be reduced by the application of

Ans. Silicon

Q.2666. Metalaxyl is effective against

Ans. Oomycetous fungus

Q.2667. Agar Agar is produced by

Ans. Red Sea algae (Nostoc)

Q.2668. Mycoplasma is sensitive for

Ans. Tetracycline

Q.2669. First transgenic plant developed in world

Ans. Tobacco

Q.2670. Most common vectors of plant viruses are

Ans. Aphids

Q.2671. The degree of infectivity of a given pathogen called as

Ans. Virulence

Q.2672. The period of time between penetration of a pathogen to the host and the first appearance of symptoms on the plant is known as

Ans. Incubation period

Q.2673. Iris famine is related to the crop

Ans. Potato (1845)

Q.2674. Disease responsible for Iris famine

Ans. Late blight of potato (*Phytophthera infestans*)

Q.2675. Disease responsible for Bengal famine

Ans. Brown spot disease of Rice

Q.2676. Causal organism of Rice Brown spot disease

Ans. *Helminthosporium oryzae*

Q.2677. When a disease is more or less constantly present from year to year in moderate to severe form in particular part of country or state it is called

Ans. Endemic Disease

Q.2678. Example of endemic disease

Ans. Early blight of potato

Q.2679. A disease which occurs widely in severe but periodically called as

Ans. Epidemic Disease

Q.2680. Example of epidemic disease

Ans. Late blight of potato, Red rot of S. cane , Wheat rusts

Q.2681. A disease occurs at very irregular intervals and locations called as

Ans. Sporadic disease

Q.2682. Example of Sporadic disease

Ans. Green ear of Pearl millet

Q.2683. A disease prevalent throughout the country, continent or the world is called as

Ans. Pandemic disease

Q.2684. Death of tissues and cells it is symptoms of

Ans. Necrosis

Q.2685. Increase in size of plant organ due to increase of cells size known as

Ans. Hypertrophy

Q.2686. Increase in the size of plant organ due to increase in numbers of cells known as

Ans. Hyperplasia

Q.2687. Composition of Bordeaux mixture

Ans. Copper sulphate + Lime + Water (5:5:50)

Q.2688. Composition of Burgundy mixture

Ans. Copper sulphate + Sodium carbonate + Water (5:6:50) (Given by Mason)

Q.2689. Copper sulphate was discovered by

Ans. Prevost (1807) and used against the bunt of wheat

NEMATOLOGY

Q.2690. Father of Nematology
Ans. Nathan Augustus Cobb

Q.2690a. Nematode name is derived from
Ans. Greek word *Nema* which mean a *thread* and include roundworms.

Q.2690b. What are entomopathogenic nematode
Ans. A group of nematode that cause death of insects.

Q.2691. Helminthology is a study of
Ans. Animal & human parasitic nematodes

Q.2692. First plant parasite nematode *Anguina tritici* discovered by
Ans. J. T. Needham (1743)

Q.2693. Mentake disease of rice caused by
Ans. Rice root nematode

Q.2694. Ufra disease of rice occurs due to
Ans. Nematode (*Ditylenchus angustus*)

Q.2695. Molya disease of wheat & barley caused by
Ans. Cereal cyst nematode *Heterodera avenae*

Q.2696. Barley Variety which is resistant for cyst nematode
Ans. RD-2052 & Rajkiran

Q.2697. Golden nematode of Potato (*Globodera rostochiensis*) was first reported by
Ans. FGW Jones (1961)

Q.2698. Resistant variety potato for Golden nematode
Ans. Kufri Swarana

Q.2699. Domestic quarantine is applied for infested potato from Golden nematode at
Ans. Nilgiri Hills (T. Nanu)

Q.2700. Root knot nematode (*Meloidogyne Svp.*) first reported by
Ans. Berkley (Cucumber Crop) in 1855

Q.2701. Nematode which transmit virus for Grape vine fan leaf disease
Ans. *Xiphinema index*

Q.2702. Nematode is responsible for tobacco rattle virus
Ans. *Trichodorus similis*

Q.2703. *Xiphinema Longidorus* is a type of nematode
Ans. Migratory Ecto parasite nematode

Q.2704. *Meloidogyne & Heterodera* are type of nematode
Ans. Sedentary Endo parasite nematode
Q.2705. *Ditylenchus* and *Anguina* is a type of nematoe
Ans. Both Ecto & Endo parasite nematode
Q.2706. Scientific name of Ring nematode
Ans. *Criconema spp.*
Q.2707. Scientific name of Pin nematode
Ans. *Paratylenchus spp.*
Q.2708. Scientific name of Needle nematode
Ans. *Longidorus spp.*
Q.2709. Scientific name of Dagger nematode
Ans. *Xiphinema index*
Q.2710. Scientific name of Lance nematode
Ans. *Haplolaimus spp.*
Q.2711. Scientific name of Citrus nematode
Ans. *Tylenchulus semipenetrans*
Q.2712. Scientific name of Stem & Bulb nematode
Ans. *Ditylenchus dipaci*
Q.2713. Scientific name of Potato root nematode
Ans. *Ditylenchus destructor*
Q.2714. Scientific name of false root knot nematode
Ans. *Naccobus spp.*
Q.2715. Scientific name of Stunt/Stylet nematode
Ans. *Tylenchorhynchus spp.*
Q.2716. Example of Univoltine (only one generation completed in a year)
Ans. *Anguina tritici* and *Heterodera avenae*
Q.2717. Example of Multivoltine (many generation are completed in a year)
Ans. *Meloidogyne, Pratylenchus, Tylenchulus spp. & Rotylenchulus*
Q.2718. Type of pesticide which are commonly used for nematode control
Ans. Soil application based Fumigant & Non fumigant type
Q.2719. Example of Fumigant type pesticide for nematode control
Ans. EDB, Methyl bromide, Nemagone (DBCP)
Q.2720. Example of Non-Fumigant type pesticide for nematode control
Ans. Aldicarb, Carbofuran, Phorate, Thionazin (Nemaphos)
Q.2720a. First plant parasitic nematode reported in India by
Ans. Barber (1901) (Root knot nematode)

DISEASES OF IMPORTANT CROPS

1. Rice

Q.2721. Blast disease is due to.
Ans. *Pyricularia Oryzae* (airborne)

Q.2722. Brown spot disease is due to.
Ans. *Helminthosporium Oryzae* (seed borne)

Q.2723. Bacterial blight disease is due to.
Ans. *Xanthomonas Campestris* pv. oryzae

Q.2724. Kresek symptom found in.
Ans. Bacterial blight (Seedling stage)

Q.2724a. Casual organims
Ans. *Xanthomonas oryzae*

Q.2725. Foot rot disease due to.
Ans. *Fusarium moniliforme*

Q.2726. Sheath blight disease is due to.
Ans. *Rhizoctonia solani*

Q.2727. Udabatta disease due to.
Ans. *Ephelis oryzae*

Q.2728. Vector of rice tungro virus.
Ans. Green leaf hopper (Nephotettik virescens)

Q.2729. Sheath rot disease is due to
Ans. *Sarocladium oryzae*

Q.2730. Stem rot disease is due to
Ans. *Sclerotium oryzae*

Q.2730a. False smut or green smut
Ans. *Ustilaginoidea virens*

Q.2731. Khaira disease of rice is due to
Ans. Zinc deficiency **(JRF, 2019, ARS 2010)**

Q.2731a. Tungro disease of rice is due to
Ans. *Nephotettix virescens (though leafhopper vector)*

Q.2731b. Ufra diseases of rice is due to
Ans. Ditylenchus angustus (Nematode)

2. Wheat

Q.2732. Black stem rust disease is due to.
Ans. *Puccinia graminis tritici*

Q.2733. Brown rust is due to.
Ans. *Puccinia graminis recondita*
Q.2734. Yellow rust is due to.
Ans. *P. graminis striiformis tritici*
Q.2735. Loose smut is due to.
Ans. *Ustilago tritici*, (Internally seed borne) (*ARS 2010*)
Q.2736. Hill bunt is due to.
Ans. *Tilletia foetida* (Externally seed borne)
Q.2737. Karnal bunt or Partial bunt is due to.
Ans. *Neovossia indica* (Soil and airborne)
Q.2738. Flag smut is due to.
Ans. *Urocystis agropyri*
Q.2739. Molya disease is due to.
Ans. *Heterodera avcnae* (nematode)
Q.2740. Vitavax & hot treatment is used for.
Ans. Loose smut
Q.2841. Ear cockle disease is due to
Ans. *anguina tritici* (nematode) *(ARS-10, Ph.D-10 & BHU-12)*
Q.2742. Powdery Mildew is due to
Ans. *Erysiphegraminis f.sp. tritici* (powdery growth)
Q.2742a. Leaf blight of wheat is due to
Ans. *Alternaria trictinia*
Q.2742b. Stem melanosis
Ans. *Pseudomonas Cichorii.*

3. Cotton

Q.2743. Wilt disease is due to.
Ans. *Fusarium oxysporum*
Q.2744. Root rot is due to.
Ans. Casual organims: *Rhizoctonia bataticola*
Q.2745. Black arm or bacterial blight is due to.
Ans. *Xanthomonas compesttris*
Q.2746. Root rot disease is due to
Ans. *Rhizoctonia bataticola*
Q.2747. Anthracnose disease is due to
Ans. *Colletotrichum capsici*
Q.2748. Leaf blight is due to
Ans. *Alternaria macrospora*

Q.2749. Grey mildew or dahiya disease is due to
Ans. *Ranularia areola*

4. Sugarcane

Q.2750. Red rot is due to.
Ans. *Colletrotrichum falcatum*
Q.2751. Smut is due to.
Ans. *Ustilago scitaminae*
Q.2752. Gumming disease is due to.
Ans. *Xanthomonas campestris, Pv. vascularum*
Q.2753. Red stripe disease is due to.
Ans. *Pseudomonas rubrilineans*
Q.2754. Grassy shoot is due to.
Ans. Mycoplasma
Q.2755. Ratoon stunting disease is due to
Ans. Ratoon stunting virus
Q.2755a. Smut in sugarcane is due to
Ans. Fungus Sporisorium scitamineum
Q.2755b. Sugarcane rust is caused due to
Ans. *Puccinia erianthi*

5. Chick Pea

Q.2756. Wilt disease is due to.
Ans. *Fusarium oxysperium sp. ciceri*
Q.2757. Blight is due to.
Ans. *Asehochyta rabei*
Q.2757a. Powdery mildew
Ans. *Oidiopsis taurica*
Q.2577b. Stem rot sclerotinia
Ans. *Sclerotinia sclerotiorum*
Q.0577c. Rust
Ans. *Uromyces cicerarietim*
Q.2758. Mancozeb, Zenab, Thiram are
Ans. *Dithio carbamates*
Q.2759. PMA (Agrosan G N) is a
Ans. Organomercurials
Q.2760. Endosulfhan is a
Ans. Chlorinated hydrocarbons

Q.2761. Aldicarb (Temik), Carbaryl is

Ans. Carbamets & their thio salts

Q.2762. Ceresan & Aretan are.

Ans. Organomercurials fungicide

6. Groundnut

Q.2763. Stem rot disease is due to

Ans. *Sclerotium rolfsii*

Q.2764. Collar rot disease is due to

Ans. *Aspergillus niger*

Q.2765. Tikka disease is due to

Ans. *Cercospora personata & C. arachidicola*

Q.2766. Hollow heart disease is due to

Ans. Boron deficiency

7. Maize

Q.2767. Seed rot seedling blight is due to

Ans. *Phythium appani dermatum*

Q.2768. Charcol rot is due to

Ans. *Macrophomina phaseoil*

8. Tobacco

Q.2767a. Blank shank caused by

Ans. *Phytophthora parasitica nicotianae.*

Q.2767b. Frog eye leaf spot caused by

Ans. *Cercospora nicotianae*

Q.2767c. Powdery Mildew caused by

Ans. *Erysiphe cichoracearum*

Q.2767d. Angular leaf spot

Ans. *Burkholderia (P) solanacearum*

Q.2767e. Mosaic

Ans. *Nicotiana Virus 1*

Q.2767f. Leaf Curl caused by

Ans. Virus

IMPORTANT RURAL DEVELOPMENT PROGRAMS IN INDIA; ORGANISATIONAL SET UP OF AGRICULTURAL RESEARCH

Q.2769. In which year was the word 'extension' used for the first time?

Ans. 1873.

Q.2770. Father of Extension education

Ans. James Stuart

Q.2771. Father of Extension in India

Ans. K. N. Singh

Q.2772. Model villages given by.

Ans. Daniel Hamilton (1903)

Q.2773. Rural reconstruction Institute started.

Ans. Shanti Niketan-1921

Q.2774. Marathandan project was given by

Ans. Spencer Hatch-1921

Q.2775. Gurgaon Experiment was given by

Ans. F. L. Brayne-1921

Q.2776. Sevagram project was given by

Ans. Mahatma Gandhi (1929)

Q.2777. Indian village service was given by

Ans. A. T. Moscher & B. N. Gupta

Q.2778. Firka Development Scheme was given by

Ans. T. Prakasham (1946)

Q.2779. Mazdoor Manzil was given by

Ans. S. K. Dey (1947)

Q.2780. Grow more food campaign was given in?

Ans. 1948

Q.2781. Etawah pilot project was started by

Ans. Albert Mayer (1948)

Q.2782. Grow more food campaign enquiry committee was founded is the year?

Ans. 1952

Q.2783. (C.D.P.) community development project was started in

Ans. 1952

Q.2784. (NES) National Extension Service.

Ans. 1953 (S. K. Dey)

Q.2785. Panchayati Raj was started.

Ans. 1957-1958

Q.2786. First state to adopt Panchayati Raj.

Ans. Rajasthan (Nagaur Dist.) 2nd Oct., 1959

Q.2787. (IADP) Intensive Ag. District programme was started in the year.

Ans. 1960

Q.2788. (IAAP) Intensive Ag. Area programme started in

Ans. 1964

Q.2789. (ICDP) Intensive Cattle Development project was started in

Ans. 1964

Q.2790. High yielding varieties programme.

Ans. 1966

Q.2791. (MCP) Multiple Cropping Programme.

Ans. 1966

Q.2792. (MKP) Minikit Programme for Rice.

Ans. 1971

Q.2793. (SFDA) Small Farmer's Development Agency.

Ans. 1970

Q.2794. (MFAL) Marginal Farmers and Agricultural Labour Programme started in the year.

Ans. 1970

Q.2795. (DPAP) Drought Prone Area Programme.

Ans. 1970

Q.2796. Administration Reform Commission

Ans. 1966 (GOI)

Q.2797. (MNP) Minimum Needs Programme.
Ans. 1974 (Fifth Five Year Plan)

Q.2798. (T & V) Training & Visit Programme was started by.
Ans. Daniel Boner (1974)

Q.2799. Name the state that adopted (T & V) first
Ans. Rajasthan **(1974)**

Q.2800. K.V.K. (Farm Science Centre) was recommended by.
Ans. Mohan Singh Mehta Committee (1974)

Q.2801. First K.V.K. was established at.
Ans. TNAU at Pondicherry (1974)

Q.2802. (CAD) Command Area Development Programme was established in?
Ans. 1974

Q.2803. (IRDP) Integrated Rural Development Programme was established/started in?
Ans. 1979

Q.2804. (TRYSEM) Training Rural Youth for self Employment programme was started in ?
Ans. 1979

Q.2805. (NREP) National Rural Employment Programme was started in?
Ans. 1980

Q.2806. (NARP) National Agricultural Research Project was started in?
Ans. 1980

Q.2807. District Rural Development Agency (DRDA) was brought up in
Ans. 1980

Q.2808. (NAEP) National Agricultural Extension Project was started in?
Ans. 1983

Q.2808a. Credit for success of KVK's goes to
And. Chandrika Prasad Yadav (AFO, 2021)

Q.2809. (RLEGP) Rural Landless Employment Guarantee Programme was started in?

Ans. 1983

Q.2810. (JRY) Jawahar Rozgar Yojna.

Ans. 28 April, 1989

Q.2811. (IVLP) Institute Village Linkage Programme.

Ans. 1994

Q.2812. (MSY) Mahila Samrudhi Yojana.

Ans. 1993

Q.2813. (PMRY) Prime Minister Rozgar Yojana.

Ans. 1994

Q.2814. (NATP) National Agricultural Technology project.

Ans. 1999

Q.2815. (JGSY) Jawahar Gram Samuridhi Yojana.

Ans. 1999

Q.2816. (SGSY) Swaran Jyanti Gram Swarojgar Yojana.

Ans. 1999

Q.2817. Jan Shri Bima Yojana was started in the year

Ans. 2000 (Providing life insurance to rural and urban people whow are living under poverty).

Q.2818. Pradhan Mantri Jal Samvardhan Yojana was started

Ans. in the year 2002

Q.2819. Pradhan Mantri Gram Sarak Yojana was started

Ans. in the year 2000

Q.2820. Vishesh Krishi Upaj Yojana to promote exports of fruit, vegetables, and Minor forest produce

Ans. 2004

Q.2821. IRDP, TRYSEM, DWCRA, SITRA merged into.

Ans. SGSY

Q.2822. Lab to land programme started.

Ans. 1 June 1979 (on the occassion of ICAR Golden Jubilee celebrations)

Q.2823. Integration of all programmes in K.V.K.
Ans. 1 April, 1992

Q.2824. First Agricultural University established in the year
Ans. 1960, Uttarakhand, Pantnagar

Q.2825. Establishment of MANAGE.
Ans. 1986, Hyderabad

Q.2826. Royal commission on Agriculture.
Ans. 1928

Q.2827. FISCAL commission.
Ans. 1949

Q.2828. Land care programme.
Ans. 1990 Australia

Q.2829. N.S.S. was founded in
Ans. 1969

Q.2830. First rural youth programme.
Ans. 1920 (Sri Niketen)

Q.2831. Indira Mahila Yozana.
Ans. 1995

Q.2832. (N.E.S.) National Extension Service.
Ans. 1953

Q.2833. (PPTD) Pilot Project for Tribal Development.
Ans. 1972–73

Q.2834. Antuoyadaya Yojana.
Ans. 2 Oct. 1977

Q.2835. (N.S.C.) National seed corporation founded.
Ans. 1963 (Delhi)

Q.2836. (DPAP) Drought Prone Area Programme.
Ans. 1970-71 (Pant Nagar)

Q.2837. Indian society of Agronomy.
Ans. 1955

Q.2838. Indian society of Soil Science.
Ans. 1934

Q.2839. First Department of Agriculture established.
Ans. 1881

Q.2840. First Irrigation commission appointed.
Ans. 1901

Q.2840a. Famine commission was appointed in year
Ans. August 1880

Q.2841. Indian central sugarcane committee.
Ans. 1944

Q.2842. Indian central cotton committee was constituted.
Ans. 1921

Q.2843. National commission of Agriculture set up by Govt. of India.
Ans. 1970

Q.2844. Imperial Lac Research Institute (Ranchi).
Ans. 1925

Q.2845. 1st five year plan started.
Ans. 1951–1956

Q.2846. Food corporation of India (F.C.I.)
Ans. Founded on 14th January, 1965

Q.2847. Nationalization of major Banks.
Ans. 1969

Q.2848. NABARD came into existence.
Ans. 12th July, 1982

Q.2849. National Agriculture Science Museum is located at
Ans. New Delhi

Q.2840. First Bharatiya Mahila Bank (BMB) was launched in
Ans. 19 Nov, 2013 (Mumbai)

Q.2841. Irish Potato famine
Ans. 1845-1849

Q.2842. Agricultural Research Services were established by
Ans. M. S. Swaminathan (in the year 1973)

Q.2842a. Village knowledge centre project implemented by
Ans. MSSRF (M. S. Swaminathan Research Foundation in 1988). First started at village Villianur (Puducherry)

Q.2843. Components of VKC (Village Knowledge Centre)
Ans. 1. Desktop computer system, 2. At least one printer, 3. Radio communication equipment, 4. Farmers and workers.

Q.2844. Gram Panchayat is
Ans. Basic unit of local administration

Q.2845. The person associated with Community Development Programmes (1952)
Ans. S. K. Dey

Q.2846. Food and Agricultural Organisation is situated at
Ans. Rome (FAO was founded on 16 Oct. 1945 in Quebec city, Quebec, Canada. In 1951 its head quarters were moved to Rome, Italy).

Q.2847. NREGA (National Rural Employment Guarantee Act) now MGNREGA
Ans. Employment Guarantee Program for the people below poverty line (ARS/NET 2009)

Q.2848. Mass schemes were started on
Ans. 2nd October 1952.

Q.2849. The right approach of agricultural extension should be.
Ans. Bottom up approach.

Q.2850. The term 'Vision' in India in 'Vision-2020' by Dr. A.P.J. Abdul Kalam mean
Ans. Articulation of desired end results.

Q.2851. The ratio of Central and State Government in NREGA scheme
Ans. 50 : 50

Q.2852. Youths are being trained every year under TRYSEM?
Ans. Two Lakh

Q.2853. Khadi Village Industry Commission was formed?
Ans. In 1968–69.

Q.2854. Marginal Framers and Agricultural Laborers Development Agency (MAFALDA) was established in
Ans. 1971–72.

Q.2855. Operation Flood is related

Ans. Milk production and is also known as the White Revolution in India

Q.2856. The rural development of the village programme?

Ans. R. N. Tagore.

Q.2857. From where the extension education originated?

Ans. Japan.

Q.2858. Intensive Agriculture District Programme was established?

Ans. In the year 1963.

Q.2859. In Farm Management most important unit is

Ans. Production unit

Q.2860. Applied Nutrition programme in india was started

Ans. 1963

Q.2861. The state which was electrified ruraly first is

Ans. Haryana.

Q.2862. ISOPOM: (Integrated Scheme of Oilseeds, Pulses, Oil palm and Maize) was started in the year

Ans. 2004

Q.2863. ICDP (Intensive Cotton Development Programme) was started in the year

Ans. 2003-04.

Q.2865. National Food Security Mission (NFSM) was stared during the year.

Ans. 2007–2008

Q.2866. Rashtriya Krishi Vikas Yojana was in the year.

Ans. 2007-2008 with 100% grant by Central Govt.

Q.2867. Which year was declared as "Food security and quality year" by Govt. of India?

Ans. 2008–2009.

Q.2868. Imperial Biological laboratory was established in year

Ans. 1889

Q.2869. Foundation of agriculture research in year

Ans. 1890

Q.2870. Foundation of agriculture research laid by

Ans. Dr. J.A. Voelcker

Q.2871. First chairman of Royal commission of Agriculture

Ans. Lord Linlithgow

Q.2872. Great Bengal famine occurred in year

Ans. 1943

Q.2873. 3 Tier Panchayat Raj given by

Ans. Balwant Rai Mehta

Q.2874. First Agriculture University pattern was based on

Ans. Land Grant System of U.S.A.

Q.2875. Blue print for establishment of SAU's was prepared by

Ans. Dr. Ralp W. Cummings

Q.2876. Father of Sociology

Ans. Auguste Comte

Q.2877. ARS was started in year

Ans. 1973

Q.2878. ASRB (Agri. Scientist Recruitment Board) was started in year

Ans. 1975

Q.2879. Which committee recommended ASRB

Ans. Gajendra Gadkar Committee

Q.2880. Food for work was started in year

Ans. 1977

Q.2881. Two Tier Panchyati Raj system was proposed by

Ans. Ashok Mehta Committee

Q.2882. Crop Insurance Scheme (CIS) was started in year

Ans. 1985

Q.2883. First hybrid of Pigeon Pea was released in year

Ans. 1991 (ICPH-8) by ICRISAT & IARI

Q.2884. National Agricultural Insurance Scheme was started in year

Ans. 1999

Q.2885. First National Conference of KVK was held in year

Ans. 2005 (New Delhi)

Q.2886. National demonstration scheme is associated with

Ans. Kalwar & Subramanyam

Q.2887. Agricultural statistics at glance is published by

Ans. Ministry of Agriculture and Farmers Welfare

Q.2888. Indian Farmer Digest is published by

Ans. Pantnagar Agricultural University. (Uttrakhand)

Q.2889. First Five year plan was started in year

Ans. 1951

Q.2890. RBI was nationalized in year

Ans. 1st Jan., 1949

Q.2891. "Krishi" television channel was launched in the year

Ans. 2004

Q.2892. Macro management of Agriculture scheme was first launched in the year

Ans. 2000-2001

Q.2893. Macro management of Agriculture scheme revised in the year

Ans. 2008

Q.2894. Khetihar Mazdoor Beema Yojana started in year

Ans. 1st July 2001

Q.2895. India's first Biotech crop technology approved for commercialization in year

Ans. 2002 (BT Cotton)

Q.2896. Pilot weather based crop insurance scheme (WBCIS) is started in the year

Ans. 2010

Q.2897. Concept of PRA (Participatory Rural Appraisal) was given by

Ans. Henry Ford

Q.2898. Person considered as father of PRA

Ans. Robert Chambers

Q.2899. PRA was introduced in India by the

Ans. NGO's

Q.2900. The term community development originated from

Ans. Cambridge (England, 1946)

Q.2901. IAAP (Intensive Agriculture Area Programme) came into operation in the year

Ans. 1964 March.

Q.2902. ICPS (Integrated Child Protection Scheme)

Ans. Started in the year 2009.

Q.2903. Village Grain Bank Scheme

Ans. Launched in the year 1996-97

Q.2904. NRHM (National Rural Health Mission)

Ans. Started on 12th April 2005 By Prime Minister.

Q.2905. First Newspaper published from India

Ans. Bengal Gadget

Q.2906. Radio Broadcasting started from India

Ans. 23rd July, 1927

Q.2907. National Bamboo mission started in

Ans. 2006-07

Q.2908. Indias first-magazine contributed to Agriculture

Ans. Kheti

Q.2909. Bringing Green Revolution to Eastern India (BGREI) 2010-11

Ans. To address the constraints limiting the productivity of rice based cropping systems in seven Eastern states of India.

Q.2909a. AGREX (Expert System) developed by

Ans. Center for Informatics Research and Advancement, Kerala to provide correct advice to farmers.

Q. 2910. Father of method demonstration-
Ans. Dr. Seeman A. Knapp

Q. 2911. Percent of innovators in India of total population is-
Ans. 2.5%

Q. 2912. Type of adopters having highest percentage -
Ans. Early majority (34%) & Late majority (34%)

Q. 2913. Most important element of communication is-
Ans. Receiver

Q. 2914. Method demonstration is a type of demonstration-
Ans. Short type

Q. 2915. General body of village panchayat in Panchayati Raj is-
Ans. Gram Sabha

Q. 2916. Basic unit of civilization is-
Ans. Family

Q. 2917. Basic unit of rural society is-
Ans. Village

Q. 2918. T & V system was introduced in Rajasthan and Madhya Pradesh for-
Ans. Canal Area and Chambal Area (1974)

Q. 2919. Basic idea and concept of CDP programme was borrowed from-
Ans. U.S.A

Q.2920. Deen Dayal Upadhayaya Grameen Kaushlya Yojna (DDU-GKY) was launched in-
Ans. 25 Sep, 2014

Q.2921. Sector for DDU-GKY Yojna-
Ans. Rural Development Sector

Q.2922. Deen Dayal Upadhayaya Gram Jyoti Yojna was launched in-
Ans. 25 July, 2015

Q.2923. Sansad Aadarsh Gram Yojna was launched in-
Ans. 11 Oct, 2014 (Rural Development Sector)

Q.2924. Soil health Card Scheme was launched in-

Ans. 17 Feb, 2015 (Agriculture Sector) from Suratgarh Rajasthan.

Q.2925. Pradhan Mantri Jan Dhan Yojna was launched in-

Ans. 28 Aug, 2015

Q.2926. Namami Gange project was launched in-

Ans. Oct, 2016 (River Devp Sector)

Q.2927. Solar Charkha mission was started in-

Ans. 27 June, 2018 by President

Q.2928. Solar Charkha mission comes under-

Ans. Ministry of MSME

Q.2929. Gobar Dhan Yojna was started on-

Ans. 30 April, 2018

Q.2930. Gobar Dhan Yojna was launched by-

Ans. Haryana's C M Manhor Lal Khattar and Uma Bharti

Q.2931. National Nutrition Mission (POSHAN Abhiyaan) was started on-

Ans. 8 March, 2018 (At Jhunjhunu, Raj. by PM)

Q.2932. National Nutrition Mission (POSHAN Abhiyaan) is a expansion of-

Ans. Beti Bachao Beti Padhao Programme

Q.2933. Integrated Scheme for Agricultural Marketing (ISAM) was effective from-

Ans. 1 April , 2014

Q.2934. National Mission of Agricultural Extension and Technology (NMAET) has started -

Ans. 1 April , 2014

Q.2935. Desert development programme was started in-

Ans. 1977-78 (By ICAR)

Q.2936. First chairman of National Farmer Commission-

Ans. Mr. Sompal

Q.2937. Young Farmer Association was formed in India in-

Ans. April, 1956 (By P.S. Deshmukh

Q.2937a. ATMA (Agricultural Technology Management Agency)

Ans. Objective to coordinate various research and extension activities at the district level and decentralization of public agriculture technology system.

ABOUT I.C.A.R. & I.A.R.I.

Q.2938. I.A.R.I. was established in Bihar.

Ans. 1905

Q.2939. I.A.R.I. was established under the vice royalty of.

Ans. Lord Curzon in 1905

Q.2940. Building of I.A.R.I. was damaged due to earthquake.

Ans. 1934

Q.2941. Transfer of I.A.R.I. to New Delhi.

Ans. 1936

Q.2942. I.A.R.I. was given the status of deemed university.

Ans. 1958

Q.2943. ICAR was established by the recommendation of

Ans. Lord Linlithgo

Q.2944. Imperial council of Agricultural Research was established on

Ans. 16th July, 1929

Q.2945. First President of ICAR was

Ans. Sir Mohammad Habibullah

Q.2946. First secretary of ICAR was

Ans. S. A. Hydari

Q.2947. ICAR was renamed on

Ans. March 1946

Q.2948. Renaming of ICAR was under the president ship of

Ans. Jogendra Singh

Q.2949. ICAR was reconstituted into full autonomous body.

Ans. 1966

Q.2950. First D. G. of ICAR

Ans. Dr. B.P. Pal

Q.2951. First Indian Director of IARI.

Ans. B. Vishwanathan

Q.2952. First vice chairman of ICAR.

Ans. Diwan Bhadur Vijayraj Acharya

Q.2953. Establishment of IRRI.

Ans. 1960 (Philipines)

Q.2954. National Bureau of Agriculturally Important Microorganism is situated at

Ans. Mau (U.P.)

Q.2955. Motto of ICAR in english

Ans. Agrisearch with a human touch

Q.2956. Vision 2030 launched in the year

Ans. January 2011

Q.2957. First Agriculture Minister of Independent India

Ans. Rafi Ahmed Kidwai

Q. 2958. Word "diorama" originated in 1823 as picture viewing device, orginated from (*ICAR, Net 2023*)

Ans. Greek words

Q.2959. SREP (Strategic Researh Exension Plan is omponent of (*ICAR NET 2023*)

Ans. ATMA

Q.2960. SFURTI

Ans. Scheme of Fund for Regeneration of Traditional Industries by GoI (MSME) in 2005

IMPORTANT REVOLUTIONS

Q.2958. Green revolution occurred in.

Ans. 1965-66 (related to crop production)

Q.2959. Term Green Revolution coined by

Ans. William Gaad

Q.2960. White revolution is related to.

Ans. Milk production (Verghese Kurien)

Q.2961. Blue revolution is related to.

Ans. Fisheries (Dr. Arun Krishnan)

Q.2962. Round revolution is related to.

Ans. Potato

Q.2963. Silver revolution is related to.

Ans. Eggs & Poultry (Promoted by Indira Gandhi)

Q.2964. Brown revolution is related to.

Ans. Leather/Non conventional products

Q.2965. Red revolution is related to

Ans. Tomato / Meat (Vishal Tewari)

Q.2966. Pink revolution is related to

Ans. Onions / Prawns (Durgesh Patel)

Q.2967. Black revolution is related to

Ans. Petroleum Products / Biofuels

Q.2968. Rainbow revolution is related to

Ans. Overall development of Agriculture (Nitish Kumar)

Q.2969. Golden revolution is related to

Ans. Horticulture (Dr. K. L. Chadha)

Q.2970. Yellow revolution is related to

Ans. Oilseeds (Sam Pitroda)

Q.2971. Protein revolution

Ans. Higher production technology driven 2nd Green Revolution (Coined by Narendra Modi and Arun Jaitley)

Q.2971. Grey revolution

Ans. Fertilizer

ELEMENTS OF STATISTICS

Q.2972. The first step of summarizing the data is.

Ans. Classification

Q.2973. The measure of central tendency to be used to study average rate of change in population is.

Ans. Geometric Mean

Q.2974. To find the average size of shoes sold in the market we should use.

Ans. Mode

Q.2975. To find the average speed of a vehicle when same distance is covered with different speeds, the suitable measure is.

Ans. Harmonic mean

Q.2976. To find the average of quantity prices, the measure of central tendency to be used is.

Ans. Harmonic mean

Q.2977. To find the average height of plants we should use.

Ans. Arithmetic mean

Q.2978. We study measure of central tendency.

Ans. To represent the whole data by only single value.

Q.2979. The suitable measure to find average speed when time for each speed is fixed would be.

Ans. Arithmetic mean

Q.2980. In industries for quality control, the most important measure of dispersion is

Ans. Range

Q.2981. What is the order of three averages Arithmetic Mean, Geometric Mean & Harmonic mean for a given data

Ans. AM $\geq$ GM $\geq$ HM

Q.2982. The formula for Geometric mean for n values x_1, x_2 .... x_n is.

Ans. $(x_1 , x_2 \text{ .... } x_n)^{1/n}$

Q.2983. If all the variate values are negative the standard deviation will be.

Ans. Positive

Q.2984. A quantitative or qualitative characteristics that varies from observation to observation in the same group is called

Ans. Variable

Q.2985. The standard deviation for the values 4, 5, 6, 7, 8 will be.

Ans. $(2)^{½}$

Q.2986. Coefficient of variation is calculated by the formula.

Ans. $\text{C.V.} = \dfrac{\text{S.D.}(\sigma)}{\text{Mean}(\mu)} \times 100$

Q.2987. The degree of freedom for error in R.B. design with 10 treatments and 4 replication will be.

Ans. 27

Q.2988. Under the assumption for analysis of variance, the parent population should be.

Ans. Normal

Q.2989. For any two values a & b the what relationship among Arithmetic mean (A) Geometric mean (G) and Harmonic mean (H) holds?

Ans. G^2 = A.H.

Q.2990. For testing the hypothesis about mean of one population we use 't' test when.

Ans. Sample size is small and S. D. is unknown

Q.2991. For testing the significance of correlation coefficient we use.

Ans. 't' test

Q.2992. To test the agreement between observed frequencies and expected frequencies we use.

Ans. Chi Square test

Q.2993. The minimum sample size for using chi square test should be.

Ans. 50

Q.2994. The analysis of variance technique is used for.

Ans. Comparing the means of more than two population

Q.2995. Latin square design is suitable for comparing.

Ans. 5 to 12 treatments

Q.2996. In field experiments the commonly used design is.

Ans. Random Block Design (R.B.D.)

Q.2997. The design where no. of replications = no. of treatments is

Ans. Latin square design

Q.2998. What is the relationship among the correlation coefficient (r) and the two regression coefficient *byx* & *bxy*?.

Ans. $r^2 = bxy \; . \; byx$

Q.2999. The regression coefficient lies between.

Ans. $-\infty$ to $+\infty$

Q.3000. The correlation coefficient lies between.

Ans. –1 to +1

Q.3001. For testing the independence of two attributes, the test used is.

Ans. χ^2 test

Q.3002. Correlation of continuity in 2 × 2 contingency table should be used when expected frequency of any cell is.

Ans. Less than 5

Q.3003. The value of (χ^2) chi square always lies between.

Ans. 0 to ∞

Q.3004. The degree of freedom to test the significance of difference between two means based on n_1 & n_2 observation is.

Ans. $n_1 + n_2 - 2$

Q.3005. For comparing 4 treatments with R.B. design, the 15 different frequencies can be obtained when the no. of replication are.

Ans. 6

Q.3006. Most commonly used measure of central tendency is.

Ans. Arithmetic Mean

Q.3007. Which measure of dispersion is considered as best ?

Ans. Standard deviation

Q.3008. The value of standard deviation may vary between.

Ans. 0 to ∞

Q.3009. Average rate of depreciation would be obtained by.

Ans. Geometric Mean

Q.3010. If *r* is the observed correlation coefficient in a sample of *n* pairs of observation then its standard error is, given by.

Ans. $S.E.\ (r) = \dfrac{1-r^2}{\sqrt{n}}$

Q.3011. Probable error of the correlation coefficient given by.

Ans. $S.\ E.\ (r) = 0.6745\ \dfrac{1-r^2}{\sqrt{n}}$

Q.3012. When two variables move in the same direction correlation is said to be.

Ans. Positive

Q.3013. When two variables move in the opposite direction correlation is said to be.

Ans. Negative

Q.3014. When the treatments are arranged randomly over the whole of a previously determined set of experimental units, the design is known as

Ans. Completely randomized design (CRD)

Q.3015. In the case of perfect negative correlation the coefficient of correlation will be.

Ans. –1

Q.3016. There is no skewness when the values of mean, median & mode are.

Ans. Equal

Q.3017. When Mean > Median > Mode, skewness will be.

Ans. Positive

Q.3018. Probability of any event is a number lying between.

Ans. 0 to 1 ($0 \leq p \leq 1$)

Q.3019. Probability is the science of.

Ans. Decision making (uncertain phenomena in science)

It deals with the occurence of random event.

Q.3020. Mode of given set of the observation in that value which occurs with the.

Ans. Maximum frequency

Q.3021. χ^2 test for goodness of fit is given by

Ans. $\chi^2 = \sum_{i=1}^{n} \frac{(O_i - E_i)^2}{E_i^{\,2}}$

Q.3022. Median is better suited for.

Ans. Positional interval series

Q.3023. Harmonic mean is the reciprocal of the.

Ans. Arithmatic mean

Q.3024. If a card is drawn from a pack of cards the probability of getting either a king or a queen is.

Ans. 2/13

Q.3025. If A & B are mutually exclusive events $P(A \cap B)$ is equal to

Ans. 0

Q.3026. The normal distribution with $\mu = 0$, and $\sigma = 1$ is known as.

Ans. Standard normal distribution

Q.3027. The distribution of χ^2 depends on the.

Ans. Degree of freedom

Q.3028. The χ^2 test should not be applied if N is less than.

Ans. 50

Q.3029. ANOVA table stands for analysis of.

Ans. Variance

Q.3030. A two way table containing two attributes is known as

Ans. Contingency table

Q.3031. In a double sampling plan the decision to accept or reject a lot is made on the basis of.

Ans. Two samples

Q.3032. Given mean 25, mode 25, the median would be.

Ans. 25

Q.3033. When data are observed over a period of time the type of classification is known as.

Ans. Chronological classification

Q.3034. Neutral scale indicates.

Ans. Absolute changes

Q.3035. Ratio scale indicates.

Ans. Relative changes

Q.3036. Quartile deviation and standard deviation are related as

Ans. Quartile deviation = $\frac{2}{3}$ × Standard deviation

Q.3037. When mean is 79, variance is 64 then coefficient of variation is.

Ans. 10.126 $\left(\because C.V. = \frac{\sigma}{\mu} \times 100\right)$

Q.3038. β_2 is a measure of.

Ans. Kurtosis

Q.3039. When the value of χ^2-static is zero ?

Ans. If observed frequency is equal to expected one.

Q.3040. The S. D. of the sampling distribution is called.

Ans. Standard error

Q.3041. The student 't' distribution test was discovered by.

Ans. W. S. Gosset (1908)

Q.3042. The student 't' test was perfected by.

Ans. R. A. Fisher (1926)

Q.3043. In describing the amount of variation in a population & measure often used is

Ans. Coefficient of variation (C. V.)

Q.3044. In an investigation when two groups or two procedures are compared, these procedures are called.

Ans. Treatments

Q.3045. Whether 't' test is parametric or non parametric.

Ans. Parametric

Q.3046. Ordinary sign test is a

Ans. Non parametric test

Q.3047. Who developed the idea of regression?

Ans. Sir Franis Galton

Q.3048. The analysis of covariance is a technique that combines the features of analysis of variance and.

Ans. Regression

Q.3049. In normal distribution.

Ans. Mean, Mode, Median are equal

Q.3050. If fertility variation in the field is in two direction at right angles, which of the experimental design is suittable

Ans. Latin Square Design (L.S.D.)

Q.3051. If the data is classified by attributes and if two or more characters or groups are to be compared within each attribute we use

Ans. Multiple bar diagram

Q.3052. Range of distribution is.

Ans. The difference between largest and smallest observation

Q.3053. Biometry deals with.

Ans. Observation with living things

Q.3054. A measure of peakedness or convexity of a curve is known as

Ans. Kurtosis

Q.3055. 'F' test can be used for testing the significance of.

Ans. Several difference

Q.3056. χ^2 test is applied for.

Ans. Testing the homogeneity of correlation coefficients.

Q.3057. Local control helps in.

Ans. Reducing Experimental error and homogeneity of experimental units.

Q.3058. When sample size is small and population S.D. is known the test used should be

Ans. 't' test

Q.3059. To reduce the experimental error with heterogeneous material we need.

Ans. More replication & use of Local control technique

Q.3060. The father of vital statistics was

Ans. Captain John Grant.

Q.3061. Pie-diagram is named as pi because

Ans. The sectors resembles the slices cut from a pie.

Q.3062. When the numerical data are presented in a descriptive form then it is called

Ans. Textual presentation.

Q.3063. Commulative frequency is also known as

Ans. Ogive.

Q.3064. The age of an individual can be considered under the category of

Ans. Continuous variable

Q.3065. The Poisson distribution is also used to represent the probability distribution of a

Ans. Discrete random variable

Q.3066. The distribution of observation in a population is described in terms of certain measure like mean, standard deviation, proportion etc. These measures are known as

Ans. Parameters

Q.3067. What is the combined geometric mean of two sizes n_1 and n_2 if their geometric means are G_1 and G_2 ?

Ans. $\log G = \dfrac{n_1 \log G_1 + n_2 \log G_2}{n_1 + n_2}$

Q.3068. What will be harmonic mean of two observations 5 and –5 ?

Ans. ∞

Q.3069. What will be mean of squares of first n natural numbers ?

Ans. $\dfrac{n(n+1)(2n+1)}{6n}$

Q.3070. If in a frequency distribution, the maximum frequency is being repeated, then the value of mode can be ascertained by

Ans. Method of grouping.

Q.3071. The lack of homogeneity in the size of distribution is called

Ans. Dispersion.

Q.3072. The measure of dispersion which is affected least by fluctuation of sampling is

Ans. Standard deviation.

Q.3073. Standard deviation is independent of ________ and depends on ________.

Ans. Change of origin, change of scale.

Q.3074. The measure of Kurtosis is

Ans. β_2 and γ_2.

Q.3075. If the Kurtosis of a distribution is less than 3, then it is called

Ans. Platycurtic curve

Q.3076. If the Kurtosis of a distribution is more than 3, it is said to be ________ curve.

Ans. Leptocurtic.

Q.3077. What is the probability of an impossible event?

Ans. Zero.

Q.3078. What is the name of the limiting relative frequency approach probability ?

Ans. Statistical probability.

Q.3079. Who gave the definition of mathematical probability?

Ans. Von-Mises.

Q.3080. If *A* and *B* are independent events then P(A ∩ B) is equal to

Ans. $P(A)\cdot(B)$.

Q.3081. By deterministic phenomena, we mean that results can be predicted with

Ans. Certainty.

Q.3082. For two independent events A and B, the conditional probability P(A/B) = ...

Ans. A

Q.3083. When do two random variables are said to be independent ?

Ans. When $cov(X, Y) = 0$ *i.e.* correction coefficient $r = 0$.

Q. 3084. The standard deviation of the sampling distribution of a statistic is called the ___of the statistic

Ans. Standard error

Q.3085. What is the maximum probability at the point of inflection of normal distribution ?

Ans. $\frac{1}{\sigma\sqrt{2\pi}}e^{\frac{-1}{2}}$.

Q.3086. The normal distribution was discovered by

And. De-Moivre.

Q.3087. If X is normally distributed with zero mean and unit variance, then mean and variance of X^2 equal to

Ans. 1 and 2.

Q.3088. What does β in the regression equation $Y = \alpha + \beta x + \gamma$ indicate ?

Ans. Slope of the line.

Q.3089. Predictor variables are also known as

Ans. Independent variable.

Q.3090. Response variables are also known as

Ans. Dependent variable.

Q.3091. What is the angle between two regression lines when correlation coefficient between two variables is zero ?

Ans. 90°.

Q.3092. Non parametric test are based on

Ans. Flexible assumptions.

Q.3093. What do you understand if the rank correlation coefficient is +1 ?

Ans. It reveals that the ranks given by two persons are the same and there is a perfect association between the ranks.

Q.3094. The Z-test does not take into account

Ans. Degrees of freedom.

Q.3095. The number of independent constraints in the set of observation is called

Ans. Degrees of freedom.

Q.3096. Which test of non parametric methods is similar to χ^2 test of goodness of fit ?

Ans. Kolmogorov - Smirnov test.

Q.3097. Ordinary sign test is based on the sign of the deviations from

Ans. Median.

Q.3098. The correction factor in the analysis of variance is equal to

Ans. $\dfrac{(\text{Grand observation})^2}{\text{Number of observations}}$

Q.3099. When a hypothesis specifies all the parameters of a probability distribution, it is known as

Ans. Simple hypothesis

Q.3100. If the hypothesis specified only some of the parameters of the probability distribution, it is known as

Ans. Composite hypothesis

Q.3102. What is the range of *t* static ?

Ans. $-\infty \leq t \leq \infty$

Q.3103. Equality of two population variances can be tested by

Ans. *F*-test.

Q.3104. The test whether theory fits well in practical can be judged by

Ans. χ^2-test.

Q.3105. If *n* = 1, then the chi-square reduces to

Ans. Exponential distribution.

Q.3106. For testing a single proportion of the population test used when the sample size is large is

Ans. Z-test.

Q.3107. The linear model in ANOVA of two way classification is

Ans. $y_{ij} = \mu + T_{ij} + e_{ij}$.

Q.3108. The criterion for one sided or two sided test depends on

Ans. Alternative hypothesis

Q.3109. For what type of trend the method of least squares is used.

Ans. It is used to know about linear and quadratic trends.

Q.3110. What is the best method for finding out seasonal variations ?

Ans. Ratio to moving average method.

Q.3111. How many control lines are in a control chart ?

Ans. Three control lines.

Q.3112. The number of defects should follow .

Ans. Poisson distribution

Q.3113. The main tools of statistical quality control are

Ans. Shewhart charts and acceptance sampling plans.

Q.3114. What does a hypothesis called under test ?

Ans. Null hypothesis.

Q.3115. What do you call the probability of type 1 error?

Ans. The level of significance.

Q.3116. To apply Mann-Whitney U test, the two samples that are to be compared should be

Ans. Independent

Q.3117. The difference between sample estimate and the corresponding population parameter is called

Ans. Sampling error.

Q.3118. Larger samples are required for getting the

Ans. Higher precision

Q.3119. If our estimates based on sample values are closer to the population parameters, the estimate can be regarded as

Ans. Reliable

Q.3120. What is the most frequently used index number formula ?

Ans. Weighted formula.

Q.3121. What is another name of Laspeyre's index number?

Ans. Base year method index number.

Q.3122. Which index number is generated if the geometric mean of Laspeyre's and Paasche's index number are taken ?

Ans. Fisher's ideal index number.

Q.3123. Laspeyre's was a

Ans. French economist.

Q.3124. The basic pay, D. A. etc. of various employees can be determined on the basis of

Ans. Cost of living index number.

Q.3125. The repeated application of the treatments under investigation is known as

Ans. Replication

Q.3126. Prof. Karl Pearson was a

Ans. Biometrician.

Q.3127. What is the number of all possible samples of size three from a population of 20 units ?

Ans. $^{10}C_3 = 120$

Q.3128. If the sample values are 1, 3, 5, 7, 9, what should be standard error S. E. of sample mean?

Ans. $\sqrt{2}\left(\because \text{S.E.} = \dfrac{S}{\sqrt{n}}\right)$.

Q.3129. What is the name of any population constant ?

Ans. Parameter.

Q.3130. What is the identification of the index of precision of an estimator ?

Ans. Standard error.

Q.3131. The stratified random sampling is applicable when the population is

Ans. Heterogenous.

Q.3132. Sampling error occur due to ________ selection of samples and ________ method of analysis.

Ans. Improper, faulty.

Q.3133. What will be the regression coefficient of *Y* on *X* and *X* on *Y*, if a constant *a* is added to each values of *X* and *Y* ?

Ans. It will remain same.

Q.3134. Given two regression lines $X + 2Y = 5$ and $2X + 3Y = 8$ and variance $X = 12$, what is the value of standard deviation of *Y* ?

Ans. 2.

Q.3135. The limit of partial correlation coefficient is

Ans. –1 to 1 (Range of co-relation coefficient).

Q.3136. In which field in the daily life the probability theory has major application.

Ans. Risk assessment and trade on commodity market.

Q.3137. If the points of scatter diagram are widely scattered a ________ between the variables is expected.

Ans. Greater amount of correlation.

Q.3138. The correlation between price and demand of a commodity is

Ans. Negative.

Q.3139. The another name of critical region is

Ans. Region of rejection.

Q.3140. When does paired *t*–test is applicable ?

Ans. When the observations are paired together.

Q.3142. Which test is often used to refer specifically to the one sample location test?

Ans. *Z*-test.

Q.3142. Sampling inspection by variables requires

Ans. Less inspection than the attributes.

Q.3143. The expected sample size required to arrive at a decision about a lot is called

Ans. Average sample number.

Q.3144. Bowley index number formula is arithmetic mean of

Ans. Laspeyre's and Paasche's index number.

Q.3144. The condition for factor reversal test in the usual notation is given by

Ans. $P_{01} \times Q_{01} = V_{01}$.

Q.3146. One of the limitations in the construction of index number is

Ans. Choice of average.

Q.3147. The errors involved in the construction of index number are

Ans. Sampling error, formula errors and errors in collection of data.

Q.3148. Which test is not satisfied by Fisher's formula?

Ans. Circular test.

Q.3149. Term 'Kurtosis' was introduced by

Ans. Karl Pearson in the year 1906.

Q.3150. For what sample size the *t* test can be applied.

Ans. When sample size < 30

Q.3151. What are the basic steps of field experimentation?

Ans. (*i*) Replication, (*ii*) Randomization, (*iii*) Local control.

Q.3152. In which fields the geometric means is commonly used?

Ans. To find average percent increase in sales, production or other economic or business series.

Q.3153. Geometric mean can be easily calculated with help of

Ans. logarithm

Q.3154. What is the relation between mean, median, and mode?

Ans. Mode = 3 median – 2 mean.

Q.3155. In variance ratio test, which variance is kept as the numerator?

Ans. Larger variance.

Q.3156. How many types of hypothesis exist?

Ans. Four simple, null, composite and statistical.

Q.3157. By whom and when the Poisson distribution was developed?

Ans. S.D. Pearson in 1837.

Q.3158. The mean and variance of random variable is given by

Ans. Mean $\mu = \Sigma P_i X_i$

Variance $\sigma^2 = \Sigma P_i X_i^2 - (\Sigma P_i X_i)^2$

Q.3159. The mean and variance of Binomial distribution is given by

Ans. Mean = *np*, Variance = *npq*

Q.3160. The mean and variance of Poisson distribution are

Ans. equal (both equal to *np*)

Q.3161. In Bernauli trails all trails have

Ans. Equal probability

Q.3162. The experimental design for fertility gradient of land in two directions is

Ans. Least square design.

Q.3163. The calculation of expected frequencies is crucial in

Ans. χ^2 test

Q.3164. Out of the two kinds of error in sampling, which one is more serious?

Ans. Type I error.

Q.3165. Positively and negatively skewed distributions are also known as

Ans. Right skewed and left skewed respectively.

Q.3166. As per law of probability the total probability of event *A* associated with events $E_1, E_2, ... E_n$ is given by

Ans. $P(A) = P(E_1) \cdot P(A/E_1) + P(E_2) \cdot P(A/E_2) + ... + P(E_n) \cdot P(A/E_n)$

Q.3167. By whom the term null hypothesis was first coined?

Ans. Ronald Fisher

Q.3168. The test whether the theory fits well in practical can be judged by

Ans. χ^2 test (Chi-square)

Q.3169. Who introduced the concept of alternative hypothesis?

Ans. Jerzy Neyman and Egon Pearson.

Q.3170. In analysis of variance, all populations from which samples have been drawn are:

Ans. Normal

Q.3171. Who defined mathematical probability?

Ans. Von Mises

Q.3172. Give a well known example of skewed distribution.

Ans. Personal wealth

Q.3173. What is the asymptotic power of Kolmogorov Smirnov test?

Ans. One

Q.3174. What is the skewness for normal distribution?

Ans. Zero.

Q.3175. What are the other names given to alternative hypothesis?

Ans. Research hypothesis or maintained hypothesis.

Q.3176. Non parametric tests are based on ______ assumptions.

Ans. Flexible.

Q.3177. For three mutually exclusive events, *A*, *B*, *C*.

$P(A \cup B \cup C) =$

Ans. $P(A) + P(B) + P(C)$

Q.3178. From where does the word Kurtosis taken?

Ans. Greek word Kyrtos mean bulging.

Q.3179. Experimental Design Principles were given by

Ans. R. A. Fisher

Q.3180. Median is also called as

Ans. 50th percentile or 2nd quartile

Q.3181. A distribution said to be symmetrical when the value are uniformly distributed around the

Ans. Mean

Q.3182. Coefficient of skewness for a symmetrical distribution is

Ans. Zero

Q.3183. F-distribution in sampling was first developed by

Ans. R. A. Fisher

Q.3184. Square of standard deviation is

Ans. Variance

Q.3185. Standard deviation of a sampling distribution of statistics

Ans. Standard error

Q.3186. An analysis of variation between two or more variable

Ans. Correlation

Q.3187. Mathematical reason of the average relation-ship between two or more variables

Ans. Regression

Q.3188. The foundation of mathematical theory of probability was laid in the mid-seventeenth century by two French Mathematicians

Ans. B. Pascal and P. Fermal

Q.3189. The measures of the direction and the degree of asymmetry are called

Ans. Measures of skewness

Q.3190. Bowley's coefficient of skewness is also known as

Ans. Quartile coefficient of skewness

Q.3191. If $\beta_2 > 3$ the distribution is said to be

Ans. Leptokurtic

Q.3192. The concept of consistency is concerned with the variability of the

Ans. Estimator

Q.3193. The non-parametric alternative to paired *t-test* is the

Ans. Wilcoxon signed-rank test

Q.3194. An example of non-parametric test for comparison of several related samples is

Ans. Quade test

Q.3195. The Karl Pearson's coefficient of correlation is often referred as ________ in order to distinguish it from other measure of relationship.

Ans. Product moment correlation

Q.3196. The value $1 - r^2$, where r is correlation coefficient is called

Ans. Coefficient of alienation

Q.3197. The multiple correlation is a measure of strength of relationship between the dependent variable and

Ans. All the independent variable *X*'s Jointly

Q.3198. The error due to data entry, coding, classification etc. are known as

Ans. Processing errors

Q.3199. When all the treatments have equal changes of being allocated to different experimental units it is known as

Ans. Randomization

Q.3200. The minimum size of an experimental plot for a given degree of precision is known as

Ans. Optimum lot size.

Q.3201. The two important methods that are used to determine optimum plot size are

Ans. Maximum curvature method and Fairfield Smith's variance law.

Q.3202. The analysis of variance can be regarded as a special development of

Ans. Regression analysis

Q.3203. The functional relationship, $\sigma^2 = \mu\ (1-\mu)/n$ occurs when distribution is

Ans. Bionomial

Q.3204. In Latin Square Design (LSD) each row and each column is a

Ans. Complete block or replication

Q.3205. When the treatments are arranged randomly over the whole of a previously determined set of experimental units, the design is known as

Ans. Completely Randomized Design (CRD)

Q.3206. The square root transformation is used when the observations follows a

Ans. Poisson distribution

Q.3207. Modern concepts of experimental design are due primarily to

Ans. R. A. Fisher

Q.3208. Relationship between arithmetic mean (A) harmonic meant (H) the geometric means (G) is

Ans. $A \times H = G^2\ (A > G > H)$

AGRICULTURAL ECONOMICS AND FARM MANAGEMENT

Q. 1. Who is known as father of economics?

Ans. Adam Smith

Q. 2. In which year the 'AGMARK' act was enforced?

Ans. 1937 and amended in 1986 *[ARS NET 2001]*

Q. 3. What is wealth according to Keynes?

Ans. Wealth consists of potentially exchangeable means of satisfying human wants.

Q. 4. Central Agmark Laboratory situated at

Ans. Nagpur

Q. 5. Full form of AGMARK

Ans. Agricultural Produce Grading and Marketing introduced in 1937

Q. 6. Types of Wealth are?

Ans. (*i*) Individual wealth, (*ii*) Social wealth, (*iii*) National wealth, (*iv*) Cosmopolitan wealth, (*v*) Negative wealth.

Q. 7. Fixed cost includes?

Ans. Taxes, insurance, cess, depreciation on machinery, tools building, land rent etc.

Q. 8. Where does the dumping activity seen?

Ans. Monopoly

Q. 9. What is the formula for average variable cost?

Ans. $AVC = \frac{\text{Total variable cost}}{\text{Output}} = \frac{TVC}{Q}$

Q. 10. Define marginal cost?

Ans. It is the change in the total cost due to change in one unit of output, *i.e.*, $MC = \frac{\Delta TVC}{\Delta Q}$

Q.10a Marginal cost always zero

Ans. When fixed cost does not change with change in output

Q.10b For a competitive firm profit maximizes when

Ans. MC = MR

Q. 10c. Prime cost is also known as

Ans. Variable cost

Q. 11. When two products are substituting each other, the optimum level of these inter prices can be worked out through. *[JRF 2001, NET 2001]*

Ans. Production possibility curve

Q. 12. Elasticity of supply is expressed as

Ans. Elasticity of supply (E_S) =

$$\frac{\text{percentage change in quantity of good supplied}}{\text{percentage change in good supplied}}$$

Q. 13. When the supply of commodity increase to infinite quantity or unlimited quantity even though there is a minute rise in price the elasticity of supply is said to be

Ans. Perfectly elastic

Q. 14. When percentage change in quantity supply equals percentage change in price, then the supply is called

Ans. Unitary Elastic supply.

Q. 15. As per Dalton, the optimum production is that which gives the

Ans. Maximum income per head.

Q. 16. When does a firm should close down the operation?

Ans. If the price of its product is less than average variabie cost (AVC).

Q. 17. Name two types of wages?

Ans. (*i*) Nominal wages or money wages (*ii*) Real wages

Q. 18. What is the full form of GNF?

Ans. Gross National Product

Q. 19. Demand for wheat is

Ans. Inelastic

Q. 20. Relationship between cost function and production function is

Ans. Negative

Q.20. BIRD (Bankers Institute of Rural Development) was promoted by

Ans. NABARD in 1983 for Training, Research and Consultancy at Lucknow (U.P).

Q. 21. When marginal pro·· ··tion is equal to ··erage production, then elas··city of production ·· equal to:

Ans. One [*ARS, NET 2002*]

Q. 22. GNP through income method can be expressed by the formula

Ans. GNP = wages and salaries + rents + interest + profit of uncorporated firms + dividends + undistributed corporate profit + corporate taxes + indirect taxes + depreciation – transfer payment.

Q. 23. What is the full form of NNP?

Ans. Net National Product or National income at Market price.

Q. 24. The amount of money available with the private individual to spend is called?

Ans. Disposable Income (DI)

Q. 25. Which is the most effective way to overcome the defects of agriculture marketing?

Ans. Regulated Marketing [*JRF 1998*]

Q. 26. What are the five important functions of money?

Ans. (*i*) Medium of exchange.

(*ii*) A limit of Account.

(*iii*) Standard of deffered payment

(*iv*) Store of value.

(*v*) Transferable.

Q. 27. The ratio of normal GNP to real GNP which can be interpreted as comprehensive price index is called

Ans. GNP Deflator

Q. 28. A sustained rise of less than 3 percent of price per annum is called

Ans. Creeping inflation.

Q. 29. What is the full form of MRTS?

Ans. Marginal Rate of Technical Substitution.

Q. 30. What is the formula for marginal rate of substitution?

Ans. $$MRTS = \frac{\text{Number of units of replaced product}}{\text{Number of units of replacing product}}$$

Q. 31. The presence of a single buyer for the products produced by the firm is called?

Ans. Monopsony

Q. 32. The nature of agricultural economics is

Ans. Ideal science and Art.

Q. 33. The number of divisions of agricultural economics?

Ans. Five

Q. 34. One of the method to deal with seasonal unemployment in villages is

Ans. Development of Kutir and local Industries.

Q. 35. The reason for poverty of people in prosperous India is

Ans. Lack of proper use of natural resources.

Q. 36. The 'Poor people live in prosperous India' was given by

Ans. Dr. Veer Anstey

Q. 37. Liquidity preference theory was given by

Ans. J. M. Keynes

Q 38. When was Planning Commission established ?

Ans. In the year 1950.

Q. 39. First Indian all Women Bank

Ans. Syndicate Bank, Bengaluru, 1962, (Seshadripuram)

Q.40. First Regional Rural Bank to launch ATM

Ans. Pallavan Grama Bank

Q. 41. 'Key Village Scheme' was started in five year plan?

Ans. First plan

Q. 42. One who gets profit from money inflation is

Ans. The person taking loan.

Q. 43. In India maximum agricultural area is irrigated by the

Ans. Canals.

Q. 44. The maximum number of tube well or pumping set in India are found in the state.

Ans. Tamil Nadu.

Q. 45. The wholesale business of wheat in India was nationalised in the year

Ans. 1973

Q. 46. When was Community Development Programme started?

Ans. 2nd October 1952.

Q. 47. Levels in Cooperative Organisation in India?

Ans. Three.

Q. 48. The bank in India having largest number of branches is

Ans State Bank of India.

Q. 49. The Rural Bank takes loan from

Ans. Lead Bank.

Q. 50. The agricultural cost is equal to

Ans. Sum of fixed and variable cost.

Q. 51. Agricultural cost and price commission was established

Ans. In the year 1965

Q. 52. When was Food Corporation of India (FCI) established ?

Ans. In January 1965.

Q. 53. Regional Rural Banks were established in the year

Ans. 2nd Oct. 1975 (Muradabad, Gorakhpur, Bhiwani, Jaipur and Malda).

Q. 54. When was Indian Reserve Bank established ?

Ans. On First April in the year 1935. First Governor of RBI- Sir Osborne Smith (1935-37). First Indian Governor of RBI- Sh. CD Deshmukh (1943-1949). First Governor of Free India- Sir Bengal Rama Rau (1st July 1949 to 14th Jan., 1957).

Q.55. First Indian Bank to offer ATM

Ans. Bank of India at Mumbai

Q. 56. When were 14 main banks nationalised ?

Ans. 19th July 1969 (Bank having about 85% of deposit in the country)

Q. 57. Where is World Bank located ?

Ans. Washington D.C. (World Bank Group constituents IDA, IFC, IBRD, MIGA and ICSID)

IDA: (International Development Association), IFC: (International Finance Corporation), IBRD: (International Bank for Reconstruction and Development), MIGA: (Multilateral Investment Guarantee Agency), ICSID: (International Centre for Settlement of Investment Disputes).

Q. 58. If the price of a commodity increases its demand will

Ans. Decrease

Q. 59. Welfare definition of Economics was given by

Ans. Alfred Marshal (*ADA, Indore 2009*)

Q. 60. Agriculture Insurance company of India Ltd. started on

Ans. 20th Dec., 2002

Q. 61. Asian Development Bank (ADB) was established in the year

Ans. 19th December, 1966 at Manila (Philippines)

Q. 62. A line or curve connecting the least cost combination of inputs for all output levels is known as

Ans. Isocline

Q. 63. Inflation is a situation

Ans. When money supply grows at higher rate than GDP in real terms

Q. 64. Measure of inflation from supply side

Ans. Increase in supply of products or commodities

Q. 65. Present RBI Governor

Ans. Shakti Kanta Das (Since 12th Dec. 2018)

Q. 66. Sum of market value of all final goods and services produced within domestic territory of the country during an accounting year both by Indians and Foreigners residing within Indian boundary is called.

Ans. Gross Domestic Product (GDP)

Q. 67. The full form of PPP is

Ans. Purchasing Power Parity

Q.68. First co-operative bank in India

Ans. Greater Bombay Co-operative of Bank Ltd., in 1952

Q. 69. The full form of NDC is

Ans. National Development Council

Q. 70. The full form CACP is

Ans. Commission for Agricultural Costs and Prices

Q. 71. What is Pigou effect

Ans. It is a term in economics referring to relationship between consumption, wealth, employment and output during periods of deflation. Pigou effect states that when there is deflation of prices, employment (and thus output) will be increased due to an increase in wealth (and thus consumption).

Q. 72. What is the full form of SFAC?

Ans. Small Farmers' Agri-Business Consortium

Q. 73. What is the full form of NAFED?

Ans. National Agricultural Cooperative Marketing Federation of India Limited.

Q. 74. The first five year plan draft outline was issued by the planning commission in the year

Ans. 1951

Q. 75. In which branch of economics the study of farm management comes?

Ans. Microeconomics

Q. 76. The three broad classifications of marketing function are

Ans. 1. Function of exchange 2. Function of physical supply, 3. Facilitating functions.

Q. 77. The full form of the PACS is

Ans. Primary Agricultural Credit Societies

Q. 78. The full form of DCCB is

Ans. District Central Co-operative Bank

Q. 79. The average size of an self help group (SHG) is

Ans. 15 people

Q. 80. SHG Bank linkage programme was started on

Ans. 31/12/2005

Q.81. First ISO-9002 certified regional rural bank

Ans. Devi Patan Kshetriya Gramin Bank of Gonda (UP) (Sponsored by PNB)

Q. 82. The maximum net revenue is obtained when marginal cost equals the

Ans. Price of product

Q. 83. Three physical factors determining the type of farming are

Ans. Climate, soil, topography

Q. 84. Two basic classification of farm labour are

Ans. 1. Unpaid labour 2. Paid labour (Hired)

Q. 85. Three kind of farm wages are

Ans. 1. Time basis 2. Price basis 3. Share basis

Q. 86. Labour efficiency in agriculture refers to

Ans. The amount of productive work accomplished per man on the farm per unit of time.

Q. 87. Four types of co-operating farming are

Ans. 1. Co-operative better farming
2. Co-operative joint farming
3. Co-operative collective farming
4. Co-operative tenant farming

Q. 88. The ordinary labour employed for manual work, which does not need any training of specialised nature is called ________

Ans. Unskilled labour

Q. 89. A farm on which no single product source of income equals as much as the totals receipts is called as ________

Ans. Diversified or general farm

Q. 90. Five types of farming are

Ans. 1. Specialised farming, 2. Diversified farming, 3. Mixed farming, 4. Ranching, 5. Dry farming

Q. 91. First organised study in the field of farm cost accounting was initiated in ________ by the ________

Ans. 1923-24, Punjab Board of Economic Enquiry

Q. 92. The Agro-economic research centres was established in ________

Ans. 1954-55

Q. 93. First credit card for Farmers

Ans. Pattabhi Agricard on 30th Aug. 1998. (Andhra Bank supported by SBI)

Q. 94. Average fixed costs (AFC) = Total fixed cost (TFC) + ________________

Ans. Cost of number of output units

Q.95. First ATM on wheels for villages

Ans. Karnataka Vikas Grameena Bank launched on wheel service GVG with ATM on 4th July 2017 with association from NABPRD

Q. 96. Total variable cost (TVC) = Total Cost (TC) – (________________)

Ans. Total fixed cost (TFC) or sum of variable cost

Q. 97. When demand curve is flatter then demand is

Ans. Relatively elastic

Q. 98. Profit is maximum when marginal product = __________

Ans. Factor product price ratio

Q. 99. A line or curve connecting the least cost combination of inputs for all output levels is known as

Ans. Isoclines

Q. 100. The border lines which separate complementarity from substitution is called

Ans. Ridge lines

Q. 101. The rate of change in quantity of one product as a result of an unit increase in the other product when the amount of input used remains constant is called

Ans. Marginal Rate of Product Substitution (MRPS)

Q. 102. The percentage decrease in output of one product associated with a given percentage increase in second product is called

Ans. Elasticity of Product Substitution.

Q. 103. The line defining all possible combinations of two commodities which would yield an equal revenue is called

Ans. Iso-revenue line

Q. 104. If marginal rate of substitution is small than zero the enterprice relationship is

Ans. Competitive

Q.105. First Institution for Agricultural Credit

Ans. Agricultural Refinance Corporation 1963

Q.106. First RuPay Kisan Card

Ans. Bank of Maharashtra at Pune on Nov., 2012.

Q. 107. Antagonism is the opposite of ______ relationship

Ans. Complementary

Q. 108. The size of farm is generally understood in terms of

Ans. Physical area, volume of production and value of production

Q. 109. Cropping pattern is expressed at a ______ level *i.e.* __________, ______ or ________ level.

Ans. micro, district, taluk, village

Q. 110. A specialized farm is one where 50% or more of income is derived from

Ans. One single source

Q. 111. The income that could have been received, if the input have been used in its most profitable alternative use is called

Ans. Opportunity cost

Q. 112. The market condition where in the entire supply of a commodity is concentrated in the hands of a single firm is called

Ans. Monopoly

Q. 113. The market situations in which there are only two sellers is called

Ans. Duopoly

Q. 114. The presence of a few firms in the market, producing either a homogeneous product or products which are close but not perfect substitution to each other is called

Ans. Oligopoly

Q. 115. The rent received exclusively from use of land only is called
Ans. Economic rent

Q. 116. The proponent of Liquidity Preference Theory is
Ans. J. M. Keynes

Q. 117. Dynamic theory of profits was postulated by
Ans. J.B. Clark (an American economist)

Q. 118. Innovation theory was postulated by
Ans. Joseph A. Schumpter, (an American Economist).

Q. 119. Wage theory of profits was advocated by
Ans. Prof. Taussig, (an American economist)

Q. 120. Which economist awarded Nobel prize for microcredit
Ans. Muhammad Yunus

Q. 121. Population theory of demographic transition was given by
Ans. C.P. Blacker and Frank Notestein

Q. 122. Fully employed situation is seen in
Ans. Recession

Q. 123. State first accorded status of Industry to Agriculture
Ans. Maharashtra in 1997

Q. 124. Mahalwari system of land tenure during British period was started by
Ans. Lord William Bentinck

Q. 125. Law of substitution was propounded by
Ans. Austrian economist Hermann Heinrich Gossen

Q. 126. Agriculture census is carried out
Ans. Once in 5 years (First census carried out in the year 1970)

Q. 127. Percent population under poverty line in India
Ans. 26%

Q. 128. Gini coefficient is a measure of
Ans. Income inequality among population

Q.129. First Govt. owned women bank
Ans. Bhartiya Mahila Bank 2013

Q. 130. Theory of profit was given by-

Ans. A. Walker

Q. 131. Father of cooperative movement in India was-

Ans. F. Nicholson

Q. 132. First mortgage bank was established in-

Ans. 1920, (Jhang, Punjab)

Q. 133. First crop loan was given to-

Ans. Cotton (Berar, Maharashtra)

Q. 134. Period for short term loan-

Ans. 1 to 1.5 Years

Q. 135. Period for long term loan-

Ans. 5 to 30 Years

Q. 136. Loan which is given to purchase for Agri Equipments & Animals is-

Ans. Medium term loan

Q. 137. Tractor Loan is a type of loan-

Ans. Medium term loan

Q. 138. Loan which is given to purchase for expensive Agri Equipments is-

Ans. Long term loan

Q. 139. Crop loan is termed as a-

Ans. Short term loan

Q. 140. Committee on Rural Banks was headed by-

Ans. M. Narasimhan (1975)

Q. 141. Land holding for marginal farmer is-

Less than 1 ha

Q.142. Land holding for large farmer-

Ans. 10 ha or above

Q. 143. Land holding for small farmer-

Ans. 1-2 ha

Q. 144. Land holding for Medium farmer-

Ans. 4-10 ha

Q. 145. Land holding for semi- medium farmer-

Ans. 2-4 ha

Q. 146. Cost which is known as sunk cost-

Ans. Fixed cost

Q.147. Short term credit is given for the purpose of-

Ans. Crop production

Q. 148. Which price is announced by Govt. on the recommendation of CACP before sowing of crops.

Ans. MSP (Minimum Support Price)

Q. 149. Which price is announced by Govt. on the recommendation of CACP before harvesting of crops-

Ans. Procurement Prices (PP)

Q.150. Which mark is given to ensure that product is environment friendly-

Ans. ECO MARK (1991)

Q. 151. Formula of Net Capital Ratio-

Ans. $= \dfrac{\text{Total Assets}}{\text{Total Liabilities}}$

Q.152. Formula for Working Capital Ratio-

Ans. $= \dfrac{\text{Working assets and Current assets}}{\text{Intermediate \& Current Liabilities}}$

Q.153. Formula for Current Ratio-

Ans. $= \dfrac{\text{Current Assets}}{\text{Current Liabilities}}$

Q.154. Formula for Debt- equity ratio-

Ans. $= \dfrac{\text{Deferred liabilities}}{\text{Net worth}}$

Q.155. Formula for Net Income per ha.-

Ans. $= \dfrac{\text{Total returns from fixed resources}}{\text{Total area in hectare}} \times 100$

Q.156. Formula for Rate of Capital Turn Over-

Ans. $= \dfrac{\text{Gross Income}}{\text{Total Farm Income}} \times 100$

Q.157. Formula for Gross Ratio-

Ans. $= \frac{\text{Total Expenses}}{\text{Gross Income}} \times 100$

Q.158. Formula for Fixed Ratio-

Ans. $= \frac{\text{Fixed Expenses}}{\text{Gross Income}}$

Q.159. Formula for Operating Ratio-

Ans. $= \frac{\text{Operating Expenses}}{\text{Gross Income}}$

Q.160. Formula of Production efficiency-

Ans. $= \frac{\text{Yield per acre}}{\text{Yield of locality}} \times 100$

Q.161. Formula for Rate of Turn Over-

Ans. $= \frac{\text{Gross Income}}{\text{Total Assets}} \times 100$

Q.162. Formula for Operating Cost Ratio-

Ans. $= \frac{\text{Total Operation Cost}}{\text{Total Profit}}$

Q.163. BIS (Bureau of Indian Standards) was set up on-

Ans. 1st April, 1987

Q.164. Organization which promote a universal standard for facilitating international exchange of goods & Services is-

Ans. ISO (International Organization for Standardization) Estd. on 23rd Feb. 1947.

Q.165. A National Bank which provide indirect finance to rural communities/ farmers-

Ans. NABARD

Q. 166. Organization which is responsible to develop the system of forest product marketing for tribal communities is-

Ans. TRIFED (1987)

Q. 167. Cooperative movement was started in India in-
Ans. 1904

Q. 168. Important fixed cost on farm is-
Ans. Land rent

Q.169. Most important principle in farm management is-
Ans. Law of diminishing return

Q.170. Risk bearing ability depends on-
Ans. Net worth

Q.171. The share of capital is referred as a-
Ans. Interest

Q.172. In case of Fruit & Vegetables the marketed surplus is nearly about-
Ans. 80%

Q.173. Law of diminishing returns operates when
Ans. One input is variable

Q.174. As per RBI norms how much percent net bank credit should be deployed in agriculture-
Ans. 18%

Q.175. Most common method in regulated market is-
Ans. Open auction system

Q.176. Technique which protect the trades from extreme falls in price is called-
Ans. Hedging

Q.177. The market where permanent/ durable commodities are traded known as-
Ans. Secular market

Q.178. Example of complementary enterprises is-
Ans. Crop farming + Milk production

Q.179. Agriculture produce (Grading & Marketing) act was passed in-
Ans. 1937

Q.180. Co- operative credit societies act was passed in-
Ans. 1904

Q.181. The tax which is collected by State Govt. on an Intra- State sale is-
Ans. SGST

Q.182. The tax which is collected by Central Govt. on an Intra- State sale is-
Ans. CGST

Q.183. The tax which is collected by Central Govt. for Inter- State sale is-
Ans. IGST

Q.184. The income elasticity of demand for inferior goods is
Ans. Negative

Q.185. In India Capitalistic farming is restricted for the crops-
Ans. Tea, Coffee & Rubber

Q.186. Farming having major object for research, demonstration and quality seeds known as-
Ans. State Farming

Q.187. Major Corporate farming states in India are-
Ans. Maharashtra & Tamilnadu

Q.188. Agricultural operation method in Cooperative better farming is-
Ans. Individual type

Q.189. Organization which is responsible for quality control of Agri commodities-
Ans. DMI (Directorate of Marketing & Inspection, Set up 1935)

Q.190. (NCUI) National Co-operative Union of India established in the year
Ans. 1929

Q.191. National Co-operative Development and Warehousing Board was established in
Ans. 1956

Q.192. (NCDC) National Cooperative Development Corporation was established in the year
Ans. 1963 (Head Quarter. New Delhi)

Q.193. National Agricultural Cooperative Marketing Federation (NAFED) was established in the year
Ans. 1958 (H.Q. New Delhi)

Q.194. Central Warehousing Corporation was started in the year
Ans. 1957 (Corporate office New Delhi).

ENVIRONMENTAL SCIENCE AND METEOROLOGY

Q. 1. The region inhabited by living organism on land, ocean and atmosphere.

Ans. Biosphere also called ecosphere (term coined by Edward Suess in 1875)

Q. 2. Name the plants which produce food themselves through photosynthesis—

Ans. Autotrophs

Q. 3. In which eco-system, maximum number of producers in an unit area

Ans. Forest ecosystem

Q. 4. Primary consumers in aquatic system are—

Ans. Zooplankton

Q. 5. The first major human source of air pollution was:

Ans. Fire

Q. 6. Movement of matter within or between ecosystem is called:

Ans. Biochemical cycle

Q. 7. Most poisonous pollutant in drinking water is

Ans. Arsenic

Q. 8. Swine flu disease (in humans) is caused by virus

Ans. H1 N1 virus

Q. 9. Colourless odourless and non-corrosive air pollutant is

Ans. SO_2 (Sulfur dioxide)

Q. 10. Sphere of Water is called.

Ans. Hydrosphere (70% of globe)

Q. 11. The micro-bacteria (organism) which live only in presence of free oxygen:

Ans. Aerobic

Q. 12. The microbial count of potable water should be

Ans. < 100Cfu/ml

Q. 13. Microorganism responsible for the conversion of CO to CO_2

Ans. *Acetobacter*

Q. 14. Sources of soil pollutants are–

Ans. Herbicides, Pesticides and Weedicides

Q. 15. Soil erosion can be prevented by

Ans. Contour farming, Strip farming and Terracing

Q. 16. Following is responsible for ozone layer depletion?

Ans. CFC, CO_2, SO_2 and Methane

Q. 17. DDT is an —

Ans. Nondegrable agro-chemical used as pesticide (Its use is banned in India in 1985)

Q. 18. Term ecology was first defined by

Ans. German biologist Ernest Haeckel in the year 1866

Q. 19. Maximum concentration of ozone is found in

Ans. Lower stratosphere (*JRF, 2011*)

Q. 20. Global warming is mainly due to

Ans. Increase in Green House Gases (GHG) and deforestation

Q. 21. Kind of forest found maximum in India?

Ans. Tropical dry deciduous forest

Q. 22. Forest conservation act was made in the year

Ans. 1980

Q. 23. Name the test which has self purification capacity of water body

Ans. BOD test

Q. 24. Salt efflorescence

Ans. Accumulation of excessive salt in the soil system which makes soil unfit for cultivation.

Q. 24a. Fluoride pollution mainly affects?

Ans. Teeth (Tooth decay)

Q. 25. Hydro-carbons present in the air are :

Ans. Primary air pollutants (Hydro-carbon are : aromatics, napthenes, olefines and paraffins).

Q. 26. Name metal toxic in drinking water even at low concentration

Ans. Chromium

Q. 27. Bacteria that obtain both energy and material from organic sources

Ans. Heterotrophs

Q. 28. The specific gravity of sewage is

Ans. Slightly more than 1

Q. 29. Most commonly used disinfectant in water purification

Ans. Chlorine.

Q. 30. Instrument used for measuring the concentration of ozone in air is

Ans. Ozonometer

Q. 31. Wind speed is measured by

Ans. Anemometer

Q. 32. The term ecosystem was coined and developed by

Ans. Arthur G,. Tansley (1935)

Q. 33. The organisms who eat dead organic matter are

Ans. Decomposers (Bacteria and Fungi).

Q. 34 Upright pyramid of biomass in forest eco-system is

Ans. Producer-herbivores-carnivores

Q. 35. Upright pyramid of numbers of grassland ecosystem

Ans. Producer-herbivorous-secondary consumers-top consumer (carnivorous)

Q. 36. Concept of ecological pyramid was first introduced by :

Ans. Charles Elton in 1927.

Q. 37. World environment day is celebrated on

Ans. 5th June

Q. 38. Gases responsible for acid rain. *(ARS 2005)*

Ans. SO_2 and NO_2

Q. 39. Reduction of nitrates and nitrites to nitrogen by microorganisms

Ans. Denitrification

Q. 40. A wooded wetland in which water is near or above ground level.

Ans. Swamp

Q. 41. Rapid growth of algae in surface waters due to increase in inorganic nutrients.

Ans. Algal bloom

Q. 42. A chemical or physical agent that causes cancer

Ans. Carcinogen

Q. 43. Accumulation of nutrients in a lake or pond due to human intervention or natural process is

Ans. Eutrophication

Q. 44. Agriculture in which several plant species are grown simultaneously to reduce insect infestation and disease

Ans. Heteroculture

Q. 45. Troposhere

Ans. It is upto 15 km from the earth surface upwards

Q. 46. Stratosphere

Ans. 15–45 km from the surface of the earth upwards

Q. 47. Mesosphere

Ans. Is the layer between stratosphere and thermosphere.

Q. 48. Ionosphere (Thermosphere)

Ans. It rest above the mesosphere, upto the height of about 130 km earth's surface.

Q. 49. Intertidal zons

Ans. Area of shoreline between low and high

Q. 50. Pollutant free alternative to petrol for automobiles.

Ans. Propane

Q. 51. The gas does not pollute air.

Ans. Nitrogen Dioxide

Q. 52. The core of Lithosphere mainly consists of

Ans. Ca and Mg

Q. 53. Microorganism responsbile for formation of H_2S

Ans. *Desulfovibrio vulgaris*

Q. 54. Which gas contributes least in green house

Ans. NO_2 (*JRF, 2012*)

Q. 55. The gas responsible for Bhopal gas tragedy.

Ans. Methyl isocyanate (Tragedy occurred 2-3 Dec. midnight 1984)

Q. 56. Fungi are plant that lack.

Ans. Chlorophyll pigment

Q. 57. Biodegradable water can be converted to useful substances with the help of

Ans. Earthworms

Q. 58. Land that can be cultivated to grow crops

Ans. Arable land

Q. 59. Species way of life in a community and includes everything that affects its survival and reproduction

Ans. Niche or ecological niche

Q. 60. Algae found in almost all terrestial environment have potential agricultural applications

Ans. They reduce soil erosion reclamation of soil P, biocontrolly agricultural pests etc.

Q. 61. Place or type of place where an organism or a population of organism lives.

Ans. Habitat

Q. 62. In a polluted lake Index of pollution is

Ans. Daphnia

Q. 63. What is climate forcing

Ans. It refers to a change in the earth's energy balance, leading to either a warming or cooling. Due to increase in atmospheric concentrations of greenhouse gases produces a positive climate forcing (warming).

Q. 63a. Greenhouse gases (According to WMO)

Ans. Carbon Dioxide (CO_2), Methane (CH_4), Nitrous Oxide (N_2O), Halocarbons and SF_6, and Molecular Hydrogen (H_2)

Q. 64. Seamounts are

Ans. Submerged oceanic volcanoes.

Q. 65. *Euphoribia lathyris* a potential source petroleum

Ans. Gasoline plant

Q. 66. Founder of Kenyan greenbelt movement

Ans. Wangari Mathai

Q. 67. Author of famous book 'Silent spring'

Ans. Rachel Carson. (Book Published on 27th Sep. 1962)

Q. 68. Author of famous book 'Small is beautiful'

Ans. Earnest Schumacher (1911–1977)

Q. 69. Mitti Bachao Andolan launched at

Ans. Hoshangabad (M. P.)

Q. 70 Ultra violet radiations are absorbed by

Ans. Ozone layer (*ARS 2005*)

Q. 71. A tropical grassland, usually scattered with trees or shrubs.

Ans. Savanna.

Q. 72. Carbon cycle

Ans. The process of removal and uptake of carbon on a globai scale.

Q. 73. Photochemical smog consists of

Ans. O_3, PAN (peroxyacetyl nitrate) and NOx (NO_2 and NO)

Q. 74. Plants which mostly emits terpenes are

Ans. Eucalyptus and Oak

Q. 75. Mean residence time of methane gas is about

Ans. 7-10 years

Q. 76. CO_2 content in atmosphere is

Ans. 0.03% (*JRF 2012*)

Q. 77. Green house gases are (*ARS/NET, 2009*)

Ans. CO_2, CH_4 (methane), O_3 and CFC's

Q. 78. Chilika lagoon lake is situated in the state

Ans. Odisha (*ARS/NET 2009*)

Q. 79. Unlimited resources are

Ans. Climate with wind energy, and solar energy

Q. 80. Eutrophication is the cause of

Ans. Excessive nutrients in water body by human beings

Q. 81. The first stage of soil erosion is

Ans. Sheet erosion

Q. 82. National Environmental Awareness Campaign (NEAC) or National Environmental Month (NEM) is observed

Ans. Nov. 19 to Dec., 18 every year since 1986

Q. 83. World Environment Day is observed

Ans. June 5th every year. (JRF 2010)

Q. 84. Permissible limit of iron in drinking water is

Ans. 1 ppm

Q. 85. Cadmium pollution is associated with disease

Ans. Itai-Itai (*ARS*)

Q. 86. Maximum limit of chromium in drinking water recommended by WHO

Ans. 0·05 mg/litre

Q. 87. Amount of copper essential in small quantity for plant, animal and human beings is

Ans. 2 mg per day

Q. 88. World's Driest place is

Ans. The Atacama desert in Chilli. It has only 0.51 mm of rain in a year.

Q. 89. El-Nino (associated with a warning of central and eastern tropical pacific) and La-Nina are

Ans. Reverse process

Q. 90. Kyoto protocol is related to

Ans. Green House Gas minimization

Q. 91. Which organism is responsible for production of methane from biogas

Ans. Methanobacteria (*JRF 2012*)

Q. 92. National Diversity Board is situated at

Ans. Chennai (*JRF 2011*)

Q.92a. Blue baby syndrome is due to

Ans. NO_3 Toxicity

Q.92b. International Panel for Climate Change (IPCC)

Ans. Year 1988

Q. 93. Biological Diversity Act

Ans. 2002

Q.93a. Wildlife Protection Act.

Ans. 1972

Q.93b. National Environment Policy

Ans. 2006

Q.93c. World Watch Institute started by

Ans. Lester Brown (An Agricultural Economist)

Q.94. Balram yojna is related to

Ans. Water management

Q.95. How much noise can be tolerated by human ear

Ans. 80 decibels

Q.96. Developing technology of waste clean up applicable to leaking storage tanks and spills waste is

Ans. Bioremediation

Q.97. Pyramid upright in all ecosystems

Ans. Pyramid of energy

Q.98. Tree best suitable for wetlands

Ans. Eucalyptus

Q.99. Fluoride pollution mainly affects

Ans. Teeth

Q.100. Biochemical oxygen demands for?

Ans. Dissolved oxygen needed by the microbes to decompose the organic waste

Q.101. The increased rate of skin cancer and mutation is due to?

Ans. Ozone depletion

Q. 102. Name the disease which is caused by eating fish which inhibits mercury contaminated water?

Ans. Minamata Disease (A disease caused by mercury pollution in Japan in 1953.)

Carbon dioxide compensation point

The point where uptake of CO_2 through photosynthetic pathways is equal to respiratory release of CO_2 at this point plant does not absorb CO_2 from surrounding.

Q. 103. Which plant is used as an indicator of SO_2 pollution in air?

Ans. Epiphytic lichens

Q. 103a. National Tree of India

Ans. Banyan tree

Q.104. Which of the following built plants contribute to the maximum risk of the environment?

Ans. Nuclear Plants

Q.105. Iti-Iti disease is caused due to

Ans. Cadmium

Q.106. Which green house gas is being generated in agricultural fields?

Ans. Nitrous Oxide/Methane

Q.107. Contamination of water due to the presence of coliform bacteria is because of?

Ans. Mammalian excreta

Q.108. HPLC stands for?

Ans. High performance liquid chromatography

Q.109. How is phosphate removed through precipitation from waste water?

Ans. Lime or Alum

Q.110. Name the country in which the first plant based remediation system was installed over 300 years ago?

Ans. Germany

Q.111. Which country is the highest per capita emitter of carbon dioxide in the world?

Ans. Qatar

Q.112. In Nitrogen Cycle, soil nitrates are transformed into free nitrogen by?

Ans. Denitrifying bacteria.

Q.113. Greatest sink of carbon on the earth

Ans. Oceans

Q.114. Green wave stands for

Ans. Youth programme to create awarness about biodiversity and reduce its loss in future.

Q.115. El-Nino phenomenon leads to rise in

Ans. Ocean temperature and dry weather and drought like situation is created

Q.116. Greenhouse effect coined by the scientist

Ans. Joseph Fourier in the year 1824

Q.117. Photo chemical smog sometimes called as

Ans : Brown Air

Q.118. pH of rain water is about

Ans; 5-6

Q.119. Green House effect given by

Ans. Joseph Fourier (1824)

Q.120. Mitigation measures

Ans. Carbon capture and storage

Q.121. Koyoto Protocol

Ans Agreement between countries for reducing green house gases (UNFCCC in 1997) India signed in year Aug., 2002)

Q.122. Working group on climate change

Ans; IPCC (Intergovernmental Panel on Climate Change) (Setup in 1988)

Q.123. What are sinks

Ans. Reservoir that uptakes chemical elements from another part of the its natural cycle

Q.124. Bioaccumulation

Ans. Process by which a substance or chemical first enter through food chain

Or Process by which certain toxic substances (such as heavy metals and polychlorinated biphenyls) accumulate and keep on accumulating in living organisms, posing a threat to health, life, and to the environment

Bioaccumulation occurs when an organism absorbs a substance at a rate faster than that at which the substance is lost by catabolism and excretion.

Bio-magnification: Steady increase in the concentration of a substance from one level of food chain to next

Q.125. The substances that biomagnifies and that is resistant to biodegradation and thus remains in the environment for a long period of time.

Ans : Persistent Organic Pollutant

125a. NICRA (National Innovation on Climate Resilient Agriculture)

Ans. Launched in (Feb. 2011) by ICAR

Meteorology

Q. 126. IMO (International Meteorological Organization) established in year

Ans. 1873

Q. 127. NMM (National Monsoon Mission) rolled out in

Ans. 2012 (By IMD to improve monsoon forecast

Q. 128. IMD (Indian Meteorological Department) was established in the year

Ans. 1875

Q. 129. Indian Institute of tropical meteorology (IITM)

Ans. Pune (1962)

Q. 130. Meteorology is derived from

Ans. Greek word meteros means high in the sky

Q. 131. Famous book "book of signs" on weather forecasting was compiled by

Ans. Greek scientist Theopharstus

Q. 132. What is Albedo

Ans. Reflected solar radiations without any change in quality

Q. 133. Solar radiations are measured by

Ans. Pyranometer

Q. 134. Atmospehric pressure is measured by

Ans. Aneroid Barometer

Q.135. Relative humidity is measured by

Ans. Psychrometer

Q. 136. Rainfall is Measured by

Ans. Raingauge

Q.136a. India's first recorded rainfall data was obtained at

Ans. Kolkata (1784)

Q. 137. Wind speed is measured by

Ans. Anemometer

Q. 138. Strongest recorded wind
Ans. In USA in April 1934. The record wind speed was 371 kmph (231 mph)

Q. 139. Chemical used for artificial rainfall
Ans. Silver Iodide (Ag I) for *cold clouds seeding* while common salt, Urea and ammonium nitrate for *warm cloud seeding.* (*ARS 2010*)

Q. 140. Blue colour of sky & red colour of sunset is due to
Ans. Dispersion

Q. 141. PAR (Photosynthetic active radiation) measured in
Ans. Einstein units

Q.142. PAR is the radiation of wavelength
Ans. 400 to 700 lanometre. (*ARS 2010*)

Q.143. Albedo is highest for
Ans. Transpiration

Q.144. Atmospheric phenomenon like fog and dew require a minimum percent relative humidity
Ans. 75%

Q.145. Average albedo of the earth
Ans. 0.30

Q. 146. Satellites operate at the height above earth surface
Ans. 36,000 km

Q.147. Father of agricultural meteorology in India
Ans. Dr. L.A. Ramdas (1900-1971)

Q.148. NCMRWF
Ans. National Centre for Medium Range Weather Forecasting

Q.159. CO_2 concentration in the atmosphere is.
Ans. 0.03% (*JRF, 2012*)

Q.160. Invisible radiant energy
Ans. Electromagnetic radiation with longer wavelength than visible light

Q.161. Ionizing radiation
Ans. Radiation that carries enough energy to liberate electrons from atoms or molecules

Q.162. Hydrology

Ans. The branch of science concerned with the properties of the earth's water, and especially its movement in relation to land.

Q.163. Evapotranspiration

Ans. Evapotranspiration is the plant's total water use, it includes, the actual water loss in transpiration as well as water evaporated from the surface of the leaves or from the soil surrounding the plant.

Typical ET values of some plants in a year

1. Alfalfa–	79-91 cm	2. Soybean–	51-64 cm
3. Sorghum–	46-58 cm	4. Sugar beet–	61-66 cm

New Additional Questions

Q. 1. Lower most layer of atmosphere (Nearest to earth)

Ans. Troposphere

Q. 2. Extending height of Troposphere form earth surface-

Ans. 8 to 18 Km.

Q. 3. How many ^{0}C temp decrease with increase altitude per km in troposphere-

Ans. 6 ^{0}C/Km

Q. 4. Which atmosphere layer influence earth climate-

Ans. Troposphere

Q. 5. In which atmosphere layer all weather phenomena occurs-

Ans. Troposphere

Q. 6. Atmosphere layer which is warmest layer and having uniform temp with altitude-

Ans. Stratosphere

Q. 7. Atmosphere which is also known ozonosphere-

Ans. Stratosphere

Q. 8. Lower layer and outer most layer of Thermosphere is called as-

Ans. Ionosphere

Q. 9. Outer most layer of Thermosphere is called as-

Ans. Exosphere

Q. 10. Value of solar constant

Ans. 1.94cal/cm^2/min

Q. 11. In cyclone atmospheric pressure is lowest at

Ans. Center region

Q. 12. Which climate classification method adopted by ICRISAT for classify semi arid tropic in India

Ans. Troll method

Q. 13. Climate classification method adopted by AICRP on dryland is

Ans. MDI (Moisture Deficit Index) method

Q. 14. MDI values for Sub humid area

Ans. 0 to 33.3

Q. 15. MDI values for Semi Arid areas

Ans. – 33.3 to – 66.6

Q. 16. MDI Value for Arid areas

Ans. > – 66.6

Q. 17. Unit of absolute humidity is

Ans. g/m^3

Q. 18. Unit of specific humidity is

Ans. g/kg.

Q. 19. Drop size of > 0.5 to 6mm called as

Ans. Rain

Q. 20. Drop size of < 0.5 mm called as

Ans. Drizzle

Q. 21. Melting and boiling point of water in Fahrenheit is

Ans. 32 °F and 212 °F

Q. 22. Absolute temperature is measured in

Ans. Kelvin units

Q. 23. Melting and Steam point of water in Kelvin is

Ans. 273° K and 373 °K

Q. 24. Golden yellow colour tag

Ans. Breeder seed

Q. 25. White tag

Ans. Foundation seed

Q. 26. Purple tag shows the seed category of-

Ans. Registered Seed

Q. 27. Truthfully labelled seed

Ans. Opal green

Q. 28. Blue tag shows the seed category of-

Ans. Certified Seed

Q.29. Indole acetic acid inhibits growth of the

Ans. Roots

Q.30. Height of sugarcane plant can be increased by application of

Ans. Gibberellins

IMPORTANT METEOROLOGICAL INSTRUMENTS AND THEIR USES

Instruments		*Uses/Measures*
Altimeter	:	Height
Anemometer	:	Wind speed/velocity
Aneroid barometer	:	Atmospheric pressure
Auxanometer	:	Growth of plant
Barograph	:	Continuous atmospheric pressure
Cambel stokes recorder	:	Sunshine duration
Crescograph	:	Growth of plant
Drosometer	:	Dew
Hygrometer/Psychrometer	:	Relative humidity (RH)
Evaporometer	:	Evapotranspiration
Irrometer	:	Water stress, soil moisture tension
Infiltrometer	:	Infiltration
Lactometer	:	Fats % in milk
Lysimeter	:	Evapotranspiration
Pyrheliometer	:	Direct solar radiation
PAR	:	Quantum sensor
Psychrometer	:	Leaf water potential/RH
Porometer	:	Transpiration rate (ET)
Pyranometer	:	Total incoming solar radiation
Peizometer	:	Depth of water table
Rain gauge	:	Amount of rainfall
Tensiometer	:	Soil moisture tension (0.8 bar)
Wind vane	:	Wind direction
Tensiometer	:	Soil moisture tension (0.8 bar)
Wind vane	:	Wind direction

FISHERIES

Q. 1. Two main zones of a sea

Ans. 1. Benthic Zone 2. Palegic Zone

Q. 2. What is Benthic Zone ?

Ans. It is the deepest part of Ocean or its area is from sea floor to the tidal effected areas.

Q. 3. Divisions of Benthic Zone

Ans.1) *Supralittoral* : The region is the area above the spring high tide line on esturies and coastline that is regularly splashed.

2) *Littoral* : Region between low tide and high tide

3) *Sub Littoral* : Region upto 200–1000m

4) *Abyssal* : Region from 1000–4000m

5) *Hadal* : 4000m onwards

Q. 4. What are mangroves ?

Ans. These are diverse group of salt tolerant trees that grow along warm, calm marine coasts.

Q. 5. What are esturies ?

Ans. These are bays where rivers empty into the sea.

Q. 6. What are marshs?

Ans. These are wetlands without trees

Q. 7. What are swamps?

Ans. Wetlands with trees.

Q. 8. Pelagic Zone

Ans. It is the part of the ocean or sea that is not near the coast or sea floor

Q. 9. Divisions of Pelagic Zone

Ans. 1. The neretic zone 2. Oceanic zone

Q. 10. What is luminescence

Ans. Source of Light in dark depth of sea produced by marine animals and fishes

Q. 11. Importance of luminescence

Ans. To attract prey or meet opposite sex, to keep together in large shoals in case of fishes. As recognition signs and means of communication to serve as an aid to frighten or confuse attackers.

Q. 12. Name zone between high tide and low tide

Ans. Intertidal zone

Q. 13. Important physical properties of sea water

Ans. Salinity, temperature, light, pressure, density, dissolved oxygen,Carbon dioxide and pH.

Q. 14. Important Characteristics of fishes.

Ans. 1. Cold blooded vertebrates with body suited to aquatic life.

2. Locomotion possible with the help of paired (pectoral and pelvic) and unpaired fins(dorsal, ventral and caudal).
3. Body is covered with dermal scales.
4. Respiration is possible through the Gills.
5. Skull & visceral arches are well developed.
6. Heart two chambered venous. Circulation is single type. Venous/deoxygenated blood flows through the heart.
7. Nerves are medullated
8. Lateral lines are sense organs and are well developed.
9. Middle ear is absent and internal ear is provided with the semicircular pinnas
10. Notochord mis partially replaced by vertebrate column.
11. Ventral and dorsal roots of spinal nerves are united at the base.
12. Kidney of most of the marine fish is mesonephric/ metanephric. But in some cases non excretory pronephric.
13. Paired nostrils

Q. 15. Name important group of fishes

Ans. Two important groups of fishes are : 1. Chondrichthyes (Cartilagenous) 2. Ostechthyes (Bone Fishes). *Examples* of Chondrichthyes are; Scaliodon, Rays, Rawas, Mackerel, Ribbon Fish, Bombay duck etc.

Examples of Ostechthyes are Labeo (carp) and Cat fish.

Q. 16. Cartilaginous fishes have pairs of Gills

Ans. 5 pairs

Q. 17. Name other marine organisms of economic importance

Ans. Crustaceans, molluscs and algae.

Q. 18. Crustaceans includes

Ans. prawns,crabs and lobsters

Q. 19. Largest quantity of prawns and shrimps produced by the country

Ans. Thialand.

Q. 20. Preferred for food item of carp sprawn is

Ans. Rotifer.

Q. 21. Largest fish catch species in India is
Ans. Sardin.
Q. 22. Per capita availability of fish in India is about
Ans. 9 kg (According to NFDB)
Q. 23. Maximum fish availability in the world is in
Ans. Asia
Q. 24. The National fish of India is
Ans. Mackerel
Q. 25. Indian coral reefs mostly found in the coasts of Indian state
Ans. Andman and Nicobar island
Q. 26. Rich source of Iodine is
Ans. Sea weeds.
Q. 27. The scientific name of goat fish is
Ans. *Upeneus sundaicus.*
Q. 28. Study of fresh water aquaculture is termed as.
Ans. Limnology
Q. 29. Fish production in India in the year 2021-22 was
Ans. 16.18 million metric tonnes (India become 3rd largest producer of fish and 2nd largest aquaculture nation after China)
Q. 30. Oil of fish contains high amount of
Ans. Vitamines
Q. 31. Maximum fishery producing state in India
Ans. 1. Gujrat, 2. Maharashtra, 3. Karnataka, 4. Kerala
Q. 32. Silver fish is studied under
Ans. Entomology
Q. 33. Length of coastline of India
Ans. 7516.6 km (Longest coast line state is Gujarat)
Q. 34. Percentage share of India in Fish Production (2020-21)
Ans. 7·96%
Q.34a. Percent share of Fish and aquaculture in India's GDP
Ans. About 1%
Q. 35. Fish production is higher in the _____ sea.
Ans. Arabian sea due to broader continental shelf.
Q. 36. Ichthyology
Ans. Ichthyology is the branch of biology devoted to the study of fish (commercial aspects) also known as *Fish Science*.
Q. 37. National Institute of Oceanography is situated at
Ans. Goa
Q. 37. Blue revolution, PMMSY
Ans. FIDF: Fisheries and Aquaculture Infrastructure Development Fund.
Q. 38. Most ideal fish for farming coastal areas.
Ans. Milkfish (Chanos chanos)

FACTS ABOUT THE EARTH

1.	Age of the earth (Approx.)	: 46000 million years
2.	Mass of the earth	: 5.98×10^{24} kg
3.	Diameter at the equator	: 12756 km.
4.	Diameter at the poles	: 12714 km
5.	Diameter of the Sun	: 1.39×10^{6} km
6.	Circumference at the equator	: 40030 km
7.	Circumference at the Meridian	: 40008 km. (at poles)
8.	Surface area of the earth	: 510,072000 km^2
9.	Volume of the earth	: 1.08321×10^{12} cubic km
10.	Total land area	: 1,48,300,000 sq. km
11.	Total water area	: 3,61,149,700 sq. km.
12.	Average distance from the sun	: 1,49,58, 900 km
13.	Greatest ocean depth	: Pacific-Mariana Trench
14.	Largest Ocean	: Pacific
15.	Smallest Ocean	: Arctic
16.	Fastest rotating planet	: Jupiter
17.	Earth completes one rotation on its axis	: 23 hrs. 56 minutes and 4.9 seconds.
18.	Warmest Ocean	: Indian Ocean (3rd Biggest)

Composition of the Earth's Crust (Weight percent)

Earth's crust is made up of many elements. The important elements & their pecentage composition by weight are given below:

Oxygen (O) 46.60%
Silicon (Si) 27.72%
Aluminium (Al) 8.13%
Iron (Fe) 5.00%
Calcium (Ca) 3.63%
Sodium (Na) 2.83%
Potassium (Mg) 2.78%
Magnesium (Mg) 2.09%
Others 1.30%

Four Distinct Spheres of the Earth

1. Lithosphere : Top crust which includes land surface and ocean floor

2. Hydrosphere : Water surface which includes oceans, seas, rivers, lakes and ponds

3. Atmosphere : Cover of air that envelopes the earth's surface

4. Biosphere : Sphere where life exists.

INTERNATIONAL INSTITUTES

Africa Rice Centre (Africa) Contonou, Benin

AVRDC: Asian Vegetable Research & Development Centre, Taiwan

Biodiversity International Rome, Italy (1974)

CABI : Commonwealth Agriculture Bureaux International

CABI: Centre for Agricultural Bioscience International Wallingford Oxfordshire OX10 8DE

CGIAR: Consultative Group of International Agricultural Research Centres, Montpellier France (1971)

CIAT: Centro International de Agriculture Tropical (International Centre for Tropical Agriculture), Cali-Columbia (South America) (1967)

CIDA: Canadian International Development Agency, Quebec (1968)

CIFOR: Center for International Forestry Research, Bogor Indonesia (1993)

CIMMYT: International Centre for the Improvement of Maize and Wheat, Mexico City, Mexico (1966)

CIP: International Potato Centre, Lima, Peru, South America (1971)

FAO: Food and Agriculture Organization, Rome, Italy (1945)

IBPGR: International Board for Plant Genetics Resources, Rome, Italy (1974)

IBRD: International Bank for Reconstruction and Development (the World Bank), Washington DC (USA)

IBSNAT: International Benchmark Sites Network for Agrotechnology Transfer Hawaii, USA

IPI: International Potash Institute, Zug, Switzerland (1952)

ICBA: International Center for Biosaline Agriculture, Dubai, UAE

ICARDA: International Centre for Agricultural Research in the Dry Areas, Aleppo, Syria (1977)

ICGEB: International Centre for Genetic Engineering and Biotechnology Trieste, Italy and JNU (New Delhi) (1994)

ICRAF: International Centre for Research in Agro forestry, Nairobi, Kenya (1978)

ICRISAT: International Crops Research Institute for the Semi – Arid Tropics, Hyderabad Telangana, India (1972)

IFAD: International Fund for Agricultural Development, Rome (1977)

IFDC: International Fertilizer Development Centre, Alabama, USA (1974)

IFPRI: International Food Policy Research Institute, Washington DC (1975)

IIMI: International Irrigation Management Institute, Digana Village, Sri Lanka (1985)

IIRB: International, Institute for Sugar Beet Research Rue Washington Brussels Belgium (1932)

IITA: International Institute of Tropical Agriculture, Ibadan, Nigeria (1967)

IJSG: International Jute Study Group, Dhaka, Bangladesh (2002)

ILRAD: International Laboratory for Res. on Animal Diseases, Nairobi, Kenya (1991)

ILRI: International Livestock Research Institute, Nairobi, Kenya (1994)

IPGRD: International Plant Genetic Resources Institute, Rome (1974)

IRRI: International Rice Research Institute, Los Banos, Philippines (1960)

ISNAR: International Service for National Agriculture Research, The Hague, Netherland (1979)

ISSS: International Society of Soil Sciences, Rome (1924)

IWMI: International Water Management Institute, Digana Village, Srilanka (1985)

TPRI: Tropical Pesticide Research Institute, Arusha, Tanzania (1979)

WARDA: West African Rice Development Association, Abidjan (1971)

WMO: World Meteorological Organization, Geneva, Switzerland (1950)
(World Meteorological Day is celebrated on 23rd March, every year. (Secretary-general Petteri Taalas).

ICAR INSTITUTES

Deemed Universities

1. Indian Agricultural Research Institute, Pusa Campus, **New Delhi-110012**.
2. Indian Veterinary Research Institute, **Izatnagar,** Uttar Pradesh (243 122).
3. National Dairy Research Institute, **Karnal**, Haryana (132 001)
4. Central Institute of Fisheries Education, **Mumbai**, Maharashtra (400 058).

National Institutes 65

1. Central Island Agricultural Research Institute, Port Blair
2. Central Arid Zone Research Institute, Jodhpur
3. Central Avian Research Institute, Izatnagar
4. Central Inland Fisheries Research Institute, Barrackpore
5. Central Institute Brackishwater Aquaculture, Chennai
6. Central Institute for Research on Buffaloes, Hissar
7. Central Institute for Research on Goats, Farah, (U.P.)
8. Central Institute of Agricultural Engineering, Bhopal
9. Central Institute for Arid Horticulture, Bikaner
10. Central Institute of Cotton Research, Nagpur
11. Central Institute of Fisheries Technology, Cochin
12. Central Institute of Freshwater Aquaculture, Bhubneshwar
13. Central Institute of Research on Cotton Technology, Mumbai
14. Central Institute of Sub Tropical Horticulture, Lucknow
15. Central Institute of Temperate Horticulture, Srinagar
16. Central Institute on Post-harvest Engineering and Technology, Ludhiana
17. Central Marine Fisheries Research Institute, Kochi
18. Central Plantation Crops Research Institute, Kasargod
19. Central Potato Research Institute, Shimla
20. Central Research Institute for Jute and Allied Fibres, Barrackpore
21. Central Research Institute of Dryland Agriculture, Hyderabad
22. National Rice Research Institute, Cuttack

23. Central Sheep and Wool Research Institute, Avikanagar, Rajasthan
24. Indian Institute of Soil and Water Conservation, Dehradun
25. Central Soil Salinity Research Institute, Karnal
26. Central Tobacco Research Institute, Rajahmundry
27. Central Tuber Crops Research Institute, Trivandrum
28. ICAR Research Complex for Eastern Region, Patna
29. ICAR Research Complex for NEH Region, Barapani
30. Central Coastal Agricultural Research Institute, Ela, Old Goa, Goa
31. Indian Agricultural Statistics Research Institute, New Delhi
32. Indian Grassland and Fodder Research Institute, Jhansi
33. Indian Institute of Agricultural Biotechnology, Ranchi
34. Indian Institute of Horticultural Research, Bengaluru
35. Indian Institute of Natural Resins and Gums, Ranchi
36. Indian Institute of Pulses Research, Kanpur
37. Indian Institute of Soil Sciences, Bhopal
38. Indian Institute of Spices Research, Calicut
39. Indian Institute of Sugarcane Research, Lucknow
40. Indian Institute of Vegetable Research, Varanasi
41. National Academy of Agricultural Research & Management, Hyderabad
42. National Institute of Biotic Stresses Management, Raipur
43. National Institute of Abiotic Stress Management, Malegaon, Maharashtra
44. National Institute of Animal Nutrition and Physiology, Bengaluru
45. National Institute of Research on Jute & Allied Fibre Technology, Kolkata
46. National Institute of Veterinary Epidemiology and Disease Informatics, Hebbal, Bengaluru
47. Sugarcane Breeding Institute, Coimbatore
48. Vivekananda Parvatiya Krishi Anusandhan Sansthan, Almora
49. Central Institute for Research on Cattle, Meerut, Uttar Pradesh
50. National Institute of High-Security Animal Diseases, Bhopal
51. Indian Institute of Maize Research, New Delhi

52. Central Agroforestry Research Institute, Jhansi
53. National Institute of Agricultural Economics and Policy Research, New Delhi
54. Indian Institute of Wheat and Barley Research, Karnal
55. Indian Institute of Farming Systems Research, Modipuram
56. Indian Institute of Millets Research, Hyderabad
57. Indian Institute of Oilseeds Research, Hyderabad
58. Indian Institute of Oil Palm Research, Pedavegi, West Godawari
59. Indian Institute of Water Management, Bhubaneshwar
60. Indian Institute of Rice Research, Hyderabad
61. Central Institute for Women in Agriculture, Bhubaneshwar
62. Central Citrus Research Institute, Nagpur
63. Indian Institute of Seed Research, Mau
64. Indian Agricultural Research Institute, Hazaribag, Jharkhand
65. Central Institutes for Fisheries Education, Mumbai

NATIONAL BUREAUS

1. National Bureau of Plant Genetic Resources, **New Delhi** (110 012).
2. National Bureau of Soil Survey and Land Use Planning, **Nagpur**, Maharashtra
3. National Bureau of Animal Genetic Resources, **Karnal**, Haryana.
4. National Bureau of Fish Genetic Resources, **Lucknow,** Uttar Pradesh.
5. National Bureau of Agriculturally Important Micro Organisms, **Mau, Nath Bhanjan,** Uttar Pradesh.
6. National Bureau of Agriculturally Important Insects Bengaluru

PROJECT DIRECTORATES

Agricultural Sciences

1. Project Directorate of Cropping Systems Research, **Meerut**, Uttar Pradesh
2. Project Directorate of Oilseeds Research, **Hyderabad**, Andhra Pradesh.
3. Project Directorate of Rice Research, **Hyderabad,** Andhra Pradesh.

4. Project Directorate of Wheat Research, **Karnal**, Haryana (Renamed Indian Institute of Wheat and Barley Research (IIWBR)
5. Project Directorate of Biological Control, **Bengaluru**, Karanataka.
6. Project Directorate on Maize, Cummings Laboratory, Indian Agricultural Research Institute, **New Delhi** (Renamed Indian Institute of Maize Research (IIMR).
7. Project Directorate on Soybean Processing and Utilization, **Indore,** Madhya Pradesh.
8. Project Directorate of Water Management Research (Renemed Indian Institute of Water Management (IIWM)
9. Directorate of Mushroom Research, Solan, **Himachal Pradesh**
10. Directorate of Sorghum Research, **Hyderabad**
11. Directorate of Oil Palm Research Pedavegi, West Godavari, **Andhra Pradesh.**
12. Directorate of Cashew Research, Puttur, **Karnataka.**
13. Project Director for Farming Systems Research, **Modipuram** (Renamed Indian Institutc of Farming Systems Research (IIFSR).
14. Directorate of Weed Science Research, **Jabalpur**
15. Directorate of Groundnut Research **Junagadh**

Animal Sciences

16. Project Directorate on Cattle, **Meerut,** Uttar Pradesh.
17. Project Directorate on Poultry Research, **Hyderabad**, Andhra Pradesh.
18. Project Directorate on Animal Disease Monitoring and Surveillance (PDADMS). **Bengaluru.**
19. Project Directorate on Foot and Mouth Diseases (PDFMD). **Mukteshwar.**
20. Directorate of Coldwater Fisheries Research, Bhimtal, **Uttarakhand.**
21. National Fisheries Development Board (NFDB) Rajendra Nagar, Hyderabad, **Telangana.**
22. Directorate of Knowledge Management in Agriculture, **New Delhi**

NATIONAL RESEARCH CENTRES-15

1. ICAR-National Research Centre for Banana, Trichi
2. ICAR-National Research Centre for Grapes, Pune
3. ICAR-National Research Centre for Litchi, Muzaffarpur
4. ICAR-National Research Centre for Pomegranate, Solapur
5. ICAR-National Research Centre on Camel, Bikaner
6. ICAR-National Research Centre on Equines, Hisar
7. ICAR-National Research Centre on Meat, Hyderabad
8. ICAR-National Research Centre on Mithun, Medziphema, Nagaland
9. ICAR-National Research Centre on Orchids, Pakyong, Sikkim
10. ICAR-National Research Centre on Pig, Guwahati
11. ICAR-National Research Centre on Plant Biotechnology, New Delhi
12. ICAR-National Research Centre on Seed Spices, Ajmer
13. ICAR-National Research Centre on Yak, West Kemang
14. ICAR-National Centre for Integrated Pest Management, New Delhi
15. National Research Centre on Integrated Farming (ICAR-NRCIF), Motihari

ALL INDIA COORDINATED RESEARCH PROJECT -60

1. AICRP on Nematodes, New Delhi
2. AICRP on Maize, New Delhi
3. AICRP Rice, Hyderabad
4. AICRP on Chickpea, Kanpur
5. AICRP on MULLARP, Kanpur
6. AICRP on Pigeon Pea, Kanpur
7. AICRP on Arid Legumes, Kanpur
8. AICRP on Wheat & Barley Improvement Project, Karnal
9. AICRP Sorghum, Hyderabad
10. AICRP on Pearl Millets, Jodhpur
11. AICRP on Small Millets, Bangalore
12. AICRP on Sugarcane, Lucknow
13. AICRP on Cotton, Coimbatore

14. AICRP on Groundnut, Junagarh
15. AICRP on Soybean, Indore
16. AICRP on Rapeseed & Mustard, Bharatpur
17. AICRP on Sunflower, Safflower, Castor, Hyderabad
18. AICRP on Linseed, Kanpur
19. AICRP on Sesame and Niger, Jabalpur
20. AICRP on IPM and Biocontrol, Bangalore
21. AICRP on Honey Bee Research & Training, Hisar
22. AICRP -NSP(Crops), Mau
23. AICRP on Forage Crops, Jhansi
24. AICRP on Fruits, Bangaluru
25. AICRP Arid Zone Fruits, Bikaner
26. AICRP Mushroom, Solan
27. AICRP Vegetables including NSP vegetable, Varanasi
28 AICRP Potato, Shimla
29. AICRP Tuber Crops, Thiruvananthapuram
30. AICRP Palms, Kasaragod
31. AICRP Cashew, Puttur
32. AICRP Spices, Calicut
33. AICRP on Medicinal and Aromatic Plants including Betelvine, Anand
34. AICRP on Floriculture, New Delhi
35. AICRP in Micro Secondary & Pollutant Elements in Soils and Plants, Bhopal
36. lAICRP on Soil Test with Crop Response, Bhopal
37. AICRP on Long Term Fertilizer Experiments, Bhopal
38. AICRP on Salt Affected Soils & Use of Saline Water in Agriculture, Karnal
39. AICRP on Water Management Research, Bhubaneshwar
40. AICRP on Ground Water Utilisation, Bhubaneshwar
41. AICRP Dryland Agriculture, Hyderabad
42. AICRP on Agrometeorology, Hyderabad including Network on Impact adaptation & Vulnerability of Indian Agri. to Climate Change

43. AICRP Integrated Farming System Research, Modipuram including Network Organic Farming
44. AICRP Weed Control, Jabalpur
45. AICRP on Agroforestry, Jhansi
46. AICRP on Farm Implements & Machinery, CIAE Bhopal
47. All India Coordinated Research Project on Ergonomics and Safety in Agriculture
48. AICRP on Energy in Agriculture and Agro Based Indus., CIAE Bhopal
49. AICRP on Utilization of Animal Energy (UAE), CIAE Bhopal
50. AICRP on Plasticulture Engineering and Technologies, Ludhiana
51. AICRP on PHT, Ludhiana
52. AICRP on Goat Improvement, Mathura
53. AICRP- Improvement of Feed Sources & Nutrient Utilisation for raising animal production, Bangalore
54. AICRP on Cattle Research, Meerut
55. AICRP on Poultry, Hyderabad
56. AICRP-Pig, Izzatnagar
57. AICRP Foot and Mouth Disease, Mukteshwar
58. AICRP ADMAS, Bangalore
59. AICRP on Home Science, Bhubaneshwar

NETWORK PROJECTS-20

1. All India Network Project on Pesticides Residues, New Delhi
2. All India Network Project on Underutilised Crops, New Delhi
3. All India Network Project on Tobacco, Rajahmundry
4. All India Network Project on Soil Arthropod Pests, Durgapura
5. Network on Agricultural Acarology, Bangalore
6. Network on Economic Ornithology, Hyderabad
7. All India Network Project on Rodent Control, Jodhpur
8. All India Network Project on Jute and Allied Fibres, Barrackpore
9. Network project on Improvement of Onion & Garlic, Pune

10. Network Bio-fertilizers, Bhopal
11. Network Project on Harvest & Post Harvest and Value Addition to Natural Resins & Gums, Ranchi
12. Network project on Animal Genetic Resources, Karnal
13. Network Project on R&D Support for Process Upgradation of Indigenous Milk products for industrial application Karnal
14. Network Programme on Sheep Improvement, Avikanagar
15. Network Project on Buffaloes Improvement, Hisar
16. Network on Gastro Intestinal Parasitism, Izatnagar
17. Network on Haemorrhagic Septicaemia, Izatnagar
18. Network Programme Blue Tongue Disease, Izatnagar
19. Network Project on Conservation of Lac Insect Genetic Resources, Ranchi
20. Network Project on Agricultural Bioinformatics and Computational Biology, New Delhi

OTHER PROJCTS-10

1. Technology Mission on Cotton (CICR, Nagpur)
2. Technology Mission on Jute (CRIJAF, Barrackpore)
3. Continuation, Strengthening and Establishment of Krishi Vigyan Kendras
4. Strengthening & Development of Higher Agricultural Education in India, New Delhi
5. Central Agricultutral University, Imphal
6. Strengthening and Modernization of ICAR Headquarters
7. Intellectual Property Management & Transfer/Commercialisation of Agricultural Technology (Upscaling of existing component IPR HQ)
8. Indo US Knowledge Initiative
9. National Agricultural Innovative Project, New Delhi
10. National Fund for Basic and Strategic Research, New Delhi

ATARI (AGRICULTURAL TECHNOLOGY APPLICATION RESEARCH INSTITUTE)

ATARI Name	ATARI Address	Pincode	No. of KVKs
ICAR-ATARI Zone-I Ludhiana	PAU Campus Ludhiana, Punjab	141004	72
ICAR-ATARI Zone-V Kolkata	Bhumi Vihar Complex Block- GB Sector-III, Salt Lake Kolkata, West Bengal	700097	59
ICAR-ATARI Zone-VII Barapani	ICAR Research Complex for NEH Region Barapani, Meghalaya	793103	43
ICAR-ATARI Zone-III Kanpur	Rawatpur Near Vikas Bhawan Kanpur, Uttar Pradesh	208002	89
ICAR-ATARI Zone-X Hyderabad	CRIDA Campus, Santhosh Nagar Hyderabad, Telangana	500059	75
ICAR-ATARI Zone-II Jodhpur	CAZRI Campus, Diesel Shed Road PO Krishi Upaj Mandi, Basni Jodhpur, Rajasthan	342005	66
ICAR-ATARI Zone-IX Jabalpur	PO Adhartal, JNKVV Campus, Jabalpur, Madhya Pradesh	482004	82
ICAR-ATARI Zone-XI Bangalore	M.R.S, H.A.Farm (P.O), Hebbal, Bengaluru, Karnataka	560024	48
ICAR-ATARI Zone-IV Patna	ICAR-Agricultural Technology Application Research Institute, Zone-IV, Patna.	800001	68
ICAR-ATARI Zone-VI Guwahati	ICAR Agricultural Technology Application Research Institute, ICAR CPCRI Campus Kahikuchi , Guwahati, Assam	781017	47
ICAR-ATARI Zone-VIII Pune	ICAR-Agricultural Technology Application Research Institute (ATARI), College of Agriculture Campus, Shivajinagar, Pune (Maharashtra)	411005	82
		Total:	731

PREVIOUS YEARS QUESTIONS ON MEMORY BASED (ASKED IN DIFFERENT EXAMINATIONS) LIKE J.R.F. & I.A.R.I Ph.D. EXAMS & B.H.U., Ph.D. & P.G. EXAM., (2023–1991)

JRF Paper-2023 (Memory Based)

Q. 1. Aeroponic technology is commercially used in quality seed planting material production of

Ans. Tomato

Q. 2. Tawa, Lower Bhavani, Balimala & Matatila dams are related to states

Ans. M.P., TamilNadu, Odisha & U.P.

Q. 3. The term allelopathy was coined by

Ans. H. Molisch

Q. 4. An intercropping system can be said beneficial if it has LER

Ans. >1.0

Q. 5. IWDP programme was renamed in year

Ans. 2003 (Renamed -Hariyali)

Q. 6. Kufri Bahar is a prominent variety of

Ans. Potato

Q. 7. Area of micro watershed is

Ans. 100-1000 Ha

Q. 8. The transcription factor SNACI is involved in drought stress response in which crop

Ans. Sorghum

Q. 9. Leaf relative growth rate (LRGR) can be calculated using which of the expression

Ans. LRGR= $\text{Log}LW_2 - \text{Log } LW_1 / t_2 - t_1$

Q. 10. Biochar produced by incomplete combustion of biological materials is rich in

Ans. Carbon

Q. 11. Weed seed dispersal by ants is called as

Ans. Myrmecochory

Q. 12. Example of ephemeral weed

Ans. Chickweed

Q. 13. The targeted yield concept of soil fertility evaluation was proposed by

Ans. B. Ramamoorthy

Q. 14. A sugarcane crop of 2 Ha area was irrigated 5 times with 6 cm water in each irrigation workout the total quantity of water applied in cubic meter

Ans. 6000

Q. 15. Type of soil water available for normal crop growth

Ans. Capillary

Q. 16. The country that has given name of the tropical cyclone Mocha developed in month of May, 2023 in Bay of Bengal is

Ans. Yemen

Q. 17. Planting geometry that ensures a uniform incidence of solar radiation

Ans. Square planting

Q. 18. Inhibitors of photosynthesis at photosystem I

Ans. Bipyridyliums

Q. 19. In which of the following organelles, enzyme pyruvate dehydrogenases complex and glycolytic pathway are located

Ans. Cytosol and Mitochondria

Q. 20. The Dapog method of raising rice nursery was introduced in India from

Ans. Philippines

Q. 21. Which is the minor millet

Ans. Foxtail millet or proso millet

Q. 22. The concentration of nitrogen in atmosphere up to 50 km from the ground surface is

Ans. 78%

Q. 23. International Livestock Research Institute (ILRI) is located at

Ans. Nairobi, Kenya

Q. 24. The practice of controlling water erosion by cultivation of alternate erosion permitting and erosion resistant crops is called as

Ans. Strip Cropping

Q. 25. Which is the herbicide is highly volatile

Ans. EPTC

Q. 26. Beushening is related to

Ans. Rice

Q. 27. Blind hoeing is related to

Ans. Sugarcane

Q. 28. Which is the temperate grass

Ans. White and red clover

Q. 29. First herbicides produced in world

Ans. 2, 4- D

Q. 30. The region of atmosphere having the constant temperature

Ans. Mesopause

Q. 31. Relation between plant population and yield in fodder crop is

Ans. Asymptotic

Q. 32. Prismatic soil structure feature found mostly in soil horizon

Ans. B Horizon

Q. 33. Suitable nozzles for herbicides spray

Ans. Regular Flat Fan Nozzle

Q. 34. Protein and Oil content in soyabean is

Ans. 43 and 20

Q. 35. Full form of NISAR is

Ans. NASA-ISRO Synthetic Aperture Radar

Q. 36. The interaction between legume and non legume plants in the form of supplementation is called as

Ans. Supplementary.

Q. 37. Mentha crop is commercially raised through

Ans. Stolon's

Q. 38. Quantum sensor is used for

Ans. Photosynthetically active radiation.

Q. 39. Drought avoidance mechanism is found in which crop

Ans. Sorghum

Q. 40. Fertilizer having with lower CRH and most hygroscopic is

Ans. Ammonium Nitrate

Q. 41. Kashi Amul is the variety of

Ans. Tomato

Q. 42. The largest orchid species in the world is

Ans. Grammatophyllum speciosum

Q. 43. India's rank in mango production

Ans. First Rank

Q. 44. Improper filling physiological disorder of banana is due to deficiency of

Ans. Potassium

Q. 45. The tree associated with the life of Lord Shree Krishna

Ans. Anthocephalus cadamba

Q. 46. Pant Ragini is a variety of

Ans. Fenugreek

Q. 47. Most common rootstock for budding in rose under South Indian condition

Ans. Rosa multiflora

Q. 48. Garden which is made basis on the ideology of heaven

Ans. Garden of Eden

Q. 49. Polyembryony is controlled by

Ans. Dominant gene.

Q. 50. The wheat grain can be stored well at a moisture percent of

Ans. 10%

Q. 51. Pusa sarda is variety of

Ans. Muskmelon

Q. 52. Irrigation system most useful in water scarcity and salt problem

Ans. Drip irrigation

Q. 53. Dicladispa annegera is major pest of

Ans. Rice

Q. 54. Major pest of early summer cauliflower

Ans. Spodoptera litura

Q. 55. Causal organism of early blight of potato

Ans. Alternaria solani

56. Country proposed UN for declaring the year 2023 as international year of millets

Ans. India

57. Llyod botanical garden is located at

Ans. Darjeeling, West Bengal

58. Exposing seed to beneficial microorganism is called as

Ans. Bio priming

59. What is the new name of Mughal Garden of Rashtrapati Bhavan, New Delhi

Ans. Amrit Udyan

60. Bhartiya Prakritik Krishi Padhati (BPKP) was launched on

Ans. 2019

61. Educating a farmer about the ill- effects of harmful insecticides belongs to

Ans. Non formal education

62. Bhoodan movement began in year

Ans. 1951

63. Self life of liquid bio fertilizers can be

Ans. 12-24 Months.

64. Which state having highest area under micro irrigation

Ans. Karnataka

65. Total plant quarantine stations at airports, Seaports and land frontiers are in India

Ans. 73

66. In paddy the elongation of leaf sheath after the infestation of gall fly is due to the chemical

Ans. Cecidogen

67. Symphilids are closely related to

Ans. Millipeds and Centipedes

68. Trichoderma spp recommend for rice leaf folder

Ans. Trichoderma chilonis

69. Secondary tillage is done primarily

Ans. To prepare a fine tilth seed bed

70. Parrot beaking is caused by mirid bugs in which crop

Ans. Bt. Cotton

JRF Paper-2022 (Memory Based)

Q.1. Seed rate of Peral millet to get seedling for transplanting in one ha area

Ans. 2kg/ Ha

Q.2. The most critical growth stage of moisture stress in lentil

Ans. Pod formation

Q.3. IWDP stands for

Ans. Integrated Watershed Development Project

Q.4. In Ca and S deficient soils gypsum placement should be at

Ans. 30 Days after Sowing

Q.5. In dry spell during crop period, which of the following can't be moisture stress mitigation practice

Ans. Intensive cropping

Q.6. Fertilizers having least hygroscopicity is

Ans. Ammonium Sulphate

Q.7. Which is Not a pressure unit

Ans. Newton

Q.8. Orobanche is problem weed in

Ans. Tomato

Q.9. Highest area under conservation agriculture is in which of continent

Ans. America

Q.10. Mineral having max cation exchange capacity

Ans. Smectite

Q.11. How many days in advance weather forecast are currently issued by IMD in DAMU project

Ans. 5 Days

Q.12. Terminal drought is also known as

Ans. Late season drought

Q.13. At what height geostationary satellites are placed above the earth

Ans. 36000 KM

Q.14. Which is a Sulphur containing mineral

Ans. Epsomite

Q.15. Which of the hormone deficiency causes dwarfism in plants

Ans. Gibberellin

Q.16. NRC for integrated farming is located at

Ans. Motihari (Bihar)

Q.17. The optimum temp for the ideal vegetative growth of wheat

Ans. 16-22 °C

Q.18. Vegetative organ of Cyperus rotundus (Nut glass)

Ans. Tuber

Q.19. Crops require higher amount of Calcium

Ans. Legumes crops

Q.20. The soils with more than 20 kg/ha phosphorus are rated as

Ans. Medium in P

Q.21. The suitable seed rate for row sown sesame is

Ans. 3-4 kg. /ha

Q.22. Latest addition to the list of essential nutrients is

Ans. Nickel

Q.23. Honeycomb structure found in soils

Ans. Laterites and lateritic soils

Q.24. Best suitable cereal crop for silage is

Ans. Maize

Q.25. AD$^+$ in cellular respiration act as

Ans. An electron carrier

Q.26. Ammonium sulphate fertilizer is advisable for

Ans. Potato crop

Q.27. The total age of rice seedling for transplanting is main field through double transplanting method

Ans. 60-70 Days

Q.28. In green gram nitrogen fixation starts from the age of

Ans. 12-15 Days

Q.29. Bulk density is expressed as

Ans. (Mg/m^3) & g/cm^3

Q.30. Which of the following is not a film forming type anti Transpirant

Ans. $CaHCO_3$

Q.31. The Book Nature & Properties of soils is written by

Ans. N C Brady & R.R.Weil

Q.32. ICRISAT was established in year

Ans. 1972 at Hyderabad

Q.33. Optimum spacing (Row × Plant) of hybrid cotton in irrigated area is

Ans. 120 × 60 cm

Q.34. Percentage of lignin in sugarcane bagasse

Ans. 25-32%

Q.35. Farming in humid regions with growing period more than 120 days called

Ans. Rainfed farming

Q.36. Establishment year of IIHR (Bengaluru)

Ans. Sep 5, 1967

Q.37. Which the following garden is based on the pattern of Mughal garden

Ans. Rashtrapati Bhawan Garden, New Delhi & Nishat Bag Kashmir

Q.38. Perennial plants having erect and bushy growth and attain a height of .5 to 4 meter are called as

Ans. Shrubs

Q.39. Kanji is prepared of

Ans. Carrot

Q.40. Major onion producer state of India

Ans. Maharashtra

Q.41. What is the main purpose of blanching in cauliflower

Ans. To keep the curd white

Q.42. Which of the following part of carrot contains the highest carotene level

Ans. Top part of the root

Q.43. Mango hybrid Pusa Peetambar is cross between

Ans. Amrapali × Lal Sindhuri

Q.44. Soils of B horizon are dominated by

Ans. Clay, Fe, Al & Humus

Q.45. Dr. Puskarnath is associated to

Ans. Potato breeding

Q.46. Homeostasis that's lead to adaptability and stability is NOT favored by

Ans. Self pollination

Q.47. Which portion of sugarcane is the best for use as seed

Ans. Top 1/3 to 1/2 portion

Q.48. What is the main cause of appearance of gold color flecks on tomato

Ans. Low K: Ca ratio

Q.49. Which is predator/ parasite of white fly

Ans. Brumus suturalis

Q.50. Which of the following is biological agent for suppressing Parthenium hysterophorus

Ans. Mexican Beetle

Q.51. Grape variety suitable for preparing Monukka

Ans. Anabe-e-Shai

Q.52. Zinc is responsible for the biosynthesis of which plant auxin

Ans. IAA

Q.53. ICAR Krishi Megh (NARES-Cloud Infrastructure and Services) was launched in year

Ans. Aug, 2020

Q.54. To find average height of plants we should use

Ans. Arithmetic mean

Q.55. If all the variate values are negative the standard deviation will be

Ans. Positive

Q.56. The ratio of normal GNP and real GNP which can be interpreted as comprehensive price index is called

Ans. GNP Deflator

Q.57. Chairmen of CACP is

Ans. Prof Vijay Paul Sharma

Q.58. Pradhan Mantri Kisan Samman Nidhi (PM-KISAN) Scheme was launched in

Ans. 2019.

Q.59. State got the Krishi Karman award of 2022

Ans. Tamil Nadu

Q.60. Insect which survives at Higher temperature and low humidity is

Ans. Khapra beetle

61. Nematode which transmits virus for grape vine fan leaf disease

Ans. *xiphinema index*

Q.62. The agricultural cost is equal to

Ans. Sum of fixed and variable cost

Q.63. Potato leaf curl virus is mainly transmitted by

Ans. Aphid

JRF and State Pre P. G. Papers-2021 (Memory Based)

Q. 1. Kashi Aman and Kashi Adarsh are variety of

Ans. Tomato

2. The KISAAN 2:0 (Krishi Integrated Solution for Agri Apps Navigation) App, was envisaged to help

Ans. e-agriculture

3. Karaknaath is a breed of

Ans. Chicken

4. KAU Nihara, a dwarf variety of

Ans. Cashew

5. Which is the early variety of Cauliflower

Ans. Sabour Mukta

6. Rajasthan Sheep breed which is under MSSP Project

Ans. Sonadi & Malpura Sheep

7. Pusa Decomposer is mainly used for the degradation of

Ans. Paddy Straw

8. In which maize variety first time leaf spot disease of maize was discovered in India

Ans. Kanchan (2017)

9. A new herbicide for broad leaf weed control in chickpea is

Ans. Topramezone

10. Best training system for Canopy management in Ber for more fruit yield:

Ans. An improved 'Y' shaped training system

11. 'Arka Sasya Poshak Ras' is most suitable for

Ans. Vegetables

12. Mushroom variety which has first time produced under artificial condition in India

Ans. Guchchhi Mushroom

13. Potassium help in maintaining cytoplasmic pH between

Ans. 7-8

14. No of Agricultural Universities accredited by NAEAB

Ans. 62

15. First Ranking of Agri University/Institutes under ICAR in 2020

Ans. 21 1st NDRI Karnal, 2nd PAU Ludhiana

16. Which enzyme is used by the biscuit manufacturers to lower the protein level of flour

Ans. Proteases

17. Maximum solar energy is trapped by

Ans. Growing algae in tanks.

18. Magnesium is component of

Ans. Chlorophyll

19. Example of the simplest carbohydrates;

Ans. Monosaccharides

20. Which is the correct nucleotide pair in RNA

Ans. Adenine- Uracil

21. The millennium ecosystem assessment (MEA) was launched in year

Ans. 2001 by the support of United Nations.

22. The dominant clay mineral present in red soils is

Ans. kaolinite with small quanties of illite

23. Recently developed mineral soil with no diagnostic horizons, low degree of soil development due to less time, occurs in all state of India is called

Ans. Entisols with small quantities of illite

24. Minimum, Optimum and Maximum temperature for germination of Rice is

Ans. 10-12°C, 30 to 32 °C and 30 to 40 °C respectively

25. Agro ecological zones classified by NBSS & LUP

Ans. 20 zones

26. Nitrogen concentration in atmosphere is

Ans. 78%

27. Harvesting time of Boro and Aus rice carried out during

Ans. Summer and Rabi

28. Edible oil which also used in paint and furnishing industry

Ans. Linseed oil

29. Scientific name of green gram is

Ans. *Vigna radiata*

30. Efficiency of sprinkler irrigation system

Ans. 70-85%

31. Two multiple-stress tolerant rice varieties namely Swarna Samridhi Dhan and Swarna Sukha Dhan have been released by

Ans. ICAR

32. Crop of which Husk is a economical part

Ans. Isabgol

33. Formulation of pesticide which is not diluted and supplied to ready to use is called

Ans. ULV

34. Sowing technique which is recently promoted by government in cereal crop is

Ans. Direct Seedling of Rice Method

35. Maximum permissible slope percent for furrow irrigation method is

Ans. 0.5%

36. Purpose to Reduce Urea bag weight 5 kg per bag was

Ans. To Reduce luxury consumption of Urea

37. Which crop can be grown in Kharif, Rabi and Spring

Ans. Maize

38. Command area development (CDP) programme was initiated during

Ans. Dec., 1974

39. To find out average height of plant we should work out the

Ans. Mean

40. Seed act was enacted in

Ans. 1966

41. Chemical which is beneficial for releaser but not for receiver

Ans. Allomones

42. Agency which is enveloped in enhancing export of processed agricultural products

Ans. APEDA

43. In which farming there is use of information and technology-based system

Ans. Precision Farming

44. Internationally recognized system for reducing the risk of safety hazard in crop is

Ans. HACCP (Hazard analysis critical control points)

45. The chromosome number of common Rice is

Ans. 24

46. Random allocation of treatments to the experimental unit is known as

Ans. Randomization

47. Guava is suitable for jelly making because of

Ans. It is rich in Vita C, acidity and good flavor

48. In which variety of Banana hard lump is common problem

Ans. Poovan

49. Clotting of blood is related to deficiency of Vitamin

Ans. Vitamin K

50. Wooliness usually occurs when fruits are stored at temperature of

Ans. 2-6 °C

51. Most serious post-harvest disease of Mango is

Ans. Anthracnose

52. Which fruit crop will be well performed in saline irrigation water condition

Ans. Date Palm

53. Which sticky trap can be used to detect infestation of adult white flies

Ans. Yellow

54. Which of the following is often cross pollinated vegetable crop

Ans. Chilly

55. Gross area under Cotton in 2019-20

Ans. 13.47 M.ha.

56. Which of the following breeding method is also known as modified pedigree method

Ans. S.S.D.

57. Removal of central portion of curd for flower initiation in cauliflower is known as

Ans. Scooping

58. Institute of Himalayan Bioresource Technology is located at

Ans. Palampur (H.P.)

59. Crop is not is not propagated by terminal cutting

Ans. Rose

60. India s First planned and laid out botanical garden Lal Bagh is situated at

Ans. Bangalore

61. Which is the quickest technique for lawn making

Ans. Turfing

62. State which is the largest producer of cut flowers

Ans. West Bengal

63. Zabo Indian farming system is mainly practiced in

Ans. Nagaland (Word Zabo means "impounding run-off" water and utilization).

64. Cyst nematode of Potato first reported in India in Year

Ans. 1961

65. Pure line breeds refer to

Ans. Homozygosity

65a. Barium Carbonate is used as

Ans. Rodenticide

66. Quickest method of plant breeding is

Ans. Mutation Breeding

67. The variables whose calculation is done according to the weight and length are known

Ans. Continuous Variables

68. First transgenic crop approved for cultivation in USA during year 1994 was

Ans. Tomato

69. Which pesticide is not used in storage warehouse

Ans. Dimethoate

70. In India where the Potato crop is grown round the year

Ans. Nilgiris & Palani Hills

71. As per the cost concept the value of fertilizer, seeds and pesticides comes under

Ans. Cost A_1

72. Environmental Protection Act (EPA) was passed by Govt of India In Year

Ans. May, 1986

73. Average yield Kg. /ha of Total food grains in 2019-20

Ans. 2343 kg./ha

74. The Web of digital dissemination system of Indian Agricultural Research is

Ans. AGROWEB

75. Contribution of awns in photogyuthem to grain drought conditions

Ans. 5%

76. Point at which CO_2 uptake exactly balances the CO_2 release is

Ans. Light compensation point

JRF and State Pre P. G. Papers-2020 (Memory Based)

Q. 1. Triose phosphate is formed in the reduction step of

Ans. Calvin Cycle

Q. 2. First operational Indian remote sensing satellite IRS-1A was launched on

Ans. 17th March 1988

Q. 3. Ozone concentration in the atmosphere is reduced by

Ans. CFC and Cl_2

Q. 4. The important event in the early history of agriculture started in 4400 BC was

Ans. Domestication of Sheep and Cultivation of Wheat & Barley

Q. 5. Which is the commonly used method to determine soil moisture content in laboratory conditions

Ans. Neutron Probe

Q. 6. The point at which CO_2 uptake exactly balances CO_2 released is called

Ans. Carbon dioxide compensation point

Q. 7. Silver shoot occurs in Rice due to attack of the pest

Ans. Gall Midge

Q. 8. What will be the error d.f. for a RBD field experiment with 8 treatments and 3 replications

Ans. 14

Q. 9. Which one is not a quality parameter of Rice grain

Ans. Curing Quality

Q. 10. Water requirement of Rice, Wheat, G. nut & Pulses

Ans. 1200 mm, 600 mm, 400 mm and 300 mm

Q. 11. Which one is the suitable irrigation method for wheat

Ans. Bubler Method

Q. 12. When an abnormal failure of rainfall occurs the type of drought is referred to

Ans. Contingent Drought

Q. 13. The concentration of macronutrient in dry plant tissue is

Ans. > 0.1%

Q. 14. Post emergence herbicide for weed management in Rice is

Ans. Thio bencarb (Biosyribic sodium)

Q. 15. Rate of transpiration in crops occurs maximum during
Ans. Afternoon

Q. 16. Which one is not a 1:1 type clay mineral
Ans. Nontronite

Q. 17. In which year first time ICAR NBBS & LUP categorized different Agro Ecological Zones
Ans. 1996

Q. 18. Maximum use efficiency in optimum condition of phosphorous fertilizer is
Ans. 20-25%

Q. 19. Most common method of herbicide application in India is
Ans. Spraying

Q. 20. Central Research Institute for Dry land Agriculture is located at (CRIDA)
Ans. Hyderabad

Q. 21. Useful part of Plantago (Psyllium) is
Ans. Seed husk

Q. 22. Which is not a variety of Chick pea
Ans. GNG1581

Q. 23. The important biological control insect for *parthenium hysterophorus* in India is
Ans. *Zygogramma bicolorata*

Q. 24. Chemiosmotic therapy was first proposed by
Ans. Peter Mitchell

Q. 25. NDVI of a crop can be measured by using
Ans. Leaf Colour Chart

Q. 26. Which one weed is a holo root parasite
Ans. Orobanche

Q. 27. Diara land comes under
Ans. Highly rainfall area

Q. 28. Amino acid which is also known as stress indicator
Ans. Proline

Q. 29. Under dry land farming condition which practice is most prominent
Ans. Single Cropping

Q. 30. Branching and pod formation are most critical condition of irrigation in crops
Ans. Pulse crops

Q. 31. What will be the effect of tillage on thermal conductivity of soil

Ans. Decreases

Q. 32. Nutrient index approach of soil fertility evaluation was proposed by

Ans. Liebig

Q.32a. 'Saksham' is an important variety of

Ans. Japanese Miint **(BHU 2020)**

Q.32b. Sheep has highest average wool production ?

Ans. Deccani **(BHU 2020)**

Q.32c. Potato tuber is scientifically called as

Ans. Underground Stem **(BHU 2020)**

Q. 33. For the analysis of variance of data from plot sampling in a CRD with 't' treatments 'r' replications and 's' samplings units per plot, sampling error will be

Ans. rt(s - 1)

Q. 34. Total pulses production in India in 2018-19

Ans. 22.08 MT

Q. 35. NIASM (National Institute of Abiotic Stress Management) is located at

Ans. Baramati (MH)

Q. 36. Pithiness is a physical disorder in which vegetable

Ans. Radish

Q. 37. Basmati rice which has highest share in export

Ans. Pusa Basmati-1

Q. 38. Vegetable most suitable for saline soils

Ans. Beet root

Q. 39. In which crop Solasodine is found

Ans. Brinjal

Q. 40. Grafting in vegetable crops is useful for managing

Ans. Bud necrosis virus

Q. 41. Girdling is generally practiced in which crop

Ans. Grapes

Q. 42. Flame of Forest is botanically known as

Ans. *Butea monosperma*

Q. 43. (A) Granulation, (B) Black tip, (C) Fruit cracking & (D) Pink blush is related to

Ans. (A) Citrus, (B) Mango, (C) Tomato & (D) Grapes

Q. 44. Holly hock and sweet pea are suitable annuals for

Ans. Hanging Basket

Q. 45. The optimum average daily temperature for wheat sowing is

Ans. 25-27 °C

Q. 46. Application of Remote sensing and GIS based technologies are essential for

Ans. Precision Farming

Q. 47. Which was the first hybrid Tea rose in world

Ans. La France

Q. 48. Botanical term of Banana inflorescence

Ans. Spadix

Q. 49. Persian garden is type of garden

Ans. Formal Garden

Q. 50. 2,4-D is a

Ans. Growth Regulator & Herbicide

Q. 51. Which one is known as mood food

Ans. Red Cherry

Q. 52. Most commonly used rooting hormone

Ans. IBA

Q. 53. Pusa Aditi is a variety of

Ans. Grapes

Q. 54. First Plant Quarantine Act passed in India

Ans. 1914

Q. 55. Molya disease of wheat is severe in state

Ans. Rajasthan

Q. 56. The acronym GEAC stands for

Ans. Genetic Engineering Approval Committee

Q. 57. Downward entry of water in soil is called as

Ans. Percolation

Q. 58. Which one of the following scheme is exclusively meant for self employment of rural youth

Ans. TYRESM

Q. 59. Sexual spores of fungi are

Ans. Zoospores

Q. 60. Mineral element plays important role in biological Nitrogen fixation

Ans. Molybdenum

Q. 61. Law of diminishing return operates when

One input is variable

Q. 62. Pulse having highest percent share of total cropped area

Ans. Gram

Q. 63. Foundation day of FAO

Ans. Oct. 16, 1945

Q. 64. Demand of Wheat is a

Ans. Inelastic

Q. 65. Major corporate farming states are India

Ans. Tamil Nadu & Maharashtra

Q. 66. India's Rank in world in Total Cereals production in 2018

Ans. Third

Q. 67. The National Programme for Organic Production (NPOP), started in

Ans. 2001

Q. 68. Harvest index of wheat is

Ans. 0.35 - 0.45

Q. 69. Radio broadcasting starts in India

Ans. 23rd July, 1927

Q. 70. Crossing of plants through fusion of somatic cell is

Ans. Somatic Hybridization.

Q. 71. Navstar Global Positioning System is maintained by

Ans. USA Department of Defence

Q. 72. For propagation cutting of black pepper mainly raised from **(ADO, 2022)**

Ans. Runner shoots

Q. 73. Aflatoxin are type of

Ans. Mycotoxins in food and feed

JRF and State Pre P. G. Papers-2019 (Memory Based)

Q.1. Sugarcane breeding station is located at
Ans. Coimbatore (*JRF 2019*)

Q.2. Which state is known as Garden of Spices
Ans. Kerala (*JRF 2019*)

Q.3. Probability can take values from
Ans. 0 to + 1 (*JRF 2019*)

Q.4. Where does the water vapor does exit in the plant
Ans. Stomata (*JRF 2019*)

Q.5. Khaira Disease in Rice is associated to
Ans. Zn Deficiency (*JRF 2019*)

Q.6. Seeding of cold clouds is achieved by
Ans. Silver iodide (*JRF 2019*)

Q.7. IGFRI is located at
Ans. Jhansi (*JRF 2019*)

Q.8. A barley plant in wheat crop is type of weed
Ans. Relative weed (*JRF 2019*)

Q.9. Chromosome number in common rice are
Ans. 22 (*JRF 2019*)

Q.10. Red Rose is the national flower of which of the country
Ans. England (*JRF 2019*)

Q.11. Buck wheat belongs to the family of
Ans. Polygonaceae (*JRF 2019*)

Q.12. Brown manuring is mainly used in crop
Ans. Rice (*JRF 2019*)

Q.13. Deficiency of Molybdenum in cauliflower leads to
Ans. Whiptail (*JRF 2019*)

Q.14. Pusa Peetamber is cross between
Ans. Amrapali × Lal Sundari (*JRF 2019*)

Q.15. CAN is a type of fertilizer
Ans. Neutral (*JRF 2019*)

Q.16. Optimum temperature for Nitrifying bacteria
Ans. 30-35°C (*JRF 2019*)

Q.17. Which hormone known as stress hormone in plants
Ans. Abscisic Acid (*JRF 2019*)

Q.18. Scooping operation is done in
Ans. Cauliflower (*JRF 2019*)

Q.19. Longest meiotic phase is
Ans. Prophase (*JRF 2019*)

Q.20. Which year is dead line for PMGSY
Ans. 2019 (*JRF 2019*)

Q.21. Plants growing in high salt concentration are called
Ans. Halophytes (*JRF 2019*)

Q.22. World Soil Day is celebrated on
Ans. 5th Dec (*JRF 2019*)

Q.23. *Neovossia Indica* cause
Ans. Karnal bunt in Wheat (*JRF 2019*)

Q.24. Multigerm Seed is found in crop
Ans. Sugar beet (*JRF 2019*)

Q.25. One hectare-CM water is equal to
Ans. 100000 liter (*JRF 2019*)

Q.26. Per hectare fertilizer consumption of India
Ans. 137.4 Kg/ha. (*JRF 2019*)

Q.27. Among which crop is known as Queen of cereals
Ans. Maize (*JRF 2019*)

Q.28. Fertilizer news is published by
Ans. Fertilizer Association of India (*JRF 2019*)

Q.29. Polarimeter is used to determine
Ans. Sucrose (*JRF 2019*)

Q.30. The correlation between price and demand of a commodity is
Ans. Negative (*JRF 2019*)

Q.31. What is Skewness for Normal distribution
Ans. Zero (*BHU Pre PG*)

Q.32. The limit of partial correlation coefficient is
Ans. – 1 to 1 (*Raj PRE PG*)

Q.33. Marginal cost always zero when fixed cost does not change in
Ans. Out put (*JRF 2019*)

Q.34. Most effective way to overcome the defects of agriculture marketing
Ans. Regulated Marketing (*JRF 2019*)

Q.35. A line or curve connecting the least cost combination of inputs for all output levels is known as
Ans. Isoclines (*JRF 2019*)
Q.36. NAFED was established in year
Ans. 1958 (*Pre PG Exam Raj*)
Q.37. Pusa 209 is variety of
Ans. Gram (*JRF 2019*)
Q.38. Sunflower is an indicator plant for the deficiency for
Ans. Boron (*JRF 2019*)
Q.39. State having max area and production in chickpea (Gram)
Ans. M.P. (*JRF 2019*)
Q.40. The shell of groundnut pod is morphologically known as
Ans. Pericarp (*JRF 2019*)
Q.41. Pesticides always to be applied at
Ans. ETL Level (*JRF 2019*)
Q.42. Protein responsible for baking is
Ans. Glutamine - gluten (*JRF 2019*)
Q.43. NAARM is located at
Ans. Hyderabad (*JRF 2019*)
Q.44. Agri- silviculture involves
Ans. Crops & Trees (*JRF 2019*)
Q.45. Cropping system which has max area in India
Ans. Rice- Wheat (*JRF 2019*)
Q.46. Most common bio fertilizers used in Rice crop is
Ans. Azolla (*JRF 2019*)
Q.47. Which of the following is a rich source of vitamin B12
Ans. Coconut milk and Cashew (*JRF 2019*)
Q.48. Which fruit pulp is suitable for Marmalade preparation
Ans. Citrus (*JRF 2019*)
Q.49. Seedless variety of Mango is
Ans. Sindhu (*JRF 2019*)
Q.50. Vacuum cooling is most suitable for
Ans. Leafy Vegetables (*JRF 2019*)
Q.51. Country shares major contribution in fruit and Vegetables export
Ans. U.A.E. (*JRF 2019*)
Q.52. Durgapur Kesar is variety of
Ans. Watermelon (*JRF 2019*)

Q.53. For population improvement in any crop the best method is
Ans. Recurrent selection *(JRF 2019)*
Q.54. City known as World Foliage Capital is
Ans. Florida *(JRF 2019)*
Q.55. Operating pressure required in high pressure drip system
Ans. 30 or > 30 psi *(JRF 2019)*
Q.56. Atmosphere which is best for fruits and vegetables storage life
Ans. High in CO_2 & Low in Oxygen *(JRF 2019)*
Q.57. Which is non mineral nutrient essential for plants
Ans. Carbon *(JRF 2019)*
Q.58. Pear shaped variety of tomato is
Ans. Sweet-72 or teardrop *(Pre PG MP)*
Q.59. The Arid zone has total growing area of
Ans. 90 Days *(BHU Pre PG)*
Q.60. Chasma Shai garden of Srinagar is type of garden
Ans. Mughal garden *(JRF 2019)*
Q.61. Z test is used when sample size is
Ans. Large > 30 *(JRF 2019)*
Q.62. Most influence type of demonstration is
Ans. Method demonstration- *(M.P Pre PG)*
Q.63. Cytokinesis is complete by the end of
Ans. Telophase *(Raj Pre PG)*
Q.64. Stage of Khapra beetle which is most harmful
Ans. Larvae Stage *(JRF 2019)*
Q.65. In Termite the King is developed from
Ans. Unfertilized Eggs *(BHU Pre PG)*
Q.66. International organization for Biological control is situated at
Ans. Zurich (Switzerland) *(Raj Pre PG)*
Q.67. In Groundnut pods occurs without kernels / unfiled grain due to
Ans. Ca Deficiency *(BHU Pre PG)*
Q.68. Average yield Kg./ha of total Food grains
Ans. 2235 kg./ha *(JRF 2019)*

Q.69. By product of Sugarcane which used for fuel & Paper making
Ans. Bagasses *(BHU Pre PG)*

Q.70. Area under Mango cultivation in India
Ans. 2293 Thousand ha. *(BHU Pre PG)*

Q.71. Present Chairman of ASRB
Ans. Dr. A.K. Misra *(JRF 2019)*

Q.72. Minimum support price for Pearl millet
Ans. 1950/Qtl *(Raj Pre PG)*

Q.73. Present Director General of ICRISAT
Ans. Peter Carberry *(BHU Pre PG)*

Q.74. Per capita availability of milk in India
Ans. 394 gm/Day/person *(Raj pre PG)*

Q.75. State No.1 in livestock population
Ans. U.P. *(Raj Pre PG)*

Q.76. Total area under Micro irrigation
Ans. 92.05 lac ha. *(BHU Pre PG)*

Q.77. Cereal which having highest percent share of total cropped area
Ans. Paddy *(BHU Pre PG)*

Q.78. Soil Health Card Scheme (SHC) was launched on
Ans. 19th Feb 2015 *(JRF 2019)*

Q.79. No. of AICRP under ICAR are
Ans. 60 *(BHU Pre PG)*

Q.80. State having highest Area under Wheat production
Ans. U.P. *(M.P. Pre PG)*

Q.81. Year known as International Horticulture year
Ans. 2012 *(Raj Pre PG)*

Q.82. Crop which is known as a King of Oil seed
Ans. Groundnut *(Raj Pre PG)*

Q.83. Which type of plough makes V shaped furrow
Ans. Desi Plough *(BHU Pre PG)*

Q.84. National Food Security Mission' (NFSM), was launched in
Ans. October 2007. *(JRF 2019)*

JRF and State Pre P. G. Papers-2018 (Memory Based)

Q.1. TV broadcast for rural development was started in

Ans. 1957 *(JRF 2018)*

Q.2. Mango mily bug female lays eggs in

Ans. Soil *(JNKVV, Jabalpur P.G. Exam)*

Q.3. India's Rank in world in total Cereals production in 2016-17

Ans. Third *(JRF 2018)*

Q.4. Average yield Kg./ha of total Food grains in 2016-17

Ans. 2129 kg./ha *(Raj P.G. Exam)*

Q.5. Among total livestock species which has highest population in India

Ans. Cattle *(JNKVV P.G. Exam)*

Q.6. MSP rupees increased per /Qtl for Common paddy from last year

Ans. Rs. 80/Qtl *(JRF 2018)*

Q.7. Most common widely used herbicide in soya bean is

Ans. Imazethapyr *(JRF 2018)*

Q.8. Best method for quantities estimation of changes in soil moisture

Ans. Neutron moisture meter *(AO Exam, ARS 2018)*

Q.9. Drop size of < 0.5 mm called as

Ans. Drizzle *(JRF 2018, JNKVV, Jabalpur P.G. Exam)*

Q.10. Total production of Rice in 2016-17

Ans. 110.15 MT *(ARS, 2018)*

Q.11. Lines connecting points of equal sunshine hours called as

Ans. Isohels *(JRF 2018)*

Q.12. Revised weight of Urea bag is

Ans. 45 Kg *(JRF 2018)*

Q.13. Minimum support price for Common paddy in 2017-18

Ans. 1550/Qtl *(ARS 2018)*

Q.14. Fruit having highest yield tone/Ha
Ans. Papaya *(Pre P.G. Exam Raj)*

Q.15. State having highest area under Micro irrigation
Ans. Rajasthan *(ARS 2018)*

Q.16. National Forest commission was set up in year
Ans. 2002 *(ARS 2018)*

Q.17. National Organic Farming Research Institute (NOFRI) will be set up at
Ans. Sikkim *(JRF 2018)*

Q.18. No. of KVK's under ICAR
Ans. 64 *(BHU P.G., JRF 2018)*

Q.19. Year known as International Horticulture year
Ans. 2012 *(BHU P.G.)*

Q.20. Pradhan Mantri Fasal Bima Yojana was launched in
Ans. 18 Feb., 2016 *(ARS 2018)*

Q.21. Full form of FCAC programme
Ans. Farmers Capacity Assessment & Certification *(JRF 2018)*

Q.22. Tractor Loan is a type of loan
Ans. Medium term loan *(JRF 2018)*

Q.23. Oil seed having highest percent share of total cropped area
Ans. Ground nut *(Raj Pre P.G. 2018)*

Q.24. Director of Food and Agriculture Organization (FAO)
Ans. Jose Graziano da Silva *(ARS 2018)*

Q.25. BBF method widely practiced in which crop
Ans. Grount Nut *(Raj Pre P.G. 2018, ARS 2018)*

Q.26. KCC scheme was launched in
Ans. 1998-1999 *(JRF 2018, AO Exam)*

Q.27. Which state has highest average Yield Kg/ha. of Rice
Ans. Punjab *(Raj Pre P.G. 2018)*

Q.28. Damaging stage of mustard saw flay is
Ans. Caterpillers *(Raj Pre P.G. 2018)*

Q.29. At present the farming system of India has become
Ans. Market oriented (*JRF 2018*)

Q.30. North East monsoon causes rainfall in the South Eastern part of the peninsular India during
Ans. Nov.-Dec. (*Raj Pre P.G. 2018*)

Q.31. For providing inputs like quality seed, fertilizers & pesticides the agency which is present at village panchayat samiti level is
Ans. Cooperative society (*JRF 2018*)

Q.32. Most popular extra early maturing variety of pigeon pea is
Ans. UPAS-120 (*BHU P.G., ARS 2018*)

Q.33. ICAR day is celebrated on
Ans. 16th July (*JRF 2018*)

Q.34. Seed potato from hilly regions of India is preferred because
Ans. It is disease free (*Raj Pre P.G. 2018*)

Q.35. The correct sequence of operations which are practiced in tobacco
Ans. 1. Topping, 2. Desuckering, 3. Priming (*JRF 2018*)

Q.36. Total State Agriculture Universities under ICAR
Ans. 64 Universities (*BHU P.G. 2018*)

Q.37. Most common irrigation method in India is
Ans. Surface Irrigation (*JRF 2018, AO Exam 2014*)

Q.38. Highest net cultivated area for the state
Ans. Rajasthan (*Raj Pre P.G. 2018*)

Q.39. The relation between AM, GM and Harmonic mean is
Ans. $GM^2 = AM \times HM$ (*JRF 2018*)

Q.40. EC and SAR of normal soils is
Ans. EC < 4 and SAR < 13 (*ARS 2018*)

Q.41. Rubisco enzyme play important role in
Ans. Photosynthesis (*Raj Pre PG, *)

Q.42. Which one is not a measure of central tendency
Ans. Frequency (*JRF 2018*)

Q.43. The statement being tested in test of statistical significance is called as

Ans. Null Hypothesis (*ARS 2018*)

Q.44. Drifts hazards of 2,4-D is mostly observed in

Ans. Cotton (*Raj P.G. 2018*)

Q.45. VAT is type of tax

Ans. Indirect tax (*Raj Pre P.G. 2018, AO Exam*)

Q.46. Bt was discovered by

Ans. S. Ishiwata (*JNKVV, Jabalpur P.G. Exam*)

Q.47. Suitable time of budding in rose is

Ans. Dec. - Jan. (*JRF 2018*)

Q.48. Clistogamy encourages

Ans. Self-pollination (*Raj Pre P.G. 2018*)

Q.49. Fungus associated with late blight of tomato is

Ans. *Phytophthora infestans* (*Raj Pre P.G. 2018*)

Q.50. The most important causal organism responsible for the spoilage and rotting of vegetables is

Ans. Bacteria (*Raj Pre P.G. 2018*)

Q.51. For one talk how may flash cards should be used

Ans. 10-12 cards (*BHU P.G. 2018*)

Q.52. Mycoplasma is sensitive for

Ans. Tetracycline (*JRF 2018,*)

Q.53. Formula for Rate of Turn Over

Ans. $= \frac{\text{Gross Income}}{\text{Total Assets}} \times 100$ (*JRF 2018*)

Q.54. Which one is a unit less measure of dispersion

Ans. Coefficient of variation (*JRF 2018*)

Q.55. Kerbs cycle produces

Ans. 30 ATP (*Raj Pre P.G. 2018*)

Q.56. Use of M B plough for

Ans. Primary tillage (*Raj Pre P.G. 2018, AO Exam*)

Q.57. Enzymes accelerate the speed of biochemical reactions by

Ans. Lowering activation energy (*JRF 2018*)

Q.58. Soil Ph generally increased in the season of
Ans. Winter (*JRF 2018*)

Q.59. Pinjore garden is a type of garden
Ans. Formal style garden (*JNKVV, Jabalpur P.G. Exam*)

Q.60. Which is not the attribute of wealth
Ans. Instability (*BHU P.G. 2018*)

Q.61. Potato variety which is extra early
Ans. Kufri Ashoka (*JRF 2018*)

Q.62. First stable product in Calvin cycle is
Ans. C-3 compound (*BHU P.G. 2018*)

Q.63. Which fruit required heavy pruning
Ans. Ber (*JRF 2018*)

Q.64. A cross of an inbred line with an open pollinated variety is known as
Ans. Top cross (*JRF 2018*)

Q.65. In maize Ganga Safed 2 is a
Ans. Double top cross hybrid (*JRF 2018*)

Q.66. State which is leader in floriculture business
Ans. Karanataka (*JRF 2018*)

Q.67. Which is common Hill reagent
Ans. Benzoquinone (*BHU P.G. 2018*)

Q.68. NIAM is located at
Ans. Jaipur (*Raj Pre P.G. 2018*)

Q.69. Tomato variety suitable for transport and processing
Ans. Punjab Chhuhara (*BHU P.G. 2018*)

Q.70. Micronutrient which is essential for synthesis of auxin
Ans. Zn (*AO Exam*)

Q.71. During mitosis chromosomal DNA duplication takes place in
Ans. Interphase (*BHU P.G. 2018*)

Q.72. Acid content is a very important consideration for the preparation of
Ans. Jelly (*BHU P.G. 2018*)

Q.73. Sand drown disease of tobacco is due to deficiency of

Ans. Mg *(JNKVV, Jabalpur P.G. Exam)*

Q.74. Which is a gram storage pest

Ans. Callosobruchus chinensis *(JNKVV, Jabalpur P.G. Exam)*

Q.75. Energy currency in a cell is

Ans. ATP *(BHU P.G. 2018)*

Q.76. Insecticide Act was passed in

Ans. 1968 *(JRF 2018)*

Q.77. Chlorophyll type which have green black micro-crystalline solid is

Ans. Chlorophyll b *(JRF 2018)*

Q.78. Area under bt cotton in India is approx.

Ans. 6 million ha *(JNKVV, Jabalpur P.G. Exam)*

Q.79. Initiation of milk secretion is called

Ans. Lacto genesis *(BHU P.G. 2018)*

Q.80. Optimum moisture content for wheat storage is

Ans. 12% *(JRF 2018)*

Q.81. Hybrid seeds can be produced through open pollination with the use of

Ans. Male sterile lines *(Raj Pre P.G. 2018)*

Q.82. Silver fish belong to the order of

Ans. Thysanura *(JNKVV, Jabalpur P.G. Exam)*

Q.83. Site of protein synthesis in a cell is

Ans. Robosomes *(BHU P.G. 2018)*

Q.84. Pest which attacks all parts of plants is

Ans. Termite *(JRF 2018)*

Q.85. Ooze test is done to detect for

Ans. Bacteria *(BHU P.G. 2018)*

Q.86. Specific gravity of milk fat is

Ans. 0.93 *(BHU P.G. 2018)*

Q.87. Mentake disease of rice caused by

Ans. Rice root nematode *(JRF 2018)*

Q.88. Bulk population breeding is suitable for

Ans. Small grain crops *(JRF 2018)*

Q.89. *Xiphinema Longidorus* is a type of nematode

Ans. Migratory Ecto parasite nematode

(Raj. Pre PG 2018, JNKVV, Jabalpur P.G. Exam)

Q.90. Guava is commercially propagated by

Ans. Stooling *(JNKVV, Jabalpur P.G. Exam)*

Q.91. Barbari breed of goat is native of

Ans. East Africa *(BHU P.G. 2018)*

Q.92. Famous north Indian variety of Mango which has characteristic turpentine flavor

Ans. Langra *(JRF 2018)*

Q.93. Initially in how many major states the ISOPOM scheme was launched

Ans. 14 States *(BHU P.G. 2018)*

Q.94. Kisaan Bharti periodical is published by

Ans. Pant Nagar *(Jabalpur P.G. Exam)*

Q.95. For a study of farming system the best PRA exercise is

Ans. Resource mapping *(JRF 2018)*

Q.96. Project "Mera Gaon Mera Gaurav" was launched by

Ans. I C A R *(AO Exam, Jabalpur P.G. Exam 2018)*

Q.97. Edible part of pomegranate is

Ans. Aril *(Jabalpur P.G. Exam)*

Q.98. Best Ph for semen preservation is

Ans. 6.8 Ph *(Raj Pre P.G. 2018)*

Q.99. Reactions which are purely chemical reaction and not required light called as

Ans. Blackman Reactions *(Raj P.G. 2018)*

Q.100. Which glands secretes oxytocin

Ans. Pituitary *(BHU P.G. 2018)*

Q.101. Geometric mean of a given series is always less than its

Ans. Airthmetic mean *(Raj Pre P.G. 2018)*

Q.102. Recently a insecticide which is restrict to use in India is

Ans. Endosulphan *(JRF 2018)*

JRF and State Pre P.G. Papers-2017 (Memory Based)

Q.1. Study of grasses is called

Ans. Agrostology (*JRF 2017*)

Q.2. Agro climatic zones classified by NARP of ICAR are

Ans. 131 (*JRF 2017*)

Q.3. Minimum level of water erosion

Ans. Splash erosion (*JRF 2017*)

Q.4. Power tiller is

Ans. Rotatory type (*JRF 2017*)

Q.5. Pusa kranti is variety of

Ans. Brinjal (*Raj Pre P. G. 2017*)

Q.6. State which is first in production and area for potato

Ans. Uttarakhand (*JRF 2017*)

Q.7. Cereal which having highest percent share of total cropped area

Ans. Paddy (*Raj Pre P. G. 2017*)

Q.8. Aluminum phosphate is a

Ans. Rodenticide (*Raj Pre P. G. 2017*)

Q.9. Trichoderma is a

Ans. Egg parasitoid (*BHU P. G. 2017*)

Q.10. In which year Varsha Beema Yojna has started

Ans. 2004 (*JRF 2017*)

Q.11. Fruit which is not considered in Pomology

Ans. Water melon (*Raj Pre P. G. 2017*)

Q.12. Productivity of banana is highest in state

Ans. Kerala *(JRF 2017)*

Q.13. The phenomenon of heterosis has been most commercially exploited first in

Ans. Maize *(Raj Pre P. G. 2017)*

Q.14. Total Kharif food grains production in 2015-16

Ans. 125.09 MT *(JRF 2017)*

Q.15. Country having highest productivity in Paddy

Ans. Australia *(JRF 2017)*

Q.16. State having highest area under wheat production

Ans. U.P. *(JRF 2017)*

Q.17. Vitavex (Carboxin) is a effective fungicide for

Ans. Smut disease *(BHU P. G. 2017)*

Q.18. Per capita availability of total Food grains in India

Ans. 491.20 gm/day *(Raj Pre P. G. 2017)*

Q.19. Crop which has highest consumption of (N+ P + K)

Ans. Sugar Cane *(BHU P. G. 2017)*

Q.20. N P K use ratio during 2015-16

Ans. (N) 7.2 : (P) 2.9 : (K) 1 *(JRF 2017)*

Q.21. State having highest consumption of NPK/kg/ha in 2015-16

Ans. Punjab *(Raj Pre P. G. 2017)*

Q.22. State first in highest area in drip irrigation

Ans. Maharashtra *(JRF 2017)*

Q.23. Minimum support price for grade A paddy in 2016-17

Ans. 1510/Qtl *(JRF 2017)*

Q.24. State which has highest cost of cultivation in Wheat

Ans. Punjab *(Raj Pre P. G. 2017)*

Q.25. Wilt resistant variety of gram

Ans. Avrodhi *(JRF 2017)*

Q.26. Vegetable which is highest in Area & production in India

Ans. Potato (*Raj Pre P. G. 2017*)

Q.27. Current cropping intensity of India

Ans. 138.9% (*JRF 2017*)

Q.28. Approx percentage of annual rainfall received by post monsoon

Ans. 13.30% (*BHU P. G. 2017*)

Q.29. Latest NRC has established for

Ans. NRCIF for Integrated Farming (Mothihari, Bihar) (*JRF 2017*)

Q.30. Present Chairman of National Commissions for Farmers

Ans. Dr. M.S. Swaminathan (*JRF 2017*)

Q.31. Directorate of Knowledge Management in Agriculture(DKMA) is situated at

Ans. New Delhi (*BHU P. G. 2017*)

Q.32. National mission on Micro Irrigation (NMMI) stared on

Ans. 2010 (*JRF 2017*)

Q.33. Founder Chairman of NABARD

Ans. M. Ramakrishnayy (*JRF 2017*)

Q.34. PMA chemical is used in agri plants for

Ans. Reduce Transpiration (*Raj Pre P. G. 2017*)

Q.35. Growing of field crops in between two widely spaced rows of fast growing trees is known as

Ans. Alley cropping (*BHU P. G. 2017*)

Q.36. Plants of the same crops but other variety in the field known as type of weed

Ans. Relative weeds (*BHU P. G. 2017*)

Q.37. Long term persistent herbicide is

Ans. 2,4-D (*Raj Pre P. G. 2017*)

Q.38. Rice variety suitable for direct seeding is

Ans. Bala (*JRF 2017*)

Q.39. Perling index in wheat is to determine

Ans. Hardness of grain (*Raj Pre P. G. 2017*)

Q.40. Objectionable weed of wheat is

Ans. *Convolves arvensis* (*BHU P. G. 2017*)

Q.41. Type of maize plant is-

Ans. Cross pollinated plant (*BHU P. G. 2017*)

Q.42. Effect of shallow and early sowing of gram

Ans. Increase the wilt disease attack (*JRF 2017*)

Q.43. Most critical stage of irrigation in sugarcane is

Ans. Formative stage (*Raj Pre P. G. 2017*)

Q.44. Bud necrosis in G. Nut due to

Ans. Thrips (JRF 2017)

Q.45. Upper limit fixed by Govt. for Glucosinolates in mustard oil is

Ans. 0.5% (*BHU P. G. 2017*)

Q.46. In cotton sympodial are

Ans. Reproductive branches (*JRF 2017*)

Q.47. Varalaxmi is cross of

Ans. G. hirsutum & G. barbadence (*Raj Pre P. G. 2017*)

Q.48. Crinkle leaf in cotton caused by

Ans. Mn toxicity (*JRF 2017*)

Q.49. Priming process is related to

Ans. Tobacco (*JRF 2017*)

Q.50. Spiked millet is known as

Ans. Pearl millet (Bajra) (*BHU P. G. 2017*)

Q.51. Spacing for potato

Ans. 60 × 20 cm (*BHU P. G. 2017*)

Q.52. Mustard Variety which has first developed through biotechnology

Ans. Pusa Jai Kisaan (Bio-902) (*BHU P. G. 2017*)

Q.53. Insect which damage standing potato as well as stored potato

Ans. Potato tuber moth (*JRF 2017*)

Q.54. Punjab chhuhara is variety of

Ans. Tomato (*JRF 2017*)

Q.55. The colour of onion outer skin is due to

Ans. Quecertin (*BHU P. G. 2017*)

Q.56. Vegetable which is used as vegetable as well as fodder crop also

Ans. Cow pea (*Raj Pre P. G. 2017*)

Q.57. Vegetable can be processed at temperature of

Ans. 115-121°C (*JRF 2017*)

Q.58. Mango variety Sai sugandha is cross of

Ans. Totapuri × Kesar (*JRF 2017*)

Q.59. Quarantine importance pest of mango is

Ans. Stone weevil (*BHU P. G. 2017*)

Q.60. Panama wilt is associated with

Ans. Banana (*Raj Pre P. G. 2017*)

Q.61. Nematode resistant varieties of papaya

Ans. Pusa majesty (*JRF 2017*)

Q.62. Grape is propagated by

Ans. Hard wood cuttings (*JRF 2017*)

Q.63. Hardness of citrus fruit is due to

Ans. Boron deficiency (*BHU P. G. 2017*)

Q.64. Which is not fat soluble vitamin

Ans. Vitamin "C" (*JRF 2017*)

Q.65. National Horticulture Board is situated at

Ans. Gurugram (Haryana) (*JRF 2017*)

Q.66. Percent area for very dense forest in India is

Ans. 2.61% (*JRF 2017*)

Q.67. Edible part of Papaya & Mango

Ans. Mesocarp (*Raj Pre P. G. 2017*)

Q.68. Laddu is a variety of

Ans. Mandarin (Citrus fruit) *(BHU P. G. 2017)*

Q.69. Development of embryo without fertilization is called

Ans. Apomixis *(JRF 2017)*

Q.70. Garden lanterns and dry landscape are key element of

Ans. Japanese garden *(JRF 2017)*

Q.71. Scientist who is associated with rose breeding

Ans. Dr. B.P. Pal *(JRF 2017)*

Q.72. Mineral soil has usually percent of organic matter

Ans. Less the 20% *(BHU P. G. 2017)*

Q.73. On the number basis population of microbes in soils

Ans. Bacteria>Actinomycetes>Fungi> *(BHU P. G. 2017)*

Q.74. Which nutrient availability increase in rice field due to water logging

Ans. Mn^{+} *(Raj Pre P. G. 2017)*

Q.75. Most soil erosion caused by

Ans. Saltation *(JRF 2017)*

Q.76. Latest 17th Essential nutrient was added in the list of essential nutrients is

Ans. Nickel *(BHU P. G. 2017)*

Q.77. Indicator plant for Phosphorus deficiency

Ans. Mustard *(JRF 2017)*

Q.78. Insects having Siphoning type mouth parts

Ans. Moths *(Raj Pre P. G. 2017)*

Q.79. (NC-IPM) National Centre for Integrated Pest Management is situated at

Ans. New Delhi *(BHU P. G. 2017)*

Q.80. Effective parasitoid against sugarcane pyrilla is

Ans. *Epiricania melanoleuca* *(JRF 2017)*

Q.81. Rice gundhi bug cause on the stage of

Ans. Milky stage (*Raj Pre P. G. 2017*)

Q.82. Bordeaux mixture was developed by

Ans. P.M.A. Millardet (1885) (*BHU P. G. 2017*)

Q.83. White rust of cruciferous caused by

Ans. *Albugo candida* (*BHU P. G. 2017*)

Q.84. Example of endemic disease

Ans. Early blight of potato (*Raj Pre P. G. 2017*)

Q.85. Scientific name of Pin nematode

Ans. *Paratylenchus spp.* (*JRF 2017*)

Q.86. ICAR was established by the recommendation of

Ans. Lord Linlithgo (*JRF 2017*)

Q.87. Food for work was started in year

Ans. 1977 (*JRF 2017*)

Q.88. In farm management most important unit is

Ans. Production unit (*JRF 2017*)

Q.89. ICDP was stared in year

Ans. 2003-04 (*JRF 2017*)

Q.90. For mass contact the no of person should be

Ans. > 30 person (*JRF 2017*)

Q.91. When fixed cost doesn't change with change in output than marginal cost will be always

Ans. Zero (*JRF 2017*)

Q.92. Relation between cost function and production function is

Ans. Negative (*Raj Pre P. G. 2017*)

Q.93. No. of Cereals covered under MSP

Ans. Seven (*JRF 2017*)

Q.94. Average size of operational holdings by Small farmers

Ans. 1.42 ha. (*JRF 2017*)

Q.95. When demand curve is flatter then demand is.

Ans. Relative elastic (*Raj Pre P. G. 2017*)

Q.96. Square of standard deviation is *(JRF 2017)*

Ans. Variance

Q.97. What is the asymptotic power of Kolmogorov Smirnov test *(JRF 2017)*

Ans. One

Q.98. The limit of partial correlation coefficient is *(JRF 2017)*

Ans. –1 to 1

JRF Papers- 2016 (Memory Based)

Q.99. India's Rank in world in Wheat production *(JRF 2016)*

Ans. Second

Q.100. National Farmers day celebrated on *(JRF 2016)*

Ans. 23rd Dec

Q.101. ICRAF is situated at *(JRF 2016)*

Ans. Kenya

Q.102. Which is related to asexual reproduction *(JRF 2016)*

Ans. Amphimixis

Q.103. Piezometer is used to measure *(JRF 2016)*

Ans. Water table

Q.104. Micronaire in cotton is used to measure *(JRF 2016)*

Ans. Fibre fineness

Q.105. Which is called as a Green gold *(JRF 2016)*

Ans. Bamboo

Q.106. The two national programme which has started in same year in 1999 *(JRF 2016)*

Ans. NATP & JGSY

Q.107. Country having largest stocks of Potassium *(JRF 2016)*

Ans. Canada

Q.108. The ratio of normal GNP to real GNP which can be interpreted as comprehensive price index is called *(JRF 2016)*

Ans. GNP Deflator

Q.109. Net benefit cost is calculated by

Ans. Gross return/Total coast (*JRF 2016*)

Q.110. When marginal production is equal to average production then elasticity of production is equal to

Ans. One (*JRF 2016, 2006, Raj PG 2008*)

Q.111. Place and year of Asian Development bank (ADB) establishment

Ans. 1966, Manila (*JRF 2016*)

Q.112. An agro forestry system having components of crops + Trees + pasture/ Animals called as

Ans. Agrisilvi pasture (*JRF 2016*)

Q.113. Water hyacinth vegetative multiplies through

Ans. Offsets (*JRF 2016*)

Q.114. Highest cropping intensity for the state

Ans. Punjab (189%) (*JRF 2016, BHU 2004*)

Q.115. The sphere which lies above stratosphere is known as

Ans. Mesosphere (*JRF 2016*)

Q.116. Black revolution is related to

Ans. Biofuel/Jatropha production (*JRF 2016*)

Q.117. Feed the soil not the plant is the matching word of

Ans. Eco farming (*JRF 2016*)

Q.118. Phone No. of Kisaan Call centre

Ans. 18001801551 (*JRF 2016*)

Q.119. Acidic soils is reclaimed through use of

Ans. Limestone (*JRF 2016*)

Q.120. The largest and most important soil group of India is

Ans. Alluvial soils (*JRF 2016. 2008*)

Q.121. National mission on Micro Irrigation (NMMI) stared in year

Ans. 2010 (*JRF 2016, Raj PG 2011*)

Q.122. Amino acid are building blocks of

Ans. Protein (JRF 2016. 2004)

Q.123. Plant portion which is used for virus free plant production

Ans. Meristem (JRF 2016)

Q.124. In which group of plants the endosperm is 3 N

Ans. Angiosperm (JRF 2016)

Q.125. Reductive division in cells is known as

Ans. Meiosis (JRF 2016, 2009)

Q.126. The site of B- Oxidation is

Ans. Mitochondria (JRF 2016, 2004)

Q.127. Present Chairman & M.D. of F.C.I.

Ans. Yogendra Tripathi (JRF 2016)

Q.128. AICRP on Sesame and Niger is located at

Ans. Jabalpur (M.P.), (JRF 2016)

Q.129. The Indian Journal of Agriculture Sciences published by

Ans. ICAR (JRF 2016, BHU 2004)

Q.130. State which has highest acidic soils

Ans. West Bengal (JRF 2016)

Q.131. Directorate of Knowledge Management in Agriculture (DKMA) is situated at

Ans. New Delhi (JRF 2016)

Q.132. Conversion of Rice (Cleaned) of paddy

Ans. 2/3 of Paddy production (JRF 2016.2008, Pre PG 2006)

Q.133. Fruit which covers highest area in India

Ans. Mango (JRF 2016)

Q.134. Duration of North East (Winter) monsoon

Ans. Jan-March (JRF 2016)

Q.135. Percent share of total cropped area for fruits

Ans. 2.22% (JRF 2016)

Q.136. Country first in transgenic plants

Ans. USA (JRF 2016)

Q.137. Spices having highest production in India

Ans. Chilly (*JRF 2016*)

Q.138. MSP rupees increased /Qtl for Wheat from last year

Ans. Rs. 75 (*JRF 2016*)

Q.139. Per capita availability of milk in India

Ans. 322 gm/Day/person (*JRF 2016, JRF 2003, BHU 2006,*)

Q.140. Which product contribute highest value in Agri Exports

Ans. Marine products (*JRF 2016*)

Q.141. Which State is No. 1 for Total pulses production

Ans. M.P. (*JRF 2016, 2004,*)

Q.142. India's Rank in world in Potato production

Ans. Second (*JRF 2016, BHU 2006*)

Q.143. Average yield per/ha of Wheat in India

Ans. 3145Kg./ha. (*JRF 2016, 2005*)

Q.144. Percent contribution of livestock of India in world

Ans. 15% (*JRF 2016*)

Q.145. Full form of NOFRI

Ans. National Organic Farming Research institute (*JRF 2016*)

Q.146. Sterility is due to deficiency of vitamin

Ans. Vitamin E(*JRF 2016, 2004, Raj PG 2007, BHU 2006*)

Q.147. Enzymes that attack bonds in DNA

Ans. DNAases (*JRF 2016*)

Q.148. Soils which have pH< 4.0 known as

Ans. Cat Soils (*JRF 2016, 2008, Raj PG 2009*)

Q.149. Nutrient essential for oil seeds crop is

Ans. Sulphur

(*JRF 2015, 2009, 2004, BHU 2006, Raj PG 2008*)

Q.150. Particle density off most of the soils is

Ans. 2.65 mg/cm^3

(*JRF 2016, 2008, 2005, 2002, BHU 2006, Raj PG 2009*)

Q.151. Oil percent in ground nut is

Ans. 40-45% (*JRF 2016, 2011, Raj PG 2010, BHU 2003*)

Q.152. Sugarcane tying should be done in the month of

Ans. August (*JRF 2016*)

Q.153. Best variety of gram for dry land is

Ans. C-235 (*JRF 2016, 2006, Raj PG 2009*)

Q.154. Bio herbicide used in Rice is

Ans. Collego (*JRF 2016*)

Q.155. Saline tolerant rice variety is

Ans. SR 26 B (*JRF 2016*)

Q.156. Which pattern of planting has max plant population

Ans. Cubodial pattern

(*JRF 2016, 2007, BHU 2004, Raj PG 2010*)

Q.157. White grub is type of pest

Ans. Polyphages (*JRF 2016, 2008, Raj PG 2010, BHU 2005*)

Q.158. Vector of grassy stunt disease is

Ans. Brown plant hopper

(*JRF 2016, 2006, Raj PG 2010, BHU 2003*)

Q.159. Brown rust in wheat is due to

Ans. Puccinia graminis recondite

(*JRF 2016, 2005, Raj PG 2006, BHU 2005*)

Q.160. When two variables move in the same direction correlation is called

Ans. Positive (*JRF 2016, 2005, Raj PG 2003, BHU 2004*)

Q.161. Which measure of dispersion is considered as best

Ans. Standard deviation

(*JRF 2016, 2002, Raj PG 2003, BHU 2004*)

Q.162. Natural scale indicates'

Ans. Absolute changes *(JRF 2016, 2004)*

Q.163. 'F' test can be used for testing the significance of

Ans. Several difference

(JRF 2016, 2002, Raj PG 207, BHU 2006)

JRF (Papers–2015)

Q.164. Total food grain production of country in 2014-

Ans. 264.67 MT *(BHU 2014, JRF 2015)*

Q.165. Country leader in wheat production-

Ans. China *(RAU PG 2014, JRF 2015)*

Q.166. State no. 1 in pulses & oilseeds production

Ans. M.P. *(RAU PG 2013, JRF 2015)*

Q.167. Bale size of jute-

Ans. 180 Kg. *(JRF 2011, 2015)*

Q.168. Forest research Institute (FRI) established in year-

Ans. 1900 *(JRF 2009, 2015)*

Q.169. NAFED was established in year-

Ans. Oct 1958 *(JRF 2007, 2015)*

Q.170. NAIP was established in year-

Ans. July 2006 *(JRF 2008, 2015)*

Q.171. ARS was established in year-

Ans. 1973 *(JRF 2006, 2015)*

Q.172. IARI shifted to New Delhi-

Ans. 1936 *(JRF 1997,2001, 2015)*

Q.173. Gross cropped area of India-

Ans. 193 mha *(JRF 1995,99,2001, 2015)*

Q.174. Average yield per/ha of Rice -

Ans. 2461 Kg./ha. *(RAU PG 2013, JRF 2015)*

Q.175. State having highest area under Maize Production in 2013-14

Ans. Karnataka *(JRF 2015)*

Q.176. India's percent share contribution in world production of Sugar cane

Ans. 19.60% *(JRF 2015)*

Q.177. What was Nutrient Base Subsidy (NBS) for Phosphorous in 2014-15

Ans. Rs.18.67/Kg. *(JRF 2015)*

Q.178. Minimum support price for grade A paddy in 2014-15

Ans. 1400/Qtl *(JRF 2015)*

Q.179. Per capita availability of cereals in India

Ans. 423.5 gm./Day *(JRF 2008)*

Q.180. State soils having highest % of Zinc deficiency in India-

Ans. Maharashtra *(RAU 2014, JRF 2015)*

Q.181. RBI was established in year-

Ans. 1935 *(JRF 2006, 2010, 2015)*

Q.182. Indian institute of vegetable research (IIVR) situated at-

Ans. Varanasi *(BHU 2015, JRF 2015)*

Q.183. NBAII is situated at-

Ans. Bangalore *(JRF & BHU 2015)*

Q.184. Dr. NE Borlaug was basically a-

Ans. Plant Pathologist *(JRF 2009, 2015)*

Q.185. N P K use ratio during 2013-14

Ans. (N) 8.0 : (P) 2.7: (K) 1 *(JRF 2015)*

Q.186. State first in highest area in drip irrigation

Ans. Maharashtra *(BHU 2014, JRF 2015)*

Q.187. Maize breeder who was awarded by WFO in 2000

Ans. S.K.Vashal *(BHU 2014, JRF 2015)*

Q.188. MRP fixed for DAP in 2013-14

Ans. Rs.20000 to 20123/tone *(JRF 2015)*

Q.189. Average size of operational holdings by medium farmers-

Ans. 5.76 ha. *(JRF 2015)*

Q.190. Father of golden revolution-

Ans. Dr. K.L. Chadda *(JRF 2015)*

Q.191. Percent share of total cropped area for Mango

Ans. 0.69% *(JRF 2015)*

Q.192. National Mission on Micro Irrigation (NMMI) stared on-

Ans. 2010 *(JRF 2015)*

Q.193. Total No. of institute under ICAR at present-

Ans. 60 *(JRF 2015)*

Q.194. Genetically modified rice is-

Ans. Golden Rice *(JRF 2015)*

Q.195. Cobalomine is known as-

Ans. Vitamin B12 *(JRF 2015, BHU 13)*

Q.196. Anemia is caused by-

Ans. Folic acid *(JRF 2015)*

Q.197. Part of opium poppy used for medicinal use-

Ans. Fruit *(JRF 2015, RAU 13)*

Q.198. Hormone responsible for ripening of fruits-

Ans. Ethylene *(JRF, 2008, 2015)*

Q.199. No of essential amino acid-

Ans. 10 *(JRF 2015, BHU2009,)*

Q.200. Kishan Day is celebrated on-

Ans. 23rd Dec *(JRF 2015, BHU 12)*

Q.201. Diara cultivation method is associated with-

Ans. Cucurbits *(JRF 2015)*

Q.202. Most common method of irrigation in India-

Ans. Surface irrigation *(JRF, 2008, 2015)*

Q.203. Bio fertilizer for rice-

Ans. Azolla *(JRF 2015, RAU PG 08,11)*

Q.204. Plants of no economic importance for human being called-

Ans. Absolute weed *(JRF 2015)*

Q.205. Vegetable which is highly sensitive for boron toxicity-

Ans. Cabbage *(JRF 2015, RAU PG 06,11)*

Q.206. Development of embryo without fertilization is called-

Ans. Apomixes *(JRF, 011, 2015)*

Q.207. Major disease of mango-

Ans. powdery mildew *(JRF 2015)*

Q.208. Pineapple is a varity of-

Ans. Sweet orange *(JRF 2015)*

Q.209. Punjab chhuhara is a varity of-

Ans. Tomato *(JRF 2015)*

Q.210. Red colour of chili is due to-

Ans. Capcyanthin *(JRF 2015, 2009)*

Q.211. NPV is mostly used to control-

Ans. Lepidoptera insects *(JRF2015, BHU 2014)*

Q.212. Safe insecticide for vegetable is-

Ans. Malathion *(JRF2015, BHU 2014)*

Q.213. Widely used fungicide in India is-

Ans. Mencozeb *(JRF 2015)*

Q.214. Half of the world raisins supplied from-

Ans. California *(JRF 2015)*

Q.215. Horizontal flow of water channel is called-

Ans. Seepage *(BHU 2008, JRF 2015, 2006)*

Q.216. Rice dwarf variety which was first developed in world-

Ans. TN-1 (Taiwan) *(JRF 2015)*

Q.217. Seed rate of hybrid rice-

Ans. 15 Kg./Ha *(RAU PG 07, BHU PG 09, JRF 2015)*

Q.218. Benthiocarb is a type of herbicide

Ans. Pre emergence *(RAU PG 2006, JRF 2015)*

Q.219. Transition zone between root and shoot is known as

Ans. CRI Stage *(BHU PG 2013, JRF 2015)*

Q.220. Which type of wheat is used for Suji

Ans. T. durum *(JRF 2008, 2015)*

Q.221. Rabi maize is extensively grown in state

Ans. Bihar *(JRF 2009, 2015)*

Q.222. Top most mineral horizon is

Ans. A Horizon *(JRF 2004, 07, 2015)*

Q.223. Crop having highest tolerance to Boron

Ans. Sugar beet *(BHU 2011, JRF 2015)*

Q.224. Property of soil which can't be changed

Ans. Soil Texture *(RAU PG 2004,10, JRF 2006, 2015)*

Q.225. Fertilizer which are higher leaching loss

Ans. Nitrate fertilizers *(JRF 2006, 2015 BHU 2008)*

Q.226. Potassium nitrate contains percentage of N

Ans. 13% N *(JRF 2006,15 BHU 2009,2015)*

Q.227. Red colour in tomato is due to-

Ans. Lycopine *(RAU PG 2008, BHU 2006, JRF 2015)*

Q.228. Commercial propagation of rose is done by

Ans. Shield Budding *(JRF 2003, 2015)*

Q.229. First seedless variety of Mango

Ans. Sindhu *(JRF 2011, 2015)*

Q.230. Scooping is most common practices in crop

Ans. Cauliflower *(JRF 2015)*

Q.231. Insects having Siphoning type mouth parts

Ans. Butterfly & Moths *(JRF2015, BHU 2008,11)*

Q.232. Nation center for IPM is situated at

Ans. New Delhi *(RAU PG 2013, 2015)*

Q.233. The effective new fungicide for control of Oomycetous fungus is

Ans. Metalaxyl *(JRF15,BHU 2015)*

Q.234. Percentage share of Agri imports in national imports in 2013-14

Ans. 4.09% *(JRF 2015)*

Q.235. World food day is celebrated on

Ans. 16th Oct *(JRF 2015)*

Q.236. Simplest measure of dispersion

Ans. Range *(JRF 207,09, 2015)*

Q.237. What is median of 3,4,5,6,7,8,9

Ans. 6 *(JRF 2015)*

Q.238. Which is used for calculating average speed-

Ans. Harmonic mean *(JRF2004,07,10, 2015)*

JRF (Papers–2014)

Q.239. Constitution of NCF(National Commission of Farmers) in year

Ans. 18 Nov. 2004 *(JRF 2014)*

Q.240. Director General of ICAR

Ans. Dr. S. Ayyappan *(JRF 2014)*

Q.241. No. of KVK's are in India in year 2013

Ans. 630 *(JRF 2014,RAU PG 2014)*

Q.242. How many nutrients are included under nutrient based subsidy

Ans. 4 *(JRF 2014)*

Q.243. What is Nutrient Base Subsidy (NBS) for Nitrogen in 2013-14

Ans. Rs.20.87/Kg. *(JRF 2014)*

Q.244. Average rainfall of the country is

Ans. 1194mm *(JRF 2014)*

Q.245. Element which is responsible for osmotic and ionic balance in plants

Ans. K *(JRF 2014)*

Q.246. Browning of cauliflower disease caused due to

Ans. Mo Deficiency *(JRF 2014)*

Q.247. Bean common mosaic disease caused by

Ans. Pot virus *(JRF 2014)*

Q.248. Laghaemoglobin pigment present in

Ans. Root nodules of legumes *(JRF 2014)*

Q.249. Wolf apple is common name of

Ans. Tomato *(JRF 2014)*

Q.250. KVK awarded by Best KrishiVigyan Kendra award in 2013

Ans. Dhakrani, Dehra Dun *(JRF 2014)*

Q.251. ICAR Normon Borlaug Award given in time of years

Ans. Once in 5 Years *(JRF 2014)*

Q.252. Bt cotton mainly attacked by which type group of insects

Ans. Sucking type pest *(JRF 2014)*

Q.253. Indian village scheme is associate with person

Ans. S.N. Gupta *(JRF 2014)*

Q.254. National Farmers day is celebrated on

Ans. First Friday of Dec. *(JRF 2014)*

Q.255. Plant known as a dairy plant also is

Ans. Soybean *(JRF 2014)*

Q.256. Most common bulb vegetables crops are

Ans. Onion & Garlic *(JRF 2014)*

Q.257. Record that is used for maintaining relationship between parents and offspring's is

Ans. Pedigree Record *(JRF 2014)*

Q.258. Pusa Chamtkar is variety of

Ans. Chick Pea *(JRF 2014)*

Q.259. Throne less variety of Rose is-

Ans. Chitra *(JRF 2014)*

Q.260. Most recently formed soils are

Ans. Alluvial Soils *(JRF 2014)*

Q.261. India's rank in Rice production-
Ans. Second *(JRF 2014)*

Q.262. Sugar recovery in Sugarcane in India is
Ans. 6-8% *(JRF 2014)*

Q.263. The curve of Normal Distribution is
Ans. Bell shaped curve *(JRF 2014, 2009)*

Q.264. Correlation coefficient varies between-
Ans. 1 to +1 *(JRF 2014, 2011, 2007. 2004)*

Q.265. Directorate of Mushroom Research situated at
Ans. Solan (H.P.) *(JRF 2014, 2005)*

Q.266. Father of Plant tissue culture
Ans. H.Haberlandt *(JRF 2014)*

Q.267. NREGA was started in year
Ans. 2006 *(JRF 2014)*

Q.268. 1st Vegetable crop which is under Bt approval in India is
Ans. Brinjal *(JRF 2014)*

Q.269. Bhima Super variety which has released by ICAR is a variety of
Ans. Onion *(JRF 2014)*

OLD QUESTIONS

Q.270. First Indian Director of IARI was.
Ans. Dr.B. Vishvanath *(J.R.F. 1994)*

Q.271. Father of Modern plant pathology.
Ans. Anton de Bary *(Ph.D. 1997)*

Q.272. Bordeaux mixture was developed by.
Ans. P. M. A. Millardet *(J.R.F. 1995)*

Q.273. Wart disease of potato is caused by.
Ans. Synchitriem endobioticum *(J.R.F. 1994)*

Q.274. Nitrogen fixation in rice field is carried out by which blue green algae.
Ans. Azolla *(Ph.D. 1997)*

Q.275. Total area of India is.
Ans. 32,87263 sq. km. *(J.R.F. 1994)*

Q.276. Forest area in India
Ans. 19.39% *(Ph.D. 1998) (J.R.F. 2003)*

Q.277. Which country tops in the sugar production.
Ans. India *(J.R.F. 1994)*

Q.278. Which state is highest producer of Rice.
Ans. West Bengal *(Ph.D. 1997)*

Q.279. In India which state is leading sugar producer.
Ans. Uttar Pradesh *(J.R.F. 1995)*

Q.280. The most critical stage during the growth of wheat is.
Ans. CRI stage *(J.R.F. 1994, 2003) (A.R.S. 2009)*

Q.281. Urea contains ....% of Nitrogen.
Ans. 46% *(J.R.F. 1995) (J.R.F. 2003)*

Q.282. Pusa ruby is a variety of .... crop.
Ans. Tomato *(J.R.F. 2001)*

Q.283. Net sown area of India is.
Ans. 162 millions hectare.

Q.284. Gross cropped area of India is.
Ans. 193, million hectare *(J.R.F. 1995)*

Q.285. KVK was recommended by which committee.
Ans. Mohan Singh Mehta Committee *(Ph.D. 1997)*

Q.286. Inflorescence of sugarcane is known as.
Ans. Arrow *(J.R.F.1994)*

Q.287. Red colour of tomato is due to.
Ans. Lycopersicon *(J.R.F.1994)*

Q.288. Richest source of Vitamin C is.
Ans. Barbadose cherry *(Ph.D. 1997)*

Q.289. Which Indian scientist shared world food prize for miracle maize.
Ans. Dr. Sruinder K. Vasal *(Ph.D. 1999)*

Q.290. Power tiller is most suitable for the cultivation of.
Ans. Paddy *(J.R.F.1994)*

Q.291. Harvesting of Paddy is done at moisture content.
Ans. 21-23% *(Ph.D. 1997)*

Q.292. National Seed Corporation was established in.
Ans. 1963 *(J.R.F. 1995)*

Q.293. National Seed act was passed in the year
Ans. 1966 *(J.R.F. 1994,1998)*

Q.294. Loose smut of wheat is.
Ans. Internally seed borne. *(J.R.F. 1995)*

Q.295. Tz test is done for.
Ans. Viability & Vigour *(Ph.D. 1997)*

Q.296. Establishment of NABARD.
Ans. 12 July, 1982 *(J.R.F. 2001)*

Q.297. Cereals are deficient in which amino acid.
Ans. Lysine *(J.R.F. 1994)*

Q.298. Harvest index is.
Ans. Economic yield/Biological yield×100 *(J.R.F. 1998)*

Q.299. Kresek in rice is caused by.
Ans. Xanthomonas oryzae *(Ph.D. 1997)*

Q.300. Gressy shoot disease of sugarcane is caused by.
Ans. Mycoplasma *(J.R.F. 1994)*

Q.301. Bunchy top of Banana is caused by.
Ans. Virus *(J.R.F. 1995)*

Q.302. Which oil seed crop leads in the production.
Ans. Ground nut (33%) *(J.R.F. 2001)*

Q.303. Which state has highest production of soyabean in India.
Ans. Madhya Pradesh *(J.R.F. 1998)*

Q.304. Which state in India is the leading producer of coffee.
Ans. Karnataka *(J.R.F. 1994)*

Q.305. The name of the macaroni wheat is.

Ans. Triticum durum *(Ph.D. 1997)*

Q.306. Hand refrectometer reading for sugarcane maturity is.

Ans. 20 *(J.R.F. 1995)*

Q.307. Which portion of cane is best suited to be used as seed/set.

Ans. Top 1/3rd to ½ *(J.R.F. 2001) (J.R.F. 2003)*

Q.308. In mango edible part is called

Ans. Mesocarp *(Ph.D. 1997)*

Q.309. Amino acid which is deficient in legumes

Ans. Methionine *(J.R.F. 1994)*

Q.310. The plants growing in salt water are known as.

Ans. Halophites *(J.R.F. 1995)*

Q.311. Granite is .... rock.

Ans. Ignite *(J.R.F. 2001)*

Q.312. Sugarcane is .... plant.

Ans. C_4 *(J.R.F.1994) (Ph.D. 1997)*

Q.313. Kresek symptom is present in.

Ans. Bacterial Leaf Blight *(J.R.F. 1995)*

Q.314. Pahala blight of sugarcane is caused due to deficiency of.

Ans. Mn *(J.R.F. 1995)*

Q.315. Reclamation disease due to deficiency of.

Ans. Cu *(Ph.D. 1997)*

Q.316. Contribution of livestock to agriculture GDP.

Ans. 25% *(J.R.F. 1994)*

Q.317. What is the amount of fertilizers consumed in India anually.

Ans. 19 milllon tones *(J.R.F. 2001)*

Q.318. The term Green Revolution was coined by.

Ans. William Gudd *(J.R.F. 1998)*

Q.319. R.B.I. was established in which year.

Ans. 1935 *(Ph.D. 1997)*

Q.320. Central Soil Salinity Research Institute is situated at.

Ans. Karnal *(J.R.F. 1994, 2003)*

Q.321. Late blight of potato is caused by.

Ans. Phytophthora infestans *(Ph.D. 1999)*

Q.322. Hormone related to drought tolerance is.

Ans. Absicis acid *(Ph.D. 1997)*

Q.323. Apical bud dominance is caused by which hormone.

Ans. Auxin *(J.R.F. 1994)*

Q.324. A hormone used as a herbicide.

Ans. 2.4-D *(J.R.F. 2001)*

Q.325. Proteins are made up of.

Ans. Amino acids *(Ph.D. 1997)*

Q.326. How many nutrients are termed as essential elements for plants.

Ans. 17 *(Ph.D. 2002)*

Q.327. Translocation of water and nutrients from roots to above ground parts of plants takes place through.

Ans. Xylem *(J.R.F. 1994)*

Q.328. Causal agent of mad cow disease is.

Ans. Priens *(Ph.D. 1999)*

Q.329. Dormancy breaking hormone is.

Ans. Cytokinin *(J.R.F. 2001)*

Q.330. Total registered pesticide in India are

Ans. 164 *(J.R.F. 1994)*

Q.331. ICAR day is celebrated on.

Ans. 16th July *(Ph.D. 1999)*

Q.332. ICRISAT is situated at.

Ans. Hyderabad *(J.R.F. 1996)*

Q.333. Double helical structure of DNA was given by.
Ans. Watson & Crick *(J.R.F. 1994)*

Q.334. Which wind cause rainfall in Taminadu.
Ans. North East monsoon *(J.R.F. 2001)*

Q.335. Ring worm disease is caused by.
Ans. Fungus *(J.R.F. 1994)*

Q.336. Total number of KVK in India till 2007.
Ans. 551 *(Ph.D. 2002)*

Q.337. National Agriculture insurance scheme was introduced in the year.
Ans. 1999–2000 (Rabi) *(Ph.D. 2002)*

Q.338. Who discovered Vitamin.
Ans. Funk *(J.R.F. 1994)*

Q.339. Theory of evolution was given by.
Ans. Charles Darwin *(Ph.D. 1999)*

Q.340. Laws of heredity was given by.
Ans. Gregor Mendal *(B.H.U. Pre P.G. 2002)*

Q.341. Instrument used to measure atmospheric pressure.
Ans. Barometer *(J.R.F.1994) (J.R.F. 2003)*

Q.342. Mycorrhiza is a symbiotic association between.
Ans. Fungi & roots of higher plants *(J.R.F. 1996)*

Q.343. Growing of plants under soil less condition is called.
Ans. Hydroponics *(J.R.F. 1994)*

Q.344. Stress hardening in plants can be active by.
Ans. ABA *(J.R.F. 2001)*

Q.345. Flowering hormone used in pineapple.
Ans. $H_2C = CH_2$ *(J.R.F. 1994)*

Q.346. Photo respiration is .... process.
Ans. Energy Spending *(Ph.D. 1999)*

Q.347. The major form of transport carbohydrate in higher plant is.
Ans. Sucrose *(Ph.D. 2002)*

Q.348. Element contributing to the disease & drought resistance.

Ans. Potassium *(J.R.F. 1998)*

Q.349. Khaira disease is caused by the deficiency of.

Ans. Zn *(J.R.F. 2001, 2003)*

Q.350. Little leaf of citrus is caused by.

Ans. Zn deficiency *(Ph.D. 2002)*

Q.351. Complex disease in association with nematode.

Ans. Yellow ear rot *(B.H.U. Pre P.G. 2002)*

Q.352. Solar heat treatment is recommended against.

Ans. Ustilago segetum tritici *(J.R.F. 1996)*

Q.353. The element involved in energy transfer and storage in plants is.

Ans. Phosphorus *(J.R.F. 2001)*

Q.354. Muriate of Potash (MOP) is chemically.

Ans. KCl *(B.H.U. Pre P.G. 2002)*

Q.355. Maximum allowable Biurate content of Urea is.

Ans. 1.5% *(J.R.F. 2003)*

Q.356. For the maximization of net revenue the condition is.

Ans. $\frac{\Delta YI}{\Delta XI} = \frac{PXI}{PYI}$ *(Ph.D. 1999)*

Q.357. Over the years the contribution of Agriculture to GDP of Indian economy has been.

Ans. Decreasing *(Ph.D. 2002)*

Q.358. On the indifference curve, the utility is.

Ans. Constant *(Ph.D. 1999)*

Q.359. The 'U' shape of cost curves could be best explained by the law of.

Ans. Variable properties *(Ph.D. 2002)*

Q.360. For the consumer to be in equilibrium the necessary condition is that the ratio of.

Ans. Marginal utilities is equal to price ratio *(Ph.D. 1999)*

Q.361. Indifference curve approach could be attributed to.

Ans. Hicks *(Ph.D. 2002)*

Q.362. Pineapple is a .... plant.

Ans. CAM *(J.R.F. 1995)*

Q.363. In the tropical climate .... plants are more produced.

Ans. C_4 plants *(Ph.D. 1998)*

Q.364. Mg^{+2} is component of.

Ans. Chlorophyll *(J.R.F. 2003) (Ph.D. 1999)*

Q.365. Heart rot of sugarbeet is caused by.

Ans. Boron deficiency *(J.R.F. 1995)*

Q.366. 2, 4- D is used for .... type weeds control.

Ans. Broad leaf *(B.H.U. Pre P.G. 2002)*

Q.367. The berry size of Thomson Seedless grapes increased by the hormone.

Ans. GA_3 *(Ph.D. 1998)*

Q.368. The optimum spacing for wheat is (line to line).

Ans. 22.5cm. *(J.R.F. 1995)*

Q.369. The non traditional area for cultivating wheat is.

Ans. Eastern India *(B.H.U. Pre P.G. 2002)*

Q.370. The all India average for wheat yield is.

Ans. 12-14 qt/ha *(Ph.D. 2002)*

Q.371. The haploid number of chromosome in rice is.

Ans. 12 *(Ph.D. 1998)*

Q.372. Rice is considered as a .... plant.

Ans. Short day plant *(J.R.F. 1996)*

Q.373. The optimum depth of puddling in rice is.

Ans. 5 cm *(Ph.D. 2002)*

Q.374. In wet nursery for rice, the level of water is maintained at.

Ans. 5 cm. *(B.H.U. Pre P.G. 2002)*

Q.375. In field paddy grains is harvested when percent moisture is.

Ans. 20-25% *(J.R.F. 1996)*

Q.376. Total water requirement of the sugarcane crop is.

Ans. 200-300 cm. *(J.R.F. 1995)*

Q.377. Sugarcane is irrigated every .... days during its growing period.

Ans. 8-12 days *(J.R.F. 1996)*

Q.378. In India, the predominant species of cultivated cotton is.

Ans. G. hersutum *(B.H.U. Pre P.G. 2002)*

Q.379. Regurs refers to.

Ans. Black soil *(Ph.D. 1998)*

Q.380. Vitamin C also known as

Ans. Ascorbic acid *(A.R.S. 2007)*

Q.381. N content of F.Y.M. and Urban compost respectively is.

Ans. 0.5 and 1.4% *(B.H.U. Pre P.G. 2002)*

Q.382. Major P fertilizer in India is.

Ans. DAP *(J.R.F. 1995)*

Q.383. Salatation is a type of.

Ans. Wind erosion *(J.R.F.1996)*

Q.384. Very few poor (below poverty line) people is in.

Ans. Punjab *(J.R.F. 1991)*

Q.385. Biologically active form of glucose is.

Ans. D-form *(Ph.D. 1998)*

Q.386. The bond present in sugars.

Ans. Phosphodiester bond *(Ph.D. 1999)*

Q.387. The most abundant protein present in the world is.

Ans. Rubisco *(J.R.F. 1995)*

Q.388. The form of amino acids present in living organisms is.

Ans. L-form *(B.H.U. Pre P.G. 2002)*

Q.389. The pairs is strongly bound with each other

Ans. A + T, G + C, A + C *(Ph.D. 1999)*

Q.390. Bio fertilizer more suited for sugarcane is

Ans. Azotobacter *(Ph.D. 1998)*

Q.391. Lines joining equal rainfall called

Ans. Isohyet *(J.R.F. 1995)*

Q.392. Origin of maize country

Ans. South America *(B.H.U. Pre P.G. 2002)*

Q.393. Depth of sowing of soyabean seed

Ans. 3 cm. *(J.R.F. 2003)*

Q.394. Cropping intensity of India

Ans. 140% *(Ph.D. 2002)*

Q.395. First rice variety introduced in India

Ans. (1966) IR-8 *(J.R.F. 2001)*

Q.396. Nicotine content in tobacco is related with

Ans. Nitrogen *(Ph.D. 1999)*

Q.397. Boro rice is transplanted in

Ans. Nov.-Dec. *(Ph.D. 1998)*

Q.398. Glyphosate is a .... herbicide

Ans. Non selective *(J.R.F. 1995)*

Q.399. Indian mustard is

Ans. Brassica juncea *(J.R.F. 2001)*

Q.400. Mass flow is affected by

Ans. Transpiration *(Ph.D. 2002)*

Q.401. Carbon content in organic matter

Ans. 58% *(Ph.D. 1998)*

Q.402. Desi cotton is known as

Ans. Gossipium arboreum *(J.R.F. 1995)*

Q.403. Metric suctions is measured by

Ans. Tensiometer *(B.H.U. Pre P.G. 2002)*

Q.404. Nitrogen bio fertilizer for wheat

Ans. Azotobacter *(J.R.F.1996)*

Q.405. Brix measures.

Ans. TSS *(J.R.F. 1995)*

Q.406. Verticle mulch is used in soils.

Ans. Black cotton soils *(J.R.F. 2001)*

Q.407. Design used when fertility gradient is in two direction.

Ans. L.S.D. *(Ph.D. 1998)*

Q.408. Criteria of essentiality was given by.

Ans. Arnon & Stout *(B.H.U. Pre P.G. 2002)*

Q.409. Nucleus was discovered by.

Ans. Robert brown *(Ph.D. 1999)*

Q.410. Segregation occurs during.

Ans. Meiosis only *(J.R.F. 1996)*

Q.411. The longest mitotic phase is.

Ans. Prophase *(J.R.F. 1995), (2003), (Ph.D. 2002)*

Q.412. Tetrad is seen in.

Ans. Pachytene *(Ph.D. 1998)*

Q.413. Chaysma is seen during.

Ans. Diplotene *(J.R.F. 2003)*

Q.414. Multilines in wheat are produced by.

Ans. Back cross breeding *(Ph.D. 2002)*

Q.415. The selection procedure which provides the maximum improvement over the base population is.

Ans. Pure line selection *(Ph.D. 2002)*

Q.416. Most dangerous disease of potato.

Ans. Late blight of potato *(J.R.F. 1995)*

Q.417. Phyllody disease in plants is caused by.

Ans. Mycoplasma *(Ph.D. 1998)*

Q.418. Father of plant nematology.

Ans. Bastian *(J.R.F. 1996)*

Q.419. Rice production is highest in country.

Ans. China *(B.H.U. Pre P.G. 2002)*

Q.420. Single super phosphate contains sulphur.

Ans. 12% *(J.R.F. 1995)*

Q.421. Most prominent soil group of India.

Ans. Alluvial soil *(J.R.F. 2003)*

Q.422. Murate of potash is chemically known as.

Ans. KCl *(J.R.F. 1996)*

Q.423. Highest contribution to vegetable oil.

Ans. Ground nut

Q.424. Explosive fertilizer is.

Ans. Ammonium nitrate *(B.H.U. Pre P.G. 2001)*

Q.425. Most mutation tolerant.

Ans. Polyploids *(Ph.D. 1998)*

Q.426. The person acting as a connecting link between higher department official & farming community in T & V system, known as.

Ans. Village Extension Worker *(Ph.D. 1999)*

Q.427. The pest which attacks all the parts of the plant is

Ans. Termite *(J.R.F. 1995)*

Q.428. White grubs prefer to lay eggs on.

Ans. Sandy soil *(Ph.D. 2002)*

Q.429. King of pulses .

Ans. Chickpea. *(B.H.U. Pre P.G. 2005)*

Q.430. Micro nutrient deficient in Indian soils.

Ans. Zn *(J.R.F. 1991) (A.R.S. NET 2007)*

Q.431. Mantek disease of rice is caused by.

Ans. Rice root nematode *(J.R.F. 1996)*

Q.432. International pest is.

Ans. Schistacerca gregaria *(Ph.D. 1998)*

Q.433. The first Fisheries University was.

Ans. CIFE, Mumbai *(Ph.D. 1995)*

Q.434. Indian Institute of tropical meteorology is situated at

Ans. Pune *(J.R.F. 1996)*

Q.435. The second Indian recipients of world food prize was. Also credit goes for white Revolution.

Ans. Dr. Varghese Kurien *(Ph.D. 1996) (ARS NET 2007)*

Q.436. Spike tooth harrow is a.

Ans. Secondary tillage implement *(Ph.D. 1999)*

Q.437. India ranks .... rank in fruit production.

Ans. First *(J.R.F. 1996)*

Q.438. Club root of cauliflower is caused by.

Ans. Plasmodiophara brassicae *(Ph.D. 1995)*

Q.439. Sheath blight of rice is caused by.

Ans. Rhizoctonia solani *(Ph.D. 1995)*

Q.440. Father of microbiology.

Ans. Louis Pasteur *(B.H.U. Pre P.G. 2001)*

Q.441. Bacterium was discovered by.

Ans. Anton Leewenhock *(B.H.U. Pre P.G. 2001)*

Q.442. The variety of wheat which is resistant to all the three rusts is.

Ans. Choti lerma *(Ph.D. 1995)*

Q.443. In sugarcane, taking of ratoon crop is advisable only.

Ans. One time *(B.H.U. Pre P.G. 2002)*

Q.444. Wilt disease is .... disease.

Ans. Soil borne *(Ph.D. 2002)*

Q.445. Vector of leaf curl disease in cotton.

Ans. White flies *(Ph.D. 1995)*

Q.446. 'Flared square' symptoms seen in cotton due to the.

Ans. Spotted boll worm *(Ph.D. 1998)*

Q.447. The no. of electrons required for conversion of NO_3 to NH_4

Ans. 8 *(Ph.D. 2002)*

Q.448. Conversion of fat to sugar occurs in.

Ans. Glyoxysomes *(Ph.D. 1995)*

Q.449. The net requirement for assimilation of CO_2 in C_4 plants is.

Ans. 5 *(Ph.D. 1998)*

Q.450. The primary accepter of electron in PS II is.

Ans. Pheophytin *(J.R.F. 2001)*

Q.451. Gneiss is a .... rock.

Ans. Metamorphic *(Ph.D. 1995)*

Q.452. Under the USDA system, silt is classified as having a particle size of.

Ans. 0.02 to 0.002 mm. *(J.R.F. 1997)*

Q.453. Hygroscopic water is held at a tension of .... atmosphere.

Ans. 31 or more *(J.R.F. 1997)*

Q.454. Kaolinite is a layer silicate type.

Ans. 1 : 1 type *(Ph.D. 1998)*

Q.455. Clays are .... minerals.

Ans. Secondary *(J.R.F. 1998)*

Q.456. C : N ratio of organic matter is.

Ans. 10 : 1 *(Ph.D. 1995)*

Q.457. CAN is .... fertilizer.

Ans. Neutral *(J.R.F. 1998)*

Q.458. Most widely used for correcting soil acidity

Ans. Lime *(J.R.F. 2003)*

Q.459. Largest importer of cut flowers in the world.

Ans. Germany *(J.R.F. 1997)*

Q.460. Density of water is maximum at .... °C.

Ans. 4°C *(Ph.D. 1995)*

Q.461. Diara cultivation method is followed in.

Ans. Cucurbits *(J.R.F. 1997)*

Q.462. The irrigation method which is suitable for saline soils.

Ans. Flood method *(J.R.F. 1997)*

Q.463. The role of extension education in India is performed by.
Ans. State Agricultural University (SAUs) *(J.R.F. 1997)*

Q.464. Extension is a.
Ans. Two way flow of message *(Ph.D. 1995)*

Q.465. Oldest method of selection is.
Ans. Mass selection *(Ph.D. 1998)*

Q.466. The limit of the regression coefficient is.
Ans. 0 to 1 *(J.R.F. 1998)*

Q.467. Highly salt tolerant crop.
Ans. Barley *(BHU, 2009) (J.R.F. 2001)*

Q.468. Formation of male flowers is induced by.
Ans. GA_3 *(J.R.F. 2003)*

Q.469. Cytokinin is mostly synthesized in.
Ans. Root tips *(Ph.D. 2002)*

Q.470. Recent method of control of bollworms is.
Ans. Bt. transgenic plants *(J.R.F. 2003)*

Q.471. Orabanche is a parasite associated with..........
Ans. Tobacco *(Ph.D. 1998)*

Q.472. Dr. B.P. Pal is associated with.
Ans. Breeding *(J.R.F. 1998)*

Q.473. Art of giving shape to shrubs resembling to figure is called as.
Ans. Topiary *(B.H.U. Pre P.G. 2002)*

Q.474. The dwarf variety of mango is.
Ans. Amrapali *(J.R.F. 1997)*

Q.475. Mango malformation can be checked by.
Ans. NAA Spray *(Ph.D. 1995)*

Q.476. Photo-respiration occurs in.
Ans. Chloroplast *(J.R.F. 2003)*

Q.477. Photosynthesis is an.
Ans. Oxidation-reduction process *(J.R.F. 1997)*

Q.478. Root promoting hormone is.

Ans. IBA *(J.R.F. 1998)*

Q.479. Fruit of rose is known as.

Ans. Hips *(Ph.D. 2002)*

Q.480. What conc. of sugar is used for preservation.

Ans. 60-70% *(J.R.F. 2001)*

Q.481. Fruit crop which requires the highest no. of irrigations.

Ans. Banana *(Ph.D. 1995)*

Q.482. Photo system II is absent in.

Ans. C_4 plants *(Ph.D. 1998)*

Q.483. Conversion of fat into carbohydrate is.

Ans. Glyoxylate cycle *(J.R.F. 1998)*

Q.484. Peroxisome is a.

Ans. Single membrane organelle *(J.R.F. 1997)*

Q.485. The main site for the dark reaction of photosynthesis is.

Ans. Stroma *(J.R.F. 2001)*

Q.486. The F_2 ratio for complementary interaction is.

Ans. 9 : 7 *(Ph.D. 1998)*

Q.487. The shortest phase of all the mitosis phases is

Ans. Anaphase *(J.R.F. 1997)*

Q.488. Longest phase of all thc mitosis phases is.

Ans. Prophase *(J.R.F. 2003)*

Q.489. The site of protein synthesis is.

Ans. Ribosomes *(J.R.F. 1997)*

Q.490. Operation flood is related to.

Ans. Dairy development *(J.R.F. 1997) (2009)*

Q.491. Bunch terracing is done when the slope is more than.

Ans. 15% *(J.R.F. 1998)*

Q.492. Mass per unit volume is called as.

Ans. Bulk density *(J.R.F. 1996)*

Q.493. Disk plough are used when the soil is.

Ans. Tough *(B.H.U. Pre P.G. 2002)*

Q.494. Post harvest losses for cereals account for .... of total production.

Ans. 10% *(Ph.D. 1999)*

Q.495. National seed act passed.

Ans. 1966 *(J.R.F. 1998)*

Q.496. ICAR was initiated as per recommendation of.

Ans. Royal commission on Agriculture 1925 *(J.R.F. 1997)*

Q.497. Mango variety which is suitable for high density planting is.

Ans. Amrapali *(Ph.D. 1998)*

Q.498. Malika is a cross of.

Ans. Neelam × Dashari *(J.R.F. 2000)*

Q.499. T & V was first started in.

Ans. Rajasthan *(J.R.F. 2001)*

Q.500. Indian Journal of Agriculture sciences is published by.

Ans. ICAR *(J.R.F. 1997)*

Q.501. The design to be followed for one directional fertility gradient is.

Ans. R.B.D. *(J.R.F. 1998)*

Q.502. The distribution where the *sp.* is equal to the root of mean is.

Ans. Poisson *(J.R.F. 2000)*

Q.503. Maximum contribution of the crop to the production of cereals in the country.

Ans. Rice *(J.R.F. 2001)*

Q.504. Highest sugarcane yield per hectare in the state.

Ans. Tamilnadu *(J.R.F. 2000)*

Q.505. Distribution of mean median mode is a.

Ans. Normal distribution *(Ph.D. 1995)*

Q.506. Nursery area required for seedling of rice for one hectare field is.

Ans. 0.10 ha. *(J.R.F. 1998)*

Q.507. Marble is a.

Ans. Metamorphic rock *(Ph.D. 1995)*

Q.508. Time duration for adsali crop of sugarcane.

Ans. 18 months *(J.R.F. 2000)*

Q.509. Crossing over occurs in.

Ans. Pachytene *(J.R.F. 2003)*

Q.510. First man made cereal is.

Ans. Triticale *(J.R.F. 1998) (J.R.F. 2000)*

Q.511. Triticale is a crossing of.

Ans. Wheat × Rye *(Ph.D. 1994)*

Q.512. Intervainal chlorosis occurs due to.

Ans. Fe deficiency *(J.R.F. 1996)*

Q.513. Saline tolerant fruit crop is.

Ans. Date palm *(Ph.D. 1995)*

Q.514. Optimum pH for rice.

Ans. 4–6 pH *(J.R.F. 1998)*

Q.515. Dead heart & white ear associated with the crop.

Ans. Rice *(J.R.F. 2000)*

Q.516. Formula of Urea.

Ans. $CO(NH_2)_2$ *(J.R.F. 1996)*

Q.517. M.O.P. contains K_2O.

Ans. 58–60% *(Ph.D. 1994)*

Q.518. Fruit ripening hormone is.

Ans. Ethylene *(J.R.F. 2001)*

Q.519. Water use efficiency higher in plants.

Ans. CAM plants *(J.R.F. 2003)*

Q.520. Ooze test is done for detecting.

Ans. Bacteria *(J.R.F. 1998)*

Q.521. Maximum arable land in country.

Ans. Australia *(J.R.F. 1996)*

Q.522. Malathion is a.

Ans. Systemic organo phosphate *(Ph.D. 1994)*

Q.523. Family of sugarcane.

Ans. Graminae *(J.R.F. 2000)*

Q.524. Most widely grown rabi pulse crop.

Ans. Bengal Gram *(J.R.F. 2001)*

Q.525. Unit of rural society.

Ans. Village *(Ph.D. 1994)*

Q.526. Fruit type of guava.

Ans. Berry *(Ph.D. 1994)*

Q.527. The term genetics was coined by.

Ans. Bateson *(J.R.F. 2000)*

Q.528. Sodic soils are reclaimed by.

Ans. Gypsum *(J.R.F. 1998)*

Q.529. I.I.S.R. (Indian Institute of Sugarcane Research) is situated at.

Ans. Lucknow *(Ph.D. 1994)*

Q.530. IBPGR is situated at.

Ans. Italy, Rome *(J.R.F. 2000)*

Q.531. Proportion of sand, silt, clay known as.

Ans. Soil texture *(Ph.D. 1995)*

Q.532. Cheapest N containing fertilizer is.

Ans. Urea *(Ph.D. 1994)*

Q.533. Total Land capability classes are.

Ans. 8 *(Ph.D. 1994)*

Q.534. Bacteria change nitrite to nitrate are.

Ans. Nitrobacter *(B.H.U. Pre P.G. 2002)*

Q.535. Akiochi disease is due to.

Ans. Sulphur toxicity *(J.R.F. 2000)*

Q.536. First product of Urea hydrolysis.

Ans. Ammonium carbamate *(J.R.F. 1998)*

Q.537. Azolla is a.

Ans. Blue green algae *(Ph.D. 1994)*

Q.538. Functional nutrients given by.
Ans. Nicholas *(J.R.F. 2000)*

Q.539. State leading in wheat production is.
Ans. Uttar Pradesh *(Ph.D. 2002)*

Q.540. NABARD was set up on the recommendation of.
Ans. Siva Raman Committee, Narasimhan Committee

Q.541. White rust of crucifers is caused by.
Ans. *Albugo candiada* *(J.R.F. 2000)*

Q.542. The hard fruits of citrus is due to the.
Ans. B deficiency *(Ph.D. 1999)*

Q.543. Which is the antidote of insect poisoning.
Ans. Atropin *(B.H.U. Pre P.G. 2001)*

Q.544. Tetra sonic is.
Ans. 2n + 2 *(J.R.F. 2003)*

Q.545. Precursor of IAA is.
Ans. Tryptophan *(Ph.D. 1999)*

Q.546. Fruit of okra is.
Ans. Capsule *(J.R.F. 2000)*

Q.547. Vitamin, containing cobalt as a constituent is.
Ans. Vit. B_{12} *(J.R.F. 2000)*

Q.548. Free living nitrogen fixing organism is.
Ans. Azotobacter *(Ph.D. 1994)*

Q.549. Dwarfing gene in rice is.
Ans. Dee gee woo gene *(Ph.D. 1994)*

Q.550. Centre of origin of wheat is.
Ans. Mexico *(J.R.F. 1996)*

Q.551. Food of God
Ans. Cocoa *(Ph.D. 2000)*

Q.552. Total National Bureau are.
Ans. 6

Q.553. Lab to land programme started in the year.
Ans. 1979 *(J.R.F. 2000)*

Q.554.. Disease which was discovered in Haryana.

Ans. Karnal bunt *(J.R.F. 1993)*

Q.555. Chromosomal theory of inheritance was proposed by

Ans. Sutton & Boveri *(J.R.F. 2003)*

Q.556. Ufra disease in Rice is caused by.

Ans. (Nematode) Ditylenchus sp. *(Ph.D. 1994)*

Q.557. The fruit of mustard is known as.

Ans. Siliqua *(J.R.F. 1998. 2000)*

Q.558. N content in ammonium sulphate is.

Ans. 21% *(J.R.F. 2003)*

Q.559. Herbicides are not used in the dust formulation because of.

Ans. Drift hazard *(Ph.D. 1994)*

Q.560. The fungicide which is used for smut.

Ans. Vitavax *(Ph.D. 1994)*

Q.561. First stable product formed in C_3 plants is.

Ans. PGA *(Ph.D. 1995)*

Q.562. The green house gas that's released from paddy field is.

Ans. CH_4 (Methane) *(J.R.F. 1998)*

Q.563. B.P. Pal is a variety of.

Ans. Rose *(J.R.F. 2000)*

Q.564. Over the years the percent contribution of agriculture to G.D.P. is.

Ans. Decreasing *(J.R.F. 2003)*

Q.565. I.A.D.P. programme started in the year.

Ans. 1960 *(Ph.D. 2002)*

Q.566. Person associated with Gurgoan pilot project is.

Ans. P.L. Brayne *(J.R.F. 1993)*

Q.567. Net irrigated area of India.

Ans. 57m.ha. *(Ph.D. 1994)*

Q.568. What is the price called which is fixed by government to support farmers

Ans. Minimum Support Price formulated by CACP *(J.R.F. 2003)*

Q.569. Power house of cell is.

Ans. Mitochondria *(B.H.U. Pre P.G. 2002)*

Q.570. The percent land resource of India in world is.

Ans. 2.4% *(J.R.F. 1993)*

Q.571. Head quarters of W.T.O. is located at.

Ans. Geneva *(J.R.F. 2000)*

Q.572. The highest award presented to an agricultural scientist in the country is.

Ans. Rafi Ahmad Kidwai Award *(J.R.F. 1993)*

Q.573. Protein content in Bajra

Ans. 11-12%

Q.574. Growth of plant towards light is called?

Ans. Photoperiodism

Q.575. Highest C.E.C. found in.

Ans. Vermiculate *(Ph.D. 1995)*

Q.576. E.S.P. of normal soils is.

Ans. Less than 15% *(J.R.F. 1998)*

Q.577. Sugar turn out from S.cane in India is.

Ans. 8-10% *(J.R.F. 1993)*

Q.578. Permanent wilting point is observed at.

Ans. –15 bar *(J.R.F. 2000)*

Q.579. Mango is mostly propagated through.

Ans. Veneer grafting *(Ph.D. 1994)*

Q.580. Micro organism associated with the symbiotic N_2 fixation in non legumes is.

Ans. Azolla *(Ph.D. 1994)*

Q.581. Emer wheat is.

Ans. Triticum dicoccum *(Ph.D. 1995)*

Q.582. Net capital ratio is.

Ans. $\frac{\text{Total assets}}{\text{Total liabilities}}$ *(Ph.D. 1999)*

Q.583. Catken is a inflorescence of.

Ans. Cauliflower *(J.R.F. 1998)*

Q.584. Black heart of potato is due to.

Ans. Poorly drained soil *(Ph.D. 2002)*

Q.585. The Arka series of varieties are released from.

Ans. IIHR Bangalore *(J.R.F. 2000)*

Q.586. Major agricultural import in India is.

Ans. Edible oils *(J.R.F. 1993)*

Q.587. Which programme also known as package programme.

Ans. IADP *(B.H.U. Pre P.G. 2002)*

Q.588. Whip tail of cauliflower is due to the deficiency of.

Ans. Mo *(J.R.F. 1993, 2009)*

Q.589. Gross cropped area in India is.

Ans. 193 M.ha. *(J.R.F. 2000)*

Q.590. Yellow rust in wheat is caused by.

Ans. Puccinia striformis *(J.R.F. 1993)*

Q.591. Seedless mango variety is.

Ans. Sindhu *(J.R.F. 1993)*

Q.592. Pink revolution refers to.

Ans. Onion production *(Ph.D. 2001)*

Q.593. Correlation coefficient ranges between.

Ans. –1 to +1 *(J.R.F. 1998, 2000)*

Q.594. Pusa snow ball is a variety of.

Ans. Cauliflower *(Ph.D. 1995)*

Q.595. CRIDA is located at.

Ans. Hyderabad *(J.R.F. 1998)*

Q.596. India's share in the fruit production in the world is.

Ans. 10% *(J.R.F. 2000)*

Q.597. Fertilizer's which are completly imported.

Ans. Potassic *(J.R.F. 1993)*

Q.598. The contribution of Agriculture to GDP of India is about

Ans. 17.1% *(Economic survey 2008–09)*

Q.599. Last D.G. of ICAR.

Ans. Mangala Rai *(B.H.U. Pre P.G. 2002)*

Q.600. Moisture content for safe storage of cereals is.

Ans. 12-14% *(Ph.D. 2001)*

Q.601. Lunishree is a variety of.

Ans. Super Rice

Q.602. Highest cotton production in India.

Ans. Gujarat than Maharastra

Q.603. Bad opening of cotton balls is known as

Ans. Tirak

Q.604. Densest part of atmosphere strata is

Ans. Troposphere

Q.605. Central Agricultural University is located at.

Ans. Imphal *(J.R.F. 1993)*

Q.606. Constituent of wheat affecting its baking quality is.

Ans. Glutin *(Ph.D. 2001)*

Q.607. I.G.F.R.I. is situated at.

Ans. Jhansi (U.P.) *(J.R.F. 1993)*

Q.608. CRI stage occurs in wheat in.

Ans. 21 days *(J.R.F. 1998)*

Q.609. Bulk density is high in.

Ans. Sandy soil *(Ph.D. 1999)*

Q.610. Suicidal bags of cell are.

Ans. Lysosomes *(Ph.D. 1995)*

Q.611. The fertilizer also called Nitro chalk.

Ans. CAN *(B.H.U. Pre P.G. 2002)*

Q.612. Regression coefficient varies between.

Ans. $-\infty$ to $+\infty$ *(Ph.D. 2001) (J.R.F. 1998)*

Q.613. Soil transported through wind is.

Ans. Aeolian soil *(J.R.F. 1993)*

Q.614. Formation of mRNA from DNA is called.

Ans. Transcription *(J.R.F. 1993)*

Q.615. The ion generally dominant in soil solution is.

Ans. Ca^{2+} *(Ph.D. 2001)*

Q.616. Fixation in soil is a problem in case of.

Ans. Phosphatic fertilizers *(Ph.D. 2002)*

Q.617. Luxury consumption is generally associated with.

Ans. Potassium *(Ph.D. 2001)*

Q.618. The journal of Indian farming is published by.

Ans. ICAR *(Ph.D. 1999)*

Q.619. Pride fruit of India is.

Ans. Mango *(J.R.F. 1998)*

Q.620. Solar constant is.

Ans. 1.94 $cal/cm^2/sec$ *(JRF 2006)*

Q.621. C.V. of rainfall is more in.

Ans. Thar desert *(Ph.D. 1995)*

Q.622. Striga is .... type parasite.

Ans. Semi root. *(B.H.U. Pre P.G. 2001)*

Q.623. Lady bird beetle is a.

Ans. Predator *(Ph.D. 2001)*

Q.624. Monophagus pest is.

Ans. Yellow stem borer *(B.H.U. Pre P.G. 2001)*

Q.625. Maximum production of potato in.

Ans. Uttar Pradesh *(B.H.U. Pre P.G. 2001)*

Q.626. CIPHET is situated at (Central Institute of Post Harvest Engineering and Technology)

Ans. Ludhiana *(Ph.D. 1995)*

Q.627. First project launched in India.

Ans. Etawah *(Ph.D. 2001)*

Q.628. Maximum oil seed producer crop in India.
Ans. Ground nut (J.R.F. 1998)

Q.629. Maximum nutrient uptake by plants.
Ans. K^+ (Ph.D. 2001)

Q.630. Nutrient available in combined form.
Ans. Nitrogen (B.H.U. Pre P.G. 2002)

Q.631. Sulphur is available in which form
Ans. SO_4^{-2} (J.R.F. 1993)

Q.632. Sugarcane sets required per ha.
Ans. 35,000 - 40,000 (Ph.D. 2001)

Q.633. Most efficient irrigation system.
Ans. Drip irrigation (B.H.U. Pre P.G. 2001) (AAO, 2009)

Q.634. How many deemed Universities under ICAR
Ans. 4 (J.R.F. 1993)

Q.635. Insects which attacks rice at night.
Ans. Army worm (Ph.D. 2001)

Q.636. Formula of gypsum.
Ans. $CaSO_4 . 2H_2O$ (J.R.F. 1993)

Q.637. Maximum productivity among cereals in world.
Ans. Maize (J.R.F. 1998)

Q.638. Queen of cereals known as.
Ans. Maize (J.R.F. 1993)

Q.639. Gold of America known as.
Ans. Soyabean

Q.640. Back bone of America is the crop.
Ans. Maize (B.H.U. Pre P.G. 2002)

Q.641. Hulling % in rice is.
Ans. 65% (Ph.D. 1999)

Q.642. Weight of cotton bale is.
Ans. 170 kg. (J.R.F. 1998)

Q.643. Which is not a measure of central tendency.
Ans. Range (J.R.F. 1999)

Q.644. Pungency in onion is due to presence of.
Ans. Sulphur compound (J.R.F. 1992)

Q.645. CEC of humus is.

Ans. 200–300 *(Ph.D. 1999)*

Q.646. Movement from higher concentration to lower concentration known as.

Ans. Diffusion *(J.R.F. 1999)*

Q.647. 'V' shape pattern of yellowing shows.

Ans. N deficiency *(Ph.D. 2001)*

Q.648. Fiber of cotton contains.

Ans. Cellulose *(J.R.F. 1999)*

Q.649. Available water to crop lies between.

Ans. 33 to 15 bar *(J.R.F. 1992)*

Q.650. Growing subsidiary crop between widely rowed space of main crop is called.

Ans. Intercropping *(J.R.F. 1992)*

Q.651. Least degree of freedom in ANOVA is.

Ans. 12 *(J.R.F. 1999)*

Q.652. Sugar cane is a ....

Ans. Quantative short day plant *(J.R.F. 1992)*

Q.653. Isotopes have same no. of.

Ans. Proton *(J.R.F. 1999)*

Q.654. N_2 deficiency occurs in plants on.

Ans. Lower leaves *(J.R.F. 1992)*

Q.655. Highest K_2O containing fertilizer.

Ans. KCl (60%) *(B.H.U. Pre P.G. 2002)*

Q.656. Boron is harmful for plants having more than.

Ans. 3 PPM conc. *(Ph.D. 2001)*

Q.657. Which is not a measure of dispersion.

Ans. Coefficient of variation *(J.R.F. 1992)*

Q.658. Irish famine (1845) was caused by.

Ans. Phytophthora infestans *(J.R.F. 1992)*

Q.659. Production of ATP in the presence of light is known as.

Ans. Photo-phosphorylation *(J.R.F. 1999)*

Q.660. For seed purpose carrot is grown as.
Ans. Binneal *(Ph.D. 2002)*

Q.661. Soil moisture is measured by.
Ans. Tensiometer *(J.R.F. 1999)*

Q.662. Bacteria which convert NH_4^+ **to** NO_2^-
Ans. Nitrosomonas *(J.R.F. 1992)*

Q.663. Physical condition of soil known as.
Ans. Soil tilth *(J.R.F. 1992)*

Q.664. Infra red thermometer used for.
Ans. Crop canopy temperature *(Ph.D. 2001)*

Q.665. Most serious disease in sugarcane.
Ans. Red rot *(J.R.F. 1999)*

Q.666. Ribosome found in mitochondria
Ans. 70s type *(Ph.D. 1999)*

Q.667. Black soils found in.
Ans. Maharastra *(J.R.F. 1992)*

Q.668. World's staple food is.
Ans. Wheat /Rice *(Ph.D. 1999)*

Q.669. 3M deep, more than 18M. wide gully is known as.
Ans. Small gully *(J.R.F. 1992)*

Q.670. In which fertilizer N & P present in highest amount.
Ans. D.A.P. (18%, 46%)

Q.671. Leading state in acreage of rice.
Ans. West Bengal *(J.R.F. 1999)*

Q.672. Which crop having highest percentage of irrigation.
Ans. Wheat *(J.R.F. 1991)*

Q.673. Sugarcane sowing in trench method to.
Ans. Prevent lodging *(Ph.D. 2001)*

Q.674. Dough stage means.
Ans. Milking to just maturing *(J.R.F. 2003)*

Q.675. C : N ratio of most arable soils.
Ans. 10 : 1 *(J.R.F. 1999)*

Q.676. Noble cane is.
Ans. Saccahrum officinerum *(Ph.D. 1999)*

Q.677. Re-discovery of Mendel laws by.
Ans. Devries, Correns, Trashe Mark *(J.R.F. 1992)*

Q.678. Pattern used in dry land.
Ans. Broadcasting *(J.R.F. 1992)*

Q.679. Parboiling in rice conserves.
Ans. Vitamin B *(Ph.D. 2001)*

Q.680. Maize is a.
Ans. C_4 plant *(J.R.F. 2001)*

Q.681. In 1943 Bengal famine was due to.
Ans. Blast of rice (*Helminthosporium oryzae*) *(J.R.F. 1992)*

Q.682. Maximum total porosity found in.
Ans. Clay soils *(J.R.F. 1999)*

Q.683. Most frost effected crop.
Ans. Gram *(J.R.F. 1992)*

Q.684. H.D. - 2329 is a variety of.
Ans. Wheat *(Ph.D. 2001)*

Q.685. Drip irrigation introduced in India from.
Ans. Israel *(B.H.U. Pre P.G. 2002)*

Q.686. N% in CAN is.
Ans. 25% *(J.R.F. 1999)*

Q.687. Highest amount of rainfall received in India by.
Ans. South West Monsoon *(J.R.F. 1992)*

Q.688. The river basins have more utilizable flow in.
Ans. Ganga *(B.H.U. Pre P.G. 2001)*

Q.689. First Agricultural chemist of Imperial Agricultural Research station was.
Ans. J. W. Leather *(Ph.D. 2001)*

Q.690. Pseudo cereal is.
Ans. Buck Wheat *(J.R.F. 1999)*

Q.691. King of fruits.
Ans. Mango *(J.R.F. 1992)*

Q.692. Queen of spices.

Ans. Cardamom *(J.R.F. 1999)*

Q.693. How water rises in plant.

Ans. By Transpiration Pull *(J.R.F. 1991)*

Q.694. The first fertilizer produced in India was.

Ans. S.S.P. *(Ph.D. 2001)*

Q.695. Oil content in soyabean.

Ans. 20% *(J.R.F. 1999)*

Q.696. Hand book of Agriculture published by.

Ans. ICAR *(B.H.U. Pre P.G. 2001)*

Q.697. Growth of plants towards light is known as.

Ans. Photoperiodism *(J.R.F. 2003)*

Q.698. Agmark is.

Ans. Quality of food product *(J.R.F. 1991)*

Q.699. Diameter of clay particles.

Ans. <.002 *(Ph.D. 2001)*

Q.700. Available form of N in soil.

Ans. NH_4^- & NO_3^-3 *(Ph.D. 1999)*

Q.701. E.S.P. of alkali soils.

Ans. > 15% *(J.R.F. 1999)*

Q.702. The normal ratio of rice/paddy is.

Ans. 2/3 *(B.H.U. Pre P.G. 2001)*

Q.703. Crop for attracting insects.

Ans. Trap crop *(J.R.F. 1999)*

Q.704. Wind velocity can be measured by.

Ans. Anemometer *(J.R.F. 1999)*

Q.705. Deflocculation of soil particles occurs by.

Ans. Na *(J.R.F. 1992)*

Q.706. Planting before harvest of main crop is known as.

Ans. Relay cropping *(J.R.F. 1992)*

Q.707. Which is not a green house gas.

Ans. O_2 *(J.R.F. 1999)*

Q.708. Urea is maximum marketed as.

Ans. Prills *(J.R.F. 2003)*

Q.709. Maximum metabolic activity of plant is at.
Ans. 22–30ºC *(Ph.D. 2001)*

Q.710. ICMR recommendation for pulse consumption /day/capita.
Ans. 85 gram/day

Q.711. 1 gm. glucose has.
Ans. 0.45 gram fat *(Ph.D. 2001)*

Q.712. Harvest Index in wheat is.
Ans. 0.35–0.45 *(Ph.D. 1999)*

Q.713. Leaf curl of tomato is due to.
Ans. Virus *(J.R.F. 1991)*

Q.714. The crop which prefer ammonical form of N.
Ans. Rice *(J.R.F. 1999) (ARS, 2017)*

Q.715. PUFA content is highest in.
Ans. Sunflower *(J.R.F. 1992)*

Q.716. Interospecific hybrid are more common in.
Ans. Cotton *(B.H.U. Pre P.G. 2001)*

Q.717. Shedding of plant parts is due to.
Ans. ABA *(Ph.D. 2001)*

Q.718. Soil is clay when clay separates are minimum of.
Ans. 30% *(J.R.F. 1992)*

Q.719. Maximum N loss in rice in form of.
Ans. Denitrification *(J.R.F. 1999)*

Q.720. Head of Niti Ayog is.
Ans. Prime Minister

Q.721. Climate for arhar is.
Ans. Warm tropical

Q.722. High yielding variety of wheat produced by
Ans. N. Baurlog *(Ph.D. 1994)*

Q.723. Maximum P available at pH.
Ans. 6.5 -7.0 pH *(J.R.F. 2003)*

Q.724. Chemical used for delinting of cotton.
Ans. H_2SO_4 *(Ph.D. 1994)*

Q.725. Highest nutrient content in which fertilizer
Ans. S.S.P. *(Ph.D. 1999)*

Q.726. Sowing time of cotton in north India
Ans. April–May *(Ph.D. 1994)*

Q.727. Soil erosion in India is
Ans. 16 ton/ha/year *(B.H.U. Pre P.G. 2001)*

Q.728. No. of eggs/sec. laid by termite
Ans. 60 *(Ph.D. 1994)*

Q.729. Which is a stable element
Ans. Zr *(Ph.D. 1999)*

Q.730. The Book "Nature & Properties of Soil" is written by
Ans. N.C. Brady *(Ph.D. 1994)*

Q.731. Poor man's fruit
Ans. Ber *(Ph.D. 1994)*

Q.732. Rothemsted Agricultural Research station was founded by
Ans. Lawes & Gilbert (1834) *(Ph.D. 1999)*

Q.733. Protein content in wheat
Ans. 12-14% *(Ph.D. 1994)*

Q.734. Curing is related to the crop
Ans. Tobacco *(J.R.F. 1999)*

Q.735. Which micro organism is responsible for maximum nutrient cycling in the soil
Ans. Bacteria *(J.R.F. 2003)*

Q.736. The extension programme linked from pllaned credit for farmers
Ans. IRDP *(Ph.D. 1999)*

Q.737. Ratooning is practiced in crop
Ans. Sugarcane *(Ph.D. 1994)*

Q.738. Molybdenum is required by plants because it is a cofactor for
Ans. Nitrate reductase *(Ph.D. 1994)*

Q.739. The basic unit of development under the Integrated Rural Development Programme is a

Ans. Family *(Ph.D. 1994)*

Q.740. Botanical name of green gram is

Ans. *Vigna radiata* *(Ph.D. 1994)*

Q.741. Irrigation method best for undulating topography

Ans. Sprinkler irrigation *(Ph.D. 1994)*

Q.742. Present farming system of India has become.

Ans. Market Oriented *(J.R.F. 2003)*

Q.743. The chemical widely used to treat seed of potato to break its dormancy

Ans. Thio urea *(J.R.F. 2003)*

Q.744. Wheat is a origin of

Ans. Mediterranean *(Ph.D. 1994)*

Q.745. Azotobacter fixes atmospheric nitrogen

Ans. Non symbiotically *(Ph.D. 1994)*

Q.746. Rock phosphate used in

Ans. Acidic soil *(Ph.D. 1999)*

Q.747. Cycocel is a

Ans. Growth retardant *(J.R.F. 1999)*

Q.748. Wheat is a

Ans. C_3 plant *(J.R.F. 2003, 2006)*

Q.749. S.A.R. of alkali soils is

Ans. >13 *(Ph.D. 1994)*

Q.750. Most of the plant obtain nitrogen from the soil in the form of

Ans. Nitrate *(Ph.D. 1994)*

Q.751. Total agro climatic zones in India are

Ans. 15 *(J.R.F. 1999)*

Q.752. Foundation seed is produced by

Ans. Breeder seed *(Ph.D. 1994)*

Q.753. Pulse crop doesn't fix N from atmosphere

Ans. Rajma (Kidney Beans) *(Ph.D. 1994)*

Q.754. If only three irrigations are available indicate critical stages of wheat when there will be applied.

Ans. CRI, tillering, doughing stage

(B.H.U. Pre P.G. 2001)

Q.755. Potato tuber is a modified form of.

Ans. Stem *(B.H.U. Pre P.G. 2001)*

Q.756. The most efficient use of phosphorus is achieved by.

Ans. Basal placement at the time of sowing.

Q.757. Goodness of fit is used for distribution of.

Ans. Chi-square test *(Ph.D. 1999)*

Q.758. Growing plants of different heights in the same field at the same time is known as.

Ans. Multi storied cropping *(B.H.U. Pre P.G. 2001)*

Q.759. Seed attain maximum dry weight is considered to be

Ans. Physiological maturity

Q.760. Type of soil water which is most useful for plants.

Ans. Capillary water *(J.R.F. 1991)*

Q.761. Scorching of burning on margins of bottom leaves manifests the deficiency of.

Ans. Potassium *(Ph.D. 1999)*

Q.762. Plants takes carbon from.

Ans. Air *(J.R.F. 1991)*

Q.763. Concept of plant ideotype was first propounded by

Ans. C.M. Donald *(J.R.F. 1991)*

Q.764. The nutrient plays important role in controlling the rate of transpiration.

Ans. Potassium *(J.R.F. 1991)*

Q.765. Highest contribution in G.D.P.

Ans. Milk *(B.H.U. Pre P.G. 2001)*

Q.766. IVLP is. (Started in the year 1995)

Ans. Institution Village Linkage Programme (2009)

Q.767. Very few poor (below poverty line) people is in.
Ans. Punjab *(J.R.F. 1991)*

Q.768. Micro nutrient deficient in Indian soils.
Ans. Zn *(J.R.F. 1991) (ARS-NET, 2007)*

Q.769. In waterlogged area which gas is found abundantly?
Ans. CH_4 *(J.R.F. 1999)*

Q.770. What means photo-respiration
Ans. Production of ATP *(J.R.F. 1999)*

Q.771. Scientific name of berseem is
Ans. Trifolium alexandrium *(J.R.F. 1999)*

Q.772. The insect "Boll Worm" commonly found on
Ans. Cotton

Q.773. Most destructive disease of sugarcane is
Ans. Red rot of sugarcane *(J.R.F. 2005)*

Q.774. Global warming is caused due to :
Ans. Greenhouse effect (CO_2, CFCs and CH_4
(J.R.F. 2005)

Q.775. Primary consumers are
Ans. Zooplankatons

Q.776. The edible part of sweet potato (*Ipomoea batata*) is
Ans. Adventitious root

Q.777. Potatoes are borne on
Ans. Stolons

Q.778. An example of albuminous seed showing hypogeal germination is
Ans. Maize

Q.779. The part that develops into seed is
Ans. Ovule

Q.780. Seeds with tigers claw like structures are found in
Ans. Martynia

Q.781. Study on seasonal flowering in shrubs and trees is known as

Ans. Floriculture

Q.782. Fruit Research Institute in India is located in

Ans. Sabour, Bhagalpur (Bihar)

Q.783. Die back of shoots occur due to deficiency of

Ans. Copper

Q.784. Grey spots on leaves occur due to deficiency of

Ans. Manganese

Q.785. Yellowing of tea leaves occur due to the deficiency of

Ans. Sulphur

Q.786. Indian soil is mainly deficient of

Ans. Nitrogen

Q.787. Which fertiliser is supplied to the fields poor in phosphorus ?

Ans. Bone meal

Q.788. An element not essential for plant growth is

Ans. Iodine

Q.789. The oil which is extracted from the endosperm

Ans. Coconut oil

Q.790. The fibre yielding plant is

Ans. Crotolaria juncea

Q.791. To get food containing mainly starch we grow

Ans. Rice and wheat

Q.792. The most important cereal of temperate region

Ans. Wheat

Q.793. Saffron is produced from

Ans. Styles and stigma of Crocus plant

Q.794. Sun flower is principally cultivated for its

Ans. Oil

Q.795. Long fibres of cotton seed are known as :

Ans. Lint

Q.796. Part of the plant contains lint and fuzz

Ans. Hemp

Q.797. The part of coconut produces coir

Ans. Mesocarp

Q.798. The plant product used to prepare a milk like preparation

Ans. Soyabean

Q.799. An example of CAM plant is

Ans. Pineapple *(NET ARS. 1999)*

Q.800. Quantity of water required to produce one kg of dry matter is

Ans. Consumptive use *(NET ARS. 1999)*

Q.801. Average milling recovery of the commonly grown rice cultivars is

Ans. 60 % *(NET ARS. 1999)*

Q.802. A cultivar developed by combining selected cross pollinated lines is

Ans. Hybrid *(NET ARS. 1999)*

Q.803. The NP fertilizer containing the highest amount of P_2O_5 is

Ans. Ammonium polyphosphate *(NET ARS. 1999)*

Q.804. Nutrient mobility concept was propounded by

Ans. Bray *(NET ARS. 1999)*

Q.805. Water requirement of irrigated wetland rice is about

Ans. 1500 mm *(NET ARS. 1999)*

Q.806. Measurement of sunlight intensity is expressed in

Ans. Lux units *(NET ARS. 1999)*

Q.807. True or False

(*i*) **Type 1 error is the rejection of hypothesis when it is actually true.**

(*ii*) **Quantity of water absorbed by a crop for its metabolic use is called its consumptive use.**

(*iii*) Commercial muriate of potash contains 48·5 K_2O.

(*iv*) Fraction of light that is absorbed by a crop canopy is absorptivity coefficient.

(*v*) Movement of water molecules/nutrient ions from regions of high concentration to low concentration is known as diffusion.

Ans. (*i*) True,

(*ii*) False,

(*iii*) False,

(*iv*) True,

(*v*) True,

Q.808. Fill in the blanks

(*i*) Phanerogamic parasite with chlorophyll but no roots _________.

(*ii*) A disease caused by *Trichodorus* _______.

(*iii*) A disease caused by *Anguina* _______.

(*iv*) A disease caused by *Xiphinema* _________.

(*v*) A disease caused by *Rodopholus* ________.

(*vi*) A fungus used for biological cantrol of plant diseases ________.

(*vii*) A disease caused by Pratylenchus ________.

(*viii*) Karnal bunt was first reported by ________.

(*ix*) The first report of plant parasitic nematode was by ________.

(*x*) The virus particles were seen for the first time with the electron microscope by _______ and his colleagues

Ans. (*i*) Phanerogamic parasite with chlorophyll but no roots are called *Loranthus* spp.

(*ii*) Loranthus spp is a partial parasite of tree trunks and branches with brown stem, dark green leaves but no roots which attacks mango, citrus, apple and guavas.

(*iii*) A disease caused by *Trichodorus* is called stubby roots.

(*iv*) A disease caused by *Anguina* is called Wheat Ear Cockle disease. Wheat Ear Cockle disease causes the plant to have reduced and irregular heads (ears) having the typical cockles (light to dark brown, round galls instead of healthy grains).

(*v*) A disease caused by Xiphinema is called Yellow Mosaic disease. Yellow mosaic disease is caused by the Grapevine fanleaf virus transmitted by dagger mematode.

(*vi*) A disease caused by Rodopholus is called Tunnel form of banana

(*vii*) Rodopholus causes the attaking of bananas, causing toppling or blakhead disease

(*viii*) A fungus used for biological control of plant diseases Streptomyces

(*ix*) A disease caused by Pratylenchus root lesion disease

(*x*) Karnal Burnt was first reported by M. Mitra (1931)

Q.809. Define Heritability and describe different methods to estimate heritability

Ans. Heritability is the ratio of genotypic variance to the phenotypic variance or total variance, methods to estimate are:

1) From simple trials
2) From generation mean analysis
3) Diallel analysis. *(ARS 2003)*

Q.810. Net gain of ATD during glycolysis is

Ans. 2 *(JRF 2005)*

Q.811. Plants capable of growing in rock crevices are called

Ans. Lithophytes *(JRF 2005)*

Q.812. Maize leaf develop red and purple colour due to the defficiency of

Ans. Phosphorous *(JRF 2005)*

Q.813. Oleoresin is an important product of

Ans. Chilli *(JRF 2007)*

Q.814. 't' test is applicable when number of treatments are

Ans. 2 *(JRF 2005)*

Q.815. Protoplasms constituted of the element
Ans. Calcium *(JRF 2006)*

Q.816. Kesar (saffron) belong to the family of
Ans. Iridiceae

Q.817. Dioecious are
Ans. Plants with male and female flowers *(JRF 2006)*

Q.818. End product of glycolysis is
Ans. Pyruvate *(JRF 2005)*

Q.819. Shelford introduced the
Ans. Law of tolerance *(JRF 2005)*

Q.820. Full form of IPR is
Ans. Intellectual Property Right *(ARS, NET 2007)*

Q.821. Beaufort Scale is used to measure
Ans. Wind Pressure/Strength *(ARS, NET 2007)*

Q.822. Crop known as camel crop is
Ans. Sorghum *(ARS, NET 2007)*

Q.823. National Biodiversity Board is situated at
Ans. Chennai *(ARS, NET 2007)*

Q.824. Citrus canker is caused due to
Ans. Bacteria *(Ph.D Gen. Agri. 2007)*

Q.825. Purline theory was given by
Ans. W.L.Johannsen *(Ph.D., 2007)*

Q.826. Carbon dating technique is used to determine age of
Ans. Fossils *(ARS, NET 2007)*

Q.827. Age of plants is determined by
Ans. Annual rings *(ARS, NET 2007)*

Q.828. Pashmina used in winter clothes is obtained from the animal
Ans. Goats *(ARS, NET 2007)*

Q.829. Natural genetic engineer is
Ans. Agrobacterium tumefaciences *(Ph.D. 2007)*

Q.830. Positively charged Amino acid
Ans. Lysine *(Ph. D. 2007)*

Q.831. Blue revolution is related to
Ans. Aquaculture *(ARS, NET 2007)*

Q.832. Virus which have DNA as a genetic material

Ans. Chromatin *(ARS, NET 2007)*

Q.833. Formula for Leaf Area Index is

Ans. Leaf area/ground area

Q.834. Leaf area index is proposed by

Ans. RADFORD (1967)

Q.835. What is Eugenics

Ans. Science of selective breeding to achieve a pre-determined set of genetic characteristics.

Q.836. What is Fermentation?

Ans. An aerobic break down by micro-organism of complex organic substances, especially carbohydrate to CO_2 and alcohol.

Q.837. What is fertilization?

Ans. Fusion of two gametes and of their respective nuclei to create diploid or polyploid zygote.

Q.838. EST (Expressed Sequence Tags)

Ans. Short cDNA sequences used to link physical maps of genomes

Q.839. Anisogamy

Ans. Sexual reproduction in which female gamete is large and immobile and the male gamete is small and motile.

Q.840. Endangered species of Indian Medicinal Plant is

Ans. Podophyllum

Q.841. Third generation Pesticides are

Ans. Juvenile hormone analogues

Q.842. Division of one cell cleavage into two daughter cell or a chromosome into two arms.

Ans. Fission

Q.843. A plant that can withstand and grow under severe adverse conditions.

Ans. Hardy

Q.844. Term virus was coined by

Ans. Beijerinck

Q.845. Bordeaux mixture (fungicide) was developed by

Ans. Millardet *(Ph.D. 2009)*

Q.846. In plants, gall formation is carried out by

Ans. Insects *(Ph.D. 2009)*

Q.847. Female goat is called as

Ans. Doe *(ARS/NET 2009)*

Q.848. Mycology is a branch

Ans. To study fungi *(JRF 2006)*

Q.849. Disease common in maize, toria, sarson, peas, lentil and gram

Ans. Powdery mildew

Q.850. Whip tail disease due to molybdenum deficiency is found in

Ans. Cauliflower *(ARS 2005)*

Q.851. Olericulture is the study of

Ans. Vegetables *(ARS/NET 2009)*

Q.852. Plant / Tree most suitable for growing in wetlands

Ans. Eucalyptus *(ARS/NET 2009)*

Q.853. Presence of mycorrhiza in soils increase availability of

Ans. Phosphorus *(ARS/NET 2009)*

Q.854. Element essential for Nitrogen fixation

Ans. Molybdenum

Q.855. Reciepant of world food prize in the year 2007 was

Ans. Phillip Nelson

Q.856. Reciepant of world food prize in the year 2008 was

Ans. Robert Dole and George McGovern

Q.857. Precision farming is

Ans. Using GPS and modern technology at site *(JRF 2008)*

Q.858. Synecology term was given by

Ans. Schroter & Kirchener *(Year 2009 JRF)*

Q.859. Precurson of vitamin A is

Ans. b-Carotene *(JRF 2009)*

Q.860. The plants which contain male and female flowers
Ans. Dioecious plants *(JRF)*

Q.860. Award given for wildlife conservation in India
Ans. Amrita Devi Bishnoi Award

Q.861. pF is a measure of
Ans. Soil moisture in soil *(JRF 2010)*

Q.862. Which food is designated as "boneless meat" ?
Ans. Soybean *(JRF 2010)*

Q.863. Name herbicide which can be applied before planting /sowing
Ans. Fluchloralin *(JRF 2010)*

Q.864. Percentage of lint in seed cotton is
Ans. 33% *(JRF 2010)*

Q.865. Recovery of sugar from crop sugar cane is about
Ans. 10% *(JRF 2010)*

Q.866. Ground water table is measured by
Ans. Piezometer *(JRF 2010)*

Q.867. The colour which absorb more radiation is
Ans. black. *(JRF 2010)*

Q.868. Most widely cultivated mustard type in India is
Ans. Indian Mustard *(JRF 2010)*

Q.869. The element most mobile in soil is
Ans. Nitrogen *(JRF 2010)*

Q.870. Progeny of breeder seed in
Ans. Formation seed *(JRF 2010)*

Q.871. Pheromone trap attracts the
Ans. Male moths *(JRF 2010)*

Q.872. The crop soybean was originated from the country
Ans. China *(JRF 2010)*

Q.873. Tundru disease of wheat is caused by
Ans. *Anguina tritici and clavibacter tritici* *(ARS 2010)*

Q.874. Term gene was coined by
Ans. Johannsen *(ARS 2010)*

Q.875. Length of Gunter chain is
Ans. 66 feets *(ARS 2010)*

Q.876. Lasso is the name of herbicide

Ans. Alachor (*ARS 2010*)

Q.877. Scientific name of Popcorn is

Ans. *Zea Mays everta*

Q.878. Poor man's Timber is

Ans. Bamboo

Q.879. Nutrient which is essential for pollen viability in wheat

Ans. Boron (*J.R.F. 2010*)

Q.880. Major crop of rabi in India

Ans. Wheat (*JRF 2010*)

Q.881. Marsh spot in peas is caused by

Ans. Mn deficiency (*Ph.D. 2011*)

Q.882. Food grain production during 2010-11

Ans. 235 MT (*JRF 2011*)

Q.883. CO_2 acceptor enzyme in C_4 plants

Ans. PEP carboxylase (*Ph.D. 2011, JRF 2009*)

Q.884. Chicken and hen disorder in grapes is due to deficiency of

Ans. Zinc (*ARS 1999*)

Q.885. Removal of male bud in Banana is technically known as

Ans. Denovelling (*BHU PG 2008, ARS 1999*)

Q.886. Quince is suitable root stock for

Ans. Sapota (*Raj. PG 2009, ARS 1999*)

Q.887. Longest phase in mitosis is

Ans. Prophase (*JRF 2008, 2011*)

Q.888. Mango Hybrid Ratna is crossed of

Ans. Neelam × Alphanso

(*JRF 2010, BHU PG 2008, Raj PG 2007*)

Q.889. National Research centre for Sheep & wool research is situated at

Ans. Avikanagar, Malpura (Raj.)

(*JRF 2011, Raj PG 2009, BHU PG 2006*)

Q.890. NRC for Ground nut is situated at

Ans. Junagarh (Guj.) (*JRF 2010, ARS 2005*)

Q.891. Chemical used for preventing fruit drop in grapes is

Ans. NAA 20 PPM (*JRF 2011, BHU PG 2008*)

Q.892. Maximum per capita consumption of fish in country is

Ans. Japan (*ARS 2005, JRF 2007*)

Q.893. Plough which makes V shaped furrow is

Ans. Desi plough (*JRF 2009, ARS 2005*)

Q.894. Arrowing occurs in

Ans. Sugar cane (*JRF 2011*)

Q.895. Rice gene bank is situated in

Ans. Philippines (*JRF 2006, 2011, ARS 2005*)

Q.896. Fruits are generally have nature

Ans. Acidic (*JRF 2007*)

Q.897. Central Plant Protection Engineering Institute is at

Ans. Hyderabad (*JRF 2011*)

Q.898. The gas which is responsible for acid rain is

Ans. SO_2 & NO_2 (*JRF 2008*)

Q.899. Soil order Vertisols is related to

Ans. Black Soils (*JRF 2011, 2007*)

Q.900. National Agriculture Insurance scheme was stared in

Ans. 1999-2000 (*JRF 2011, BHU PG 2009*)

Q.901. The first DG of ICAR was

Ans. Dr. B. P. Pal (*JRF 2011, 2006*)

Q.902. Statistical test used to determine the goodness of fit is

Ans. Chi-Square test (*JRF 2010*)

Q.903. Cotton leaf curl virus is transmitted by

Ans. White fly (*JRF 2004,06, 2011, Ph D. 2005*)

Q.904. Citus canker disease was introduced in India from
Ans. USA (*JRF 2009*)

Q.905. Hatch & Slack pathway is found in crop
Ans. Maize (*JRF 2010*)

Q.906. Utera cultivation is mainly practiced in
Ans. Chhattisgarh and M.P. (*JRF 2010*)

Q.907. National Biodiversity board is situated at
Ans. Chennai (*JRF 2010, ARS 2007*)

Q.908. Which soil group having highest infiltration rate
Ans. Aridisol (*JRF 2009*)

Q.909. Total soil loss of India in a per year is about
Ans. 500 million tonnes

Q.910. Apical dominance in Sugarcane is suppressed by
Ans. Cytokinin (*JRF 2011, BHU PG 2009*)

Q.911. Clove is obtained from
Ans. Bud (*JRF 2011, RAJ PG 2010*)

Q.912. FMD disease is caused by
Ans. Virus (*JRF 2007*)

Q.913. Study of Soil Shape is known as
Ans. Geodesy (*JRF 2009, ARS 2007*)

Q.914. Credit of white revolution goes to
Ans. V. Kurien (*JRF 2010, ARS 2007*)

Q.915. Bordeaux paste is mixture of
Ans. Copper Sulphate + Lime (*JRF 2008, 2011*)

Q.916. Blue revolution is related with
Ans. Aquaculture (*JRF 2009, BHU PG 2010*)

Q.917. Vitamin C is known as
Ans. Ascorbic acid (*Raj. PG 2010, JRF 2011*)

Q.918. Which form of P_2O_5 is highest available at pH 6.7
Ans. H_2PO_4

Q.919. Major pest of Maize is
Ans. Chilo Partellus (Stem Borer)
(*JFR 2009, RPSC Raj.*)

Q.920. Depth of sowing for soya bean is
Ans. 3-5 cm *(JRF 2002. 2009. Ph.D 2010)*

Q.921. Most active portion of soil is
Ans. Clay *(JRF 2010, BHU PG 2009)*

Q.922. NRC for Grapes is situated at
Ans. Pune *(JRF 2007, Ph.D 2006)*

Q.923. Most important alkaloid in opium is
Ans. Morphin *(JRF 2010, Raj. PG 2009)*

Q.924. Papen in papaya is extracted from
Ans. Immature fruits *(JRF 2011, Raj. PG 2009)*

Q.925. IPM term as given by Barlet in year
Ans. 1956 *(JRF 2008, BHU PG 2007)*

Q.926. Indian Institute of Vegetable is situated at
Ans. Varanasi *(JRF 2010)*

Q.927. A popular alcoholic drink fenny is made from
Ans. Cashew apple *(JRF 2007)*

Q.928. White Grub hibernates in stage of
Ans. Adult *(RPSC 2007, BHU PG 2008)*

Q.929. NRC for women in agriculture is situated in
Ans. Bhubaneswar *(JRF 2007)*

Q.930. Wheat protein is called as
Ans. Glutenin *(JRF 2004, 2009)*

Q.931. Project Directorate of biological control is located at
Ans. Bangalore *(JRF 2010)*

Q.932. Pusa Purple long is variety of
Ans. Brinjal *(JRF 2010, BHU PG 2008)*

Q.933. Crop which has maximum area in M.P.
Ans. Soya bean *(JRF 2009)*

Q.934. State having largest area for wheat
Ans. U. P. *(JRF 2011)*

Q.935. The Pyrite is mostly found in
Ans. Bihar *(JRF 2007)*

Q.936. The acid which contents in sorghum plants
Ans. HCN (*JRF 2010*)

Q.937. Optimum Plant population for Maize/hec
Ans. 65000 plants (*Raj. PG 2009*)

Q.938. Reserve Bank of India was set up in
Ans. 1934 (*JRF 2011*)

Q.939. Green revolution started in Indian in
Ans. 1969 (*JRF2010*)

Q.940. World environment day is celebrated on
Ans. 5th June (*JRF 2010*)

Q.941. Highest consumption of fertilizer in crop
Ans. Sugar cane (*JRF 2011*)

Q.942. First herbicide produced in world is
Ans. 2-4 D (*JRF 2009, BHU PG 2007*)

Q.943. Trade name of Pendamethalin is
Ans. Stamp

Q.944. Luxury consumption is generally associate with
Ans. Potassium

Q.945. Water requirement of Sugar cane is
Ans. 200 cm (*JRF 2009, Ph.D 2005*)

Q.946. Pulse which doesn't fix atmospheric Nitrogen is
Ans. Rajma (*JRF 2010*)

JRF (Papers–2012)

Q.947. Urea is hydrolyzed by which enzyme
Ans. Urease (*JRF 2012, JRF 2009, Raj. PG. 2008*)

Q.948. Progeny of breeder seed is-
Ans. Foundation Seed (*JRF 2012, BHU PG 2002 P.hD.2008*)

Q.949. QPM (Quality Protein Maize) was developed by gene-
Ans. Opaque-2 (*JRF 2012*)

Q.950. The permanent property of soil is-
Ans. Soil Texture (*JRF 2012, JRF 2001, RAJ PG 2006*)

Q.951. Element which is responsible for stability of protein structure

Ans. Sulphur *(JRF 2012, 2007)*

Q.952. Marsh spot of pea is caused by

Ans. Mn deficiency *(JRF 2012, BHU PG 2004)*

Q.953. Surface temperature of sun is about

Ans. 6000 °K *(JRF 2012)*

Q.954. Mentha arvensis is a type of plant

Ans. Aromatic plant *(JRF 2012, Raj. PG 2009)*

Q.955. Leading state in Isabgol production-

Ans. Rajasthan *(JRF 2012)*

Q.956. Fern leaf disease of potato is due to deficiency of

Ans. Zn *(JRF 2012, BHU Pre PG 2008)*

Q.957. Bacteria, actionomyceties, fungi and algae comes under the group of organism

Ans. Micro flora *(JRF 2012, 2002)*

Q.958. Biggest rural employment scheme is-

Ans. MGNREGA (2 Oct. 2009) *(JRF 2012)*

Q.959. In groundnut aflatoxin is produced by-

Ans. Aspergillus flavus *(JRF 2012)*

Q.960. White grub is mainly confined to which soil-

Ans. Dry sandy loam soils *(JRF 2012, 2006)*

Q.961. Per capita cultivable land availability (ha.) in India is-

Ans. 0.15 Ha. *(JRF 2012)*

Q.962. Recently terminator technology has been developed in

Ans. Cotton *(JRF 2012)*

Q.963. Tag colour of certified seed is

Ans. Blue *(JRF 2012, 2003, Raj PG 2006)*

Q.964. Protien synthesis takes place in

Ans. Rough ER *(JRF 2012, 2001, 1998)*

Q.965. Enzymes are chemically
Ans. Vitamins *(JRF 2012, 2009, 2003)*

Q.966. Country leading in pears production
Ans. China *(JRF 2012)*

Q.967. State leading in Aonla production
Ans. U.P. *(JRF 2012)*

Q.968. SRI system of paddy was started first at
Ans. Madagascar *(JRF 2012)*

Q.969. Genes are composed of
Ans. DNA *(JRF 2012, 2007)*

Q.970. Maximum hybrids developed in which crop
Ans. Maize *(JRF 2012)*

Q.971. Favr - savr is a variety of
Ans. Tomato *(JRF 2012)*

Q.972. PPV & FR authority is located at
Ans. New Delhi *(JRF 2012)*

Q.973. "Hollow heart" in groundnut is caused by
Ans. B deficiency *(JRF 2012, BHU PG 2005)*

Q.974. Pungency in mustard oil is due to
Ans. Glucosinolates *(JRF 2012, 2002, BHU PG 2007)*

Q.975. Most affected crop from stem borer-
Ans. Maize *(JRF 2012,2002)*

Q.976. Organism which is responsible for production of methane from Bio gas plant
Ans. Methanobacteria *(JRF 2012)*

Q.977. Pollination by birds is called
Ans. Ornithophily *(JRF 2012)*

Q.978. Mode of pollination in pearl millet is
Ans. Cross pollination *(JRF 2012)*

Q.979. Concept of plant ideotype was first given by
Ans. Donald *(JRF 2012)*

Q.980. Most effective herbicide to control Phalaris minor in wheat
Ans. Isoproturon *(JRF 2012, 2008)*

Q.981. Blooter test is used for detection of
Ans. Fungi *(JRF 2012, BHU PG. 2009)*

Q.982. Coverage of Bt cotton in cotton production of India is
Ans. > 90% *(JRF 2012)*

Q.983. Which oil seed crop NRC is located at Gujarat
Ans. Ground Nut *(JRF 2012)*

Q.984. Irrigation Period
Ans. No. of days allowed for Irrigation *(ARS 2013)*

Q.985. Pesticide active ingredient in cotton has dropped by Bt –
Ans. > 50% dropped *(JRF 2012)*

Q.986. Organic source of nicotine
Ans. Tobacco *(JRF 2012)*

Q.987. How many acres are in one hectare.
Ans. 2.47 Acre *(JRF 2012, Raj PG 2013)*

Q.988. Non Agricultural institute which has developed 40 varieties of crops
Ans. BARC *(JRF 2013)*

Q.989. World Food Summit was organized in year
Ans. 1996 *(PG BHU 2013)*

Q.990. Kinnauri Apple mostly produced in state
Ans. H.P. *(JRF 2013)*

Q.991. Highest palm oil production in country
Ans. Indonesia *(JRF 2013)*

Q.992. Flood tolerant Rice verity
Ans. Scuba Rice *(JRF 2013)*

Q.993. Across the world the percent use of water in agriculture
Ans. 70% *(JRF 2013)*

Q.994. India's lowest HP tractor
Ans. 15 HP Tractor *(JRF 2013)*

Q.995. Country leader in watermelon production
Ans. China *(JRF 2012)*

Q.996. Vertical revolution in Agriculture

Ans. Value addition to crops (*ARS 2013*)

Q.997. Largest producer of strawberries in world

Ans. USA (*JRF 2013*)

Q.998. The flower which is used as natural colour for Holi

Ans. Gulmohar (*JRF 2012*)

Q.999. Rice productivity during 2012-2013

Ans. 2.5 t/ha. (*ARS 2013*)

Q.1000. India's rank in potato production

Ans. Second, (*JRF 2012*)

Q.1001. Current certified organic farming area in world

Ans. 37 m. ha. (*JRF 2013*)

Q.1002. Largest producer of grape fruit in world-

Ans. China (*JRF 2013*)

Q.1003. National Horticulture Mission was set up in year

Ans. 2005-06 (*JRF 2013*)

Q.1004. First tractor bought in India in year

Ans. 1914 (*JRF 2012*)

Q.1005. First MSP was announced by Indian government in

Ans. 1966-67 for Wheat (*JRF 2012*)

Q.1006. CAGR growth of Horticulture production from last 20 years

Ans. 4.2% CAGR (*JRF 2102*)

Q.1007. Largest Producer of cashew in world

Ans. Nigeria (*JRF 2012*)

Q.1008. Percent use of cotton for globally denim production

Ans. 17 % (*JRF 2012*)

Q.1009. Growing potato by hanging the plants in air without soil method called as

Ans. Aeroponics (*JRF 2012*)

Q.1010. Largest exporter of dairy products in world
Ans. New Zealand (*JRF 2012*)

Q.1011. India's percent share in world milk production
Ans. 17 % (*JRF 2013*)

JRF (Papers–2013)

Q.1012. Yellow coloured fruits are rich in
Ans. Vitamin A

Q.1013. Bengal famine was due to which fungus
Ans. *Helminthosporium oryzae*

Q.1014. Which gas is responsible for ozone hole formation
Ans. Chlorofluorocarbon (CFCs)

Q.1015. CRIDA is situated at (Central Research Institute for Dryland Agriculture)
Ans. Hyderabad (Santosh Nagar) 1985

Q.1016. The element constituent of chlorophyll is
Ans. Magnesium (Mg)

Q.1017. Nitrogen content in anhydrous ammonia
Ans. 82%

Q.1018. Water use efficiency is highest under which method
Ans. Drip irrigation

Q.1019. Fertilizer well suited for submerged soil is
Ans. Ammonical fertilizer

Q.1020. Botanical name of alfalfa
Ans. *Medicago sativa*

Q.1021. Plasticity of soil is measured by
Ans. Atterberg Apparatus

Q.1022. Lowest infiltration rate of soil
Ans. Sodic clayey soil

Q.1023. Availability of which micro nutrient increase with high pH
Ans. Mo (Molybdenum)

Q.1024. Stomata open in plants during at night
Ans. CAM plants

Q.1025. Which nutrient is taken up largely by the diffusion process
Ans. Phosphorus

Q.1026. Readily available form of K in soil in ranges
Ans. 0.1-2%

Q.1027. Required nutrient for soil synthesis in oilseed crops is
Ans. Sulphur

Q.1028. Seed used for commercial seed production
Ans. Certified seed

Q.1029. Tag colour of certified seed is
Ans. Blue

Q.1030. Charcoal rot is due to
Ans. *Macrophomina spp*

Q.1031. ADP to ATP conversion is called
Ans. Phosphorylation

Q.1032. Element essential for osmotic and ionic balance in plants
Ans. K (Potassium)

Q.1033. Nitrogen content in anhydrous ammonia
Ans. About 82%

Q.1034. Ripening hormone for crops is
Ans. Ethylene

Q.1035. Chemical Oleoresins found in
Ans. Chillies and Capsicum

Q.1036. Average rainfall in India is
Ans. 1194 mm

Q.1037. First KVK was started at
Ans. Pondicherry

Q.1038. Precursors of Ethylene
Ans. Methionine

Table 1. Founders/Fathers of Various Fields

Justus von Liebig	–	Father of Agricultural Chemistry
Aristotle	–	Father of Biology
F. Galton	–	Father of Eugenics
Theophrastus	–	Father of Botany
Aristotle	–	Father of Zoology
George Pallade	–	Father of Modern Cell Biology
Edward Osborne Wilson	–	Father of Sociobiology
G. J. Mendel	–	Father of Genetics
T. H. Morgan	–	Father of Modern Genetics
Karl Correns	–	Father of Cytoplasmic Inheritance
Stephen Hales	–	Father of Plant Physiology
Robert Koch	–	Father of Bacteriology
Louis Pasteur	–	Father of Fermentation, Microbiology
C. Linnaeus	–	Father of Taxonomy
H. Santapau	–	Father of Systematic Botany
M. Malphighi	–	Father of Microscopic Anatomy
N. Grew	–	Father of Plant Anatomy
Sir R. A. Fisher	–	Father of Population Genetics
Anton DeBary	–	Father of Modern Plant Pathology
P. A. Micheli	–	Father of Mycology
M.W. Stanley	–	Father of Virology
G. Haberlandt	–	Father of Plant Tissue Culture
Paul Berg	–	Father of Genetic Engineering
Edward Jenner	–	Father of Immunology
Captain John Grant	–	Father of Vital Statistics
Auguste Comte	–	Father of Sociology
Pietro decresenzi	–	Father of Agronomy
Nathan Augustus Cobb.	–	Father of Nematology
Norman E. Borlaug	–	Father of Green Revolution
Alec Jeffreys	–	Father of DNA Printing
Harrison	–	Father of Tissue culture
Salim Ali	–	Birdman of India
John Snow	–	Father of Epidemiology
L. Ramdas	–	Father of Indian Agricultural Meteorology
Knapp and Leagans	–	Father of Extension Education
E.O. Wilson	–	Father of Biodiversity
Robert Hooke	–	Father of Cytology
Sir Dietrich Brandis	–	Father of Indian Forestry
A.V. Leeuwenhoek	–	Father of Microbiology
L. D. Weber	–	Father of Soil Physics
P. R. Pisharyoti	–	Indian Remote sensing
Hugo de Vries	–	Mutation theory
Beijerinck	–	Biological Nitrogen Fixation
M. S. Randhawa	–	Ornamental Gardening in India
F. Platter	–	Parasitology
Joseph Gottlieb Kölreuter	–	Polygenic Inheritance
Oscar Brefeld	–	Pure Culture Technique
H. H. Bennet	–	Soil Conservation
Dr. B. P. Pal	–	Indian Plant Breeding
R. Mishra	–	Indian Ecology
JB Boussingault	–	Field Plot Experiment

Table 2. Superlatives of Plants

Largest marine alga — *Macrocystis macrocarpa* (a brown alga, attaining a length of 600 feet)

Largest Unicellular organism — *Acetabularia* (a marine green alga)

Largest fern — *Alsophila excelsa* (a tree fern)

Tallest tree — *Sequoia gigantea* (a gymnosperm found in America)

Tallest angiosperm — *Eucalyptus* (Australian eucalyptus) (Also considered as the tallest tree)

Largest herb — *Musa paradesiaca*

Largest grass — *Bamboosa*

Largest ovule — *Cycas circinalis*

Largest seed — *Lodoicea seychellarum* (double coconut)

Largest leaf — *Victoria regia* (angiosperm)

Largest inflorescence — *Puya raimondii*

Largest flower — *Rafflesia arnoldii* (Diameter 1 meter, Weight about 8-10 kg)

Largest fruit — *Lodoicea maldivea*

Smallest bacteria — *Dialister pneumosintes*

Smallest mycoplasma — *Mycoplasma gallisepticum* (smallest unicellur organism)

Smallest pteridophyte — *Azolla* (a water fern)

Smallest gymnosperm — *Zamia pygmaea*

Smallest angiosperm — *Wolffia microscopica*

Smallest seed — *Orchids* (Diameter 0.1 mm)

Smallest flower — *Wolffia*

Minimum number of chromosomes — *Haplopappus gracilis* (2n = 4)

Lightest wood — *Ochroma pyramidale*

Sprilluna : (Small spiral) : Botanical name of a tiny bluish green variety of algae scarcely longer than half a millimetre in length.

Table 3: Famous Variety of Field and Horticultural Crops

Crops	Variety	Specification
Rice	**TN-1 (Taichung native-1)**	**First variety of rice developed in World in Taiwan in 1965** **Its cross between Dee Gee Woo × Tasai Yung Chung)**
	IR-8	**World first high yielding rice variety developed by Dr. M.Beacheli in 1966 at IRRI, Philippines.**
		Its cross between **Dee Gee Woo × Peta (from Indonasia).** This is also known as **Miracle rice at International level**
	Jaya	**TN-1 (from Taiwan) × T-141 (from Indian variety) developed by Dr. Shastry in 1968**
		First rice variety is developed under india's rice programme. It outyielded TN-1 and IR-8 both, hence called Miracle rice in India
	Pusa Basmati-1 (IARI-1989)	It is the World's first high yielding dwarf basmati rice variety. It is also called "Todal", because the flower has awns
	Pusa RH-10 (IARI-2001)	World's first superfine grain aromatic rice hybrid possesses typical Basmati quality traits. Strongly scented, long slender grains with almost twice kernel elongation on cooking.
	Lunishree	**World first Superfine rice** variety suitable for saline and alkaline condition developed in 1992 at CRRI,Cuttak
	Improved Pusa Basmati 1 (Pusa 1460) (IARI-2007)	It was developed by incorporating risistance to bacterial leaf blight by pyramiding genes Xa13 and X21 in Pusa Basmati 1 through marker assisted back cross bredding method while retaining the agronomic features method of Pusa Basmati 1.

	Phalguna, Suraksha, Sulekha	Rice variety suitable for Gall midge resistant (Phalguna is resistant for both gall midge and Fe toxicity)
	CSR-30	Salt tolerance rice variety
	CSR-10, CSR-13	Alkaline tolerance
	Indira sona	First hybrid rice variety developed at IGKV Raipur, CG Its cross between IR-58025 A X R 710-437-1-1
	C.G.Zinc rice-1	First high zinc containing (22 ppm zinc) rice variety developed in IGKV Raipur using Pedigree method its cross between Poornima X Ananda
	Tilkasturi and Korma	First Green rice identified by IGKV Scientist at Durg and Dhamtari Disstt.of CG
	Heera	First high protein containing rice variety (11.5%)
Wheat	C-306	First variety of wheat developed in India at Hissar by Dr. Dhaniram Vasudeva. (Tall variety of wheat)
	Single gene dwarf	Lerma rojo 64-A, Sonalika, Sujata (Sujata-Suitable for rainfed condition)
	Double gene dwarf	Kalyansona, Sonaro-64, Chhoti Lerma, HD 2009 (Arjun), Pratap, Janak
	Triple gene dwarf	Hera, Moti, Jawahar, Jyoti, Lal Badsah
	Mutant variety	**Sarbati Sonaro** (Sharbati Sonara obtained from Sonara - 64 with gamma rays.)
		Pusa lerma (pusa lerma obtained from lerma rojo 64-A with gamma rays.)
	Marconi wheat	Jayraj, meghdoot, Bansi, Malvika
	Bread wheat	C306, C-3, K-65, K-68

	Other important variety	(**Lok-1**-Suitable for irrigated condition), (**Pusa pachheti**-IARI variety suitable for late sown), (**Pusa suketi**-This is only variety of wheat that resistant all three rust, IARI variety), (**Poshan**-IARI variety Good for dalia,suji,pasta) **Pusa Gold-(FCI 2015)**
Maize	Ganga-1, Ganga-101, Deccan and Ranjit	First time in India double cross hybrid were released in 1961
	Ganga-2 and Highstarch	Top cross variety of maize (White seeded variety)
	Proteina (4.0% lysine), Shakti and Ratna	Lysin rich composite variety of maize released in 1971, such variety are called opaque-2
	Shaktiman-1, Shaktiman-2 and HQPM-1	This are the QPM variety of maize using opaque-2 gene
	Amber, Kisan	Composite variety of maize
	Baby corn-	Prakash, Parvati, VL-78, HM-4
	Sweet corn-	Composite madhuri, Composite priya
	Pop corn-	VL pop corn, Amber pop, Pearl popcorn
	Fodder maize	African tall, Pratap chari-6
	Flood prone area	Diara-3
Sorghum	CSH-1	First variety of Sorghum 1964
	M-35-1	Best high yielding variety of rabi sorghum
	CSH-9	Drought and salinity tolerance variety of sorghum
	IS-208, IS-28450	Low HCN containing variety
Bajra	HB-1	Release in 1965 **:Tift 23-A (male sterile) x Bill 3B**
	Pusa moti	**Mass selection variety of bajra (SADO CG VYAPAM-2016)**
	Other	Parbhani sampada, pusa safed
Balrley	BHS 380 (Pusa Losar)	IARI variety Resistant to blight and all three rusts

	BHS 352 (Himadri):-	Huskless; tolerant to yellow rust. (Dolma, Karan- Huskless)
	Vijay	Best for malt & brewing
	Kedar	Double purpose (both grain and fodder)
	HD 2052	Molya disease resistant
Oat	Kent, Coachmen	
Triticale	**DT-46**	
Gram	**Pusa chamatkar (BG 1053) (Kabuli)**	IARI variety resistant to soil borne diseases; bold seeded; good cooking quality **(IBPS, AO- 2015)**
	Avrodhi	Wilt resistant variety
	Pusa 256/BG 256	Most common and best for rain fed condition
	C-235	Widely adapted for dry land, tolerant to Ascochyta blight
	Gaurav	Resistant to Ascochyta blight
	Aparna	Grain remain green colour
	ICRISAT variety	ICCV-10 (Bharati), ICCV-2 (Sweta), ICCV-37 (Kranti)
	IIPR	Subhra, Ujjawal
	IGKV	Indira chana-1, Vaibhav
	Karnal Chana1	Recommended for salt affected areas; Wilt resistant.
	Other	JG-74, JG-315, Samrat
Pigeon pea	**ICPH-8**	World first hybrid of pigeon pea developed at ICRISAT, Hyderabad 1991 **(GMS-Prabhat × ICPL-161), Indeterminate type**
	UPAS-120	It is an early maturing (120-125 days) selection evolved at Pantnagar.
	Pusa ageti, prabhat, Manak	**Extra early maturing (110-120 days)**
	Rajiv lochan	**Developed at IGKV Raipur through pedigree method , Drought, Sterility and Phytophthora blight resistant**

	Other variety	**Asha, Bahar, Mukta (Wilt resistant), Durga, Laxmi**
Green gram	**Pusa baisakhi**	**It is selection from type 44, It has wider adaptability and is suitable for summer season**
	Pusa Vishal (IARI)	Resistant to mungbean yellow mosaic virus.
	Pusa Ratna (IARI)	Tolerant to yellow mosaic virus.
	IPM 410-3 (Shikha) (IIPR)	Highly resistant to Mungbean Yellow Mosaic disease and powdery mildew
	Pairy mung	Developed at IGKV Raipur, Cross between TARM-1 × J781
	Other Important variety	HUM-2 (Malvia Jagriti), HUM-6 (Janpriay), HUM-12 (Janchetna), HUM-16 (Jankalyani)
Black gram	PDU-1 (Basant bahar) IIPR, variety	YMV tolerance
	OBG-17 (Ujala)	
	T-9	Suitable for all season
	Indira urd-1 (IGKV, variety)	Resistant to yellow moisac and Powdery mildew
	Other	
Field pea	Rachna	First tall & powdery mildew resistant variety of pea
	Aparna (HFP-4)	First dwarf variety of pea
	Salvia	Whole pod are edible and suitable for kitchen garden
	Bonevile	Introduced from USA (suitable for dehydration)
	CG variety	Ambika, Paras, subhra
	Other important variety	Early bajer, Early December, Alankar, Pusa panna (DMR-7)
Table pea	Arkel	Introduced from England
		Pusa pragati, Lincoln

Lathyrus	Pusa 24 (IARI variety)	
	Ratan (Bio lathyrus 212)	
	Pratik	
	Mahateora	
	Nirmal	
Rajma	Pusa parvati	**Contender, Amber (IIPR)**
Ground-nut	Girnar-1	**Bulk pedigree method (developed by DOG, a high yielding bunch type for kharif cultivation)**
	Girnar -2	
		Kausal, prasuna, Jyoti, Chandra, Chitra
Soybean	Bragg, Lee, Clark, Silajit	**(Exotic variety indroduced from USA)**
	Indira soya-9	**(IGKV Raipur)**
	Ahilya, Kargil	
Rapseed & Mustard	Pusa jai kisan (Bio 902)	**1st Somaclonal variety of Indian mustard developed by V.L. Chopra in 1994 at National centre for biotechnology New Delhi, Donor parent used Varuna (T-59)**
	Pusa Agrani (SEJ-2) (IARI -1998)	**The first early maturing (110 days) variety of Indian mustard. It is a substitute of toria.**
	Pusa Bold (IARI 1985)	**Most widely adapted variety in the country. It is a bold seeded variety with > 6 g/1000 seeds weight having 40% oil content. It matures in 140 days. (Yellow sarson)**
	Pusa Jagannath, Pusa Mahak	
	Pusa Karishma (LES-39) (IARI-2005)	It is the first single zero (<2% erusic acid) variety of Indian mustard having attractive yellow seed coat colour.
	Pusa Tarak	
	NRCHB-6	1st hybrid mustard in India developed at DRM, Bharatpur in 2009

	Pusa '00'mustard	1st canola type hybrid mustard with low < 30 ppm glucosinolate and < 2.0 % Erusic acid
	PGSH-51	1st Gobhi sarson variety
	Vardan	Suitable for intercropping
	Varuna, Kranti, Rohini	
Brown sarson	Pusa kalyani, Suffla	
Yellow sarson	T-42, Benoy,Pusa Bold	
Taramira	T-27	
Karanrye	Pusa swarnima	Narendra rai -1 (Salt tolerance variety)
Toriya	Indira toria-1	(Developed by IGKV, Raipur cross between Bhavani × L.toria-1)
	C.G sarson-1 Bastar toriya-1	(IGKV Variety)
Sunflower	BSH-1 (1980)	1st hybrid variety of sunflower in India based on CMS method and it cross between 234-A × RHA-274)
	Russian, Vinmik, Armviskis	Exotic variety
	Modern, Jwalamukhi, Surya	
Safflower	DSH-129	1st commercial hybrid variety released by directorate of oilseed research based on CMS (Spiny type)
	NARI-1	1st Non-spiny hybrid variety of safflower in World level
	Bhima, Subra, Phule kusum	
Linseed	Indira alsi-32, Deepika, Kartika, Kiran	(IGKV Variety)
Castor	Aruna	Popular variety of castor
Sesamum	Kanchan	
Tobacco	GTH-1	1st hybrid variety

	Gautami, Sona, Gandak,	
Cotton	H-4	It was 1st Commercially hybrid variety of cotton. Developed in the World by Crossing G-67 × American nector less by Dr. C.T. Patel of GAU, Surat in 1970
	Varalaxmi	1st Interspecific hybrid between G.hirsutam × G.barbandense Released from UAS, Dharwad
	MCU-5	It is extra long staple variety of cotton
	Savita, Surya	Intraspecific hybrid of hirsutam
	Maru vikas	1st hybrid hirsutam in Rajasthan
	Sujata, Savin	Egyptian variety
	Bikaneri nerma, Dhanlaxmi, mahalaxmi	Boll guard -1,2
Jute	White jute-	Sabuj sona, Sonali, Shyamali
	Tosa jute-	Baisakhi tosa, Chaitali tosa
Lucern	Rambler	
Berseem	Meskavi	Diploid variety
	Pusa giant	Tetraploid variety (IARI)
	Vardan	(IGFRI, Jhansi)
Napier grass	Pusa giant	

Horticultural Crops		**Fruits Crops**
Mango	Alphonso	Best export variety but susceptible to Spongy tissue
	Chausa	Sweetest variety
	Baganpalli	Most commercial variety of Andhra Pradesh
	Bombay green	Earliest variety of North India, Pollishing variety, highest Vit-C. It is also called Malda in UP and Sehroli in Delhi
	Kesar	It has good processing quality

	Langra	It has characteristic of Turpentine flavor, Most prone to fruit drop
	Niranjan	Off season variety of mango (Bear fruit in October)
	Lal sindhuri	Powdery mildew resistant variety
	Neelam	Ideal for long transport, taken two crops in a Year
	Rossica	Mutant variety
	Rummani	Apple shaped variety
	Sindhu	1st Seedless variety of mango in World
	Edward	Antracnose resistant variety
	Gulab khas	Rosy flavor
	Olour, Goa, Mulgoa, Bellary, Chandrakaran	Poly embryonic variety
Hybrid	Amrapali (IARI)	Suitable for HDP 2.5 × 2.5 m^2, Dasheri × Neelam (DNA)
	Mallika (IARI)	Highest Vitamin content, Neelam × Dasheri (NDM)
	Ratna	Spongy tissue disease resistant, Neelam × Alphonso (NAR)
	Sindhu	1st Seedless variety of mango in World, Ratna × Alphonso (RAS)
	Arka anmol	Alphonso × Janardan pasand (AJA), Free from spongy tissue
	Arka punnet	Alphonso × Baganpalli (ABA), Free from spongy tissue
	Arka aruna	Baganpalli× Alphonso (BAA)
	Sai sugandha	Totapari × Kesar (TKS)
	IARI variety -	Pusa Lalima, Pusa Shreshth, Pusa surya, Pusa Arunima
	IGKV variety-	Chhattisgarh Nandi raj
Banana	Poovan	AAB-genome, resistant to panama wilt
	Dwarf gavendish (Barsai, Kadali)	AAA-Genome group, resistant to panama wilt (Kadali-Leading commercial variety-58 % production)

	Robusta	AAA-Genome
	Rasthali	AAB-Genome, Choicest table variety
	G-9	France, Tall mutant of dwarf gavendish
	Lady finger	Diploid variety
Lime	Sai sarbati	Tolerance to tristeza and canker
	chakradhar	Seedless variety of lime
	Vikram	Offseason and bunch bearing habit
Orange	Sathgudi	Grannulation susceptible variety, commercially grown in Andhra Pradesh, it is also known as Chinese Orange
	Jaffra	Suitable for arid region, Mid season variety
	Valencia	Late maturity variety
	Mosambi	For Maharastra
	Malta blood red	For North India
	Pineapple	
	Shamouti	Famouse seedless sweet Orange variety
Grape	Tetraploid-	Thampsom seedless, Pusa seedless, Parlette
	New parlette	Self thining
	Anab-e-shahi	Popular cultivar variety of whole India
	Arka Hans,Arka shyam	(Pusa urvashi, Pusa Navrang-IARI variety)
	Crimson seedless	Indroduced from USA, Coloured seedless
Papaya	Gynodioecious-	
	Pusa delicious -	(From IARI Regional Station, Pusa Bihar, variety)
	Pusa Majesty--	(from IARI Regional Station, Pusa Bihar), Nematode resisitant)
	Coorg honey dew-	(IIHR variety, Selection from honey dew)
	Taiwan-	Blood red colored variety
	Surya-	(Sunrise solo × Pink flesh sweet)
	Sunrise solo- Dioeciouse-	Pink flesh, highest Carotene content

	Pusa giant-	(From IARI Regional Station, Pusa Bihar). Suitable for canning
	Pusa dwarf	From IARI Regional Station, Pusa Bihar).
	PUSA Nanha	(From IARI Regional Station, Pusa Bihar an ultra dwarf dioecious variety of papaya developed through mutation breeding. Fruiting starts at a height of 30 cm and the total height of the plant is 120 cm)
	Transgenic variety-	**Sunup and Rainbow**
	Other	**Co-5-Popular variety for papain extraction, Red lady-Popular exotic variety**
Guava	**Allahabadi safeda-**	**Popular variety of UP, Suitable for HDP (3 × 2 m^2)**
	Lukhnow-49	**Selection from allahabadi safeda , Also known as Sardar, National variety of guava, Highest vit. C content variety**
	Harijha-	**Popular variety of Bihar**
	Chittidar-	**Highest TSS content variety of guava**
	Safed jam-	**Allahabadi safeda × Kohir**
	Kohir safeda	**Kohir × Allahabadi safeda**
	Pusa Srijan	**Developed from IARI New Delhi,** Extremely dwarf rootstock, hence, recommended for high density planting. Tolerant to wilt
	Behat coconut	**Seedless variety of Guava**
	Hybrid-45	**Allahabad safeda × L-49**
	Other	**Sweta, Lalit, Apple**
Sapota	Kalipatti, Murabba,	**Cricket ball, (A popular variety from Tamil Nadu) Kirthi bharati**
Pineapple	Queen-	**1st hybrid variety, Earliest variety**
	Kew	**Fibreless variety**
Jackfruit	Singapore or ceylonjack	**Introduced from Ceylon , Off season variety**
	Rudrakshi-	**Pummelo size fruit**
Litchi	Swarn roopa-	**Seedless, highly resistance to fruit cracking 1st variety developed in India**

	Shahi	**Earliar variety**
	Rose scented	**Mid season variety**
Apple	Lal amberi	**Red delicious × Amberi**
	Sunheri	**Amberi × Golden delicious**
	Amberi	**Longest shelf life and indigenous variety of India**
	Red delicious	**Most important variety of India**
Straw-berry	Pajaro-	**Most succesfull under summer season**
	Chandler-	**Most popular variety of North Indian condition**
Ber	Gola	**Early variety suitable for dryland area**
	Seb, Kaithali	**Mid maturity**
	Umaran	**Late for heavy production, Originated from Rajasthan, processed and used as Chuhara**
	Illaichi-	**Triplest variety, 90 % pollen sterility found**
	Thar Sevika-	**Seb × Katha (Early maturity)**
	Thar Bhujraj-	
	Goma Kirti	**Early maturity and high yielding variety**
	Katta paul-	**Apple's variety**
Aonla	Banarasi	**Early maturity variety, Prone to heavy fruit drop, Best for Murrumba**
	Krishna (NA-5)	
	Narendra (NA-9)	**Suitable for preservation and Candy making**
	Kanchan (NA-4)	**Mid season variety, Preferred by industries for pulp extraction**
	Francis (Hathi jhool)	**Mid Season , Highly susceptible to fruit necrosis**
	Chakaiya	**Late maturity, suitable for pickles making**
	Goma Aishwarya	**Clonal selection from NA-7, Early variety**

Pome-granate	Ganesh	
	Mridula	**Ganesh × Gul-e-Shah Red, Blood red aril and soft seeded**
	Ruby	**(Ganesh × Kabul) × Yercaud**
	Amlidana	**Hybrid from Ganesh × Nana, Most suitable for Anardana purpose**
	Phule Arakta	**Ganesh × Gul-e-Shah Red**
	Bhagwa (Sinduri)	**Most popular variety of Indian condition, Semi hard seeded variety**
Bael	Kagzi Gonda,	**Kagzi Etawa, Kagzi Banarasi, Naredra Bael-1, 2**
Jamun	Paras	**Large seeded variety**
	Narendra jamun-6	

Vegetables crops

Tomato	Pusa Ruby	**Sioux × Improved Merruti**, Most Famous variety of tomato
	Pusa early dwarf	**Improved Merruti × Red cloud**
	Pusa Rohini ,	Highly suitable for long distance
	Puas Gaurav	transportation and processing
	Pusa uphar	Tolerant to fruit borer
	Parker	Resistant to fruit borer
	Hisar lalit, S-120	Nematode resistant variety
	SL-12	**X rays Mutant of Sioux**
	Arka Rakshak	1st Triple disease resistance Variety of Tomato **to Leaf Curl Virus, Bacterial Wilt and Early blight.**
	Pusa shcetal	Suitable for low temperature (Winter season)
	Flaver savar	Delay ripening variety of tomato developed by Biotechnology in 1994 by Calgene Company
	Punjab Chhuhara, Roma, Angoorlata,	

Chilli	Pusa Jwala	NP-46 A × Puri Red , Trips resistant
	Pusa Sadabahar	Pusa Jwala × IC-31339
	Punjab lal	Parenial × Long red
	Jwala Mukhi	Suitable for HDP with low Pungency and resistant to virus
	Bharat, Bhaskar	
Capsicum (Shimla Mirch)	California Wonder, Yolo Wonder	
	Arka Mohini	Introduced from USA
	Pusa Deepti	Yolo Wonder ×Russian Yellow
Brinjal	Pusa purple long	Extra early maturity
	Pusa purple round	
	Pusa purple cluster	
	Pant Samrat	Resistant to Phomopsis blight and Bacterial wilt, Tolerance to Fruit and Shoot borer
	Annamalai	Aphid resistant
	Pant rituraj	
	Pusa Anmol (Hybrid)	PPL × Hyderpur
	Arka Navneeth (hybrid)	IIHR22-1 × Supreme
	Arka Nilkantha	Resistant to bacterial wilt
Bhindi	Pusa Makhmali	Selection from West Bengal, Suitable for Spring-Summer season
	Pusa Sawani	Pusa Makhmali × IC-1542, Tolerantce to salinity, Spinless, Day Natural variety
	Harbhajan	Selection from Perkins long green, suitable for hills only
	Parbhani Kranti	Pusa Sawani ×A. Manihot, Resistance to YVMY
Cabbage	Golden Acre	Selection from EC-6774 (Japan),
	Pride of India	Early and round variety
	September	Popular in Nilgiri hills
	Pusa Ageti	1st Tropical variety in India

	Pusa drum head	Flat head variety
	Pusa Mukta (Sel-8)	Selection from EC-24855 ×EC-10109
Cauli-flower	Pusa Himjyoti	It is only variety which can grow from April to July in Hills , Self Blanched habit
	Pusa Snowball	**Self Blanched habit**
	Pusa Deepali ,	Pusa ketki, Pusa meghma, Early Kunwari
Knol-Khol	White Vienna	Most popular Early variety
	Purple Vienna	
Cucumber	Pusa Sanyog	Japanese gynoecious line × Green long Naples, 1st Gynoecious F1 Hybrid developed in India in 1971
	Japanese long green	Introduced from Japan
	Pineset	
Musk-melon (Kharbooz)	Pusa Rasraj	M-3 × Durgapur Madhu, Utilization of Monoecious line
	Pusa Sharbati	Kutana × PMR-6 (USA)
	Pusa Madhuras	
	Durgapur Madhu	
	Arka Rajhans	
	Hara Madhu	Non Spiny variety
Watermelon (Tarbooz)	Sugar Baby	Introduced from USA
	Pusa Bedana (2n=33)	Tetra-2 (4×) × Pusa Rasal (2×), Triplod (3×) seedless watermelon
	Durgapur Meetha	Selection from local cultivar
	Durgapur Kesar	Selection from local cultivar
	Durgapur Lal	Sugar baby × K-3-566, Unlobed leaf marker
	Arka Jyoti	IIHR-20 × Crimson Sweet
	Arka Madhura	Triploid seedless, Suitable for year round cultivation under protected cultivation
Bitter-gourd	**Pusa Do mausami**	Suitable for Spring –Summer and rainy Season
	Arka Harit	
	Arka Annupmma	
	Kashi Urvasi	IC-85650B × IC-44435 A

Bottle-gourd	**Pusa Meghdoot**	**PSPL × Sel-2,** Suitable for Summer and Rainy Season, 1st hybrid variety in India
	Pusa Manjira	**PSPR × Sel-11,** Suitable for Summer and Rainy Season
	Pusa Summer Prolific long (PSPL)	
	Pusa Summer Prolific Round (PSPR)	
	Samrat	
Pointed gourd	**Kashi Alankar**	
Ashgourd	**Pusa Ujjwal**	**Ideal variety for Petha preparation**
Sponge gourd	**Pusa Nasdar**	
	Pusa Chikni	
	Pusa Supriya	
	Pusa Sneha (IARI)	
Pumpkin	Pusa Vishwas	
	Pusa Vikas	rich in vitamin A.
Summer Squash	Pusa Alankar	EC-27050 × Sel.IPI-8
French bean	Pusa Parvati	X-ray Mutant of Wax pod (EC-1906)
	Arka Komal	Photoinsenstive
	Contender	
Cluster bean	Pusa mausami	Suitable for rainy season
	Pusa Sadabahar	Suitable for Summer and rainy season
	Pusa Navbahar	Pusa mausami × Pusa Sadabahar, single stem variety of cluster bean
Cowpea	Pusa phalguni	Introduced from Philipines, Spring season variety
	Pusa Barsati	Introduced from Philipines, Rainy season variety

	Pusa Dofasali	Pusa phalguni × Philipines selection, Spring and rainy season
Onion	Pusa Ratnar	
	Arka Pitambhar	Tolerant to Purple blotch and Thrips, Suitable for Kharif and Rabi season
	Bhima Subhra ,	Bhima Shweta
	Agrifound Red	
Garlic	Agrifound White	
	Bhima Omkar,	Bhima Purple
	Pant Lohit	
Carrot	Pusa Meghali	Highest Vitamin –A content variety
	Pusa Yamadagni	EC-9981 × Nantes Half Long, Self coloured Core variety
	Imperator	Nantes × Chantenay
	Pusa Kesar	Local Red × Nantes Half Long, Tolerance to high temperature
Radish	Pusa Reshmi	Green type × Desi type
	Pusa safed	White-5 × Japanese White
	Arka Nishant	Multiple disease resistance
Potato		
Early Variety	Kufri Chandramukhi Kufri Surya	(Seedling-4485 × Kufri Kuber) Mature in 90-120 days
Medium variety	Kufri Bahar	Kufri red × Gineke Yield 200-250 quintals
	Kufri Pukhraj	Created in 1998 by CPRI, Good for home use
Late variety	Kufri Sindhuri	Suitable for Dehydrated flakes and canning
	Kufri Badshah	Kufri joyti × Alankar
Other varieties	Kufri Chipsona-1	For Potato chips and home use.
	Kufri Chipsona-2	–do–
	Kufri khyati	High yielding variety
Rose	Tushar	White flower
	Crimson Glory	Red flower
	Pusa Mohit, Suchitra	Thornless variety
Rajnigandha	Prajwala	IIHR-Mexicon single × Shrinagar
Gladiolus	Rose supreme	Vibrant colour and ruffled blooms

Table 3a: Fruit Crops

Mango	Veneer grafting(June to Sept-Oct is best for grafting) (Epicotyl grafting- The germinated seedlings of 8-15 days old are used for grafting)
Banana	Sward sucker (The weight of the suckers should be 1.0-1.5 kg, Age of suckers: 3-4 months)
Gauva	Air layering
Papaya	By Seed (Diocious-400-500gm/ha 10 % extra male plants), (Gynodiocious-250-300 gm/ha)
Spota	Commercially propagated by Inarching/Aproach grafting on Rayan or khirni (Manilkara hexandra) sroot tocks)
Phalsa	Seed
Coffee	Seed
Fig	Hardwood cutting
Grape	Hardwood cutting
Pomegranate	Hardwood cutting/Air layring
Litchi	Air layring
Tea	Softwood cutting /Leaf bud cutting
Date palm	Offshoot/Sucker
Pineapple	Sucker/Slip
Strawberry	Runner
Aonla	T-Budding/Patch budding
Bael	Patch budding (June to July is ideal time)
Ber	Ring/T-budding (best time for budding April to May)
Custard apple	T-budding
Rose	Shield budding (Best time for buuding- Nov. to Feb.)
Apple	
Gladiolus	Corm
Chrysanthemum	Root sucker

Table 4. Important Crops and Their Nativity

Crop	*Origin*	*Crop*	*Origin*
Rice	South-East Asia	Oat	Asia Minor
Sugarcane	South-East Asia	Mango	India
Wheat	South-West Asia	Sorghum	Africa
Barley	South-West Asia	Bajra	Africa
Buckwheat	South-West Asia	Sunhemp	Africa
Turmeric	India	Seasamum	Africa
Lucerne	South-West Asia	Castor	Africa
Tobacco	America	Clusterbean	Africa
Maize	South America	Groundnut	Brazil
Toesinate	Mexico	Bearseem	Egypt
Potato	Peru	Napier grass	Rhodesia
Tomato	Peru	Soybean	China
Linseed	Afghanistan	Rapeseed	China
Sunflower	USA	Black pepper	India
Arhar	India	Tea	China
Mung	India	Coffee	Ethiopia
Urd	India	Chilli	Mexico
Cotton	India	Cow pea	Africa
Jute	India	Mustard	China
Kutki	India	Red gram	Africa & India
Bengal gram	South-West Asia	Banana	Southern asia

Table 5. Plant Growth Substances

Substances	*Function*
Auxins	Stimulates elongation of cells of stems and coleoptiles Cell division & Root formation are functions of auxins. Can be used as herbicide: Enhances fruit set and fruit ripening. (Discovered by Darwin 1880)
Cytokinins	To stimulate cell division Promotes orderly development of embryos of seed Break dormancy of seeds and buds Delays senescence.
Gibberellins	Increases cell division & cell elongation Increase in size of leaf, flower, fruit Dormancy is broken & flowering is induced
Abscisic Acid	Dormancy is induced Facilitates stomata closure
Ethylene	Ripening of fruits (or) Iso-diametric growth of stems and roots

Table 6. Plant Growth Substances used in Agriculture and Horticulture

Plant growth parameter	*Use*
2, 4 -D	Herbicide
Glyphosine, Ethepon	Sugarcane ripener
Abscisic acid	Cotton defoliant
NAA	Increasing number of flowers
Gibberllins	Production of seedless grape, increase in size of fruits.
Ethylene	Regulating ripening of fruits.

Table 7. Crops and Oil Percentages

Castor	—	40-50%
Indian Mustard	—	38-40%
Toria	—	42-44%
Yellow brown Sarson	—	43-48%
Soybean	—	18-20%
Linseed	—	33-48%
Cotton seed	—	33-40%
Peanut	—	44-55%
Ricebran	—	18-20%
Mesta	—	17-22%
Maize	—	4%
Sesame	—	48-55%
Sunflower	—	45-50%
Groundnut	—	44-50%
Safflower	—	24-36%
Niger	—	35-45%
Coconut	—	55-60%

Table 8. Chemical Properties of Different Soils

Type of soils	*Parameter*		
	pH	*EC*	*ESP*
1. Saline	< 8.5	> 4	<15
2. Sodic	< 8.5	> 4	>15
3. Alkali	> 8.5	< 4	>15

Table 9. Thermal Properties of Common Soils and Soil Components

Component	*Density (g/cm^3)*	*Specific heat (cal/g/degree)*	*Thermal conductivity (cal/cm/degree/second)*	*Thermal diffusivity (cm^2/second)*
Clay	1.8	0.8	0.0028	0.002
Humus soil	0.3	0.3	0.00027	0.003
Wet sandy soil	1.6	0.4	0.0064	0.01
Air	0.0012	0.24	0.00005	0.1667
Water	1.0	1.0	0.0014	0.0014
Ice	0.917	0.505	0.005	0.0108

Table 10: Pungency and Colours of Crops

Yellow colour of Cassava	- **Carotene**
Yellow colour of Onion	- **Quercetin**
Pink colour of Onion	- **Anthocynin**
Red colour of Tomato	- **Lycopene**
Purple colour of Brinjal	- **Anthocynin**
Pungency of Chilli	- **Capsaicin alkaloid**
Colouring pigment of Chilli	- **Capsanthin**
Pungency of cole crops (Cabbage)	- **Sinigrin**
Anticancerous principle in Broccoli	- **Sulphoraphane**
Pungency of Onion	- **Allyl propyl disulphide**
Pungency of Garlic	- **Diallyl disulphide**
Pungency of Radish	- **Isothiocynate**
Anticancerous principle in Tomato	- **Lycopene**
Anticancerous principle in Cabbage	- **Indole-3-carbinole**
Alkaloids present in Brinjal	- **Solasodine**
Yellow pigment in Papaya	- **Caricaxanthin**
Muscat flavour of Grape	- **Methyl anthranilate**
Enzyme present in papain	- **Pepsin**
Active compound in Fig	- **Ficin**
Enzyme present in Pineapple	- **Bromelain**
Principle gylcoside in Peach	- **Prunacin acid**
Principle compound in Coffee	- **Caffeine**
Principle compound in Cocoa	- **Theobromin**
Red violet colour of Beet root	- **ß-cyanin**
Bitterness in almond	- **Imailedin**
Bitterness in pepper	- **Marmelosin**
Odour in pepper	- **Oleriosin**
Bitterness in cucumber	- **Cucurbitacin**
Acidity in arbi	- **Calcium oxalate**
Acidic taste in gram leave	- **Malic acid/Oxalic acid**
Pungency of toria	- **Allyl iso-thiocyanate**
Pungency of mustard oil	- **Glucosinolates**
Bitterness in aonla	- **Polyphenole and tannin**
Red colour in carrot	- **Anthocyanin**
Red colour in chilli	- **Capsanthin**
Green colour in potato	- **Solanine and Chaconine**
Orange red colour of saffron	- **Carotenoid (crocin)**
Bitterness in saffron	- **Picrocrocin**
Antioxidant, anti-ageing, antimicrobial ingredient in turmeric, yellow colour in turmeric	- **Curcumin**
Bitterness in fenugreek	- **Trigonelline and Choline**
Orange colour of carrot	- **Carotene**

SOME FACTS ABOUT SOILS

Vasily Vasil'evich Dokuchaev is commonly regarded as the father of soil science, he was the first person to make wide geographical investigations of different soil types.

Soil refers to upper biochemically weathered portion of the regolith.

Weathering is a process by which rocks are chemically or physically disintegrates into fragments.

Pedon is the smallest volume which can be called as a "soil".

Soil viscosity is the shear stress required for a fluid to pass from one layer to another for specific distance away from the original position of the fluid with unit velocity

- Clay soils have higher pore space than sandy soils.
- Moisture holding capacity of clayey soils is maximum about 13%.
- Soil class represent the texture of the *soil*
- Bulk density of soil is always smaller than its particle density. Bulk density of sub soils is higher as compared to surface soils. Bulk density of normal soils is 1.33 g/cm^2
- Particle density also called true density.
- Acid soils can be reclaimed by $CaCO_3$.
- Aluvial soil group is dominant in India.
- *Soil profile* is a vertical section of the soil from the surface through all its horizons or Soils are characterized by a definite sequence of horizons known as *soil profile.*

  Zone of maximum leaching - A Horizon (Eluviated horizon)

  Zone of maximum accumulation – B. Horizon

  Zone of least weathering and desiccation - C Horizon
- For irrigated agriculture best suitable soil is red soil.
- Soil which are most suitable for most of the crops are sandy loams.
- Soil most suitable for plantation crops – *Laterite soils*
- Element responsible for flocculation in soil is – *Calcium* and Mg
- For dry land agriculture black soils are best suitable.
- (C.N.) Carbon nitrogen ratio of normal soils is 10-12 : 1.
- Humus of soils is a product of plant and animal residues decomposition.
- Soil structure is the arrangement of aggregates in soil.
- *Granular* & *Crumb* is most preferable soil structure for agriculture.
- Most important soils for agriculture -Alluvial Soils
- Best Agricultural soils in the world are- Mollisol

- Sodic soils also called as alikali soils
- Sodic soil is a soil with an exchangeable sodium of greater than 6% of the cation exchange capacity.
- Optimum bulk density of soils preferable is 1.2 to 1.6 g/cc
- Diameter of clay particle is less than 0.002 mm.
- Diameter of fine sand particle is 0.02-0.2 mm.
- Diameter of silt particle is 0.02-0.002 mm.
- Taxonomically organic soil are *Histosols*.
- *Organic soils* are extensively composed of plant remains peat and muck.
- *Peat* is a partially decomposed plant accumulated either in water or in a water saturated soil.
- *Muck* is decomposed of plant remains.
- For determining soil colour, Munsell colour chart is used it contains 175 colour chips
- pF of soil is the logarithmic height of water column in cm representing total stress with which water is held in the soil.
- Negative logarithm of soil water tension is measured by pF scale.
- Highest area of acidic soils in West Bengal
- *Alfisols soils* are mostly high in respect of their fertility and are used either as crop land or forest or range land.
- *Texture* refers to the size distribution of soil particles in a soil. It is important in determining the water holding capacity of soil.
- *Porosity* is the ratio of the volume of pores to the total soil volume (Porosity is highest in clays).
- Father of soil physics – *Hillel*
- **Edaphology** is the study of the soil properties in relation to higher plant.
- **Pedalogy** refers to the study of origin of the soils in their natural environment, its classification and description.
  Pedosphere: It is the sum of the planet's all soil ecosystems.
- **Largest soil group of India:** Alluvial soils
- Yellow colour of soil is due to higher degree of hydration of ferric oxide or silica.
- Red colour of red soils is due to presence of *iron oxide.*
- Pore space is space present between the soil particles in a given soil sample.

- Brown colour of soil is due to organic matter.
- Who reported in 1840 that crop yields are increased by adding minerals or inorganic elements to the soil.

Ans. Justin von Leibig (German Scientist)

- Nature takes about 300 years to form 2.4 cm (1 inch) of soil.
- Clay and humus have large surface area per unit weight due to their extremely small particle size.
- Most of the plants are likely to suffer if the soil volume occupied by the air is less than 10-12%.
- Soil moisture content at which plant remains wilting day and night (or in humid chamber) is called the *permanent wilting point* or *wilting coefficient*. Concept proposed by **Briggs and Shantz (1912)**
- Three recognized types of soil water movement are: 1. Saturated flow 2. Unsaturated flow and 3. Vapour movement.
- Resistance to a change in pH of the soil is called Buffering.
- Higher the exchange capacity of a soil greater is the buffering capacity.
- Plasticity index given by **Albert Atterberg (1911)**

Petrology: The branch of science concerned with the origin, structure, and composition of rocks.

Carbon nitrogen ratio: The ratio of weight of organic carbon to the weight of total nitrogen in soil or in organic material.

Adhesion: Molecular attraction that holds surfaces of two substances in contact. (Like soil particles and water).

Soil Fertility: Ability of a soil to supply essential elements for plant growth without a toxic concentration of any element.

Soil productivity: Soils capacity to produce certain yield of crops or other plants with optimum management.

***Cation exchange capacity*:** It is the amount of exchangeable cations a soil can sustain. The soils which have high cation exchange capacity hold nutrients better and fertile. Cation exchange capacity of clay, sand and humus soils are about 30, 9 & 200 milliequivalents per 100 grams. Exchangeable cations available in soil particles include Alluminium (Al^{3+}), with the strongest attachment followed by hydrogen (H^+), calcium (Ca^{2+}), magnesium (Mg^{2+}), potassium (K^+), ammonia (NH_4^+), and sodium (Na^+). Cation exchange capacity is important because it allows us to understand the necessary supplemental soil fertility requirements.

Electrical conductivity: It is measure of soluble salts present in a soil. Soil having high EC has higher soluble salts in it and vice versa.

Tilth: It is the physical condition of soil related to the ease of tillage, fitness of soil as seabed and desirability for seedling emergence and root penetration.

Tillage: Mechanical manipulation or modification of soil conditions for plant growth purposes of tillage are (*i*) to manage crop residues (*ii*) to kill weeds and (*iii*) to alter soil structure especially preparation of soil for planting seeds or seedlings.

Horizon: A layer of soil approximately parallel to the surface, that has distinct characteristics produced by soil forming process.

Sodic soils: Sodic soils are characterized by a disproportionately high concentration of sodium (Na) in their cation exchange complex. A non saline soil.

Epipedon: These are diagnostic surface horizons of mineral soils forms near soil surface.

SOIL GROUPS (12 orders)

Alfisols: Soils with gray to brown surface horizons, medium to high supply of bases and B-horizons of alluvial clay accumulation usually fertile.

Andisols: Soils that have formed in volcanic ash or other volcanic ejecta. They differ from those of other soil orders in that they are typically dominated by glass and shortrange order colloidal weathering products (Allophane, imogolite and ferrihydrite minerals). It accounts for 1% of the ice free land area. These are also known as Andosols

Aridisols: Soils of dry regions/climate. Low in organic matter. Have pedogenic horizons.

Entisols: Soils not having any pedogenic horizons. These soils may found in any types of climate on very recent geomorphic rocks.

Gelisols: Soils of very cold climates that contain permafrost within 2 meters of the surface. Gelisols estimated to occupy about 9% of the earth's ice free land area.

Histosols: Organic soils formed from organic matter and do not have clay.

Inceptisols: Mineral soils with beginning or inception of profile development.

Mollisols: Soils with nearly black organic rich surface horizons and high supply of bases.

Oxisols: Soils with residual of accumulation of non-activity clays, free oxides, kaolin and quartz, found in tropical climates.

Spodosols: These are acid soils characterized by a sub- surface accumulation of humus that is complexed with Al and Fe. Globally these soils occupy 4% of ice free land surface.

Ultisols: Soils low in bases and have subsurface horizons of illuvial clay accumulations. Usually moist.

Vertisols: Clayey soils with high shrink-swell potential having wide, deep cracks when dry. These soils have distinct wet and dry period throughout the year.

SOME TERMS DEFINED

1. **Agro ecosystem:** The land or area used for crops, pasture and livestock, the adjacent uncultivated land that supports other vegetation and wildlife, and the associated atmosphere the underlying soils, groundwater, and drainage networks.
2. **Abiotic stress:** Conditions or factors which can cause harmful effects to plants such as soil conditions, drought or extreme temperatures.

2a. **Aerosol:** A container in which a liquid substance is kept under pressure. When you press a button the liquid comes out in a fine spray

3. **Anabolism:** The synthesis of complex molecules from simpler ones together with the storage energy. All biosynthetic reactions in a living organisms.
4. **Alley cropping:** Planting of crops in strips with rows of trees or shrubs on each side.
5. **Achene:** A small, dry seeded fruit that does not open to release the seed.
6. **Amino Acids:** These are fundamental building blocks of proteins. There are 20 naturally occurring amino acids in animals and around 100 or more found in plants.
7. **Annual:** Plant that grows, sets seed and dies in one growing season.
8. **Arable land:** Land that can be cultivated to grow crops
9. **Aridity:** A measure of dryness or water deficiency. An index based on water deficiency over water need.
10. **Azotobacter:** Azotobacter is defined as a genus of free living soil microbes that are responsible for nitrogen fixing in the atmosphere. These microbes play a very significant role by synthesizing some biologically active substances which helps in stimulating plant growth. Azotobacter are mainly used in agriculture, most importantly in nitrogen biofertilizers such as azotobacterin.
11. **Biodiversity:** The variability among living organisms from all sources and the ecological complexes of which they are part and include diversity within species or between species and of ecosystems.
12. **Biochemical pathways:** Various steps involved in any bioconversion process.

13. **Bioremediation:** Process of using microorganism to degrade or consume to remove pollutants from environment.
14. **Biotic stress:** A stress that occurs in plants as a result of damage done by other living organisms, such as bacteria, viruses and fungi, parasites, weeds, insects and other native or cultivated plants.
15. **Biofuel:** Gas (as methane) or liquid fuel (ethyl alcohol) made from plant material (biomass).
16. **Biofertilizers:** These are organisms that enrich the nutrient quality of soil. Main sources of biofertilizers are bacteria, fungi and cyanobacteria (blue green algae).
17. **Biome:** terrestrial regions characterized by certain types of vegetation and other forms of life.
18. **Carbon credit:** A form of permit which allows a country or organization to produce a certain amount of carbon emissions and which can be traded if the full allowance in not utilized.
19. **Carbohydrates:** They are the compounds either aldehydes or ketones having an empirical formula (CH_2O), *i.e.* they are composed of carbon, O_2 and H.
20. **Chemosynthesis:** Process of carbohydrate synthesis in which the organisms use chemical reactions to obtain energy from inorganic compounds. *Example* bacteria of the genus Nitrosomon as oxidise ammonia to nitrite.
21. **Chromosomes:** Thread or rod like structure in eukaryotes found in nucleus; made up of nucleoprotein and contain all the nuclear genes. The paired, self-replicating genetic structures of cells that contain the cellular DNA; the nucleotide sequence of the DNA encodes the linear array of genes.
22. **Chromatid:** A thread of DNA formed during cell division when a chromosome splits along its length.
23. **Cistron:** A unit function in DNA. One DNA cistron specifies one polypeptide chain in protein synthesis.
24. **Clone:** All individuals derived by vegetative propagation from a single original individual.
25. **Coenzymes:** These are organic molecules that are required by certain enzymes to carry out catalysis. They bind to the active site of the enzyme and participate in catalysis but are not considered substrates of the reaction. They often function as intermediate carriers of electrons, specific atoms or functional

groups that are transferred in the overall reaction. An example is the role of NAD in the transfer of electrons in certain coupled oxidation reduction reactions. Some examples are: Nicotine adenine dinucelotide (NAD), nicotine adenine dinucelotide phosphate (NADP), flavine adenine dinucelotide (FAD), Thiamine pyrophosphate (thiamine vit. B_1) and biotin.

26. **Cofactor:** It is a non-protein chemical compound that is bound (either tightly or loosely) to a protein and is required for the protein's biological activity. These proteins are commonly enzymes and cofactors can be considered as "helper molecules/ions" that assist in biochemical transformations. Cofactors are often classified as inorganic substances that are required for, or increase the rate of, catalysis. Some examples are Zn^{++}, Fe^{+++} or Fe^{++}, Cu^{++} or Cu^{+}, K^{+} and Mg^{++}.

26.a **Condensation:** Joining of monomers by covalent bonding with the accompanying loss of water molecules.

26.b **Cutin:** A waxy water-repellent substance in the cuticle of plants, consisting of highly polymerized esters of fatty acids.

27. **Embryogenesis:** Embryogenesis is a process whereby a plant is formed from one cell in a tissue, or callus by the normal zygotic pathway. Process is initiated by hormones or in the absence of hormones (*e.g.*, carrot, brome grass).

28. **Coniferous trees:** Cone bearing trees, mostly evergreens, that have needle-shaped or scale like leaves.

29. **Contour farming:** Ploughing and planting across the changing slope, rather than in straight lines to help retain water and reduce soil erosion.

30. **Crust** solid outer zones of the earth. It consists of oceanic crust and continental crust.

31. **Dioxins:** Family of 75 chlorinated hydrocarbon compounds formed by unwanted by-products in chemical reactions involving chlorine and hydrogen, usually at high temperature.

32. **Ecological footprint:** Amount of biologically productive land and water needed to supply a population with the renewable resources it uses and to absorb or dispose of the wastes from such resource use. It measures the average environmental impact of populations in different countries and areas.

33. **Ecological tipping point:** Point in the development of an environmental problem where a threshold level is reached, causing an irreversible shift in the behaiour of a natural system.

34. **Enzymes:** These are macromolecular biological catalysts. They help in catalyzing the biochemical and biological reactions by binding to the substrate at the substrate.
35. **Explant:** A plant organ or piece of tissue used to initiate a tissue culture.
36. **Fallow land:** It is the land which is kept for one or more growing seasons to kill the weeds or making the soil richer with nutrients and moisture.
37. **Famine:** Widespread malnutrition and starvation in a particular area because of a shortage of food, usually caused by drought, war, flood, earthquake, or other catastrophic events that disrupt food production and distribution.
38. **Fermentation:** It is the process of energy production in a cell under anaerobic conditions without oxygen.
39. **Fishprint:** Area of ocean needed to sustain the consumption of an average person, a nation, or the world.
40. **Food web:** Complex network of many interconnected food chains and feeding relationships.
41. **Functional agro-biodiversity:** Those elements of biodiversity on the scale of agricultural fields or landscapes; which provide ecosystem services that support sustainable agricultural production and can also deliver benefits to regional and global environment and to public at large.
42. **Fuzz:** Non separable part of fiber attached to cotton seeds.
43. **Gene:** A chromosomal segment that codes for a single functional polypeptide chain or RNA molecule.
44. **Gene duplication:** Genetic alteration in which a segment of DNA is repeated. Duplications may appear anywhere, but where the duplicated segment is adjacent to the original one is called *tandem duplication*.
45. **Genetic drift:** It is the change in the relative frequency in which a gene variant occurs in a population due to random sampling and chance rather than natural selection. In genetic drift, a population experiences a change in the frequency of a given allele; prompted by random luck rather than a need for adaptation.
46. **Gibberellins:** A group of plant hormones responsible for stimulation of the growth of leaves and shoots in general and internodal length of genetically dwarf plants in particular.

47. **Gossypol ($C_{30}H_{30}O_8$):** It is a polyphenol derived from the cotton plant. It acts as a inhibitor for several dehydrogenase enzymes. It is a yellow pigment.
48. **Green revolution:** Popular term for the introduction of scientifically bred or selected varieties of grain (including rice, wheat, and corn) that, with adequate inputs of fertilizer and water, can greatly increase crop yields.
49. **Guttation:** Loss of water in the liquid state from uninjured parts of plants is known as guttation. It usually occurs from tips and margins of leaves during night or early morning when there is high atmospheric humidity.
50. **Herbarium:** It is a collection of preserved plant specimens. There specimens may be whole plants or plant parts; these are usually in dry form, mounted on sheet or may be kept in alcohol or other preservative.
51. **Hormones:** Organic substance secreted by plants and animals that functions in the regulation of physiological activities and in maintaining homeostasis.
52. **Homokaryon:** A bi or multinucleate cell having nuclei of the same kind in a common cytoplasm.
53. **Humus:** Slightly soluble residue of undigested or partially decomposed organic material in topsoil. This material helps retain water and water-soluble nutrients, which can be taken up by plant roots.
54. **Hydrolysis:** A process that occurs when a large molecule is broken down into smaller parts by the addition of water.
55. **Hybrid:** Offsprings received from two genetically dissimilar individuals, strains, purelines etc.
56. **Inoculum:** A biological material (like a virus or toxin or immune serum) that in injected to increase immunity to a particular disease.
57. **Inter cropping:** Growing two or more different crops at the same time on a plot. Example a carbohydrate-rich grain that depletes soil nitrogen and protein rich legume that adds nitrogen to the soil may be intercropped.
58. **Jhum cultivation:** Also called slash and burn agriculture, is the process of growing crops by first clearing the land of trees and vegetation and burning them there after. The burnt soil contains potash which increases the nutrients contents of the soil for

better yield. This is practiced in the states like Arunachal Pradesh, Meghalaya, Mizoram and Nagaland in India.

59. **Leaching:** In agriculture, leaching refers to the loss of water soluble nutrients from the soil due to rain or irrigation.

60. **Lichen:** These are composite organisms consisting of a association of a fungus with a photosynthetic partner usually either a green alga (commonly *Trebouxia*) or (commonly *Nostoc*). The morphology, physiology and biochemistry of lichens are very different from those of the isolated fungus and alga in culture. Lichens occur in some of the most extreme environments on Earth, hot deserts, rocky coasts and toxic slag heaps.

61. **Lipids:** Compounds composed of C, O_2 and H which are insoluble in water. These are a class of organic compounds that tend to be soluble in non-polar solvents such as alcohol includes fats and oils. Fat and oils are used as long term storage compounds in plants and animals.

62. **Lenticel:** It is a porous tissue consisting of cells with large intercellular space in the periderm of the secondary thickened organs and the bark of woody stems and roots of dicotyledonous flowering plants allow gas exchange between the atmosphere and internal tissues.

63. **Loams:** Soils containing a mixture of clay, sand, silt and humans. Good for growing most crops.

64. **Lysis:** Process of cell disintegration or membrane rupturing.

65. **Microtubules:** Submicroscopic, hollow tubes of protein that function throughout the cytoplasm to provide structural support and enable movement.

66. **Micronutrients:** Chemical elements that organism need in smaller trace amounts to live, grow or reproduce like Ferrous, copper, chlorine and Iodine, zinc etc.

67. **Molecular farming:** The synthesis of certain chemicals through transgenic plants is known as molecular farming. The genes confirming quality characters are being transferred to get increased synthesis of some chemicals *viz.* Mannital dehydrogenage gene from *Escherichia coli* is transferred to tobacco for increased production of mannital.

68. **Monoculture:** Cultivation of a single crop, usually on a large area of land.

69. **Mycorrhiza:** This term refers to symbiotic association between the fungus and the roots of higher plants. Some of these fungi are

present on the root surface only whereas others enter into roots as well. These fungus solubilizes phosphorus, produce plant growth promoting substances and protect host plant from soil pathogens. Mycorrhiza is classified into two types (*i*) ectotrophic and (*ii*) endotrophic

70. **Mycotoxin:** A deadly poison produced by fungi.
71. **Nitrogenase: Nitrogenases** are enzymes that are produced by certain bacteria, such as cyanobacteria (blue-green algae). These enzymes are responsible for the reduction of nitrogen (N_2) to ammonia (NH_3).
72. **Nucleic acid:** Complex molecules that store and transfer information within a cell. They are constructed of funda-mental monomers known as nucleotides.
73. **Nucleoli:** Nuclear structures composed of completed or partially completed ribosomes and the specific parts of chromosomes that contain the information for their construction.
74. **Nucleoplasm:** The liquid matrix of the nucleus composed of a mixture of water and the molecules used in the construction of the rest of the nuclear structure.
75. **Nucleotide:** A fundamental subunit of nucleic acid constructed of a phosphate group, sugar, and an organic nitrogenous base.
76. **Oogenesis:** The specific name given to the gametegenesis process that leads to the formation of eggs.
77. **Organelles:** Cellular structures that perform specific functions in the cell.
78. **Orthologues:** Homologous proteins that perform the same function in different species.
79. **Perennial:** Crops that can live for more than 2 years.
80. **Photoperiodism:** It is the ability of plants to detect and respond to the relative length of day and night to which plant is exposed.
81. **Phytochrome:** A pigment which is a protein molecule, located within the plasma membranes of the leaf cells that is extremely sensitive to certain forms of light.
82. **Plasmids:** Small and circular DNA molecule which has self replicating property in the host cell or an extrachromosal genetic plant found in bacterial cell and can replicate independently of chromosomal DNA.

83. **Protandrous plants:** Where male reproductive organism (anthers) mature before the female reproductive organism (stigma) examples are; Maize, sunflower
84. **Protogynous plants:** The plant flower in which female reproductive organism (stigma) mature before the males ones (anthers) thereby ensuring that self fertilization does not occur . Examples are *plantain* and *figwort, plearlmillet*.
85. **Proteins:** Proteins are large biomolecules or macromolecules consisting of one or more long chains of amino acids residues. They are one of the building blocks of the body tissues and can serve as a fuel source. There are about 20 different amino acids that occur naturally in proteins.
86. **Polymerase chain reactions (PCR):** A method for amplifying a DNA base sequence using a heat stable polymerase and two primers, one complementary to the (+) strand at one end of the sequence to be amplified and other complementary (–) strand at the other end. The faithfulness of reproduction of the sequence is to the fiedility of the polymerase. PCR was developed by Karl Mullis in the year 1980.
87. **Polyculture:** Complex form of inter cropping in which a large number of different plants maturing at different times are planted together.
88. **Protoplast** is a plant, bacterial or fungal cell that had its cell wall completely or partially removed using either mechanical or enzymatic means.
89. **Rhizobium:** It is a symbiotic bacterium, and occurs in the root nodules of legumes. Rhizobium fixes atmospheric nitrogen in the presence of red pigment called leghae-moglobin. Rhizobium was first time isolated by Biejerinck in the year 1888.
90. **Salinity:** Amount of various salts dissolved in a given volume of water.
91. **Salinization:** Accumulation of salts in soil that can eventually make the soil unable to support plant growth.
92. **Scavenger:** Organism that feeds on dead organisms that were killed by other organisms or died naturally. Examples include vultures, flies, and crows.
93. **Strip cropping:** Planting regular crops and close-growing plants such as hay or nitrogen-fixing legumes in alternating rows or bands to help to reduce depletion of soil nutrients.

94. **Succulent plants:** Plants such as desert cacti that survive in dry climates by having no leaves, thus reducing the loss of scarce water. They store water and use sunlight to produce the food they need in the thick, fleshy tissue of their green stems and branches.

95. **Sustainable agriculture:** Agriculture and agrobased food system that are economically viable and meet society's need for safe and nutritious food while conserving or enhancing natural resources and the environment for sustainable life for future generations.

96. **Synecology:** The branch of ecology that deals with the structure and development of entire ecological communities and the interrelationship of the plants and animals with them. It is the study of entire ecosystem.

97. **Transgenic plants:** These are genetically modified plants which have been created by mobilizing genes of interest from one species to another species by means of biotechnology.

98. **Turnover number:** The number of molecules of substrate that a single molecule of enzyme can react with in a given unit of time.

99. **Virgin soil:** The soil, which has not been used for cultivation.

100. **Viruses:** Infectious disease causing agents, smaller than bacteria, which always require intact host cells for replication and which contain either DNA or RNA as their genetic component.

101. **Wetland:** Land that is covered all or part of the time with saltwater or freshwater, excluding streams, lakes, and the open ocean.

102. **Windrower:** A machine used for the purpose of crop harvesting.

103. **Yellowing:** It is the result of diminishing rate of chlorophyll synthesis from plants.

104. **Xerophyte:** Plant adapted to survive long periods without moisture. *Example-* Cacti.

105. **Xylan:** Polysaccharide found in the cell walls of plants, where it forms the bulk of the hemicelluloses component.

106. **Vernation:** Arrangement and folding of leaves in a bud before its opens.

107. **Zeatin:** Naturally occurring cytokinin derived from adenine which plays a role in the growth and development of plants.

SOME FACTS ABOUT CENSUS

- India is seventh largest country in the world according to area (3,287,263 square km).
- 2011 Census in India is the 15th Census.
- The first India census was conducted in the year 1872. First regular census was done in 1881 on regular basis
- The estimated global population in 2000 was 6055 million. World population is increasing 80 million per year and Asian population is 60.8% of total population of world.
- World population is increasing Approx. 80 million per year
- By 2050 India will most likely to over take China to become the most populous country on the earth.
- Largest Population of any country is that of China and lowest is that of Vatican City.
- Total Population of the Country India is 121crore (2011 census)
- Nagaland led the states with lowest population growth rate of –0·47% whereas Meghalaya showed the highest growth rate at 27·82%.
- Least populous state is Sikkim– 6,07,688
- Most populous state—Uttar Pradesh- 19,95,81,477
- Kerala is the state with highest women Population rate (1084 females/1000 males)
- Urban population of India has increased from 25·85 million in 1991 to 377.10% million 2011.
- State with lowest women population is Haryana 877/females/1000 males.
- State with lowest birth rate Kerala— 19·8/1000 person
- State with highest birth rate UP— 97/1000 person
- State with lowest death rate Kerala— 5·9/1000 person
- State with highest death rate Madhya Pradesh — 12·8/1000 person
- Least densely populated UT Andaman and Nicobar Islands (46 person/km^2).
- Least densely populated state is Arunachal Pradesh (17 person/km^2)
- 2001-2011 has registered sharpest decline since independance for 2001-2011 decadal growth has become 17.64% a decrease of 3.9% from last decade.
- Thane district in Maharastra is most populated district (11054131)
- Puducherry has highest sex ratio of 1037 females per 1000 males.

- India is 7th largest country in the world according to area.
- Geographical area of India — 328.75 Mha
- Geographical area of World — 134.90 Mha
- **Total agricultural labourers: 14,43,29,833**

Rural : 1,19,42,619
Urban: 63,88,688
Males: 8,27,40,351
Females: 6,15,89,482

- **Total cultivators : 11,86,93,640**

Males: 8,27,06,724
Females : 3,59,85,916

- **Current population of India has surpassed China and stand at 1,42.86 crore India has 7.74% of world population with population density 464 per sq. km.**

CENSUS (2011) AT A GLANCE

Population	1,21,05,69,573	(Male: 62,31,21,843) (Female: 58,74,47,730)
Sex Ratio (per 1000 men)	940	
Literacy rate (%)	74·04	
Male Literacy (%)	82·14	
Female Literacy (%)	65·46	
Population Density	382/km^2	
Highest Population Density State (Bihar.)	1102/km^2	
Lowest Population Density State (Arunachal Pradesh)	17 persons/km^2	
Highest Population Density, Union Territory. (Delhi)	11,320/km^2	
Decadal growth (%)	17·84%	
Annual Growth (%)	–3·9 %	
Lowest Literacy	Bihar (63·8%)	
Highest Literacy	Kerala (93·9%)	

STATES WITH THE HIGHEST POPULATION (CENSUS 2011)

Rank	State	Area (1000 km^2)	Density/km^2	Population
1.	Uttar Pradesh	243.3	823	19,95,81,477
2.	Maharashtra	307.7	365	11,23,72,972
3.	Bihar	94.1	1102	10,38,04,637
4.	West Bengal	88.8	1029	9,13,47,736
5.	Andhra Pradesh	275.608	308	8,46,65,533
6.	Madhya Pradesh	308.0	236	7,25,97,565
7.	Tamil Nadu	130.1	555	7,21,38,958
8.	Rajasthan	342.2	201	6,86,21,012
9.	Karnataka	191.8	319	6,11,30,704
10.	Gujarat	196.0	308	6,03,83,628

STATES 1ST IN PRODUCTION OF VARIOUS CROPS IN INDIA

Crop	States
Rice	U.P. > West Bengal > Punjab
Wheat	Uttar Pradesh > M.P. > Punjab
Jowar	Maharashtra
Maize	Karnataka > MP > Bihar
Pulses	Madhya Pradesh > Rajasthan
Oil Seeds	Rajasthan > Gujrat > MP
Black Gram	Madhya Pradesh > AP
Green Gram	RJ > AP.
Ground nut	Gujrat > Rajasthan
Rapeseed and Mustard	Rajasthan > Haryana
Sugarcane	Uttar Pradesh > Maharashtra
Cotton	Gujrat > T.N.
Tobacco	Andhra Pradesh
Rubber	Kerala
Pototo	Uttar Pradesh
Onion	Maharastra > Gujarat
Sorghum	Maharastra > KR
Pearl millet	Rajasthan > UP
Pigeon Pea	Karnataka > Maharashtra
Mustard	Rajasthan
Total Cereals	UP. > Punjab
Soybean	M.P. > Maharashtra
Jute	West Bengal > Assam
Tea	Assam
Coffee	Karnataka
Banana	Tamilnadu
Saffron	Jammu & Kashmir
Black Pepper	Kerala

IMPORTANT ABBREVIATIONS

ADB	Asian Development Bank (Founded in 1966)
AFC	Agricultural Finance Corporation
AIC	Agricultural Insurance Company of India Ltd.
AIDEA	Association for Innovation Development of Entrepreneurship in Agriculture
APEDA	Agriculture and Processed Food Products Export Dev. Authority
AGMARK	Agricultural Produce Grading and Marketing
AICRP	All India Coordinated Research Project
AIIB	Asian Infrastructure Investment Bank
AKIS	Agriculture Knowledge and Information System
AMA	Agricultural Co-Marketing Advisor
APC	Agricultural Price Commission
APHNET	Animal Production and Health Informatics Network
ARDC	Agriculture Refinance and Development Corporation
ARYA	Attracting and Retaining Youth in Agriculture
ASEAN	Association of South East Asian Nations
AEC	Atomic Energy Commission
ARS	Agricultural Research Services
ARYA	Attracting and Retaining Youth in Agriculture
ARISNET	Agricultural Research Information System Network
ATMA	Agricultural Technology Management Agency
Bt	*Bacillus thuringiensis*
BMC	Bulk Milk Cooler
BOD	Biological Oxygen Demand
CAD	Command Area Development
CAPART	Council for Advancement of Peoples Actions and Rural Technology
CACP	Commission for Agricultural Costs & Prices
CASE	Commission on Alternative Sources of Energy
CBA	Capture Based Aquaculture
CB	Co-operative Bank
CCEA	Cabinet Committee for Economic Affairs
CCI	Cotton Corporation of India
CDAC	Centre for Development of Advanced Computing

CDD	Cattle and Dairy Development
CDP	Community Development Programme
CeRA	Consortium for e-Resources in Agriculture
CFDO	Central Fodder Development Organization
CFSPF	Central Fodder Seed Production Farm
CGIAR	Consultative Group of International Agricultural Research
CIAT	International Centre for Tropical Agriculture
CITES	Convention on International Trade in Endangered Species
CIRDAP	Centre on Integrated Rural Development for Asia and the Pacific
CMTP	Central Minikit Testing Programme
CSC	Central Seed Committee
CSO	Central Statistical Organisation
CWWG	Crop Weather Watch Group
DAC	Department of Agriculture and Cooperation
DADF	Department of Animal Husbandry, Dairying & Fisheries
DARE	Department of Agricultural Research and Extension
DDP	Desert Development Programme
DDT	Dichloro-Diphenyl-Trichloroethane
DNA	Deoxyribo Nucleic Acid
DMI	Directorate of Marketing and Inspection, Faridabad
DPAP	Drought Prone Area Programme
DRM	Disaster Risk Management
EBRD	European Bank for Reconstruction and Development
EDP	Entrepreneurship Development Programmes
EPA	Eicosapetaenoic Acid / Environmental Protection Agency
EDP	Electronic Data Processing
EEC	European Economic Community
FAO	Food and Agricultural Organisation
FTC	Farmers Training Centre
FCI	Food Corporation of India
FSS	Farmers Service Society
FYM	Farm Yard Manure
FYP	Five Year Plan
GAPs	Good Agricultural Practices
GDP	Gross Domestic Product
GHGs	Green House Gases

GIR	Geographical Indications Registry
GIS	Geographic Information System
GNP	Gross National Product
GPS	Global Positioning System
G-8	Group of Eight (Canada, France, Germany, Italy, Japan, Russian Federation, United Kingdom and United States
HAD	Hill Area Development Programme
HOFED	Horticultural Cooperative Marketing Federation
HT-Crop	Herbicide Tolerant Crop
HYVP	High Yielding Varieties Programme
IAAP	Intensive Agricultural Area Programme
IADA	Intensive Agricultural District Programme
IAEA	International Atomic Energy Agency
IBRD	International Bank for Reconstruction and Development
IMD	Indian Meteorological Department
ICDP	Intensive Cattle Development Programme
ICFA	Indian Council for Food and Agriculture
ICRISAT	International Crop Research Institute for Semi Arid and Tropics, Hyderabad
ICT	Information and Communication Technology
IDAD	Integrated Dryland Agricultural Development
IDBI	Industrial Development Bank of India
IFAD	International Fund for Agriculture Development
ICRAF	International Council for Research in Agroforestry
IFOAM	International Federation of Organic Agriculture Movement
IFS	Integrated Farming System
IFFCO	Indian Farmers Fertilizer Co-operative Ltd.
IMF	International Monetary Fund
IODP	Intensive Oil Seed Development Programme
IPM	Integrated Pest Management
ISO	International Standards Organisation
IPCC	Intergovernal Panel on Climate Change
IPGRI	International Plant Genetic Resource Institute, Rome, Italy
IPR	Intellectual Property Rights
IRDP	Integrated Rural Development Programme
IUCN	International Union for Conservation of Nature and Natural Resources

IVLP	Institute Village Link Programme
JRY	Jawahar Rozgar Yojana
KCC	Kissan Call Centre
KCCS	Kishan Credit Card Scheme
KVK	Krishi Vigyan Kendra
LBS	Lead Bank Scheme
LDB	Land Development Bank
LISNET	Land Information System Network
MFAL	Marginal Farmers and Agricultural Labour Programme
MCP	Multiple Cropping Programme
MMT	Mean Monthly Temperature
MNAIS	Modified National Agricultural Insurance Scheme (2010)
MSP	Minimum Support Price/Minimum Standard Protocol
MSY	Mahila Samridhi Yojana
NABARD	National Bank for Agriculture and Rural Development (1982)
NADAMS	National Agriculture Drought Advisory and Management Systems
NAEP	National Agriculture Extension Project
NAFED	National Agriculture Co-operative Marketing Federation
NAARM	National Academy of Agricultural Research Management
NADRS	National Animal Disease Reporting System
NAFED	National Agricultural Cooperative Marketing Federation
NAIP	National Agricultural Innovative Project
NAIS	National Agricultural Insurance Scheme
NAMA	Non Agricultural Market Access
NAPCC	National Action Plan on Climate Change
NASA	National Aeronautics & Space Administration (USA)
NSAP	National Social Assistance Programme
NBAGR	National Bureau of Animal Genetic Resources
NBSSLUP	National Bureau of Soil Survey and Land Use Planning
NCA	National Commission on Agriculture
NCAER	National Council for Agricultural Economics and Research.
NCAP	National Centre for Agricultural Planning.
NCDC	National Co-operative Development Corporation
NCF	National Commission on Farmers
NCIPM	National Center for Integrated Pest Management

NDDB	National Dairy Development Board
NHB	National Horticulture Board
NIC	National Informatics Centre
NIN	National Institute of Nutrition
NISAGENET:	National Information System on Agricultural Education Network in India
NES	National Extension Service
NFSM	National Food Security Mission.
NHM	National Horticulture Mission.
NIAM	National Institute of Agricultural Marketing
	National Institute of Agricultural Management
NIPGR	National Institute of Plant Genome Research New Delhi
NITI	National Institute for Transforming India.
NOAA	National Oceanic & Atmospheric Administration (USA).
NOVOD	National Oilseeds and Vegetable Oils Develo-pment Board.
NRDWP	National Rural Drinking Water Supply Programme
NREGA	(Mahatma Ghandhi) National Rural Employment Guarantee Act
NSC	National Seed Corporation
NSRTC	National Seeds Research and Training Centre
OECD	Organization for Economic Cooperation and Development
OPDP	Oil Palm Development Programme
PD	Poultry Development
PDS	Public Distribution System
PKVY	Paramparagat Krishi Vikas Yojana
PPV&FR	Protection of Plant Varieties and Farmers' Rights
PPTD	Pilot Project for Tribal Development
PQSs	Plant Quarantine Stations
PRAP	Participatory Action Research Programme
PRA	Participatory Rural Appraisal
PURA	Provision of Urban Amenities in Rural Areas
RKVY	Rashtriya Krishi Vikas Yojana
RADP	Rainfed Area Development Programme
READY	Rural and Entrepreneurship Awareness Development Yojana
RLEGP	Rural Landless Employment Guarantee Programme

RMK	Rashtriya Mahila Kosh
RRB	Regional Rural Banks
RUCHI	Rural Centre for Human Interests
SAARC	South Asian Association for Regional Cooperation
SDGs	Sustainable Development Goals (United Nations)
SEBI	Security and Exchange Board of India
SFAC	Small Farmers Agri-Business Consortium
SFDA	Small Farmers Development Agency
SGSY	Swaran Jyanti Gram Swarojgar Yojana
SMS	Short Message Service
SPAC	Soil Plant Atmosphere Continuum
SRI	System of Rice Intensification.
TAD	Tribal Area Development
TERI	Tata Energy Research Institute
TTC	Trainer's Training Centre
TRIFED	Tribal Co-operative Marketing Federation of India
TRYSEM	Training of Rural Youth for Self Employment
UTI	Unit Trust of India
UNCTAD	United Nations Conference on Trade and Development
UNDP	United Nations Development Programme
UNEP	United Nations Environment Programme
UNESCO	United Nations Economic Social and Cultural Organization
UNFPA	United Nations Fund for Population Activities
USDA	United States Department of Agriculture
VAT	Value Added Tax
VOIP	Voice Over Internet Protocol
WADA	World Anti-Doping Agency
WBCIS	Weather Based Crop Insurance Scheme
WEF	World Economic Forum
WFC	World Food Council
WFP	World Food Programme
WVDP	Whole Village Development Programme
WHO	World Health Organization
WTO	World Trade Organisation
WSIS	World Summit on Information Society
ZSI	Zoological Survey of India

SOME INSTRUMENTS AND THEIR USES

Altimeter	: Height above sea level
Aneroid Barometer	: Atmospheric Pressure
Albodometer	: Solar radiation (Reflectivity)
Assmann's psychrometer	: Relative humidity in plant canopy
Crescograph	: Growth of Plant
Clinometer	: Measuring the slope angle
Barograph	: Continuous atmospheric pressure
Beaufort's scale	: Estimation of wind speed
Disdrometer	: Size, speed and number of rain drop.
Gypsom block	: Soil moisture in situ.
Filanimeter	: Leaf area
Lactometer	: Density of milk
Lysimeter	: Percolation and leaching loss
Portable evaporimeter	: Measuring evaporation for short period
Hydrograph	: Relative humidity continuously
Sunshine recorder	: Number of hours of sunshine per day
Ozonometer	: Concentration of ozone in air
Piezometer	: Depth of water table
Pycnometer	: Density of soil
Psychrometer	: Relative humidity
Pyranometer	: Solar radiation
Pyraheilometer	: Direct solar radiation
Net radiometer	: Net radiation
Infrared thermometer	: Crop canopy temperature
Quantum sensor	: PAR
Tensiometer (Surface tension)	: Surface tension of liquids
Tensiometer (Soil)	: Matric water potential
Saccharimeter	: Concentration of sugar solution.
Seismograph	: Intensity of earthquake
Venturimeter	: Rate of flow of liquids
Viscometer	: Viscosity of liquid
Wind vane	: Wind direction
Polarimeter	: Sucrose content estimation

FAMOUS BOOKS AND AUTHORS

1. Soil Fertility and Fertilizers —*S.L. Tisdale & W.L. Nelson*
2. Nature and Properties of Soils —*N.C. Brady and Weil*
3. Garden Flowers —*Vishnu Swaroop*
4. Vegetables —*B. Chaudhry*
5. Fruits —*Ranjeet Singh*
6. Floriculture in India —*M.S. Randhawa*
7. Handbook of Horticulture —*K. L. Chadha (ICAR)*
8. Entomology and Pest Management —*L. P. Peddigo*
9. Extension Education —*A. Reddy*
10. Spices and Condiments —*J.S. Pruthi*
11. General Texbook of Entomology (Vol.I and II) —*Imm's*
12. Principles of Biochemistry —*Lehninger*
13. Plant Pathology —*G. N. Agrios*
14. Principles of Agricultural Engg. (Vol.I and II) —*A.M. Micheal and T.P. Ojha*
15. Wings of Fire —*A. P. J. Abdul Kalam*
16. India 2020 —*A. P. J. Abdul Kalam*
17. Ignited Minds —*A. P. J. Abdul Kalam*
18. Indomitable spirit —*A. P. J. Abdul Kalam*
19. The Wonder that was India —*Bhasam*
20. History of Indian Agriculture Vol.I-II —*M. S. Randhawa*
21. It's Always Possible —*Kiran Bedi*
22. Indian Philosophy —*Dr. S. Radhakrishanan*
23. Long Walk to Freedom —*Nelson Mandela*
24. India's Struggle for Independence —*Bipin Chandra*
25. The God of Small Things —*Arundhati Roy*

26. You Can Win —*Shiv Khera*
27. Discovery of India —*Jawahar Lal Nehru*
28. Glimpses of World History —*Jawahar Lal Nehru*
29. My Experiment with Truth —*M. K. Gandhi*
30. The Augmentative Indian —*Amartya Sen*
31. The Good Earth —*Pearls Buck*
32. The Name Shake —*Jhumpha Lahiri*
33. The Song of India —*Sarojini Naidu*
34. The Time Machine —*H. G. Wells*
35. The White Tiger —*Arvind Adiga*
36. Origin of Species —*Charles Darwin (ARS 2010)*
37. Wealth of Nations —*Adam Smith*
38. Rose in India, —*B. P. Pal*
39. Horticulture at a Glance (Vol. I, II, III) —*A. S. Salaria & Babita Salaria*
40. Insects Structure and Function —*Chapman*
41. Revolutions: to Green the Environment to Glow the Human Heart — *M.S. Swaminathan*
42. Unaccoustomed Earth —*Jhumpha Lahiri*
43. India after Gandhi —*Ramachandra Guha*
44. Makers of Modern India —*Ramachandra Guha*
45. Three Billion —*A. P. J. Abdul Kalam*
46. Reimagining India: Unlocking the Potential of Asia's Next Superpower —*McKinsey*
47. India After Independence —*Bipin Chandra*
48. General and Applied Entomology —*B.V. David and T.N. Ananthakrishnan*
49. Fundamentals of Ecology —*E.P. Odum*
50. Diseases of Crop Plants in India — *G. Rangaswami*

51. Introductory Mycology —*C. J. Alexopoulos*
52. Remote Sensing and Image Interpretation—*I.M. Lilliesand*
53. Outlines of Biochemistry —*E.E. Conn and P K Stumpf*
54. Weed Science: Basic and Applications —*T. K. Das*
55. Field Crop Production I and II -- *Dr. Rajendra Prasad*
56. Plant Breeding theory and Practice —*V. L. Chopra*
57. Plant Diseases — *R. S. Singh*
58. Int. Principles of Plant Pathology — *R. S. Singh*
59. Elements of Economic Entomology —*B. V. David*
60. Modern Techniques of Raising Field Crops —*Chhidda Singh*
61. Economics of Farm Production and Management —*V. T. Raju*
62. Introduction to spices, plantation crops, medicinal and aromatic plants —*N. Kumar*
63. Agricultural Economics —*S. Subba Reddy*
64. Textbook of Soil Science —*T. D. Biswas*
65. Introduction to Plant Biotechnology —*H. S. Chawla*
66. You are Unique —*A. P. J. Abdul Kalam*
67. Principles of Insect Physiology by – *Wiggles Worth*
68. Text Book of Animal Husbandry – *G. C. Banerjee*
69. Soil Microbiology – *N. S. Subba Rao*
70. Irrigation: Theory and Practice. – *A. M. Michael*
71. Fundamentals of Soil/Science – *Indian Society of Soil Science*
72. Textbook of Soil Science – *Susanta Kumar Pal*
73. Textbook of Soil Science – *Mehra*
74. Textbook on Introductory Plant Nematology – *R. K. Walia & H. K. Bajaj*
75. Planned Economy of India – *M. Visvesvaraya*
76. Das Capital – *Karl Marx*
77. The One Straw Revolution (Introduction to Natural Farming) – *Masanobu Fukuoka*
78. First Principle of Agriculture – *Henry Turner*
79. Feeding a World of 10 Billion – *Norman E. Borlaug*
80. I too had a dream – *V. Kurien (Milkman of India)*
81. Plowman's Folly – *E. H. Faulkner*

IMPORTANT/SIGNIFICANT DAYS

National Youth Day (Swami Vivekanand)	January, 12
National Tourism Day	January, 25
Indian Republic Day	January, 26
World Wetland Day	February, 2
World Cancer Day	February, 4
National Deworming Day	February, 10
National Science Day	February, 28
International Women's Day	March, 8
World Kidney Day	March, 10
World Consumer Rights Day	March, 15
Water Resources Day	March, 11
World Consumer Day	March, 15
World Wildlife Day	March, 3
World Forestry Day	March, 21
World Water Day	March, 22
World Meteorological Day	March, 23
World Health Day	April, 7
Earth Day	April, 22
National Panchayati Raj day	April, 24
Workers Day (Int. Labour Day)	May, 1
World Red Cross Day	May, 8
National Technology Day	May, 11
Agri-Tourism Day	May, 16
Endangered Species Day	May, 17
Commonwealth Day	May, 24
World Milk Day	June, 1
World Environment Day	June, 5
World Food Safety Day	June, 7
World Ocean Day	June, 8
International Yoga Day	June, 21
World Zoonoses Day	July, 6
World Mango Day	July, 7
International Day of Co-operative	July, 7
World Population Day	July, 11
NABARD Foundation Day	July, 12
ICAR Foundation Day	July, 16
International Mango Day	July, 22
International Biodiesel Day	August, 1

World Biofuel Day	August, 10
India's Independence Day	August, 15
National Sports Day	August, 29
World Coconut Day	September, 2
Teacher's Day	September, 5
World Literacy Day	September, 8
Hindi Diwas	September, 14
Engineers Day	September, 15
Ozone Day	September, 16
World Biosphere Day	September, 31
World Tourism Day	September, 27
World Habitat Day	October, 2
International Day for Non-Violence	October, 2
World Animal Welfare Day	October, 4
International Day for Girl Child	October, 11
U.N. Int. Day for Natural Disaster Reduction	October, 13
International Day of Rural Women	October, 15
Mahila Kisan Diwas Day	October, 15
World Food Day	October, 16
World Fruits and Vegetables Day	October, 18
United Nations Day	October, 24
National Integration Day	November, 9
World Fisheries Day	November, 12
Children' Day/World Diabetes Day	November, 14
Int. Day for Elimination of Violence Against Women	November, 25
World Environment Protection Day	November, 26
World Aids Prevention Day	December, 1
World Pollution Prevention Day	December, 2
World Disability Day	December, 3
National Agricultural Education Day	December, 3
Chemical Disaster Prevention Day/Indian Navy Day	December, 4
World Soil Day	December, 5
Girl, Child Day (SAARC)	December, 9
Int. Human Rights Day	December, 10
National Energy Conservation Day	December, 14
Minorities Right Day	December, 18
Farmer's Day (Kisan Divas) (Birthday of Ch. Charan Singh)	December, 23

IMPORTANT YEARS FOR INDIAN AGRICULTURE

No.	Event	Year
1.	Indian Meteorological Department established	1875
2.	First Department of Agriculture established	1881
3.	Forest Research Institute at Dehradun	1900
4.	First Irrigation Commission	1901
5.	IARI at Pusa Bihar (Imperial Agricultural Research Institute)	1905
5a.	Sugarcane breeding Institute at Coimbatore	1912
6.	First Tractor bought for Agriculture use	1914
7.	Destructive Insect and Pests Act passed	1914
8.	First Livestock Census Conducted	1919
9.	Central Cotton Committee	1921
10.	Royal Commission on Agriculture	1928
11.	ICAR at New Delhi	1929
12.	Indian Statistical Institute at New Delhi	1931
13.	Great Bengal famine caused by *Helminthosporium oryzae*	1943
14.	Indian Standards Institute (Renamed Bureau of Indian Standards (BIS) (in year 1986), CRRI at Cuttack	1946
15.	Planning Commission Established	1950
16.	Indian Institute of Sugarcane Established at Lucknow	1952
17.	Central Soil Conservation Research Development and Training Institute, Karnal	1954
18.	Indian Society of Agronomy/State Bank of India	1955
19.	First Five Year Plan started/Life Insurance Corporation of India	1951+1956
20.	CDP (Community Development Project) started	1952
21.	National Extension Service	1953
22.	T.V. Broadcast for Rural Development started	1957
23.	Panchayati Raj started	1957-58
24.	NAFED (National Agricultural Cooperative Marketing Federation)	1958
25.	Doordarshan Services Inaugurated by the President	1959
26.	First Agricultural University established at Pantnagar	1960
27.	Intensive Agriculture District Programme started	1960
28.	National Seed Corporation Founded at Delhi	1963
29.	Intensive Agriculture Area Programme started/Unit trust of India	1964
30.	Food Corporation of India	1965
31.	Agriculture Price Commission constituted	1965
32.	Krishi Darshan Programme Inaugurated	1967
33.	High yielding varieties Programme started	1966
34.	Insecticides Act passed	1968
35.	Green Revolution started	1969
36.	National Commission on Agriculture setup by GOI	1970
37.	Small Farmer Dev. Agency started	1970
38.	Accelerated Rural Water Supply Schemes/ ICRISAT at Hyderabad	1972
39.	Fertilizer subsidy started (Retention Price Scheme)	1972
40.	Agriculture Research Services started (DARE created)	1973
41.	First KVK established in Tamilnadu Agri. University., Pondicherry	1974
42.	Command Area Development Authority	1974

43.	National Seed Project	1974
44.	Drought Prone Area Program	1974
45.	Regional Rural Banks Established	1975
46.	NAARM (National Academy of Agricultural Research Management) established at Hyderabad	1976
47.	NBPGR (National Bureau of Plant Genetic Resources) established	1976
48.	NDC (National Data Centre) created at IMD Pune /HDFC Bank	1977
49.	Desert Development Programme	1977-78
50.	Farmer's Agriculture Service centres	1983-84
51.	CRIDA (Central Research Institute for Dryland Agriculture) at Hyderabad	1985
52.	MANAGE (National Institute of Agricultural Extension Management) established at Hyderabad	1987
53.	IRDP (Integrated Rural Dev.. Programme) started	1980
54.	Plant Quarantine Regulations passed	1981
55.	National Bank for Agriculture and Rural Development (NABARD)	1982
56.	Crop Insurance scheme was started	1985
57.	APEDA (Agricultural and Processed Food Products Export Dev. Authority)	1986
58.	TRIFED (Tribal Co-operative Marketing Dev. Federation of India Ltd.)	1987
59.	NIAM (National Institute for Agricultural Marketing), Jaipur	1988
60.	NAAS (National Academy of Agri. Sciences) established	1990
61.	Agriculture Research Information Systems Established	1995
62.	National Social Assistance Programme	1995
63.	National Gene Bank established at New Delhi	1996
64.	Village Grain Bank Scheme started	1996-97
65.	Indian Agriculture Started on line by Ministry of Agriculture	1997
66.	Kisan Credit Card Scheme started	1998
67.	National Agriculture Insurance Scheme started	1999
68.	National Agriculture Technology Project (NATP)	1999
69.	International Seed Network Initiative started	1999
70.	Macro-management of Agriculture (Central Govt.) Scheme started	2000-2001
71.	Agricultural Technology Management Agency (ATMA)	2000
72.	India's First Biotech Crop. technology approved for commercialization	2002
73.	Biodiversity Act Passed	2002
74.	Intensive Cotton Development Programme	2003-04
75.	National Rural Health Mission	April 2005
76.	PPV and FR were Implemented	2005
77.	National Floriculture Mission started	2006
78.	National Agricultural Innovation Project (NAIP)	2006
79.	National Bamboo Mission	2006-07
80.	National Food Security Mission	2007-08
81.	Rashtriya Krishi Vikas Yojana	2007-2008
82.	Food Security and quality year	2008-09
83.	Mahatma Gandhi National Rural Employment Guarantee Act (MGNREGA)	2 Oct., 2009
84.	National Mission for Sustainable Agriculture launched	2011-12
85.	NMAET (National Mission on Agricultural Extension and Technology)	2014
86.	Paramparagat Krishi Vikas Yojana	2015

Range of pH for Different Crops

Crop	*pH range (optimum)*
Paddy	5.0 – 6.5
Sorghum	5.5 – 7.5
Maize	5.5 – 7.5
Wheat	4.5 – 6.5
Barley	6.5 – 8.0
Rye	5.0 – 7.0
Oats	5.0 – 7.5
Leguminous crops.	
Chickpea	6.0 – 7.5
Pea	6.0 – 7.5
Field beans	6.0 – 7.5
Lima beans	6.0 – 7.0
Velvet beans	5.5 – 7.0
Soybean	6.0 – 7.0
Groundnut	5.3 – 6.6
Pipeonpea	5.5 – 7.0
Cowpea	5.0 – 7.0
Lucerne	6.2 – 7.5
Berseem	6.0 – 7.5
Red clover and White clover	5.7 – 7.0
Lupine	5.5 – 6.5
Other crops:	
Cotton	6.0 – 7.5
Sugarcane	6.0 – 8.0
Tobacco	5.5 – 7.5
Potato	5.0 – 5.5
Cabbage	5.0 – 7.0
Redbeet and Sugarbeet	6.5 – 8.0
Sunflower	6.0 – 7.5
Linseed	5.0 – 7.0
Strawberries	4.8 – 5.8
Blue grass	5.2 – 6.5
Turnip	5.5 – 6.8
Sweet potato	5.8 – 6.0

Critical Growth Stages of Different Crops for Irrigation

Crop	*Critical stages*
Cereals:	
Kharif	
Rice	Tiller initiation, banicle initiation, heading, flowering
Sorghum	Seedling, pre-flowering, flowering, milky, dough stage
Maize	Seedling, tasseling, silking, dough stage
Pearlmillet	Maximum tillering, flowering, grain development
Fingermillet	Panicle initiation, flowering, grain development
Minor millets	Maximum tillering, flowering, grain development
Rabi	
Wheat	Crown root initiation, late tillering, late jointing, boot/flag leaf, flow ering, milky stage, dough stage Active tillering, flag leaf/booting, grain filling
Barley	Active tillering, ear emergence, grain filling
Oat	Active tillering, ear emergence, grain filling
Grain Legumes:	
Kharif	
Pigeonpea	Flower initiation, flowering, pod filling
Urdbean	Pre-flowering, pod development (filling)
Mungbean	Pre-flowering, pod filling (devel-opment)
Cowpea	Pre-flowering, pod development
Clusterbean	Pre-flowering, pod development
Soybean	Vegetative phase, flowering, pod filling
Rabi	
Chickpea	Branching, pre-flowering, seed development
Fieldpea	Flower initiation, pod development
Lentil	Branching, flowering, pod develop-ment
Horsegram	Flowering, grain development

Oilseed crops	
Kharif	
Groundnut	Flowering, pegging, pod development
Seasame	Flower initiation, capsule (pod) formation
Castor	First or second or third spike development
Rabi	Branching, flowering (general)
Rape & mustard	Preflowering, silique filling
Sunflower	Buttoning, flowering, seed setling, seed development
Safflower	Elongation, branching, flowering, grain filling
Linseed	Pre-flowering, Pod filling
Toria	rosette, pod formation
Rocket *salad*	rosette, pod formation
Potato	Stolonization, tuberization, bulking,
Tomato	Flowering, fruit development, Fruit ripening
Onion	Bulb formation, bulb enlargement
Cauliflower	Seedling stage, curd growth
Cabbage	Head formation, head enlargement
Carrot	Root initiation, bulking (enlarg-ement)
Radish	Root formation, bulking
Turnip	Root initiation, root enlargement
Broccoli	Seedling, curd growth, branching
Sweet potato	Seedling, leaf development, root enlargement
Commercial and other crops:	
Sugarcane	Emergence, tillering, elongation
Tobacco	Complete growth period especially topping
Cotton	Sympodial branching, flowering, Boll development
Citrus	Flowering, fruit setting fruit growth
Banana	Early vegetative phase, flowering, fruit formation
Sugarbeet	Formative, leaf development, root development

Water Requirement of Crops

Crop	*Water requirement (mm)*
Cereals and millets:	
Rice	600-4000
Light soils	2000-2500
Medium soils	1500-2000
Heavy soils	1000-1500
Meduirn to heavy soils	1600
Sorghum *rabi*	450-650
kharif	300-450
Maize	500-800
Pearlmillet *rabi*	400-350
kharif	150-200
Fingermillet	450-500
Italian millet	250-300
Minor millets	200-300
Wheat	250-600
Northern India	250-400
Central India	500-600
Barley	250-600
Oat Grain	250-600
Fodder	350-600
Triticale	250-400
Grain legumes:	
Pigeonpea	300-350
Cowpea	250-300
Mungbean	150-250
Urdbean	200-250
Soybean	450-700
Clusterbean	200-250
Chickpea	250-300
Fieldpea	350-500
Lentil	150-200

Oilseed crops:	
Groundnut	500-700
Sesame	300-350
Castor	500-550
Rapeseed-mustard	300-350
Toria	200-250
Rocket salad	150-200
Safflower	200-250
Sunflower	350-400
Linseed	350-500
Miscellaneous	250-1500
Sorghum (Fodder)	450-650
Berseem	800-900
Lucerne	1800-2000
Vegetable crops	500-1500
Tomato	600-950
Potato	500-700
Cabbage	300-500
Pea	350-500
Radish	230-300
Turnip	250-300
Carrot	225-300
Beet root	330-350
Onion	350-550
Garlic	300-500
Commercial crops:	
Cotton	500-1300
Jute	500-600
Tobacco	400-600
Sugarcane	1500-2500
Andhra Pradesh	1600-1700
Bihar	1400-1500
Karnataka	2000-2300
Maharashtra	2800-3000
Punjab and Haryana	1700-1800
Uttar Pradesh	1800-2000
Sugarbeet	800-1000

VARIOUS ACTS IN AGRICULTURE

1904 : First Co-operative Credit Society Act
1912 : Second Co-operative Credit Society Act
1948 : Minimum Wages Act
1950 : Agriculture Pest and Disease Act
1954 : Prevention of Food Adulteration Act
1955 : Essential Commodities Act
1956 : Fruit Products Order Act
1966 : Asian Development Bank Act
1966 : Indian Seeds Act
1967 : Vegetable Oil Product Act
1968 : Insecticides Act
1972 : Agriculture Produce Market Act
1973 : Meat Food Products Order
1974 : The Water Prevention and Control of Pollution Act
1975 : Vegetable Oil Product (standard of quality) Order
1976 : Regional Rural Banks Act
1976 : National Commission on Agriculture Act
1976 : Standards of Weights and Measure Act
1982 : Sugar Development Fund Act
1985 : Fertilizer Control Order
1986 : Consumer Protection Act
1987 : Dairy Development Board Act
1992 : Milk and Milk Product Order
1992 : Central Agricultural Universities Act
1992 : Destructive Insects and Pests (Amendment and Validation) Act
2001 : Farmers Rights Act
2002 : Biodiversity Act
2002 : Agricultural Employees Protection Act
2005 : Mahatma Gandhi National Rural Employment Guarantee Act
2005 : Right to Information Act
2005 : PPV and FR (Plant Protection Varieties and Farmers Right)
2006 : Agriculture Produce Markets (Amendments) Act
2013 : National Food Security Act.

COMPOSITION OF AIR

Components	Concentration (ppm by volume)
Nitrogen (N_2)	78.009
Oxygen (O_2)	20.95
Argon (Ar)	0.93
Water vapour (H_2O)	–
Carbon dioxide (CO_2)	0.037
Neon (Ne)	0.0018
Krypton (Kr)	0.00017
Methane (CH_4)	0.00017
Helium (He)	0.00052
Ozone (O_3)	0.000006
Xenon (Xe)	0.000009
Nitrous dioxide (N_2O)	0.00002
Carbon monoxide (CO)	0.00001
Hydrogen (H_2)	0.00005
Ammonia (NH_3)	0.0000006
Nitrogen dioxide (NO_2)	0.0000001
Nitric oxide (NO)	0.00000006
Sulphur dioxide (SO_2)	0.00000002
Hydrogen sulphide (H_2S)	0.00000002

ATMOSPHERE
(Altitude and temperature range)

Region	Altitude range (k.m.)	Temperature (°C)	Important chemical species
Troposphere	0-11	15 to –56	N_2, O_2, H_2O (Vapour), CO_2
Stratosphere	11-50	–56 to – 2	O_3
Mesosphere	50-85	– 2 to – 92	O^+, NO^+

Lower layer of atmosphere where chemical composition of air is uniform– *Homosphere*.

Highest layer of atmosphere where chemical composition changes– *Heterosphere*.

DIFFERENCES BETWEEN C_3 AND C_4 PLANTS

Characteristics	C_3 plants	C_4 plants
Leaf Anatomy	No distinct bundle sheath of photosynthetic cells	Well-organized bundle sheath, rich in chloroplasts
Enzyme (Carboxylating)	Rubisco	Phosphoenolpyruvate (PEP) carboxylase, followed by Rubisco
Transpiration ratio (kg H_2O per kg dry matter produced)	450-950	250-350
Photorespiration	Yes	No
Optimum temperature for photosynthesis	(15-25°C)	(30-47°C)
Minimum ppm CO_2 required for photosynthesis	30-70	0-10
		Proposed by Hitch and Slack (1970)

Salisbury and Ross (1992)

Phosphoenolpyruvate carboxylase (also known as *PEP carboxylase*) is an enzyme in the family of carboxy-lyases found in plants and some bacteria that catalyzes the adition of bicarbonate (HCO_3^-) to phosphoenolpyruvate (PEP) to form the four-carbon compound oxaloacetate and inorganic phosphate.

$$PEP + HCO_3^- \longrightarrow \text{oxaloacetate} + Pi$$

This reaction is used for carbon fixation in CAM and C_4 organisms, as well as to regulate flux through the citric acid cycle (also known as Krebs or TCA cycle) in bacteria and plants.

ESSENTIAL MACRONUTRIENT ELEMENTS AND ROLE IN PLANTS

Element	Role/Functions	Conc. in Tissue (dry weight) %
1. Carbon (C)	Basic molecular component of proteins, carbohydrates, and lipids etc.	45.0%
2. Hydrogen (H)	Plays a key role in plant metabolism	6.0%
3. Oxygen (O_2)	Occurs in all organic compounds required for growth and development of plants	1.5%
4. Nitrogen (N)	Constituent of all proteins, chlorophyll, and in coenzymes, and nuclei acids.	1.5%

5. Phosphorous (P)	Important in energy transfer as part of adenosine triphosphate. Constituent of many proteins, coenzymes, nuclei acids, and metabolic substrates	0.2%
6. Potassium (K)	Hexokinase	1.0%
7. Calcium (Ca)	Cell wall component. Plays role in the structure and permeability of membranes.	0.5%
8. Magnesium (Mg)	Constituent of chlorophyll and enzyme activator.	0.2%
9. Sulfur (S)	Important constituent of plant proteins.	0.1%

Essential Micronutrient Elements and Their Role in Plants

		ppm
10. Boron (B)	Important in sugar translocation and carbohydrate metabolism. Essential for cell division	20
11. Iron (Fe)	Chlorophyll synthesis and in enzymes for electron transfer	100
12. Manganese (Mn)	Controls several oxidation-reduction systems, formation of O_2 in photosynthesis	50
13. Copper (Cu)	Catalyst for respiration, enzyme constituent.	6
14. Zinc (Zn)	In enzyme systems that regulate various metabolic activities	20
15. Molybdenum (Mo)	In nitrogenase needed for nitrogen fixation.	0.1
16. Chlorine (Cl)	Activates system for production of O_2 in photosynthesis	100
17. Nickel (Ni)	Deficiency produces an array of effects on growth and metabolism of plants, including reduced growth, and induction of senescence, leaf, and meristem chlorosis, alternations in N metabolism, and reduced Fe uptake.	0·5

Other Elements Needed Sometimes

1. Sodium (Na), 2. Silicon (Si), 3. Cobalt (Co), 4. Vanadium (V)

Cardinal Temperature for the Germination of Seeds of Certain Crops

Crop	Cardinal temperature		(°F)*
	Minimum	Optimum	Maximum
Rice	50-93	86-90	97-101
Maize	46-50	89-95	104-111
Sorghum	46-50	89-95	104
Wheat	39-40	77	86-90
Barley	39-40	68	82-86
Oats	39-41	77	86
Rye	34-36	77	86
Peas	34-36	86	95
Lentil	39-41	86	97
Sugarbeet	39-41	77	82-86
Carrot	39-41	77	86
Pumpkin	54	89-93	104
Melon	54-59	95	104
Alfalfa	34	86	99
Flax	35-37	77	86
Hemp	34-36	95	113
Vetch	34-36	86	95

* $\frac{C}{100} = \frac{F-32}{100} = \frac{K-273}{100}$

Recommended Optimum Moisture Content of Seed at Harvest of a Seed Crop

Crop	Moisture content per cent
Wheat and Barley	18-20
Rice	18-25
Maize	15-20
Sorghum and Pearlmillet	25-30
Field pea	25
Soybean	20-25
Groundnut	30-40

Normal, Deficient and Toxic Levels of Supply of Essential Nutrient Elements

Nutrient	*Concentration (ppm)*		*Toxic level*
	Deficiency	*Optimum*	
Nitrogen	35-70	140-350	
Phosphorous	2-10	30-60	Growth depressed
Potassium	2-20	100-200	at 5-10 times
Calcium	2-20	100-200	concentration that
Sulfur	1-10	25-50	for the normal
Magnesium	0.5-5	10-40	growth
Copper and Zinc	0.0001-0.005	0.05-1.0	2-50
Iron	0.05-2.0	5-10	30
Molybdenum	0.00001-0.00001	0.2-1.0	20-100
Manganese	0.005-0.02	0.55-5.5	10-55
Boron	0.0001-0.2	0.1-0.5	1-10

Vascular Plants

Division	*Class*	*English name*	*Group name*
Spermatophyta	Angiospermae (Covered seeds)	—	—
	sub-class		
	Monocotyledonease (one seed leaf)	Monocotyledons	Flowering plants
	Dicotyledonease (Two seed leaves)	Dicotyledons	
	Conopsida (cone)	Conifers	Gymnosperms
	Ephedropsida (ephedra)	Ephedra	
	Gnetopsida (Gnetum)	Gnetum Welwitschia	
	Cycadopsida (cycad)	Cycads	
Pteridophyta	Pterosida (feather)	Ferns	Pteridophytes
Sphenophyta	Sphenopsida (wedge)	Horsetails	
Lycophyta	Lycopsida (wolf)	Club mosses	
Psilophyta	Psilopsida (bareness)	Psilositum	

COMMON NAME OF CROPS

Wheat	: King of cereals
Maize	: Queen of cereals, Drosophila of crop plants
Potato	: King of Vegetables, Poor man's friend
Soybean	: Poorman's meat, Wonder crop, Yellow jewel, Gold crop of America
Millets	: Famine reserves
Sorghum	: Camel crop, King of coarse Cereals
Groundnut, Mustard	: King of oilseeds
Sesame	: Queen of Oilseeds
Lucene	: Queen of Fodder crops
Berseem	: King of Fodder Crops
Cowpea or Mushrrom	: Vegetable meat
Seasamum	: Poorman's ghee
Congress Grass	: King of weeds
Pearl Millet	: Poorman's food
Mango	: King of fruits
Mangosteen	: Queen of fruits
Gram	: King of Pulses
Pea	: Queen of Pulses
Cardamom	: Queen of Spices
Dahlia	: King of Flowers
Cocoa	: Food of God
Tea	: Queen of Beverages
Date Palm	: Century plant
Jojhoba	: Dessert Gold
Litchi	: Queen of Fruits
Guava	: Poorman's apple
Ber	: Poorman's Fruit
Cotton	: White gold
Dahlia	: King of flowers
Jamun	: Indian blackberry, Black plum, Java plum
Loquat	: Japan plum, Japanese maldar
Aonla	: Indian Gooseberry, Malacca tree, Myrabalan
Papaya	: Melon tree
Coconut	: Kalpavriksha
Walnut	: King of nuts
Apple	: King of Temperate Fruits
Congress Grass	: King of Weeds

Table : Fruits, Their Kinds and Edible Parts

Fruit	*Kind*	*Edible part*
Apple	Pome	Fleshy thalamus
Banana	Berry	Mesocarp, endocarp
Cashewnut	Nut	Peduncle, cotyledons
Coconut palm	Fibrous drupe	Endosperm
Cucumber	Pepo	Mesocarp, endocarp, placentae
Custard apple	Berries	Fleshy pericarp
Date palm	Berry (single seeded)	Pericarp
Fig	Syconus	Fleshy receptacle
Jack	Sorosis	Bracts, perianth, seed
Grape	Berry	Pericarp, placentae
Guava	Berry	Thalamus, pericarp
Indian plum	Drupe	Mesocarp, epicarp
Litchi	Nut (single seeded)	Fleshy aril
Wheat, rice, Maize	Caryopsis	Starchy endosperm
Mango	Drupe	Mesocarp
Melon	Pepo	Mesocarp
Orange	Hesperidium	Juicy placental part
Palmyra palm	Fibrous drupe	Mesocarp
Papaya	Berry	Mesocarp
Pea	Legume or pod	Cotyledons
Pear	Pome	Fleshy thalamus
Pineapple	Sorosis	Bracts, perianth, outer receptacle
Pomegranate	Balausta	Aril, Juicy outer seedcoat
Pummelo or shaddock	Hesperidium	Juicy placental hair
Strawberry	Achenes	Succulent thalamus
Tomato	Berry	Pericarp and placenta
Woodapple	Amphisarca (special)	Inner endocarp and placenta

AGRICULTURAL METEOROLOGY

Q. 1. Which was the first weather satellite

Ans. TIROS launched by United States in the 1960. It sent the first images from space of clouds moving around the Sun.

Q. 2. When was the first weather forecast made

Ans. In the year 1861 in USA

Q. 3. What is a barograph ?

Ans. It is an instrument that makes a continuous records of the changes in air pressure

Q. 4. What is an Ice bow?

Ans. An Ice bow is a rainbow

Q. 5. What is a blizzard?

Ans. A blizzard is caused when a very strong wind combines with falling snow or whips it up from the ground. It may blow snow into deep drifts that can bury vehicles and buildings

Q. 6. Who first classified clouds ?

Ans. Meteorologist Luke Howard in 1803

Q. 7. How many types of clouds are there ?

Ans. There are many different clouds but only 10 (Ten) are officially classified.

Q. 8. Cirrus clouds

Ans. These form at height above 6000m, it is so cold at that the water inside the clouds is frozen into crystals of ice. They have a feathery, wispy appearance and sometimes called "Mares Tails'. A large number of cirrus clouds will occasionally form a complete layer of white clouds.

Q . 9. Cumulus Clouds

Ans. These form at different heights 1000m - 2000m, although they are most often seen in the middle of the cloud layer. Fluffy in appearance, cumulus clouds are often grey on the bottom and a very bright white at the top. Sometimes known as cauliflower clouds. They have a flat base.

Q.10. Altocumulus clouds

Ans. These are small flattened cumulus compounds, grey or white in color. They may appear after a long period of hot weather, before a thunderstorm

Q.11. Stratus clouds

Ans. These form at the lowest levels of cloud layer 500 m (1600 ft) They form in layers that can build up across the whole sky. These clouds produce light rain and dizzle and in hilly areas will often produce wet fog and mist over the ground

Q.12 Altostratus

Ans. A thin watery sheet of grey clouds is altostratus clouds. Rain often follows its appearance

Q.13. Stratoculumbus clouds

Ans. These are formed when a sheet of cumulus clouds that almost join together They are probably the most common type of cloud.

Q. 14. Cumulonimbus

Ans. These clouds are huge, flat topped clouds that often bring heavy storms, rain and thunder. Because of their shapes, they are sometimes called anvil clouds and may stretch to a very great height

Q.15. Cirrocumulus clouds

Ans. These are combination of cirrus and cumulus clouds They are rows of icy particles and indicate a period of unsettled weather

Q. 16. Nimbostratus clouds (Rain clouds)

Ans. These are very thick and grey. They bring rain or snow and block out the Sun completely.

Q.17. First forecast published in Newspaper

Ans. "The Times" August 1, 1861

Q.18. First News Bulletin in India

Ans. 19th Jan., 1936 (All India Radio)

Q.19. Krishi Darshan started

Ans. 26th Jan., 1967

Q.20. International cloud atlas by WMO

Ans. New version in 2017

Agro-meterological Advisory Service (AAS) rendered by IMD and Ministry of Earth Sciences under Gramin Krishi Mausam Sewa scheme is a mechanism to apply relevent meterological information to help the farmer to make most efficient use of natural resources. Currently GKMS is being carried out through 130 Agro Met Field Units (AMFUs) is each Agro climatic zones.

District Level Agromet Service (DAAS)– IMD has started issuing quantitative district level (612 districts) weather forecast upto 5 days from 1st June 2008. The products comprise quantitative forecasts for 7 (seven) weather parameters. Viz. Rainfall, maximum and minimum temperatures, wind sped and direction, relative humidity and cloudiness, besides weekly cumulative rainfall forecast.

ADDITIONAL QUESTIONS

Q. 1. Ground water table is measured by

Ans. Piezometer

Q. 2. C_4 plant normally produce more biological yield than C_3 plant because of:

Ans. Less photorespiration

Q. 3. Element most mobile in soil ?

Ans. Nitrogen

Q. 4. The IARI was established in the year

Ans. 1905

Q. 5. The present average level of carbon-dioxide in atmosphere is

Ans. 421 ± 0.1 ppm

(As per Maura Loa Observatory, Hawaii of NOAA (National Oceanic and Atmospheric Administration USA), all time record CO_2 daily average on May 27, 2022.

Q. 6. Important constituent element of protoplasm

Ans. Calcium, iron, magnesium, chlorine, phosphorus, potassium and sulphur.

Q. 7. Criteria for the essentiality of nutrients for plants was given by

Ans. Arnon

Q. 8. The process of use of microorganism to remove salts from soil is referred as

Ans. Bioremediation

Q. 9. World Earth Day is celebrated on

Ans. March 22

Q. 10. The protoplasm contain water

Ans. About 67-75%

Q. 11. The outer layer of a seed is called

Ans. Testa

Q. 12. The CO_2 content of soil air is

Ans. 0.3 %

Q. 13. Solar constant is equal to (in cal/cm²/min) *(JRF-2006)*

Ans. 1.96/cal/cm²/min.

(It is measure of flux density. It is the average amount of solar radiation received by the earth's atmosphere per unit area. When the earth is at its mean distance from the Sun)

Q. 14. Which monsoon is the main cause of rainfall in India ?

Ans. South west monsoon

Q. 15. A plant which is salt resistant and can grow in waters of high salinity

Ans. Halophytes

Q. 16. The condition favourable when the ratio between Auxine : Cytokinin is less than one.

Ans. Shoot growth

Q. 17. Plant grow in acidic soils are

Ans. Oxylophytes

Q. 18. Leaf Area Index is computed as

Ans. Land area/ leaf area

Q. 19. Lime ($CaCO_3$) is added to reclaim ________.

Ans. Acid soil

Q. 20. The study of soil fertility with respect to crop is known as

Ans. Edaphology

Q. 21. Plant can grow on the surface of the rocks

Ans. Lithophytes

Q. 22. Hill reaction occurs in:

Ans. Presence of ferricyanide

Q. 23. Cu excess indicator plant is ________.

Ans. Sugarbeet

Q. 24. Mycorrhizae is the association of?

Ans. Higher plants & fungi

Q. 25. Plant can survive to dry conditions

Ans. Xerophytes

Q. 26. Crossing back F_1 plants to either of the two parental types is called

Ans. Test cross

Q. 27. Who has developed alley cropping

Ans. B. T. Kang

Q. 28. Neem Day is being celebrated on

Ans. 1st Day of Chaitra (Gudi Padwa)

Q. 29. In C_3 plants, the first stable compound formed after CO_2 fixation is

Ans. 3-phosphoglycerate

Q. 30. During guttation water is lost through
Ans. hydathodes

Q. 31. The term 'azofication' stands for ________.
Ans. Non symbiotic N-fixation by *Azotobacter*

Q. 32. Sowing of male and female parents on different day for the purpose of synchronisation in flowering is called
Ans. Staggering

Q. 33. The world 'Mutation' was first used by ?
Ans. Hugo de vries

Q. 34. Coastal soils are rich in
Ans. K

Q. 35. Transpiration is least in
Ans. High atmospheric humidity

Q. 36. Stomata open at night and close during day time in
Ans. Succulents

Q. 37. Increasing power of osmotic pressure in plants
Ans. Xerophytes > mesophytes > hydrophytes > halophytes. **Osmotic pressure** is the minimum pressure required to stop the flow solvent molecules across the semipermeable membrane

Q. 38. Those soil which contain low salt and high amount of exchangeable sodium is known as
Ans. Non saline alkaline soil

Q. 39. How many element occurs in plant body ?
Ans. 20

Q. 40. Evapo–transpiration expressed in ________.
Ans. cm/hr

Q. 41. Soil bulk density is greatest in which soils?
Ans. Sand

Q. 42. The father of green revolution is
Ans. N. E. Borlaug

Q. 43. Crop rotation is useful for
Ans. Improving fertility

Q. 44. N-fertilizer suitable for water logged soils
Ans. Ammonium sulphate

Q. 45. Which one is the principal 'N' fixing blue green algae

Ans. Anabaena

Q. 46. Khaira disease of rice can be controlled by spraying

Ans. Zinc sulphate

Q. 47. Zoo geographical distribution of animals was proposed by

Ans. Alfred Russel Wallace (1878)

Q. 48. Who gave the statement 'water is the sole nutrient of plants'?

Ans. J. B. Helmont

Q. 49. The test has self purification capacity of water body ?

Ans. BOD test

Q. 50. Plants neither survive in wet conditions nor in dry conditions

Ans. Mesophytes

Q. 51. To predict earthquake, the following is the easiest way

Ans. Release of Radon gas

Q. 52. UNEP-led Green Economy initiative means

Ans. Analysing the macroeconomic; sustainability and poverty reduction implications of green investment by the country.

Q. 53. Vision of Digital India is

Ans. 1. Digital infrastructure as a utility to every citizen. 2. Electronic delivery of services and 3. Digital empowerment of citizens.

Q. 54. Skill India programme aims to

Ans. Skill the youth in such a way so that they get employment and also improve entrepreneurship

Q. 55. Plants can grow in the saline environment

Ans. Halophytes

Q. 56. National park of India, also declared by a UNESCO world Heritage site

Ans. MANAS

Q. 57. Microbial type culture collection centre is situated at

Ans. Chandigarh

Q. 58. Decreasing productivity in various systems is sequenced as

Ans. Mangroves > Grasslands > Lakes > Oceans

Q. 59. The first Airport in the world that would be running fully on solar power is the

Ans. Cochin International Airport, Kerala

Q. 60. World Global Risk Report 2016 is published by

Ans. World Economic Forum (WEF)

Q. 25c. Plants grow in aquatic conditions

Ans. Hydrophytes

Ans. Australia

Q. 62. Term smog was first coined and used by

Ans. Des Voeux (1905)

Q. 63. Term Acid rain was introduced by

Ans. Robert August (1872)

Q. 64. Which gas binds Haemoglobin 200 times more than oxygen

Ans. Carbon monoxide (CO)

Q. 65. Salt content in fresh water is about

Ans. 0·5%

Q. 66. Sphere of earth covered by ice and snow is

Ans. Cryosphere

Q. 67. Process of frozen soil expansion is called

Ans. Soil heaving

Q. 68. Most important form of water available to plants

Ans. Capillary water (It is held by capillary forces between the soil particle in the micropores after the removal of runoff water and gravitational water).

Q. 69. Maximum water holding capacity is found in

Ans. Clay soil

Q. 70. Important and central to soil health is

Ans. Available soil carbon

Q. 71. India's first fully organic state is

Ans. Sikkim

Q. 72. Target Temperature rise in global temperature since 1850 set by the participating countries at Paris climate Agreement.

Ans. 2 °C maximum rise by the year 2100 (Since 1850 about 17 °C temperature already risen on earth)

Q. 73. Hottest place in India during the year 2016

Ans. Phalodi (Rajasthan) (51 °C) in May 2016

Q. 74. Hottest place in the world

Ans. Mitribish (Kuwait) 54.9 °C in July 2016

Q. 75. What will happen when all plants will die

Ans. Animals will also die due deficiency of oxygen

Q. 76. Total carbon stock in country's forest cover

Ans. 7044 million tonnes

Q. 77. Full form of MINAS

Ans. Minimum National Standards

Q. 78. What is slogan of Brudtland report our common future

Ans. Investing in our common future

Q. 79. Term Biodiversity named and coined by

Ans. Walter G. Rosen in 1985

Q. 80. Indian grain storage management institute is situated at

Ans. Hapur (U.P).

Q. 81. Central Insecticide Laboratory is situated at

Ans. Faridabad (Haryana)

Q. 82. Term Integrated pest management given by

Ans. Bartlett (1956)

Q. 83. NAIS (National Agricultural Insurance Scheme) launched in the year

Ans. 1999

Q. 84. Term vernalization was coined by

Ans. T. D. Lysenko (1928)

Q. 85. The term self incompatibility was originally coined by

Ans. Stout (1917)

Q. 86. Khaira disease first reported by scientist
Ans. Y. L. Nene in IR-8 at Pantnagar

Q.87. Percentage of water available at wilting point and field capacity
Ans. At wilting point : 10%. At field capacity : 75%

Q. 88. Transpiration ratio of crop plants usually ranges of
Ans. 200-1000 (Means crop plants transpire 200-1000 kg of water for every kg of dry matter)

Q. 89. Solar energy required to change water into vapour form is
Ans. 560/cal/g/25°C

Q. 90. Father of soil testing is
Ans. M. L. Troug

Q. 91. Most salt tolerant grass is
Ans. Karnal grass (*Diplachne fusca*)

Q. 92. Tarai soils are deficient in
Ans. Zinc

Q. 93. Chemical formula of SSP is
Ans. $Ca(H_2PO_4)$

Q. 94. Munsell colour chart contains about............ coloured paper on chips
Ans. 175

Q.95. Climacteric fruits are those
Ans. Which release ethylene gas with increased rate of respiration

Q. 96. Non-climacteric fruits are those which
Ans. Absorb ethylene gas and do not response to ethylene treatment

Q. 97. CIPHET (Central Institute of Post Harvest Engineering and Technology) is situated at
Ans. Ludhiana (Punjab)

Q. 98. Study of Animal behaviours is known
Ans. Ethology

Q. 99. Directorate of plant protection quarantine and storage is at
Ans. Faridabad (Haryana)

Q.100. Survival of insects ranges from
Ans. 04-70°C

Q.101. Who founded the Indian Statistical Institute
Ans. PC Mahalanobis (1931)

Q.102. Bank founded on May 28, 1938 under the name Divakaran Nanjee Banking Co. having headquarters at Mumbai,
Ans. Dena Bank

Q.103. The components of organic farming are
Ans. Organic manure, Bio-fertilizers, Bio-pesticides

Q. 104. Bitterness in bitterguard is due to
Ans. Momordicines compound

Q.105. Central Institute of Subtropical Horticulture is located at
Ans. Lucknow (U.P.)

Q.106. In which pulse crop, the first hybrid variety was developed in India?
Ans. Arhar

Q.107. The mutation theory was given by
Ans. Hugo de Vries (1901)

Q.108 Chemical formula of Gypsum is
Ans. $CaSO_4\ 2H_2O$

109. Mar soils are found in the region
Ans. Bundelkhand region

Q.110. Which crop is more responsive to Zinc application?
Ans. Paddy

111. Nitrification is a process of
Ans. Oxidation

Q.112. Which fertilizer is called 'Sona'?
Ans. Urea

Q.113. Which microorganism grows well in water logged soil?
Ans. Anaerobic microflora

Q.114. Micro-organisms grown at temperature below 5 °C are called
Ans. Psychrophiles

Q.115. Enzymes are
Ans. Proteins

Q.116. Microbial conversion of nitrate into chemical nitrogen is called
Ans. Denitrification

Q.117. In year 1943, Bengal famine was due to
Ans. Brown spot disease of rice

Q.118. The manure is considered as bio-fertilizer?
Ans. Blue green algae

Q.118. Which crop is used for green manure?
Ans. Sunhemp, Dhaincha

Q.119. Sulphur content in single super phosphate is
Ans. 11-12%

Q.120. Which colour tag is used for breeder seed?
Ans. Yellow/golden

Q.121. Term allelopathy was given by
Ans. Hans Molisch (1937)

Q.122. Major type of bacteria that carries photosynthesis is
Ans. cyanobacteria

Q123. Kind of plant that does not have seeds
Ans. Ferns

Q124. Oxygen in today's earths atmosphere is result of the
Ans. Photosynthesis

Q125. The first genetic material was probably
Ans. DNA

Q.126. Hypogeal crops (germinate below ground germination) are
Ans. chick pea and pea

Q.127. Epigeal crops (germinate above ground) are
Ans. blackgram, sunflower and common beans

Q.128. Who gave the idea of flowering hormone florigen in 1937
Ans. Mikhail Chailakhyan

Q.129. Giffen good

Ans. It is typically a inferior product that does not have easily available substitutes, as people consume more as price rises and vice versa violating the basis law of demand

Q.130. Herbicide used for pre-plant incorporation is

Ans. Fluchloralin

Q.131 The permissible limit for sulphur dioxide in jam is ?

Ans 40 ppm

Q.132 India's rank in coconut production in the world

Ans. 3rd rank

Q.133 National research centre for oil palm is located at ?

Ans. Pedavegi (ELURU) (A.P.) Pin-534450

Q.134 The crossing between Mandarine × Grape fruit is?

Ans. Tangelo

Q.135 The principle glycoside present in pulp of peach is ?

Ans. Prunasin

Q.136. The mechanical injury to the epidermal or oily cells of rind of citrus fruit is called as ?

Ans. Oleocellosis

Q.137. The example for bio preservation used to preserve for fruit and vegetables is called as

Ans. Niacin and Pimaricin (Antibiotics)

Q.138. The absorbing organ originating on hypha of a parasite and penetrating into host cell is called as ?

Ans. Haustorium

Q.139. The club root rot of cabbage (*Plasmodiophora brassicae*) is favoured in which type of soil?

Ans. Acidic soil.

Q.140. The example for heteroecism is ?

Ans. *Puccinia graminis tritici* (Black stem rust of wheat) (Heteroecism is the development of different stages of a parasitic species on different host plants)

Q.141. Dominant Mineral in earth's crust is

Ans Feldspar

Q.142. The capacity of storage structure for Kangi 1-3 tonnes and Kothar type of structure is?

Ans. 9-35 tonnes

Q.143. The capacity for ultra-volume sprayer is less than ?

Ans. 5 litres spray/ ha

Q.144. The capacity of vertical conveyor reaper for tractor mounted is ?

Ans. 0.4 to 0.6 ha/hr.

Q.145. An instrument commonly used for measuring angles of slope is called as ?

Ans. Clinometer

Q.146. The length of Gunter chain is ?

Ans. 66 feet

Q.147. The preparation of yoghurt by using bacterial cultures was Lactobacillus bulgaricus and ?

Ans. *Streptococcus thermophilus*

Q.148. Casein is in the form calcium caseinate-phosphate and it is 80 % of total?

Ans. Milk protein

Q.149. The yield of milk for Sahiwal breeds of cow is?

Ans. 2200– 2500 litres per lactation.

Q.150. Which committee were appointed for KVK scheme in the year 1974 ?

Ans. Mohan Singh Mehata committee

Q.151. Which mango variety is free from malformation.

Ans. Sai Sugandha (Totapuri x Kesar)

Q.152. Agricultural price commission came into existence in the year ?

Ans. 1965

Q.153. In No tillage system, the surface layer of the soil has ?

Ans. higher bulk density

Q.154. Who suggested the natural supply of nitrogen to the plant

Ans. Bartholomew (1971)

Q.155. The typical soil derived from Deccan trap is called as regular or black cotton soil and having high CEC ?

Ans. 40-60 meq / 100 gm. of soil

Q.156. Cotton Technological Research Laboratory is located at Matunga (Mumbai) was established in the year ?

Ans. 1924

Q.157. The respiratory roots are called as ?

Ans. Pneumatophores

Q.158. The total amount of water present in the soil is called as ?

Ans. Holard

Q.159. DNA content of the variant / feulgen stained interphase nuclei can be measured by ?

Ans. cytophotometer

Q.160. In which state is the first blood bank for cattle going to be set up?

Ans. Odisha under NADP programme

Q.161. What is the name of revamped National Mission on Food Processing that the government is going to launch at an outlay of Rs. 6000 crore?

Ans. SAMPADA (Period 2016-2020)

Q.162. A phenomenon when insect become quiescent due to adverse climatic condition and shows no visible sign of metabolic activity is called as ?

Ans. Cryptobiosis

Q.163. An important practice in sugarcane, hoeing after planting and before crop emergences for the purpose of weed control is known as ?

Ans. Blind hoeing

Q.164. For weed control in pea crop recommended herbicide is ?

Ans. Propanol

Q.165. Government and the jute sector have set out a programme to increase the supply of raw jute called Jute ?

Ans. I-CARE project

Q.166. Indian Journal of Dry land Agriculture Research and Development was published by Indian society of Dry land Agriculture is released ?

Ans. Half-yearly

Q.167. The degraded lands in India occupy approx an area (mha) of ?

Ans. 175

Q.168. Terms like dust, oolong and brick are varieties of the crop

Ans. tea

Q.169. Cotton fibres mainly contains

Ans. cellulose

Q.170.. Jute fibres detoriate quickly due to

Ans. lignin content is high

Q.171. Cocoa is a plant that has been used for chocolate making, so name the part of the plant that is used for this process

Ans. Seeds

Q.172. Genetic Engineering is

Ans. Introduction of the foreign genes for improving the genotypes.

Q.173. Cyanobacteria helps farmers for

Ans. reducing the acidity of the soil.

Q.174. Major source of sugar in the world is

Ans. **sugarcane** (Saccharum spp.) and sugar beets (**Beta vulgaris**).

Q.175. Method by which the amount of DNA, RNA and protein can be known at a time

Ans. Autoradiography (Agarose Gel Electrophoresis)

Q.176. Headquarter of European Union (EU) at

Ans. Brussels (Belgium)

Q.177. Catalytic converters in automobiles are used to convert

Ans. Carbon Monoxide to CO_2.

Q.178. Most of the plants cultivated in agriculture belong to which succession

Ans. Late secondary succession.

Q.179. Quick lime or burnt lime is

Ans. CaO and hydrated or slaked lime $Ca(OH)_2$

Q.180. Largest forest area country in the world

Ans. Russia

Q. 181. India's rank in forest area in the world

Ans. 10th rank (24.4% area)

Q.182. India's state has maximum forest area

Ans. M.P. > Arunachal Pradesh > Chattishgarh

Q.183. Level of pH reduces incidence of scab in Potato

Ans. 5-6.5

Q.184. 731 KVK, in the country are integral part of

Ans. NARS (National Agricultural Research System), aims at assessment of location specific technology modules in agriculture and allied enterprises, through technology assessment, refinement and demonstrations.

Q.185. Damage to crop due to low temperature is called

Ans. Heaving

Q.186. Damage due to low temperature and windspeed is called

Ans. Chilling injury

Q.187.. G.M. crop area, India's rank in the world

Ans 5th rank

Q.188. One farmer one transformation scheme launched by

Ans. Maharashtra state.

Q.189. India's First specialized hospital for elephants started on

Ans. 17th Nov. 2018 at Mathura (U.P.).

Q.190. PHM-2 is a variety of

Ans. Maize crop.

Q.191. Annamalai is a variety of

Ans. Brinjal crop.

Q.192. Udaipur-102 is a variety developed MPUAT for the crop

Ans. Onion (Bulb size 4·5 - 6·5 cm maturity 120 days)

Q.193. Plant Ideotype concept was given by

Ans. Donald

DIFFUSION OSMOSIS, VERNALISATION & IMBIBITION

Term Diffusion : Refers to movement of molecules or ions of a substance from an area of higher concentration to an area of lower concentration by virtue of their kinetic energy in order to equalize the concentration of two areas. The exchange of gases like CO_2 and O_2 between the aerial organs of the plant and the outer atmosphere, movement of gases, liquids and solutes inside the plant body are largely governed by the process of diffusion. The rate of diffusion of a substance depends upon its free energy or chemical potential.

Term Osmosis was discovered by Jean Antoine Nollet in 1748. Osmosis is the spontaneous net movement of solvent molecules through a semi permeable membrane into a region of higher solute concentration, in the direction that tends to equalize the solute concentration on two sides.

(*i*) Osmosis explains how plant roots are able to overcome the resistance of soils.

(*ii*) It explains how roots are able to absorb water from the soils.

(*iii*) It explains how the flow of gases into and out of leaves is regulated (Osmosis acting through stomata and guard cells). This is used in regulation of photosynthetic and respiration.

(*iv*) Osmosis explains how organelles in the cell and the cell itself can maintain a consistent chemical environment

Term vernalisation was coined by Trofim Lysenko in 1928 by his works on the effect of cold on cereals seeds.

Vernalisation is used to explain the ability of flowering in some plants after a chilling period due to physiological changes and external factors. It is the acquision or acceleration of the ability to flower by a chilling treatment. Or induction of flowering by prolonged exposure to low temperatures as during winters.

Typical temperature for vernalisation 5-10°C

Imbibition: It is a special type of diffusion when water is absorbed by solid-colloids causing an enormous increase in volume, examples are absorption of water by seeds and dry-wood. Due to imbibition pressure seedlings emerge out of soil into open. This term was given by Julius Sachs in 1865 a German botanist. Imbibition pressure is also known as Matric Potential.

AGRICULTURAL HYDROGELS

Agricultural hydrogels are synthetic polymers, generally made from petroleum products. They are able to absorb many times their weight water, and the same water can be distributed into dry regions in order to improve the soil ability to absorb water.

Potassium polyacrylate is the principal material used in SAD (Super Absorbent Polymers) industry and marketed as hydrogel for agricultural use because of its longer retention and high efficiency in soil with nill toxicity. They are prepared by polymerizing acrylic acid with a cross linker.

Agriculture hydrogels can change the physical properties of soils by:

1. Increasing their capacity to hold water
2. Reduce frequency of irrigation , 3. Reducing erosion and runoff
4. Increase soil permeability and infiltration
5. Reduce the tendency of the soil to get compacted
6. Increase the efficiency of the water being used
7. Help plant performance.

Hydrogel in agriculture is non-toxic pottasium based super absorbent polymer acting in drought/prone areas in reducing stress in plants, making better use of irrigation and fultilizer. They can also influence soil permeability, density, structure, texture, evaporation and infiltration.

Hydrogels are hydrophilic crosslinked molecular networks which can absorb and hold great amount of water.

Hydrogel agriculture technology uses insoluble gel forming polymers to improve the water-holding properties of different soils, such as clays and sandy loams.

Higher cost of hydrogel has limited their use in India.

Pusa Hydro-gel: A granulated product developed and patented by IARI (Indian Agriculture Research Institute) New Delhi.

It saves water 40-60% and nutrients used in farming. It improves soil fertility and soil texture.

It can perform well even at 50-60% soil temperature.

It helps for better germination (20% increase in performance), yields also increase 10-30%. It is very effective for one year in the soil than decompose.

NEW COMMITTEES FORMED DURING 2016-2022

1. **Ashok Dalwai Committee:** Blueprint for Doubling Farmers Income by 2022.
2. **T. Haque Expert Group:** For Land leasing
3. **Deepak Nayak Committee:** Revival of Godavari River
4. **Arvind Subramanyam Committee:** For tackling of pulse shortage and to review MSP and bonus for pulses
5. **Madhav Chitale Committee:** To prepare Guidelines for desiltation of river ganga
6. **Kaddiyam Srihari:** To look into issues of girl education in the country
7. **S. Ayyapan:** To draft National Policy on marine fishery.
8. **A.K. Sinha Expert Committee:** Study the silt in the rivers
9. **Pradeep Kumar Sinha:** GST council.
10. **N. Gopalan Committee** to select 20 educational institute of eminence.
11. 15th Finance commission constituted under Sh. N.K. Singh
12. PM's Science, Technology and Innovation **K. Vijay Raghavan.**
13. Research Advisory Committee- **Dr. Sankar De**
14. Committee to strengthen CIC framework headed by **Tapan Ray**
15. **Ratan P. Watal Committee** – to boost digital payments in India
16. **Vulture Conservation Committee** - Scientific management of carcass dumps

SOME OTHER IMPORTANT COMMITTEES

1. **Cotton Committee** : 1921
2. **Jute Committee** : 1936 (Technological Research Institute, Barrackpore, Kolkata, (West Bengal)
3. **Sugarcane Committee : 1946**
4. **Coconut and Tobacco Committee: 1945**
5. **Oilseed Committee : 1947**
6. **Arecanut Committee : 1949**
7. **Spice and Cashnew Committee: 1958**
8. **Khusro Committee (1986):** For Agricultural Credit System
9. **Purshotam Das Committee:** For Agricultural Finance and Co-operative societies.
10. **Abhijit Sen Committee (2002):** For long term food security
11. **Shankar Lal Gori Committee:** Agricultural Marketing
12. **UK Sharma Committee:** Role of NABARD in Rural Regional banks
13. **Mohan Singh Mehta Committee:** KVK Scheme (1974)
14. **Malhotra Committee:** Insurance Reforms (1991)
15. **Bhanu Pratap Singh Committee:** Agriculture
16. **Chakravarty Committee:** Development Monetary System (1982)
17. **Hanumantha Rao Committee:** Fertilizers (Revision of Fertilizer Policy)
18. **Mahajan Committee:** Sugar Industry (1997)
19. **Marathe Committee:** Recommendations for Urban Co-operative Banks
20. **Narasimham Committee-I:** Financial System (1991)
21. **Narasimham Committee-II:** Banking Reforms (1998)
22. **Raja Chelliah Committee:** Tax Reforms (1991)
23. **R.V. Gupta Committee:** Agricultural Credit (1997)
24. **S. Tendulkar Committee:** Redefining Poverty Line and its calculation formula

NATIONAL AND OTHER COMMISSIONS

1.	First Famine Commission	– 1880
2.	First Irrigation Commission	– 1901
3.	Royal Commission on Agriculture	– 1926
4.	Central Water commission	– 1945
5.	Finance Commission	– 1951
6.	Central Forestry Commission	– 1965
7.	Khadi and Village Industry Commission	– 1957
8.	Commission for Agricultural Cost and Prices	– 1965
9.	Second Irrigation commission	– 1972
10.	Agricultural Commission	– 1976
11.	National Flood Commission	– 1976
12.	National Commission for Women	– 1992
13.	National Commission on Backward Classes	– 1993
14.	National Commission on Cattle	– 2002
15.	Commission for Enterprise in Unorganized Sector	– 2004
16.	National Commission on Farmers	– 2004
17.	National Statistical Commission	– 2005
18.	National Forest Commission (Review of Forest Policy)	– 2013

National Action Plan for Climate Change (NAPCC)
(Action Plan Released on 30th June, 2008)

Comprises of Eight Core National Missions

(*i*) National Solar Mission.

(*iii*) National Mission for Enhanced Energy Efficiency.

(*iii*) National Mission on Sustainable Habitat.

(*iv*) National Water Mission.

(*v*) National Mission for Sustaining the Himalayan Ecosystem.

(*vi*) National Mission for a "Green India".

(*vii*) National Mission for Sustainable Agriculture.

(*viii*) National Mission on Strategic Knowledge for Climate Change.

TERMS ASSOCIATED WITH VARIOUS CROPS

Crops	Term	Crops	Term
Oil palm	- Stripping	Tobacco	- Desuckering, Topping Rubbing, Priming
Rubber	- Tapping	Maize	- De-tasseling
Tea	- Pruning and Skiffing	Jute	- Retting, Stripping
Coffee	- Stripping	Cotton	- Ginning, Topping, Nipping
Cauliflower	- Blanching, Scooping	Sugarcane	- Ratooning and Propping Arrowing, Wrapping, Detrashing
French bean	- Stalking	Rice	- Parboiling
Tomato	- Stalking	Lucerne	- Lopping
Tobacco & Tea	- Curing	Potato	- Earthing up
Gram	- Nipping	Groundnut	- Pegging
Banana, Sugarcane	- Propping	Linseed	- Retting
Sunhemp	- Degumming		

Toxins Produced in Plants

Plants	Toxins
Cotton	Gossypol
Lathyrus	Neurotoxin- ß-oxalyl-L-a,ß-diaminopropionic acid (ODAP)
Rapeseed & Mustard	Erucic acid- High erucic content is associated with myocardial fibrosis in adults and lipidosis in children.
Soybean	Goitrogen
Lucerne/Alfalfa	Saponin and plant estrogen
Groundnut	Aflatoxin-High moisture, Fungi (Aspergillus flavus & A.Niger)
Sorghum	HCN glucosides/Prussic acid/Dhurin
Sweet clover	Coumarin
Potato	Steroidal alkaloid
Yam	Alkaloids
Mango	Resin/ Urushiol
Field pea	Antivitamin E factor
Safflower, Sunflower	Polyphenolics
Pigeon pea / French bean	Tripsin inhibitors
French bean	Heamaggutinin
White clover	Glucosides
Cucurbits	Cucurbitacins
Cassava	CN Glucosides
Pearl millet, Napiergrass	Oxalic acid
Gram leaf	Malic & Oxalic acid
Cole crops	Sinigrin
Celery	Appin
Tomato	Tomatine
Brinjal	Salasodine
Pumpkin	Choline esterase

WELFARE PROGRAMMES & SCHEMES IN 2014-24

1. Swachh Bharat Abhiyan: A national campaign by Indian Government launched officially on **2nd Oct., 2014** at Rajghat by Prime Minister Narendra Modi. India's biggest cleanliness drive covering 4,041 towns, to clean the streets, roads and infrastructure of the country. This campaign aims to achieve vision of **clean India** by 2nd Oct. 2019. (150th Birth Day of Mahatma Gandhi).

Gramin area: Since Oct., 2014 number of household toilets bult - 10.24 crores.

Open defecation free villages: 6,03,175

Urban area: Since Oct., 2014. Number of household toilets built- 62,60606. **Number of community and public toilets: 6,15,864 toilets.**

Number of cities open defection free: 3,574 cities.

Swachh Bharat mission (Urban) is now continued till 2025-26. Swachh Bharat 2.0 has been launched to make all the cities garbage free and ensure gray and black water management.

2. Sansad Adarsh Gram Yojana: Launched by Prime Minister of India Sh. Narendra Modi on birth anniversary of Jayprakash Narayan on 11th Oct. 2014. It is rural development programme focusing upon the development of people of the village. This programme also initiated to bring MP (Member of Parliament) of all the political parties under the same roof for taking the responsibility of developing physical and institutional infrastructure in villages and turn them in to model villages. Each MP has to choose one village in his/her constituency that they represent, fix parameters and make it a model village.

Gram panchayat identification :

SAGY I (2014-19) 1440 No. of states covered-24

SAGYII (2019-24) 1921 No. of Adarsh gram declared 7405

3. Pradhan Mantri Jan Dhan Yojana: This scheme launched by Prime Minister Sh. Narendra Modi on 28th August 2014. It is a scheme for comprehensive financial inclusion run by department of Financial Services Ministry of Finance. Scheme is started with target to provide universal access to banking facilities starting with Basic Banking Accounts with Overdraft facility of Rs. 5,000 after 6 months and Rupay debit card with inbuilt accident cover of

Rs. 1 lakh and Rupay Kisan Credit Card. In the next phase, micro-insurance pension etc. will also be added. On the inauguration day 1.5 crore bank accounts were opened and around Rs. 10,590 crores were deposited under this scheme. This scheme launch completed one year on 28th August 2015 with 174.5 million accounts and a total deposit of Rs. 20,769.33 crore. 49.03 crore beneficiaries so far with a Rs 197,193.68 crore balance in beneficiaries account. 8.50 lakh Bank Mitras delivery branchless banking services.

More than 53.14 crore beneficiaries banked under PMJDY since inception.

- Total deposit balances under PMJDY Accounts stand at Rs. 2,31,236 crore.
- PMJDY accounts grow 3.6 -fold from 15.67 crore in March 2015 to 53.14 crore as on 14-08-2024.
- Around 55.6% Jan-Dhan account holders are women and around 66.6% Jan Dhan accounts in rural and semi-urban areas
- 36.14 crore RuPay cards issued to PMJDY account holders

4. Pradhan Mantri Suraksha Bima Yojana: It is a low-cost personal accident cover to increase the insurance penetration level for poor in the country. Scheme was announced by Finance Minister in budget 2015-2016. This insurance policy covers death or disablement of the policy holder caused due to accident or any accidental injuries. This insurance cover is available for a premium of just Rs. 12 per year to people in the age group of 18-70 years. The scheme is offered by all public sector general insurance companies. This scheme ensures social security through convenient modes like auto debit facility.

Benefeciaries as on April 2023. Gross enrolment reported by banks subject to verification for **eligibility** - 34 crore.

Total number of **claims:** 1.15 lakh **claims disbursed:** 2,302 crore.

5. Pradhan Mantri Jeevan Jyoti Bima Yojana: The scheme is to be a one year cover, renewable from year to year insurance Scheme offering life insurance cover for death due to any reason.

Scope of coverage: All savings bank account holders in the age 18 to 50 years in Bank will be entitled to join. In case of multiple saving bank accounts held by an individual in one or different banks, the person would be eligible to join the scheme through one saving bank account only. Premium Rs. 436 per annum.

Benefits: Rs. 2 lakhs is payable on member's death due to any reason. If the scheme member dies during insurable membership, subject to policy being in force and all due premiums, Service Tax and other levies (if any) having been paid and subject to any restrictions or qualifications referred to in these clauses, the amount specified as the Sum Assured for such Scheme Member shall become due to the Nominee of the Scheme Member. No maturity benefit is payable under the policy. No surrender benefit is payable under the policy.

Benefeciaries as on April 2023. Gross enrolement reported by banks subject to verification for eligibility - **16 crore.**

Total number of claims received - **6.64 lakh claims.**

Total claims amount disbursed- **13,294.46 crore.**

6. Pradhan Mantri Krishi Sinchai Yojana: The major objective of PMKSY is to achieve convergence of investments in irrigation at the field level, expand cultivable area under assured irrigation, improve on-farm water use efficiency to reduce wastage of water, enhance the adoption of precision-irrigation and other water saving technologies (More crop per drop), enhance recharge of aquifers and introduce sustainable water conservation

Components

Accelerated Irrigation Benefit Programme (AIBP) by MoWR,RD &GR PMKSY (Har Khet ko Pani) by MoWR,RD &GR. **PMKSY (Watershed development) (IWMP)** by Dept. of Land Resources (DoLR), MoRD

PMKSY(Per drop more crop) by Dept. of Agriculture & Welfare Farmers (GoI on 15th July 2015). The scheme mainly focuses on enhancing water use efficiency at the farm level through Micro Irrigation (Drip and Sprinkler Irrigation System). Besides, it also supports micro-level water storage, and water conservation/ management activities (Other Interventions) to supplement source creation for Micro Irrigation. MoA OFWM - DAC

Allocation for the financial year 2022-23-Rs. 4043.38 crore and for financial year 2023-24 Rs. 1,25,036 crore has been allocated under More Crop Per Drop (PDMC) Scheme.

District Irrigation Plans (DIPs) shall be the cornerstone for planning and implementation of PMKSY.

7. Pradhan Mantri Awas Yojana

I. Pradhan Mantri Awas Yojana – Urban (PMAY-U): A flagship Mission of Government of India being implemented by Ministry of Housing and Urban Affairs (MoHUA), was launched on 25th June 2015. The Mission addresses urban housing shortage among the EWS/LIG and MIG categories including the slum dwellers by ensuring a pucca house to all eligible urban households. The minimum carpet area for a house under the Pradhan Mantri Awas Yojana (Urban) (PMAY-U) scheme is 30 square meters (sqm). However, states and union territories (UTs) can increase the size of the house up to 60 sqm, in consultation with the Ministry of Housing and Urban Affairs (MoHUA) Based on the proposals submitted by States/ UTs, a total of 118.64 lakh houses have been sanctioned under PMAY-U by the Ministry as on 15.07. 2024. Of the sanctioned houses, 114.33 lakh have been grounded for construction; of which 85.04 lakh have been completed/delivered upto 29 Jul 2024.

II. Pradhan Mantri Awaas Yojana – Gramin

Launched on 1st April 2016, Pradhan Mantri Awaas Yojana Gramin (PMAY-G) is centre's flagship mission by the Ministry of Rural Development (MoRD), implemented by the Ministry of Housing and Urban Affairs (MoHUA). PMAY-G aims at providing a pucca house, with basic amenities, to all houseless households and those households living in kutcha and dilapidated house. PMAY-G addresses the rural housing shortage and bridges the housing deficit in rural areas of India, contributing significantly to the mission of "Housing for All". The minimum size of the houses under PMAY-G is 25 sq m including a dedicated area for hygienic cooking. As of September 27, 2022, 2 crore houses were constructed under the Pradhan Mantri Awaas Yojana - Gramin (PMAY-G). The total target was 2.72 crore houses.

8. Nai Manzil: Nai Manzil, a scheme launched on 8th August 2015, aims to benefit the minority youths who do not have a formal school leaving certificate in order to provide them formal education and skills, and enable them to seek better employment and livelihoods in the organised sector.

The Scheme has been approved with the cost of Rs.650.00 Crore for five years with 50% funding from the World Bank. This is the first World Bank supported programme for minority welfare being managed by the Ministry of Minority Affairs. The scheme is also significant as it combines education with skills for school dropouts which will significantly enhance their employability.

Scheme Objectives: The objective of the scheme is to provide education and market-driven skills training to youth between the ages of 17-35 years from minority community families who are **Below Poverty Line (BPL).** The scheme offers non-residential integrated education and skills training for 9 to 12 months, of which a minimum of 3 months are devoted to skills training. The scheme intends to provide placements to at least 70% of the trained youth in jobs which would earn them basic minimum wages and provide them with other social protection entitlements like Provident Funds, Employee State Insurance (ESI) etc.

9. Soil Health Card Scheme (Swasth Dhara, Khet Hara.): Prime Minister Narendra Modi launched **'Soil Health Card Scheme'** for 14 crore farmers to check the excessive use of fertilizer in Agriculture in India on 19th Feb. 2015 from Suratgarh (Raj). Under this scheme card will be issued for testing of soil and the estimated savings for the farmers with land holding of 3 acres shall be about Rs. 50,000 annually by adopting recommendation of crop-wise fertilizer requirement and reduction in amount and quality of fertilizer usage for crop production.

SHC is a printed report that a farmer will be handed over for each of his holdings. It will contain the status of his soil with respect to 12 parameters, namely N,P,K (Macro-nutrients); S (Secondary-nutrient); Zn, Fe, Cu, Mn, Bo (Micro - nutrients); and pH, EC, OC (Physical parameters). Based on this, the SHC will also indicate fertilizer recommendations and soil amendment required for the farm.

Swasth Dhara Khet Hara Slogan: (Healthy Earth Green Farm) It is a tagline and logo for Soil Health Card Scheme given in 2015 for improving the fertility of agricultural land which has been increased by adopting scientific techniques and about 52 new technologies have been developed in last 3 years, on soil testing, organic farming packages and 12 mini-labs made accessible to farmers to test soil parameters. Through advanced technologies and new varieties of paddy, mustard and wheat are developed which can be irrigated by the saline waters. Vulnerability Atlas of India has been made available to 572 rural districts of country and 75 ideal watershed also developed to provide irrigation water.

10. Shyama Prasad Mukherjee Rurban Mission (SPMRM) for integrated project based infrastructure in Rurral areas of the country. Mission aims to create 300 Urban growth clusters over the next (launched in 2016) years. It follows the vision of "Development of a cluster of villages that preserve and nurture the essence of rural community life with focus on equity and inclusiveness without compromising with the facilities perceived to be essentially urban in nature, thus creating a cluster of "Rurban Villages".

The Shyama Prasad Mukherji Rurban Mission (SPMRM) ended on March 31, 2022, and there is no budgetary allocation for it in 2023–2024. The Government of India did approve a one-time financial sanction to release funds for ongoing projects until March 31, 2023.

11. Deen Dayal Upadhyaya Grameen Kaushalya Yojana (DDU-GKY)

➢ **Ministry of Rural Development's** placement linked skill training programme announced on 25th September, 2014, It is a part of NRLM'S skilling component so as to build the capacity of rural poor youth to address the needs of both national and global skill requirements.

➢ The scheme is expected to promote **'Make in India'** campaign by preparing skilled workforce required for it.

➢ DDU-GKY has been aligned with the, Common Norms notified by Government.

➢ A total of 544 projects are being undertaken in the country by 268 Project Implementing Agencies (PIAs) selected in term.

12. Deen Dayal Upadhyaya Gram Jyoti Yojana (launched in 2015) is an Indian government scheme intending to provide an uninterrupted power supply to the rural parts of India. Nodal agency is Rural Electrification Corporation. Under DDUGJJY 2023 will be able to provide better connectivity of TV, Telephone, Internet and Mobile in rural India.

13. Pradhan Mantra Gramin Sadak Yojana is launched by the Govt. of India to provide connectivity to unconnected habitations as part of a poverty reduction strategy. Govt. of India is endeavoring to set high and uniform technical and management standards and facilitating policy development and planning at State level in order to ensure sustainable management of the rural roads network. (Launched on 15th Aug., 2000).

PMGSY-IV: The Cabinet approved the implementation of PMGSY-IV for the period FY 2024-25 to 2028-29. The total outlay for the scheme is Rs. 70,125 crore. The scheme will provide all-weather connectivity to 25,000 rural habitations and construct or upgrade bridges on new connectivity roads.

14. Pradhan Mantri Mudra Yojana: Pradhan Mantri Mudra Yojana (PMMY) is a flagship scheme of Government of India. The scheme facilitates micro credit/Loan up to Rs. 10 lakh to income generating micro enterprises engaged in the non farm sector in manufacturing, processing, trading or service sector. MUDRA supports Financial Intermediaries to extend loans to the non-corporate, non-farm sector income generating activities of micro and small entities.

The loans under MUDRA scheme can be availed only through banks and lending institutions which include:

Public Sector Banks, Private Sector Banks, State operated cooperative banks, Rural banks from regional sector. Institutions offering micro finance. Financial companies other than banks. (Launched in 2015)

Interest rate: Interest rates are charged as per the policy decision of the bank. However, the interest rate charged to ultimate borrowers shall be reasonable.

Loan limit increase: The loan limit for the PMMY has been increased from Rs. 10 lakh to Rs. 20 lakh for individuals who have previously availed and repaid loans under the Tarun category.

15. Pradhan Mantri Fasal Bima Yojana (PMFBY): Pradhan Mantri Fasal Bima Yojana (PMFBY) scheme was launched in India by Ministry of Agriculture & Farmers welfare, New Delhi from Kharif 2016 season onwards. National Insurance Company started participating in PMFBY from Rabi 2016 onwards and covered 8 States and 2 Union Territories during the past 5 seasons namely, Rabi 2016-17, Kharif & Rabi 2017 and Kharif & Rabi 2018 covering 70,27,637 farmers. Farmers share of premium is Rs. 453 crores and with subsidy from State/Central Govts RS.1909 Crores, gross Premium is Rs.2362 Crores for the 5 seasons together. Whilst the claims of Kharif 18 and Rabi 18 are under process, we have closed the first 3 seasons with a gross premium amount of Rs.1804 Crores collected from 35,22,616 farmers and claims to the tune of Rs.1703 crores were paid, by which 17,66,455 no of farmers benefited, which shows almost 50% of the farmers insured were benefitted.

Revised Operational Guidelines issued by Ministry of Agriculture & Farmers welfare, New Delhi are binding on all the stake holders which gives complete insight about the Scheme.

There are Two Schemes under the said operational Guidelines.

1. Pradhan Mantri Fasal Bima Yojana (PMFBY)
2. Revised Weather Based Crop Insurance Scheme (RWBCIS)

Objective of the Schemes

Pradhan Mantri Fasal BimaYojana (PMFBY) aims at supporting sustainable production in agriculture sector by way of

Providing financial support to farmers suffering crop loss/damage arising out of unforeseen events

Stabilizing the income of farmers to ensure their continuance in farming

Encouraging farmers to adopt innovative and modern agricultural practices

Ensuring flow of credit to the agriculture sector which will contribute to food security, crop diversification and enhancing growth and competitiveness of agriculture sector besides protecting farmers from production risks.

All farmers who have been sanctioned Seasonal Agricultural Operations (SAO) loans (Crop Loans) from Financial Institutions (FIs), *i.e.* loanee farmers, for the notified crop(s) season would be covered compulsorily.

The Scheme is optional for non-loanee farmers.

The insurance coverage will strictly be equivalent to sum insured/hectare, as defined in the Govt. notification or /and on National Crop Insurance Portal multiplied by sown area for notified crop.

Coverage;Food crops (Cereals, Millets and Pulses),

Oilseeds: Annual Commercial / Annual Horticultural crops.

Rate of Premium to be paid by the farmer to Insurance Company

Type of Crop Kharif Rabi: Food grains including Cereals, Pulses and Oilseeds 2% and 1.5 %

Annual Horticulture and Commercial Crops 5%

Coverage of Risks and Exclusions under PMFBY Scheme:

Following stages of the crop risks leading to crop loss are covered under the Scheme. Addition of new risks by the State Govt other than the one mentioned below, by the State Govt. is not permitted.

Prevented Sowing/Planting/Germination Risk: Insured area is prevented from sowing / planting / germination due to deficit rainfall or adverse seasonal/weather conditions. 25% of the sum insured will be paid and the Policy will be terminated.

Standing Crop (Sowing to Harvesting): Comprehensive risk insurance is provided to cover yield losses due to non-preventable risks, viz. Drought, Dry spell, Flood, Inundation, widespread Pests and Disease attack, Landslides, Fire due to natural causes ,Lightening, Storm, Hailstorm and Cyclone.

Post-Harvest Losses: Coverage is available only upto a maximum period of two weeks from harvesting, for those crops which are required to be dried in cut and spread / small bundled condition in the field after harvesting against specific perils of Hailstorm, Cyclone, Cyclonic rains and Unseasonal rains.

Localized Calamities: Loss/damage to notified insured crops resulting from occurrence of identified localized risks of Hailstorm, Landslide, Inundation, Cloud burst and Natural fire due to lightening affecting isolated farms in the notified area.

Add-on coverage for crop loss due to attack by wild animals: The States may consider providing add- on coverage for crop loss due to attack by wild animals wherever the risk is perceived to be substantial and is identifiable.

General Exclusions: Losses arising out of war and nuclear risks, malicious damage and other preventable risks shall be excluded.

The Government of India has released the **PMFBY Status 2024.** All the farmers of India who have applied for the Pradhan Mantri Fasal Bima Yojana can now **visit the official website** to check the application status released by the authorities. With the help of the online system of checking the application status the farmers and the government can both save a lot of time and effort. The applicants just need to have their receipt number to check the Pradhan Mantri Fasal Bima Yojana status online on the official website. The official website to check the PMFBY scheme application status online is pmfby.gov.in Portal.

16. Rashtriya Gokul Mission: (Dec. 2014) The "Rashtriya Gokul Mission" aims to conserve and develop indigenous bovine breeds in a focused and scientific manner for improving milk production and productivity.

Key Features of the Rashtriya Gokul Mission

1. The mission is to protect cow breeds. So, we can expect to see a rise in the number of local cow breeds.
2. The government will spend a fund of 150 crore on the protection of the cows.
3. Many new gaushalas may be constructed across the cities. I-net gaushalas, the government will construct cattle care centers.
4. If a farmer takes the scheme seriously and helps the government to preserve these local cow breeds, he will get recognition. The government is planning to provide "Gopal Ratna" awards to such exemplary farmers.

Gokul Gram: Under this component, it is proposed to establish Integrated Indigenous Cattle Centres or Gokul Grams in the breeding tracts of indigenous breeds. Gokul Grams will be established in:

(*a*) The Native Breeding Tract of an indigenous bovine breed and.

(*b*) Suburbs of Metropolitan and large cities (urban cattle)

(*c*) Setting up of biogas plant also included as one of the component.

17. National Programme for Bovine Breeding is being implemented for enhancing productivity of milch animals through extension of Artificial Insemination (AI) coverage. This is done through establishment of Multi Purpose AI Technicians in Rural India (MAITRIs); strengthening of existing AI centres; monitoring of AI etc.

18. National Mission on Bovine Productivity has been launched in November 2016 with the aim of enhancing milk production and productivity and thereby making dairying more remunerative to the farmers. The scheme is being implemented with following components (*a*) **Pashu Sanjivni-** this component includes identification of animals in milk using UID, issuing health cards to all animals in milk and uploading data on INAPH data base; (*b*) Advance reproductive Technique- under the component sex sorted semen production facility is being created at 10 A graded semen stations and 50 Embryo Transfer Technology Labs with IVF facilities are being created in the country; (*c*) Creation of E **Pashu Haat Portal-** The e-pashu haat portal has been launched in November 2016 for linking farmers and breeders of indigenous breeds and (*d*) Establishment of National Bovine Genomic Centre for Indigenous Breeds (NBGC-IB): The NBGC-IB is being established for enhancing milk production and productivity through genomic selection among indigenous breeds.

1. *Two National Kamdhenu Breeding Centres* are being established one in the State of Andhra Pradesh for southern region and other in Madhya Pradesh for northern region of the country with the aim of development and conservation of indigenous breeds in a scientific manner and thereby enhancing milk production and productivity.

2. *National Dairy Plan-I* a world Bank assisted project being implemented in 18 major dairy States with aim of enhancing milk production and productivity in order to meet demand of milk in the country through strengthening of semen stations; bull production programme (progeny testing and pedigree selection), ration balancing programme etc.

3. Government has also established three subordinate organizations namely (*i*) *Central Cattle Breeding Farms* (CCBFs) (*ii*) *Central Herd registration Scheme* and (*iii*) *Central Frozen Semen Production & Training Institute*. These organizations are also undertaking genetic upgradation of milch animals through supply of disease free high genetic merit bulls for semen production and natural service for use in the breeding programme being implemented by the States.

In the last one year, the **Indian Council of Agricultural Research (ICAR) has registered 10 new breeds of livestock species**, including cattle, buffalo, goat and pig. This has taken the total number of indigenous breeds to 212 as of January 4, 2023.

19. Pandit Deendayal Upadhyay Unnat Krishi Shiksha Yojana (2016): Launched in 2016. With a vision of transformational change in rural development processes by leveraging knowledge institutions to help build the architecture of an Inclusive India and to connect institutes of higher education with local communities to address the development changes of rural India through participatory processes and appropriate technologies for accelerating sustainable agricultural growth it is felt that it is befitting to start a new scheme "Pandit Deen Dayal Upadhyay Unnat Krishi Shiksha Yojana" under of Ministry of Agriculture and Farmers Welfare, implemented by Education Division of ICAR. Pt. Deen Dayal Upadhyay encouraged self-reliance in food grains. He emphasized decentralization of the economy to empower local communities to make economic and developmental choices. He created a model for rural development based on self-reliance to boost economic prosperity in several hundred villages. 100 farmers training centers across the country have been identified on the basis of knowledge, skill, ability and experiences related to Organic Farming, Natural Farming and Cow based Economy and having facilities to conduct training programmes.

The scheme has the following objectives:

1. To build skilled Human Resource at village level relevant to national needs towards organic farming and sustainable agriculture.
2. Provide rural India with professional support in the field of Organic Farming/Natural Farming/Rural Economy/Sustainable Agriculture.
3. To extend other activities of Pandit Deen Dayal Upadhyay Unnat Krishi Shiksha Yojana at village level through these established centres.

20. Atal Pension Yojna (APY): It is a guaranteed pension scheme and is administered by the pension fund regulatory and development authority (PFRDA). Under this scheme GOI will co-contribute to each eligible subscriber, for a period of 5 years who joins the scheme in the period of June 1 to December 2015. The scheme was made available for those who are not covered by any statutory social security and is not an Income Tax payer. This scheme is open to all citizens between 18 and 40 years of age.

21. Deen Dayal Antyodaya Yojana or **DDAY.** The scheme is integration of NULM and NRLM launched by Ministry of Rural Development. It is a scheme for the helping the poor by providing skill training. It replaces Aajeevika. The Government of India has provisioned Rs. 500 crore for the scheme. The objective of the scheme is to train 0.5 million people in urban areas per annum from 2016. In rural areas the objective is to train 1 million people by 2017. Further, in urban areas, services like SHG promotion, training centres, vendors markets, and permanent shelters for homeless will be provided for. The aim of the scheme is skill development of both rural and urban India as per requisite of international standards. **Rural Component** Deen Dayal Upadhyaya Grameen Kaushalya Yojana. (Earlier Scheme (SGSY). (Scheme launched under Ministry of Housing and Urban Poverty Alleviation (MHUPA). (Budget 3,000 crore, 2019-2020).

22. Aajeevika Grameen Express Yojna (Launched 2017)

The Government of India launched a new sub-scheme named "Aajeevika Grameen Express Yojana (AGEY)" **as part of the**

Deendayal Antyodaya Yojana – National Rural Livelihoods Mission (DAY-NRLM). The Self Help Groups under DAY-NRLM will operate road transport service in backward areas. This will help to provide safe, affordable and community monitored rural transport services to connect remote villages with key services and amenities (such as access to markets, education and health) for the overall economic development of backward rural areas.

AGEY will be initially implemented in 250 blocks in the country on pilot basis for a period of 3 years from 2017-18 to 2019-20 for a period of 3 years with each Block provided upto 624 vehicles to operate the transport services from 30th Nov., 2018. During the current year implementation of the scheme has been so far approved for 520 Blocks in 8 States namely Andhra Pradesh, Jharkhand, Maharashtra, Tamil Nadu, Telangana, Uttarakhand and West Bengal with a total provision of Rs.16.06 Crore of which the Government of India share would be Rs.10.16 Crore. The balance funding would be provided by the respective States. About 1,811 vehicles are plying on road under this scheme.

23. Pradhan Mantri Kisan SAMPDA: The Central Sector Scheme - SAMPADA (Scheme for Agro-Marine Processing and Development of Agro-Processing Clusters) was approved by the cabinet in May 2017 for the period of 2016-20 coterminous with the 14th Finance Commission cycle. The scheme has now been renamed as the "Pradhan Mantri Kisan Sampada Yojana (PMKSY)". The continuation of the Central Sector Scheme – Pradhan Mantri Kisan SAMPADA Yojana (Scheme for Agro-Marine Processing and Development of Agro-Processing Clusters) with an allocation of Rs. 4600 crore is approved till 31.03.2026.

It is an umbrella scheme incorporating ongoing schemes of the Ministry like **Mega Food Parks, Integrated Cold Chain and Value Addition Infrastructure, Food Safety and Quality Assurance Infrastructure,** etc. and also new schemes like **Infrastructure for Agro-processing Clusters,** Creation of Backward and Forward Linkages, Creation / Expansion of Food Processing & Preservation Capacities.

Objective: The objective of PMKSY is to supplement agriculture, modernize processing and decrease Agri-Waste.

Schemes to be implemented

Under PMKSY the following schemes are to be implemented.

Integrated Cold Chain and Value Addition Infrastructure

Creation/ Expansion of Food Processing and Preservation Capacities (Unit Scheme)

Infrastructure for Agro Processing Clusters

Food Safety & Quality Assurance Infrastructure

Human Resources and Institutions

The component schemes of Mega Food Park, Creation of Backward & Forward Linkages, Human Resource & Institution - Promotional Activities, Skill Development & HACCP, a component of Food Safety and Quality Assurance Infrastructure have been discontinued in the 15th FC cycle with the provision of committed liability.

Financial Allocation

PM Kisan SAMPADA Yojana is expected to leverage investment of Rs. 11,095.93 crore, benefiting 28,49,945 farmers and generating 5,44,432 direct/indirect employment in the country by the year 2025-26.

24. **AYUSHMAN BHARAT:** National Health Protection Mission will have a defined benefit cover of Rs. 5 lakh per family per year.

- Benefits of the scheme are portable across the country and a beneficiary covered under the scheme will be allowed to take cashless benefits from any public/private empanelled hospitals across the country.
- **Ayushman Bharat** - National Health Protection Mission will be an entitlement based scheme with entitlement decided on the basis of deprivation criteria in the SECC database.
- Beneficiaries can avail benefits in both public and empanelled private facilities.
- To control costs, the payments for treatment will be done on package rate (to be defined by the Government in advance) basis.
- One of the core principles of Ayushman Bharat - National Health Protection Mission is to co-operative federalism and flexibility to states.
- For giving policy directions and fostering coordination between Centre and States, it is proposed to set up Ayushman Bharat National Health Protection Mission Council (AB-NHPMC) at apex level Chaired by Union Health and Family Welfare Minister.

- States would need to have State Health Agency (SHA) to implement the scheme.
- To ensure that the funds reach SHA on time, the transfer of funds from Central Government through Ayushman Bharat - National Health Protection Mission to State Health Agencies may be done through an escrow account directly.
- In partnership with NITI Aayog, a robust, modular, scalable and interoperable IT platform will be made operational which will entail a paperless, cashless transaction.
- **1 Lakh Ayushman Mitra Jobs in PMRSSM Empaneled Hospitals – Rs. 15000 Salary**

Central government is going to deploy 1 lakh Ayushman Mitra directly in Govt. / Private hospitals under Pradhan Mantri Rashtriya Swasthya Suraksha Mission (PMRSSM). Ministry of Health expects that a total number of 10 lakh jobs will be created under Ayushman Bharat – National Health Protection Mission (AB-NHPM). All the employed youths under Ayushman Bharat Yojana will get monthly salary of Rs. 15,000.

- Prime Minister Narendra Modi launches the aspirational district programme as well as India's first wellness centre under **Ayushman Bharat** on April 14, 2018 at Bijapur, Chattisgarh. The two programmes are key initiatives of the NITI Aayog. Under the "Transformation of Aspiration Districts" programme, the government aims to quickly and effectively transform the 115 districts based on real time monitoring since 2018. Now it is digital health incentive scheme.

25. MP Bhavantar Bhugtan Yojana: Madhya Pradesh govt. is going to restart online registrations for Bhavantar Bhugtan Yojana 2018. Under this scheme, govt. will pay the entire price deficit (bhav + antar) to farmers who incur losses while selling their agriculture produce below MSP. Govt. will start this scheme for 13 Kharif Crops for crops which are being sold below minimum support prices within and outside the state. Farmers can make online registration from 28 July to 31 August 2018 at mpeuparjan.nic.in The primary objective of this Bhavantar Bhugtan Yojana is to raise farmers income in case of distress sale. In case the kharif crops are being sold at lower price than MSP in markets, then they can apply for BBY Scheme to get their loss covered.

Market Prices of 13 Kharif Crops are found to be lower than MSP this FY 2018, so the govt. has taken this step to restart MP BBY Scheme 2018. In 2024 also Govt. offers scheme to farmers.

Bhavantar Bhugtan Yojana for horticulture farmers has been announced in **Punjab state Budget (2023-24)**

In 2023, Haryana state also started the scheme for benefit of farmers.

26. Swajal Scheme: The Ministry of Drinking Water and Sanitation has launched the Swajal scheme in around 115 aspirational districts in India to provide clean drinking water. It will involve an outlay of Rs 700 crore through flexi funds under existing National Rural Drinking Water Programme (NRDWP) budget. Swajal is community owned drinking water supply. The operations and management of the project is taken care by the local villagers. Launched in 2018.

27. Krishi Kalyan Abhiyan:The Ministry of Agriculture and farmers' welfare in line with the Hon'ble Prime Minister's vision of doubling farmers' income by 2022 has launched the Krishi kalyan Abhiyaan from 1st June, 2018 till 31st July, 2018 so as to aid, assist and advice farmers on how to improve their farming techniques and raise their incomes. The Krishi kalyan Abhiyaan undertaken in 25 Villages with more than 1000 population each in Aspirational Districts identified in consultation with Ministry of Rural Development as per directions of NITI Ayog. In districts where number of villages (with more than 1000 population) is less than 25, all villages will be covered.

An Action Plan comprising specifically identified activities under various departments of the ministry namely, Department of Agriculture, Cooperation & Farmers Welfare (DAC&FW), Animal Husbandry Dairying & Fisheries (DAHD&F) and Department of Agricultural Research & Education (DARE-ICAR) will be implemented to saturate these 25 villages in each district with these activities.

Various activities to promote best practices and enhance agriculture income are being undertaken under this plan such as:-

- Distribution of Soil Health Cards to all farmers
- 100% coverage of bovine vaccination for Foot and Mouth Disease (FMD) in each village
- 100% coverage of Sheep and Goat for eradication of *Peste des Petits ruminants (PPR)* a viral disease of domestic and wild ruminants.
- Distribution of Mini Kits of pulses and oilseeds to all
- Distribution of Horticulture/Agro Forestry/Bamboo plant @ 5 per family(location appropriate)

- Making 100 NADAP Pits in each village
- Artificial insemination saturation
- Demonstration programmes on Micro- irrigation
- Demonstrations of integrated cropping practice
- So far two phases has been completed in which 11.05 lakh farmers were trained in third phase about 17 lakh farmers and trained on diversified farming practices. ICAR also sensitised and mobilised the farmers for water conservation measures. More than 7.1 lakh tree saplings planted.

28. Jal Shakti Abhiyan Launched by GOI

- Union Jal Shakti Minister recently launches Jal Shakti Abhiyan.
- The Jal Shakti Abhiyan is a **campaign for water conservation** and water security in the country.
- The campaign will run through citizen participation in two phases, firstly, during the monsoon season, from 1st July, 2019 to 15th Sep., 2019 and then from 1st October, 2019 to 30th November, 2019 for states receiving the North East retreating monsoons.
- It is a collaborative effort of various ministries of the Government of India and state governments, being **coordinated by the Department of Drinking Water and Sanitation (DDWS)** that comes under the Jal Shakti Ministry.
- The focus of the campaign is on water stressed districts and blocks. The teams of officers from the central government will visit and work with district administration in **1592 water stressed blocks in 256 districts,** to ensure five important water conservation interventions.
- The **five important water conservation interventions** are:
  - ❖ Water conservation and rainwater harvesting,
  - ❖ Renovation of traditional and other water bodies/tanks,
  - ❖ Reuse of water and recharging of structures,
  - ❖ Watershed development and
  - ❖ Intensive afforestation.
- The water conservation interventions will also be supplemented with special interventions including the development of block and district water conservation plans, promotion of efficient water use for irrigation and better choice of crops through Krishi Vigyan Kendras.

29. Har Medh Par Pedh: It is a sub mission of agroforestry to increase farmers income.

Mission on Agroforestry (Har Medh Par Ped) Scheme was launched in 2016-17 to encourage tree plantation on farm land along with crops/ cropping system to help the farmers get additional income and make their farming systems more climate resilient and adaptive.

Presently, the scheme is being implemented in 23 States/UTs including Maharashtra. Under the scheme, assistance to farmers is given through State Govt. for nursery development, boundary plantation and block plantation of prominent tree species to promote, inter-alia, fruit bearing trees, tree borne oilseeds, medicinal & aromatic plants, silk & lac rearing host plants in addition to timber species, so that farmers get early returns.

In case of promotion of horticulture, the **Mission for Integrated Development of Horticulture (MIDH),** a Centrally Sponsored Scheme is being implemented w.e.f. 2014-15, for holistic growth of the horticulture sector covering fruits, vegetables, root and tuber crops, mushrooms, spices, flowers, aromatic plants, coconut, cashew, cocoa and bamboo.

30. National Mission on Oilseeds and Oil Palm (NMOOP)

➢ The NMOOP was launched in **2014-15** restructuring the Centrally Sponsored Scheme of Integrated Scheme of Oilseeds, Oil Palm and Maize (ISOPOM) and Oil Palm Area Expansion (OPAE).

➢ The strategy of the Mission includes increasing Seed Replacement Ratio (SRR) with focus on varietal replacement; increasing irrigation coverage under oilseeds, diversification of area from low yielding cereals crops to oilseeds crops; inter-cropping of oilseeds with cereals/pulses/sugarcane; use of fallow land after paddy/potato cultivation; expansion of cultivation of Oil Palm, increasing availability of quality planting materials of Oil Palm.

31. Mera Gaon Mera Gaurav National Innovative initiative of ICAR started in 2015-16 with 10,712 villages under this programme, number now increased to 15,436 villages in 2017-18.

❖ The scheme was **launched in 2015.**

❖ The scheme envisages scientists to select villages as per their convenience and remain in touch with the selected villages and provide information to the farmers on technical and other related aspects of agriculture in a time frame through personal visits or on telephone.

❖ They may function with the help of **Krishi Vigyan Kendras (KVKs)** and **Agriculture Technology Management Agency (ATMA).**

***Objective*:**

❖ To promote the **direct interface of scientists** with the farmers to hasten the **"lab to land"** process.

Agriculture Technology Management Agency

❖ It is a **registered society responsible for technology dissemination at the district level.** It is a focal point for integrating research extension and marketing.

32. Pradhan Mantri Kisan Samman Nidhi Yojana: Launched in December 2018, this scheme provide an annual income support of Rs. 6,000 to all farmer families in the country.

33. Pradhan Mantri Ujjwala Yojana: Launched on April 20, 2018.

Saubhagya: Pradhan Mantri Sajhaj Bijli Har Ghar Yojana: Launched on April 28, 2018.

Ujjwala Yojana to distribute 50 million LPG connections to women of BPL families and Saubhagya Yojana (Pradhan Mantri Sahaj Bijli Har Ghar Yoja) to provide electricity to all households have transformed lives of every rural family, dramatically improving ease of their living. Total connections released under PMUY as on 30, Jan., 2023, 95870,19. The Union Cabinet chaired by Prime Minister Shri Narendra Modi approved the continuation of targeted subsidy of Rs. 300 per 14.2 kg cylinder (and proportionately pro-rated for 5 kg cylinder) for up to 12 refills per year to be provided to the beneficiaries of Pradhan Mantri Ujjwala Yojana (PMUY) during FY 2024-25

34. Mission Indradhanush: Launched from April 23–27, 2018.

Mission Indradhanush is a health program by the Government of India to vaccinate children and pregnant women who are unvaccinated or partially vaccinated against vaccine-preventable diseases. The program was launched on December 25, 2014 by Union Health Minister.

35. Samagra Shiksha (National Education Mission): Launched in 2018.

Objectives: The Government of India launched this scheme under the Ministry of Human Resource Development. The Department of School Education and Literacy looks after its administration.

1. The scheme was launched with a view to improving the quality of school education.
2. The Scheme represents a paradigm shift in the conceptual design of school education, by treating 'school' holistically as a continuum from pre-school, primary, upper primary, and secondary to the senior secondary levels.
3. The scheme was drafted with the broader goal of improving school effectiveness measured in terms of equal opportunities for schooling, and also equitable learning outcomes.

In June 2021, the Ministry of Education has launched the NIPUN Bharat Programme that intends to cover the learning needs of children between three and nine years. The programme is being run under the aegis of **Samagra Shiksha Abhiyan (SSA).**

36. Samagra Shiksha Abhiyan Genesis

The Samagra Shiksha Abhiyan was formed by subsuming the following three schemes to ensure an integrated and holistic school education:

1. **(RMSA)** Rashtriya Madhyamik Shiksha Abhiyan
2. **(SSA)** Sarva Shiksha Abhiyan
3. **(TE)** Teacher Education

The goal is to treat schooling as a **smooth transition** from pre-school, primary, upper primary, secondary and senior secondary level. It focuses on improving the quality of education at all levels by integrating the two T's – **Teachers and Technology.**

· **Sustainable Development Goal 4.1:** Aims to ensure that all boys and girls complete free, equitable, and quality primary and secondary education leading to relevant and effective learning outcomes.

·**Sustainable Development Goal 4.5:** Aims to eliminate gender disparities in education and ensure equal access to all levels of education.

37. PM Jan Vikas Karyakaram (PMJVK, PM People Progress Programme): Launched in 2018.

Pradhan Mantri Jan Vikas Karyakram (PMJVK) is a Scheme designed to address the development deficits of the identified Minority Concentration Areas. Prior to PMJVK, the Ministry implemented Multi-Sectoral Development Programme. The areas of implementation, under PMJVK, have been identified on the basis of minority population and socio-economic and basic amenities data of Census 2011 and will be known as Minority Concentration Areas.

PMJVK will continue to support the States/UTs in creating infrastructure to improve the quality of life of people and reduce the imbalances in the identified minority concentration areas to be at par with the rest of the country. To further strengthen the programme and ensure the benefit reaches the intended beneficiaries, project proposals will also be accepted from Central Government Departments/Organizations, Central Public Sector Enterprises/ Undertakings, Central/State Universities and Armed Police Forces, apart from the State/UT governments.

Under PMJVK, 80% of the resources would be utilized for projects related to education, health and skill development of which at least 33-40% will be earmarked for creation of assets/facilities for women/ girls.

The detailed guidelines and identified areas for implementation of PMJVK are available at Ministry of Minority Affairs.

Objective of the Scheme/ Programme

- The restructured Programme will be implemented as "Pradhan Mantri Jan Vikas Karyakram (PMJVK)" during the remaining period of 14th Finance Commission i.e. upto 2019-20. It will be implemented as a Centrally Sponsored Scheme.
- The PMJVK will continue to support the projects sanctioned under erstwhile Multi-sectoral Development Programme (MsDP) for completion of the sanctioned and ongoing projects.
- The PMJVK will continue to be implemented in Minority Concentration Blocks (MCBs), Minority Concentration Towns (MCTs) and Clusters of Minority Concentration Villages (COVs). However, certain number of population and area coverage have been specified for MCTs and COVs.
- A new area for implementation of PMJVK has been included namely Minority Concentration District Headquarters (MCD Hq).
- The PMJVK has used the selected socio-economic, basic amenities and population data of Census 2011 and have identified MCBs, MCTs and MCD Hqs. COVs will be identified by using the backwardness

parameters and population data of Census 2011 in consultation with the State government/UT administration concerned.

• The identified areas namely MCBs, MCTs, MCD Hqs and COVs for implementation of PMJVK will be known as Minority Concentration Areas (MCA).

• The PMJVK aims at improving socio-economic conditions of the minorities and providing basic amenities to them for improving quality of life of the people and reducing imbalances in the identified MCA.

Parameters for identification of Minority Concentration Areas (MCA)

The backwardness parameters are:-

Religion-specific socio-economic indicators at the district level

• Literacy rate;
• Female literacy rate
• Work participation rate
• Female work participation rate
• Basic amenities indicators at the district level
• Percentage of households with pucca walls
• Percentage of households with safe drinking water
• Percentage of households with electricity
• Percentage of households with latrine facility within the premise

Area for implementation

• Backward Minority Concentration Blocks (MCBs)
• Backward Minority Concentration Towns (MCTs)
• Backward Minority Concentration Districts Headquarters (MCD Hqrs)
• Backward Clusters of Minority Concentration Villages (CoVs)

38. Rashtriya Gram Swaraj Abhiyan (RGSA, National Village Swaraj Campaign): Launched in 2018. RGSA aims to strengthen the capacity of rural local governance institutions to achieve the Sustainable Development Goals (SDGs).

39. Pradhan Mantri Shram Yogi Maandhan Yojana Government of India in February 2019 introduced a pension scheme for unorganized workersunder named of Pradhan Mantri Shram Yogi Maandhan (PM-SYM) to ensure old age protection for unorganized sector workers.

40. Pradhan Mantri Suraksha Bima Yojana (PMSBY): Launched 9th May, 2015. This scheme provides risk coverage of Rs. 2 lakh for accidental death and full disability.

41. Pradhan Mantri Kisan Mandhan Yojana: This scheme provides an assured pension of Rs. 3,000 per month for small and marginal farmers.

42. PM CARES for Children: This scheme was launched on May 29, 2021 to support children who lost a parent or legal guardian to COVID-19.

43. The Pradhan Mantri Poshan Shakti Nirman (PM POSHAN) scheme was approved by the Government of India in September 2021. The scheme was previously known as the Mid-Day Meal Scheme and was launched in 1995 as the National Programme of Nutritional Support to Primary Education (NP-NSPE).

The scheme's goals include:

Improving nutritional levels in children

Increasing enrollment, retention, and attendance in schools

The scheme provides a hot cooked meal to children in government and government-aided schools, Special Training Centres (STC), Madarsas, and Maqtabs.

44. Pradhan Mantri Vishwakarma Scheme: A scheme that provides support to artisans and craftspeople who work with their hands and tools.

PM Vishwakarma, a Central Sector Scheme, was launched on 17th September, 2023 by the Prime Minister to provide end-to-end support to artisans and craftspeople who work with their hands and tools. The Scheme covers artisans and crafts people engaged in 18 trades, viz. Carpenter (Suthar/Badhai), Boat Maker, Armourer, Blacksmith (Lohar), Hammer and Tool Kit Maker, Locksmith, Goldsmith (Sonar), Potter (Kumhaar), Sculptor (Moortikar, stone carver), Stone breaker, Cobbler (Charmkar)/ Shoesmith/Footwear artisan, Mason (Rajmistri), Basket/Mat/Broom Maker/Coir Weaver, Doll & Toy Maker (Traditional), Barber (Naai), Garland maker (Malakaar), Washerman (Dhobi), Tailor (Darzi) and Fishing Net Maker.

Benefits to the Artisans and Crafts Persons

The Scheme envisages provisioning of the following benefits to the artisans and crafts persons:

· **Recognition**: Recognition of artisans and craftspeople through PM Vishwakarma certificate and ID card.

· **Skill Upgradation**: Basic Training of 5-7 days and Advanced Training of 15 days or more, with a stipend of Rs. 500 per day.

· **Toolkit Incentive**: A toolkit incentive of upto Rs. 15,000 in the form of e-vouchers at the beginning of Basic Skill Training.

· **Credit Support**: Collateral free 'Enterprise Development Loans' of upto Rs. 3 lakh in two tranches of Rs. 1 lakh and Rs. 2 lakh with tenures of 18 months and 30 months, respectively, at a concessional rate of interest fixed at 5%, with Government of India subvention to the extent of 8%. Beneficiaries who have completed Basic Training will be eligible to avail the first tranche of credit support of upto Rs. 1 lakh. The second loan tranche will be available to beneficiaries who have availed the 1st tranche and maintained a standard loan account and have adopted digital transactions in their business or have undergone Advanced Training.

· **Incentive for Digital Transaction**: An amount of Re. 1 per digital transaction, upto maximum 100 transactions monthly will be credited to the beneficiary's account for each digital pay-out or receipt.

· **Marketing Support**: Marketing support will be provided to the artisans and crafts people in the form of quality certification, branding, onboarding on e-commerce platforms such as GeM, advertising, publicity and other marketing activities to improve linkage to value chain.

45. National Programme for Youth and Adolescent Development (NPYAD): A scheme that was formulated by merging four 100% central sector grants-in-aid schemes.

The National Programme for Youth and Adolescent Development (NPYAD) was introduced in 2008-09. It was created by merging four central sector grants-in-aid schemes from the Ministry of Youth Affairs and Sports.

46. Promotion of Drone Technology under SMAM: A program that uses drones for pesticide and nutrient application in agriculture.

47. Scholarships for Higher Education for Young Achievers Scheme (SHREYAS): A scheme for OBC and other students. The 11th Agriculture Census of India, which began in August 2022, is the most recent agricultural census in the country. The census was launched by the Union Minister of Agriculture and Farmers Welfare on July 28, 2022.

48. Muft Bijli Yojana: PM Surya Ghar: Muft Bijli Yojana is a government scheme that aims to provide free electricity to households in India. The scheme was launched by Prime Minister Narendra Modi on February 15, 2024. Under the scheme, households will be provided with a subsidy to install solar panels on their roofs. The subsidy will cover up to 40% of the cost of the solar panels. The scheme is expected to benefit 1 crore households across India. It is estimated that the scheme will save the government Rs. 75,000 crore per year in electricity costs.

49. Pradhan Mantri Gramin Digital Saksharta Abhiyaan (PMGDISHA) is the scheme to make six crore persons in rural areas, across states/UTS, digitally literate, reaching to around 40% of rural households by covering one member from every eligible household.

The scheme would empower the citizens in rural areas by training them to operate computer or digital access devices (like tablets, smart phones etc.), send and receive e-mails, browse internet, access government services, search for information, undertake digital payment etc. and hence enable them to use the information technology and related applications especially digital payments to actively participate in the process of nation building.

Objective: Making one person in every family digitally literate is one of the integral components of the prime minister's vision of "digital india".

Beneficiaries were provided with 20 hours of training consisting of 5 modules.

(*i*) Introduction to digital devices, (*ii*) Operating digital devices, (*iii*) Introduction to the internet, (*iv*) Communications using the internet, (*v*) Application of internet (includes citizen centric services) and use of financial tools for undertaking digital cashless transactions.

Total certified canditates in all Indian states stand to 4,78,36,188 out of traine candidates 6,39,41,718 as on 26th July, 2024.

ADDITIONAL TOPICS

1. GeoMGNREGA: Ministry of Rural Development achieved a milestone by successfully geotagging one crore assets created under flagship rural employment program MGNREGA. The geotagged assets have been shared in public domain.

Each year a huge number of rural assets are created under MGNREGA. Since it started in 2006-07, more than 2.8 crore assets have been reportedly created. That makes for an annual average of about 30 lakh new assets annually which includes a variety of works such as water harvesting structures, plantations, rural infrastructure, flood control measures, individual assets for sustainable livelihood, community infrastructure and so on. The process of geotagging is going on and **all assets under MGNREGA will be geotagged.** Special focus is being given to geotag Natural Resource Management works especially the water related works.

GeoMGNREGA is a unique endeavor of the Ministry of Rural Development and Panchayati Raj in association with National Remote Sensing Centre (NRSC), ISRO and National Informatics Centre. A Memorandum of Understanding (MoU) was signed by Ministry of Rural Development with NRSC on 24th June 2016 for geo-tagging the assets created under MGNREGS in each gram panchayat. Strength of Space technology has been leveraged.

Government in 2021 launched a new geospatial planning portal Bhuvan Yuktdhara which will held in facilitating MGNAREGA assets using remote sensing and GIS system data based data. This portal will also serve as a repository of geotags created under various National Rural Development Prógrammes,, *i.e.* MGNAGERA, integrated watershed programming Per Drop More Crop and Rashtriya Krishi Vikas Yojana.

2. NITI Aayog: The **NITI Aayog (National Institution for Transforming India)** serves as the apex public policy think tank of the Government of India, and the nodal agency tasked with catalyzing economic development, and fostering cooperative federalism and moving away from bargaining federalism through the involvement of State Governments of India in the economic policy-making process using a bottom-up approach. Its initiatives include "15-year road map", "7-year vision, strategy, and action plan", AMRUT, Digital India, Atal Innovation Mission, Medical Education Reform, agriculture reforms (Model Land Leasing Law, Reforms of the Agricultural Produce Marketing Committee Act, Agricultural Marketing and Farmer Friendly Reforms Index for ranking states), Indices Measuring States' Performance in Health, Education and Water Management, Sub-Group of Chief Ministers on Rationalization of Centrally Sponsored Schemes, Sub-Group of Chief Ministers on Swachh Bharat Abhiyan, Sub-Group of Chief Ministers on Skill Development, Task Forces on Agriculture and up of Poverty, and Transforming India Lecture Series.

It was established in 2015, by the NDA government, to replace the Planning Commission which followed a top-down model. The NITI Aayog council comprises all the state Chief Ministers, along with the Chief Ministers of Delhi and Puducherry, Lieutenant Governors of all UTs, and a vice-chairman nominated by the Prime Minister. In addition, temporary members are selected from leading universities and research institutions. These members include a chief executive officer, four ex-officio members, and three part-time members.

* **National Nutritional Strategy drafted by NITI Ayog aimed at Kuposhan Mukt Bharat.** Scheme for children, adolescent Girls and Women upto the year 2022. It was aimed to decrease kuposhan by 1 to 3% per year. As on 14th Oct., 2021. Total Malnutrition children are about 33,23,322.
* On the occasion of International Women's Day, the Prime Minister, Shri Narendra Modi launched the **National Nutrition Mission** and Pan India Expansion of **Beti Bachao Beti Padhao (BBBP)** covering all 640 districts of the Country at a mega event held at Jhunjhunu in Rajasthan on 8th March, 2018. National Nutrition Mission was launched with the budget of Rs. 9046 crore.

*** SATH (Sustainable Action for Transforming Human Capital) Launched by NITI Ayog (Launched in 2017)**

Furthering the agenda for cooperative federalism, NITI Aayog has launched SATH, a program providing 'Sustainable Action for Transforming Human capital' with the State Governments. The vision of the program is to initiate transformation in the education and health sectors. The program addresses the need expressed by many states for technical support from NITI Ayog.

SATH aims to identify and build three future 'role model' states for health systems. NITI will work in close collaboration with their state machinery to design a robust roadmap of intervention, develop a program governance structure, set up monitoring and tracking mechanisms, hand-hold state institutions through the execution stage and provide support on a range of institutional measures to achieve the end objectives. The program will be implemented by NITI along with McKinsey & Company and IPE Global consortium, who were selected through a competitive bidding process.

Three states selected by NITI Aayog for SATH are: 1. MP, 2. Jharkhand and 3. Odisha.

* SATH-E 2.0, was commenced by NITI Aayog for two years, from October 2020 for three states : 1. U.P., 2. Assam, 3. Karnataka

* **Odisha** Govt. has identified around 8000 schools in 15 districts of the state which will be merged with other schools due to low enrolment (less than 20 student). The merger is being carried out under the SATH (Oct.2020). First phase was completed in March, 2021.

NITI AYOG "s initiatives in 2018/ 2019/2020/2021/2022

(*i*) Measuring performance and ranking States on outcomes in critical sectors

(*ii*) Sustainable Action for Transforming Human Capital (SATH)

(*iii*) Ek Bharat Shrestha Bharat

(*iv*) Development Support Services to States (DSSS) for Development of Infrastructure

(*v*) Public-Private Partnership in Health

(*vi*) Resolution of pending issues of States with Central Ministries

(*vii*) State Human Development Report

(*viii*) Transforming of 115 identified Aspirational Districts

(*ix*) Shoonya Podcast

(*x*) Indian Innovation Index 2021.

(*xi*) Promoting Millets in diets.

3. National Green Tribunal: The National Green Tribunal has been established on 18.10.2010 under the National Green Tribunal Act 2010 for effective and expeditious disposal of cases relating to environmental protection and conservation of forests and other natural resources including enforcement of any legal right relating to environment and giving relief and compensation for damages to persons and property and for matters connected therewith or incidental thereto. It is a specialized body equipped with the necessary expertise to handle environmental disputes involving multi-disciplinary issues. Current Chairman Justice Prakash Shrivastava appointed on 21st Aug. 2021. Goel (Retired Judge) Since July, 09,2018.

4. Zero Budget Natural Farming (ZBBF) means raising crops without using any fertilizers and pesticides or any other external materials. The word Zero Budget refers to the zero cost of production of all crops.

ZBNF guides the farmers towards sustainable farming practices thus helps in retaining soil fertility, to ensure a chemical free agriculture and ensure low cost of production(zero cost) and thereby enhancing the farmers income. In short, ZBN, is a farming method that believes in growing crops in tune with nature.

The concept was promoted by an agriculturist, Padmashri. Subhash Palekar, in the mid-1990s as an alternative to the Green Revolution's methods driven by chemical fertilizers and pesticides and intensive irrigation.

Principles of ZBNF: No external inputs.

Local seeds (use of local varieties) Microbial Seed treatment.

Microbial inoculants for soil health

Cover crops for biomass mulching and bio-mass incorporation for creating a suitable micro-climate for maximum beneficial microbial activity

Mixed cropping.

Integration of trees into the farm.

Integration of livestock, especially of native breed for cow dung and cow urine as essential inputs for several practices.

Water and moisture conservation.

Four main Elements and Model of ZBNF

(*i*) **Bijamrita** -Seed Treatment using local cow-dung and cow urine The seeds are treated with formulations prepared using cow dung and cow urine from native cow species as they have higher adaptability to our local climatic conditions and easy to maintain by the small and marginal farmers. While Bijamrita is used for seed treatment, extracts of neem leaves and pulp, tobacco and green chillies are used for insect and pest management.

Benefits: The seeds sown in the field may be affected by fungus and other seed born/soil borne diseases. The seed treatment using "Bijamrita "protects the seeds from diseases.

(*ii*) Jiwamrita/Jeevamrutha: Cow dung is a natural source used to revive the fertility and nutrient value of soil. One gram of cow dung may have between 300 to 500 crore beneficial micro-organisms. These microorganisms assists in decomposing the biomass present in the soil and convert it into available/ready-to-use nutrients for crop. Jiwamrita is prepared using Cow dung and cow urine. It is used as an input for the plants. It is a fermented microbial culture obtained from cow dung, urine, jaggery, pulse flour and uncontaminated soil. This fermented microbial culture when applied to soil, adds nutrients to the soil besides acting as a catalytic agent to promote the activity of microorganisms and earthworms in the soil. About 500 litres of jeevamrutha should be sprayed twice a month per hectare of land; after three years, the system is supposed to become self-sustaining. One indigenous cow is sufficient for 30 acres of land. Benefits: This culture will be applied in soil instead of chemical fertilizers. This culture stimulates microbial activity in the soil and enhances nutrient availability for the plants, protects the crops against soil pathogens and increases carbon content of the soil.

(*iii*) Mulching: Mulching is the process of covering the top soil with crop wastes/organic waste or with cover crops Benefits: Mulching materials decomposes and produces humus which conserves top soil, increases water retention capacity of the soil, decreases evaporation loss, encourages soil fauna besides enriching soil nutrient status and controlling weed growth.

(*iv*) Waaphasa (Soil Aeration). Good aeration is required in the soil for plant growth and development. ZBNF concept opposes the use of vermi compost as it introduces the most common composting worm the European Red Wiggler to Indian soil, which absorbs toxic metals and poisons ground water and soil. Benefits:

Due to the application of Jiwamrita and mulching, the aeration of the soil increases, thus improves humus content, water availability, water holding capacity and soil structure which is most suitable for crop growth especially during drought periods .

Government Initiatives to support ZBNF:

Government of India has been encouraging organic farming through various schemes like Paramparagat Krishi Vikas Yojana (PKVY) since 2015-16 and also through Rashtriya Krishi Vikas Yojana (RKVY). In the revised guidelines of PKVY scheme during the year 2018, various organic farming models like Natural Farming, Rishi Farming, Vedic Farming, Cow Farming, Homa Farming, Zero Budget Natural Farming (ZBNF) etc. have been included and flexibility is given to state governments to adopt any model of Organic Farming including ZBNF depending on farmer's choice. In June 2018, Andhra Pradesh Government rolled out an plan to become India's first state to practice 100 % natural farming by 2024 by converting state's 60 lakh farmers to ZBNF method.

5. National Livestock Mission: The National Livestock Mission (NLM) has commenced from 2014-15. The Mission is designed to cover all the activities required to ensure quantitative and qualitative improvement in livestock production systems and capacity building of all stakeholders. The Mission will cover everything germane to improvement of livestock productivity and support projects and initiatives required for that purpose subject. This Mission is formulated with the objective of sustainable development of livestock sector, focusing on improving availability of quality feed and fodder. NLM is implemented in all States including Sikkim.

NLM has 4 submissions as follows:

- **The Sub-Mission on Fodder and Feed Development** will address the problems of scarcity of animal feed resources, in order to give a push to the livestock sector making it a competitive enterprise for India, and also to harness its export potential. The major objective is to reduce the deficit to nil.
- **Under Sub-Mission on Livestock Development,** there are provisions for productivity enhancement, entrepreneurship development and employment generation (bankable projects), strengthening of infrastructure of state farms with respect to

modernization, automation and biosecurity, conservation of threatened breeds, minor livestock development, rural slaughter houses, fallen animals and livestock insurance.

- **Sub-Mission on Pig Development in North-Eastern Region:** There has been persistent demand from the North Eastern States seeking support for all round development of piggery in the region. For the first time, under NLM a Sub-Mission on Pig Development in North-Eastern Region is provided wherein Government of India would support the State Piggery Farms, and importation of germplasm so that eventually the masses get the benefit as it is linked to livelihood and contributes in providing protein-rich food in 8 States of the NER.
- **Sub-Mission on Skill Development, Technology Transfer and Extension:** The extension machinery at field level for livestock activities is very weak. As a result, farmers are not able to adopt the technologies developed by research institutions. The emergence of new technologies and practices require linkages between stakeholders and this sub-mission will enable a wider outreach to the farmers.All the States, including NER States may avail the benefits of the multiple components and the flexibility of choosing them under NLM for a sustainable livestock development.

6. Mission Organic Value Chain Development for North Eastern Region (MOVCD-NER). This scheme of GOI from 2015 to 2018 was started on 11th January 2016. So far 2321 farmers groups and 8 farmers produce organizations have been made under this programmes. It is a submission under National Mission for Sustainable Agriculture. This scheme is being implemented in all 8 states of NE India Phase -IV (2023-24).

7. Scheme of Fund for Upgradation and Regeneration of Traditional Industries' (SFURTI): Common Facility Centres (CFCs) to be setup under SFURTI.CFCs will facilitate cluster based development for making traditional industries more profitable, productive and capable for generating sustained employment clusters. During 2019-20, 100 new clusters to be setup with special focus on Honey, Bamboo and Khadi. This will enable 50,000 artisans to join economic value chain.

8. A Scheme for Promotion of Innovation. Rural Industry and Entrepreneurship' (ASPIRE): It was launched to set up a network of technology centres and in.agro based startups industry. It is now to be consolidated with setting up 80 **Livelihood Business Incubators** (LBIs) and 20 Technology Business Incubators (TBIs) during 2019-2020 and also 75,000 entrepreneurs to be skilled in agro-rural industry sectors.

9. Zero Budget Farming in which presently few states' farmers are already being trained to be replicated in other states. India's water security: Newly formed Jal Shakti Ministry to look at management of country's water resources and water supply in an integrated and holistic manner. Jal Jeevan Mission to achieve Har Ghar Jal (piped water supply) to all rural households by 2024. Under JJM, objective is to integrate demand and supply side management of water at local level.

For the Jal Shakti Abhiyan, 1592 critical and over exploited Blocks spread across 256 District being identified.

Additional funds available under the Compensatory Afforestation Fund Management and Planning Authority (CAMPA) fund can be used for this purpose.

10. Bharat Net Project: The Bharat Net project is the world's largest rural broadband connectivity program and also a flagship mission implemented by **Bharat Broadband Network Ltd. (BBNL)** which aims to connect all Gram Panchayats (GPs) in India with Optical Fibre Cable (OFC) connectivity. The project's objectives include delivering e-governance, e-health, e-education, e-banking, and other services to rural India

National Optical Fibre Network (NOFN) was launched in October 2011 and was renamed as **Bharat Net Project in 2015.**

BBNL is a Special Purpose Vehicle (SPV) set up by the Government of India under the **Companies Act, 1956.**

It is being implemented by the Department of Telecommunication under the Ministry of Communications.

This project involves altering the **execution strategy and employing Village Level Entrepreneurs (Udyamis)** to provide fiber connections to the last mile, thereby **accelerating the connectivity process** over the next 2.5 years.

oIt is financed by the **Universal Service Obligation Fund (USOF).** It was created under the Ministry of Communications in 2002.

It aims to **connect all 640,000 villages** across India with high-speed internet access.

It seeks to bring broadband internet connectivity to each of the more than 2.5 lakh gram panchayats across the country.

The government intends to provide a minimum of **100 Mbps bandwidth at each Gram Panchayat through BharatNet** so that everyone, especially those in rural India, can access online services.

oUnder this approach, the **government will bear the cost of extending the infrastructure** to homes, while the entrepreneurs **will contribute to the maintenance** and operation of home connections.

·This partnership will work on **a 50:50 revenue-sharing basis.**

Progress and Milestones of BharatNet Project

Phase I: Completed by December 2017, this phase connected 100,000 village councils covering 300,000 villages.

Phase II: Completed by March 31, 2023, this phase connected the remaining 150,000 village councils covering 325,000 villages in 16 states.

Amended BharatNet Program

The Union Cabinet approved the Amended BharaNet Program on August 4, 2023, to strengthen the network and connect the remaining GPs. The construction phase for the network under this program is three years from the award of the project.

Here are some milestones and progress of the BharatNet project as of 2024:

Gram Panchayats connected

As of June 30, 2024, 82.1% of the 2.6 lakh Gram Panchayats (GPs) in India have been connected to the BharatNet project, or 2,13,398 GPs.

Internet subscribers: Out of 104293 Wi-Fi hotspot installed under BharatNet, 53600 Wi-Fi hotspot are active at present. The work of providing broadband facility (Wi-Fi hotspots and Fibre to the Home connections) in Phase-I GPs, is implemented with funding from Universal Service Obligation Fund (USOF). For BharatNet Phase-II, the respective Implementing Agencies are responsible for the utilisation of the created network. Regular review & monitoring of the project is done with the Implementing Agencies for expeditiously making the Wi-Fi hotspots active/operational.

Connectivity issues: Some areas have experienced poor service quality and interrupted internet access due to connectivity issues. These issues are attributed to challenging terrain, lack of proper roads, and logistical difficulties.

Technical challenges: The BharatNet project has also faced technical challenges such as signal quality, bandwidth limitations, and network downtime

Competition from private operators: Private telecom operators like Jio and Airtel have established their own network infrastructure and services in some rural areas.

On 30.06.2021, Union Cabinet has accorded in principle approval for extending BharatNet to cover all inhabited villages in the States and UTs. The revised strategy includes creation, upgradation, operation, maintenance and utilization of the BharatNet. The asset created under BharatNet can be utilised to provide Broadband Services such as Wi-Fi Hotspots, Fibre to the Home (FTTH) connections, leased lines, Dark fibre, backhaul to mobile towers etc. Timeline of the BharatNet Project is upto 2025.

All Telecom Service Providers (TSPs) are encouraged to increase fiberisation of Towers/ Base Transceiver Station (BTSs) by utilising the existing fibres laid by PSUs, fibre laid under BharatNet and fibre sharing amongst TSPs.

Optical Fibre Cable Network of approximately 10 Lakh Route Kms laid by PSUs [Bharat Sanchar Nigam Limited (BSNL), Bharat Broadband Nigam limited (BBNL), RailTel, Mahanagar Telephone Nigam Limited (MTNL), GAIL (India) Limited] and approximately 5.73 lakh telecom towers installed by Telecom Service providers (TSPs) have been mapped on the PM GatiShakti National Master Plan Platform.

MOBILE APPLICATIONS FOR AGRICULTURE AND RURAL DEVELOPMENT

One of the areas with the greatest potential impact is in the contribution that mobile applications can make to agricultural and rural development (ARD), by providing access to information, markets, and services to millions of rural inhabitants. For both agricultural supply and demand, mobile phones can reduce waste, make delivery more efficient, and forge closer links between farmers and consumers. M-apps are software designed to take advantage of mobile technology and can be developed for technology besides mobile phones. M-apps are markedly different in developing countries because they typically run on second-generation (2G) phones rather than smartphones, which are far more common in developed countries. One of the main findings is that an enabling platform (or platforms) is probably the most important factor for the development of m-ARD apps. Platforms can facilitate interactions among ecosystem players, increase access to users, provide technical standards, and incorporate payment mechanisms. Agriculture app development

- Agri app
- Kisan Suvidha
- AgriCentral
- Cattle management app
- Crop insurance
- Fieldcheck App
- GPS tracking
- IFFCO Kisan agriculture
- Pusa Krishi
- Smart Farm Mobile

Some Abbreviations

2G : Second generation (mobile communications)

3G : Third generation (mobile communications)

4G : Fourth generation (mobile communications)

5G : Fifth generation (mobile communications)

API : Application Programming Interface

B2B : Business to Business

CSR : Corporate Social Responsibilit

DFID: Department for International Development

EDGE: Enhanced Data rates for GSM Evolution

FMCG: Fast Moving Consumer Goods
GGS : Govi Gnana Seva
ICT: Information and Communication Technology
ICTA: Information and Communication Technology Agency
IFC: International Finance Corporation
IKSL : IFFCO Kisan Sanchar Ltd
LTE : 3GPP Long Term Evolution
m-apps Mobile applications
m-ARD apps: Mobile Applications for Agriculture and Rural Development
m-money: Mobile Money
NGO: Nongovernmental Organization
NHS: National Healthcare System
OECD: Organisation for Economic Co-operation and Development
OLT: Ovi Life Tools
PE/VC: Private Equity/Venture Capital
PPP: Private-Public Partnership
PWC: Pricewaterhouse Coopers
RML: Reuters Market Light
SME: Small and medium-size Enterprise
SMS: hort Message Service
SWORB : Strengths, Weakness, Opportunities, Risks, Benefits
TOR: Terms of Reference

MOBILE APPS FOR FARMERS

1. PUSA Krishi launched at Krishi Unnati MELA 2016. The App will provide with information about weather conditions to make farmer to take measures to save the crop and also give information related to new varieties of crops developed by ICAR, resource conserving cultivation practices as well as farm machinery and its implementation will help increasing returns to farmers, With the use of mobile app farmer can get easy solution to their problems. The Agricultural Ministry has developed this App and described it as an omnibus for quick and relevant information. But farmers must have smart phones to use this App. At present information is available in

Hindi and English only, but Ministry is looking forward to add more languages to this APP.

2. e-NAM- National Agricultural Marketing (Portal) Founded 2016. This networks the existing APMC mandis to create unified national markets for farmers. Agri Market Mobile: Farmers can use this app to get the information about the market price of crops in nearby mandi within 50 km radius and also the mandi in the country. Allowing him to make right decision to sell his produce. It has connected 585 mandis and 17 states already fully or partially amended their Agri. marketing acts to allow for electronic trading. New 25 commodities are added to the NAM on 29-08-2020 for the trading. Now a total of 175 commodities are available for the trade. Currently there are 1000 markets that are linked to the e-NAM network from 18 states and 3 UT's.

Benefits of e-NAM

- Real time information on trade and price
- Transparency in trade through better price discovery
- Extended access to markets across the state
- Information on quality of commodities
- Transparent e-bidding process
- Direct online payments

Tamil Nadu Dept. of Agriculture marketing & Agri-Business, has amended TN-APMC Act, wherein now all traders of e-NAM States/ UT having Unified license, can trade at TN e-NAM mandis to all crops.

e-NAM Platform of Platforms (PoP): Launched on 14th July, 2022

New Commodities added on e-NAM: Total 209 Commodities : Food Grains/Cereals - Little Millet, Kodo Millet, Barnyard Millet, Browntop Millet, Proso Millet, Amaranth Seed

3. IFFCO Kisan Agriculture App, launched in **2015** and managed by IFFCO Kisan is a subsidiary of Indian Farmers' Fertilizer Cooperative Ltd. Its aim is to help Indian farmers make informed decisions through customized information related to their needs. Moreover, the user can access a variety of informative modules including agricultural advisory, weather, market prices, and agriculture information library in the form of text, imagery, audio, and videos in the selected language at the profiling stage. The app also offers helpline numbers to get in touch with Kisan Call Centre Services.

4. Krishi Jagran App: This recently launched app, Krishi Jagran, provides trending agri news, cultivation guides, crop calendar, complete information on crop protection, pest, and disease management, subsidies, a career in agriculture, and all the information regarding farm mechanization as well. Through this platform, farmers can get all crop-related information at the tip of their fingers. This easily accessible agri platform connects farmers to the world and provides them with all relevant agriculture-allied information. Click here to download

5. Agri App: Agri App is a total farmer-friendly app that provides **complete information on Crop Production, Crop Protection, and all relevant agriculture-allied services**. It also enables farmers to access all the information related to **"High value, low product"** category crops from varieties, soil/ climate, harvesting, and storage procedures. Moreover, an option to chat with experts, video-based learning, the latest news, and online markets for fertilizers, insecticides, etc. are also available on this app.

6. Bijak App : Bijak is an online agriculture trading platform that connects verified agri traders (farmers, loaders, commission agents) across India. It ensures suppliers and buyers (commission agent/ adati) receive end-to-end services like advance payment guarantee, limit facility, secured payments, mandi rates etc.

7. e-Pashuhaat portal launched by Union Agriculture Minister Sh. Radha Mohan on the occasion of National Milk Day 26th Nov. 2016. Portal has been developed for connecting breeders and farmers regarding availability of bovine germ -plasm. Through the portal breeders and farmers can sell/ purchase breeding stock. Information on all forms of germ-plasm including , semen, embryos and living animals with all agencies and stake holders in the country has been uploaded on the portal for the benefit of the farmers / breeders to get information about the availability of quality and disease free bovine germ plasm. The portal will lead to propagation of high genetic merit germ-plasm.

8. BHIM (Bharat Interface for Money) app. Developed by National Payments Corporation of India (NPC) BHIM gives an easy method of payments and fund transferring. User can transfer funds to any bank account and need not to register the payee before-hand.

9, KISAN Crop Insurance using space Technology and geo-informatics to fasten payment of crop insurance claims to farmers. The project envisages use of space technology and geo-informatics (GIS, GPS and smart phones) technology along with high resolution data from satellite UAV/Drone based imaging and other geospatial technology for improvement in yield estimation and better planning of crop cutting experiments (CCES), needed for crop insurance programme. There has been always a problem in getting timely and accurate data, due to which payment of claims to farmers was getting delayed. KISAN is launched to address this issue by providing timely data on crop yields.

This project shall be implemented by Mahalanobis National Crop Forecast Centre (MNCFC) an attached office of Department of Agriculture, Co-operation and FW in collaboration with ISRO Centres, IMO (India Meterological departments State agricultural departments and state remote sensing centres.

10. Pashu Poshan Mobile App: National Dairy Development Board (NDDB) launched a mobile application that will recommend a balanced diet for cows and buffaloes to help boost dairy farmers' income by raising milk yield and cutting feed cost.

The mobile application, named 'Pashu Poshan', was launched by the Union Agriculture

11. AgriCentral: Purpose: AgriCentral offers a comprehensive platform for farmers to get real-time updates on crop care, weather conditions, and market prices.

Key Features: Crop Care: Tailored recommendations to enhance crop protection and management.

Market View: Latest information on market prices for crops.

Weather Forecasts: Hyperlocal weather updates.

Community Support: Farmers can engage with fellow farmers and experts to share knowledge and tips.

Benefits: AgriCentral helps farmers improve their productivity by providing actionable insights and relevant data.

12. DeHaat: Purpose: DeHaat is a comprehensive agricultural platform that provides end-to-end solutions and services to farmers. It covers all aspects from providing seeds, fertilizers, and equipment to market linkages and advisory services.

Key Features: • Access to quality inputs like seeds and fertilizers

- Crop advisory services
- Market linkages for better price realization

Benefits: Assists farmers in enhancing their productivity and profitability through access to high-quality inputs and expert advice, while also providing a direct link to markets.

13. Govi Gnana Seva (GGS) is a not-for-profit organization that collects and distributes information on agricultural produce prices and is also known as the Farmer Intelligence Service. It was started by researchers from LIRNEasia, an information and communication technology policy and regulation think-tank in the Asia Pacific region.

14. Swachhata App: The Swachhata application is a fourth generation complaint redressal mobile and web platform. It is a quantum leap in how complaints and grievances are being redressed by Municipal Corporations in India. This solution is for all the 4041 towns and cities of India.

The core of the Swachhata application is to use citizen participation and civic engagement to help resolve the Swachh Bharat complaints. It is an initiative of the Swachh Bharat Mission in association with the Ministry of Housing and Urban Affairs.

Benefits to citizens

- The app can pinpoint the location of the complaint with accuracy using the geo-location of the picture, which will lead to faster resolution of the complaint.
- Regular complaint status updates - Citizens will get regular updates on the status of the complaint in the form of a push notification.
- Push notification when resolved - Citizens will get a push notification with the 'Resolved' image uploaded by the sanitary inspector or engineer when they change the status to Resolved.
- Reopen unresolved complaint - Citizens can reopen the complaint if they are not satisfied with the resolution.
- Users can also vote up on complaints in their locality or complaints uploaded by friends and neighbours.

The key to the working of this mobile application is large scale citizen participation. Sanitation problems in urban India cannot be left for the government alone to resolve. This is where the application comes in, with its inputs from citizens to monitor and resolve issues in real time. The app amplifies citizen voices and helps build communities that can stay engaged during the course of the Swachh Bharat movement.

The system consists of the following modules.

- **Citizen application-** A mobile application on Android and iOS for citizens to upload Swachh Bharat complaints.
- **The Engineer Application :** A mobile application in Android for the sanitary engineer/inspector to see the complaints uploaded by the citizen and take action on the ground.
- **Municipal Administrator Website :** A website for Municipal Commissioners and their digital teams to monitor the complaints received in their cities and towns.
- **Ranking & Rating Website** : This is a website open to all citizens, it will provide a rating and ranking for all the cities on the Swachhata platform based on the resolution rate. It will provide a break down of the performance of each ward in the city.

Once a citizen posts a complaint using the citizen app, Sanitary inspectors look into the issue and get it resolved. A photo of the resolved status is also uploaded for the citizen to check and provide feedback. Action on the complaints may vary depending on the categories. Action for most complaints will be initiated from 6 hours of registration.The categories and the service level agreement is as follows.

- Dead animals - 48 hours
- Dustbins not cleared - 12 hours
- Garbage dumps - 12 hours
- Garbage vehicle not arrived - 12 hours
- Sweeping not done - 12 hours
- No electricity in public toilets - 12 hours
- No water in public toilets - 12 hours
- Toilets not cleaned / Blockage - 12 hours

15. AgroStar: Kisan Agridoctor App

Purpose: AgroStar provides farmers with personalized advice and a platform to purchase high-quality agricultural inputs such as seeds and fertilizers.

Key Features: • *Personalized Agronomy Advice*: Expert guidance on farming based on the farmer's crop and location. • *Agricultural Inputs*: Access to a wide range of seeds, fertilizers, and other farm essentials.

• *Farmer Community:* A forum where farmers can connect with each other and exchange knowledge and experiences.

ANDROID APPS DEVELOPED BY ICAR-CCARI
(Central Coastal Agricultural Research Institute)

1. **Fertilizer Calculator Goa App:** The Fertilizer Calculator Goa App is completely offline app for making soil test based fertilizer recommendations to important crops of Goa. The recommendations are given on the basis of the crop type, area or number of trees/plant, soil testing report values and desired fertilizer grades. It works for both macro and micronutrients. The results would help to use the fertilizer in appropriate amount and in a balanced way. **(Prepared by – Dr. Gopal Mahajan and Mr. Vishwajeet Prajapati and Designed by – Mr. Aditya Naik)**
2. **The Fertilizer Calculator App** is completely offline app for making soil test based fertilizer recommendations. The recommendations can be obtained by feeding inputs as area or number of trees, recommended doses of fertilizers for given area, soil testing report (optional) and desired fertilizer grades. It works for both macro and micronutrients. The results would help to use the fertilizer in appropriate amount and in a balanced way. **(Prepared by – Dr. Gopal Mahajan and Mr. Vishwajeet Prajapati and Designed by – Mr. Aditya Naik)**
3. **Nutrient to Fertilizer Calculator app:** The n2f Calculator App (nutrient to fertilizer calculator app) is completely offline app for calculating amount of fertilizers for a given amount of nutrients. The calculations can be obtained by feeding the amount of nutrients and by selecting the desired fertilizer grade. It works for both macro and micronutrients. **(Prepared by – Dr. Gopal Mahajan and Mr. Vishwajeet Prajapati and Designed by – Mr. Aditya Naik)**
4. **Plant Population calculator App :** The Plant Population calculator App can be used for calculation of the number of plant that can be accommodated for the given area (square meters) and desired spacing (m). The results could be used to decide exact and precise amount of inputs for crop cultivation.

It would indirectly reduce the cost of cultivation, if used properly. **(Prepared by – Dr. Gopal Mahajan and Mr. Vishwajeet Prajapati and Designed by – Mr. Aditya Naik)**

5. **Seed Calculator App :** The Seed Calculator App can be used for calculation of exact amount of seed that is required for a given area (ha or acre) and recommended seed rate (g/ha, kg/ha, g/acre, kg/acre). This could be useful to avoid extra expenditure on seed by knowing exact amount of seed required. **(Prepared by – Dr. Gopal Mahajan and Mr. Vishwajeet Prajapati and Designed by – Mr. Aditya Naik)**
6. **Pesticide Calculator App :** The Pesticide calculator App can be used for calculation exact amount of pesticide that is required for a given recommended dose (g/L, mL/L) and active ingredient (%). This could be useful to avoid the wastage of the pesticides by knowing exact amount of pesticide required. **(Prepared by – Dr. Gopal Mahajan, Dr. R. Maruthdurai and Mr. Vishwajeet Prajapati and Designed by – Mr. Aditya Naik)**

Mobile Apps at Indian Institute of Oilseeds Research (IIOR)

1. Mobile App "ICAR IIOR"
2. ICAR-IIOR CASTOR
3. ICAR IIOR SUNFLOWER
4. ICAR IIOR SAFFLOWER
5. ICAR IIOR SESAME
6. ICAR IIOR BIOCONTROL
7. E-THILHAN
8. ICAR IIOR Seed Production Castor
9. ICAR IIOR Seed Production Safflower
10. ICAR IIOR Seed Production Sunflower
11. ICAR IIOR Sesame Seed Production

AGRICULTURAL CURRENT AFFAIRS 2016-2024

* On April 24th 2017 GOI launched a **New Model Mandi Act. 2017.** This act is expected to improve agricultural marketing system with benefits to the farmers to get higher value of their produce.

* **Bureau of Indian Standards (BIS) Act 2016** implemented replacing BIS 1986. (It aims to help in ease of doing business in the country enhancing **Make In India** campaign and ensure availability of quality products to consumers).

* **Samproon Bima Gram Yojana** (launched by GOI in 2017 with expansion of PLI: Postal Life Insurance). A scheme to provide affordable life insurance to people particularly living in rural India. With expansion of PLI will now also cover professionals apart from govt. employees. At least one village (having a minimum of 100 households) will be identified in each of the revenue districts of the country.

* **Panjabrao Deshmukh Award** for women scientist was bagged by Dr. Archana Mukherjee Director of CTCRI Thiruvananthapuram. (Foundation year of award 1995). Award is given annually.

  Panjabrao Deshmukh woman scientist award for **2017** shared by **Shelly Praveen** of Biochemistry Division IARI, New Delhi and **Seema Jaggi** of Agricultural Statistics Research Institute, New Delhi. The other award is shared by **Neeru Bhushan** of Zonal Technology Management and Business Planning, IARI and **PS Vimla Devi** of Institute of Oilseeds Research, Hyderabad. This award consists of Rs. 1 lakh in Cash and Certificate and Sponsored by ICAR. Dr. Gurinderjit Randhawa. Head, NBPGR receives 2020 award. For 2021, **Dr. (Ms) Kambham Madhavi Reddy** (ICAR-IIHR) Bengaluru was awarded.

* **Genetically modified variety of cotton developed by Punjab Agricultural University,** LUDHIANA. The seeds of crop could be reused by farmer's to save repeated cost. PAU **Bt-1** and F1861. (March, 2017). Now in 2020 BT-II is introduced which is improved version of BT-1 and is resistant to pests such as american bollworm, pink bollworm and tobacco caterpillar.

* **Mission Fingerling** (2017) with the total expenditure of about Rs.52 lakh in association with the **blue revolution** launched to

recognize the potential and possibilities in fisheries sector. Fish fingerling production is the single most important critical input visualised to launched in 2017 to achieve goals mentioned in National Action Plan 2020. Fish seed of fingerling size is used to be stocked in natural water and water bodies. Fingerling is a small fish expecially upto 1 year of age.

* New Variety of Pomegranate- **Solapur Lal** released by Dr. Roshan Kishan Lal of Purlia (West Bengal) at National Research Center on Pomegranate, Pune.

This variety contains:

1. Iron content – 5.6 to 6.1 mg/100 gm.
2. Zinc content – 0.64 to 0.69 mg /100 gm.
3. Vitamin C – 19.4 to 19.8 mg/100 gm.
4. Fruit yield – 23-27 tones/hectare
5. Adaptation semi-arid regions of the country

* **Indian Seed Congress:** With theme **"Seed of Joy"** held on February 2017 at Kolkata. (Annual Flagship event) of NSAI)

  10th edition of Indian Seed Congress 2020 held at Aerocity, New Delhi (15-17 Feb., 2020). It is a annual event. 13th edition will be organized in New Delhi in 2023-25 Feb., 2025.

* **World Food Programme (WFP) Founded in 1961.** Country Sweeden signed strategic partnership agreement with UN and committed $370 million dollars to UN world food programme for the next four years (2018-2021). Biggest ever contribution made by the donor to World Food Programme.

  WFP is a food assistance branch of UN and established in 1961 and aims to removing hunger and promoting food security. It works to help people who cannot produce or obtain enough food for themselves and their families. It provides food assistance to an average 80 million people in 125 countries each year. 160 million people received assistance and 20 million children received nutrition meals schools, snackes and takes home rations from WFP in 2022. (Awarded Nobel peace prize in 2020).

The **Jatayu Nature Park** housing the world's largest bird sculpture is located in the state of Kerala. Park sprawls over 65 acres of land and is home to giant sculpture of Jatayu. (Nearest city Kollam (Quilon) 38 km.

Rythu Bandhu scheme also Farmers' Investment Support Scheme (FISS) (2018) is a welfare program to support farmer's investment for two crops a year by the **Government of Telangana.** The government is providing 58.33 lakh farmers, Rs. 5000 per acre per season to support the farm investment, twice a year, for rabi and kharif seasons. This is a first direct farmer investment support scheme in India, where the cash is paid directly.

About 15 Lakhs farmers get benefit under the scheme\ (Who have maximum 2 acre land)

India's first corn festival held in state of MP.

Chhindwara (M.P.) on 29th and 30th September, 2018.

COMPRESSED BIOGAS

Union Minister of Petroleum and Natural Gas launched an innovation initiatives title SATAT (Sustainable Alternative Towards Affordable Transportation) in Oct., 2018

The initiative would benefit vehicle users and farmers both. It has been launched by PSU oil marketing companies (MCS) to setup compressed biogas (CBG) production plant and make available CBG fuel in the market for use as automotive fuel.

CBG is the purified form of biogas, without other gaseous impurities. It is very similar to natural gas in its composition and energy potential. It is a non-renewable energy source eco-friendly and can replace LPG and CNG easily. In CBG, biogas is cleaned of hydrogen and carbondioxide to produce 95% methane gas. This pure gas is compressed and bottled for transportation and other usage India plans to start 5000 CBG plant in 4 year time. CBG has calorific value of (5200 kJ/kg) and other properties are similar to CNG. This initiative will give better use of agricultural residue, cattle dung and municipal solid waste as well as provide additional income source to farmers. India have 46 BCG plants with an overall capacity of 272 tonnes/day as on 2nd Nov., 2023.

- **Cabinet Approves DNA Technology (Use and application) regulation Bill 2019 (Introduced in Loksabha on July 08,2019)**

  It allows enforcement agencies to collect DNA samples, create "DNA profiles" and special databanks for forensic-criminal investigations. It creates DNA Profiling Board (DPB) that will be final authority and authorise creation of state-level DNA databanks. **Approve the methods of collection and analysis of**

DNA technologies. It will make accreditation and regulation mandatory for DNA laboratories.

- **The National Board for Wildlife (NBWL)** recently added four species (July, 2018)

1. *Northern River Terrapin*: It is riverine turtle found in rivers that flow in Eastern India. It is native of Bangladesh, Cambodia, India, Indonesia and Malaysia.
2. *Clouded Leopard*: It is found in Himalayan foot hills.
3. *Arabian Sea Humpback Whale*: It is a species found in all of the major oceans and migrates from the Oman coast to Arabian sea, along the Indian coasts till the Sri Lankan coast.
4. *Red Panda*: It is closely associated with mountain forests with dense bamboo thicket. It is found in Sikkim, West Bengal and Arunachal Pradesh.

- **Queen variety of Pineapple** of Tripura state declared as state Fruit of Tripura (Tripura is largest pineapple glowing state in country.
- Queen pineapple is spiny, golden yellow in colour. Have pleasant aroma and flavour. It has received GI tag in 2015. It is being exported to West Asian countries.

- **GOI Ministry of New and Renewable Energy has declared target for off-shore wind power medium** to 20 GW by 2026 and 30 GW by 2030. This will add new element to (long term) already existing basket of renewable energy of the country. (India has 195 GW of offshore wind potential)

- **WEP Women's Entrepreneurs platform** was launched by NITI Ayog on the occasion of Int. Women Day on March 8, 2018. The initiative is aimed building an ecosystem for women across india to realize their entrepreneurial aspiration scale-up innovative initiatives and chalk out sustainable long term strategies for their businesses. The initiative is aimed building an ecosystem for women across india to realize their entrepreneurial aspiration scale-up innovative initiatives and chalk out sustainable long term strategies for their businesses. Launched in 8th Global Entrepreneurship Summit (GES) in Hyderabad in 2017.

- Sikkim's *khangchendzonga biosphere reserve* which is one of the highest ecosystem in the world (1,220 metres above sea level) added to UNESCO list.

- **Giftmilk Programme:** First initiative of NFN (National Foundation of Nutrition) and NDDB wherein 200 ml of flavoured milk fortified with vitamin A&D (As per FSSAI Standards) is distributed among govt. school children on all school working days through a dairy co-operation under CSR (Corporate Social Responsibility) allocations Target group is mainly school children aged 5-15 years in primary and under primary classes from government schools or schools run for underprivileged children by NGO and Charitable organization.

  GiftMilk by Jharkhand as a State Policy and initiated supply of milk in the entire State.

- **CRISIL's DRIP (Deficient Rainfall Impact Parameter)** provides a better assessment of deficiency because it considers the irrigation buffer available. The higher the DRIP score, the more adverse the impact of deficient rains. (Developed in 2007) , Latest DRIP scores show stress in some states -Assam, West Bengal, Andhra Pradesh ,Tamil Nadu and Haryana. This Drip index is able to provide effective information regarding impact of deficient monsoon on agriculture. It is better indicator than percentage deviation of rainfall from normal as it captures both shock (deviation from normal) as well as vulnerability of a region (measured as percentage of unirrigated area).

- **PARIVESH** (Proactive and Responsive Facilitation, by Interactive and Virtuous Environmental Single-Window Hub"). launched by PM of India on 10th August 2018. It is a single window integrated environmental management system (for Forest, Wildlife and CRZ clearances) developed in live with digital india initiative. It also follows the Minimum Government & Maximum Governance ideology.

ICAR body NBAGR (National Bureau of Animal Genetic Resources) approves 15 new breeds of livestock and poultry in the year 2018. Making total to 212 indigenous breeds. For the period 2014-2023 (40) breeds. The 15 breeds included in 2018 were;

1. **Two cattle breeds:** Ladakhi (J&K) and Konkan Kapila (Maharastra & Goa)
2. **Three Buffalo breeds**: Luit (Assam & Manipur), Bargur (T.N.), Chhttishgarhi (Chhattisgarh)
3. **One sheep breed**: Pnchali (Gujrat)

4. **Six Goat Breeds**: Kahmi (Gujrat), Rohilkhndi (UP), Assam hill (Assam & Meghalya) Bidri and Nandidurga (Karnatka)
5. **One donkey breed**: Halari (Gujrat)
6. **On pig breed**: Ghurrah (UP)
7. **One chicken breed**: Uttara (Uttarakhand)

The 10 new breeds included in 2022-23; three new cattle breeds (Kathani, Sanchori, Masilum), one buffalo breed (Purnathadi), three goat breeds (Sojat, Karauli, Gujari) and three pig breeds (Banda, Manipuri Black, Wak Chambil).

GEF assisted Green – Ag Project to transform Indian Agriculture

The Government has launched the "Green – Ag: Transforming Indian Agriculture for global environmental benefits and the conservation of critical biodiversity and forest landscapes" in association with Global Environment Facility (GEF).

About the Project:

- The project would be implemented in collaboration with the Food and Agriculture Organisation (FAO) in high-conservation-value landscapes of five States namely, (*i*) **Madhya Pradesh:** Chambal Landscape, (*ii*) **Mizoram:** Dampa Landscape, (*iii*) **Odisha:** Similipal Landscape, (iv)Rajasthan: Desert National Park Landscape and (*v*) **Uttarakhand:** Corbett-Rajaji Landscape. The Green-Ag project seeks to integrate biodiversity, climate change and sustainable land management objectives and practices into Indian agriculture.
- The project aims to catalyze a transformative change of India's agricultural sector to support the achievement of national and global environmental benefits and conservation of critical biodiversity and forest landscapes.

About GEF:

- The Global Environment Facility was established on the eve of the 1992 Rio Earth Summit to help tackle our planet's most pressing environmental problems like biodiversity, climate change. International waterways, ozone depletion, land degradation and persistent organic pollutants (POPs).
- It is an international partnership of 183 countries, international institutions, civil society organizations and the private sector that addresses global environmental issues.
- GEF funds are available to developing countries and countries with economies in transition to meet the objectives of the international environmental conventions and agreements.

- The World Bank serves as the GEF Trustee, administering the GEF Trust Fund.
- India is both a donor and a beneficiary of GEF funds.

- **Country's second tulip garden** established at Pithoragarh (Uttarakhand). It is developed by ONGC (Oil and Natural Gas Commission). First tulip garden is situated at Srinagar (J&K). 31 Jan 2019.

- **India's "100 per cent organic state" Sikkim** has won the "Oscar for best policies", conferred by the Food and Agriculture Organisation for the world's best policies promoting agroecological and sustainable food systems in the state and increasing tourism. To make Sikkim an organic state 75,000 hectares of land was converted into organic.
- Sikkim won the Future Policy Award 2018, beating 51 nominated policies from 25 countries, according to a statement. Policies from Brazil, Denmark and Quito (Ecuador) bagged silver awards. The award is co-organised by the Food and Agriculture Organisation of the United Nations (FAO), the World Future Council (WFC) and IFOAM – Organics International.

- **Punjab Launches 3 Mobile Apps To Check Stubble Burning** (Oct. 11, 2018) The 3 apps will help farmers access equipment for crop residue management, monitor tree plantation.

  These are **i-Khet** Machine for facilitating farmers to have access to the agriculture machinery or equipment for in-situ management of crop residue; **e-PEHaL** for monitoring tree plantation; and **e-Prevent** to have prompt and accurate information about incidents of crop residue burning.

- **Integrated Farming System (IFS)** is a promising approach for increasing productivity and profitability through recycling the farm by-products and efficient utilization of available resources. Further it generates employment opportunities to the farming communities round the year and provide a better economic and nutritional security. It also maintains environmental quality and ecological stability.

Components of Integrated Farming System (IFS)

• Piggery • Poultry • Duckery • Fishery • Plantation crops • Apiary • Mushroom Cultivation • Vermicomposting • Fruit cultivation and vegetables cultivation and many othes as per conditions and regions.

Model 1 : Horticulture + Piggery + Fisheries + Plantation Crops

Model 2 : Horticulture + Duckery + Fishery + Plantation crops + Vermicomposting + Apiary

Model 3 : Agriculture + Horticulture + Poultry + Fishery Fruits crop + Vegetables and Spices Crop + Field crops + Plantation Crops Fishery + Vermicompost unit + Piggery

Model 4 : Agriculture + Horticulture + Poultry + Fishery + Azolla + Mushroom

- **A paper sensor** that can **detect freshness of milk.** Scientists at Indian Institute of Technology, Guwahati, have developed a simple **paper** kit that can test **freshness of milk** and tell how well it has been pasteurized. Aided with a smart phone app, the kit can help ensure that **milk** is consumed before it turns too sour.

- **ICAR - Indian Agricultural Research Institute (ICAR - IARI), PUSA- Delhi,** is manufacturing Digital Soil Test and Fertilizer Recommendation (STFR) Kit (Digital Soil Testing Mini Lab). This soil testing kit helps farmers in testing 14 parameters of soil & recommends crop-specific fertiliser dose for 100 crops within a short time at a centre nearby, and thus helps them to achieve higher yields, increase their income and keeps soil fertile.

Pusa STFR Meter Tests Following 14 Parameters of Soil:

1. AVAILABLE ORGANIC CARBON (OC)
2. AVAILABLE NITROGEN (N)
3. AVAILABLE PHOSPHOROUS (P)
4. AVAILABLE POTASSIUM (K)
5. AVAILABLE ZINC (Zn)
6. AVAILABLE SULPHUR (S)
7. AVAILABLE BORON (B)
8. AVAILABLE COPPER (Cu)
9. AVAILABLE IRON (Fe)
10. AVAILABLE MANGANESE (Mn)
11. ELECTRICAL CONDUCTIVITY 12. pH
13. Lime Requirement Test For Acid Soil
14. Gypsum Requirement Test For Alkaline Soil

This soil testing kit can be used for generating soil health card. It is widely used, licensed and few firms have brought it to market commercially. Cost of Technology Rs. 30,000

• **Water technology centre, IARI, New DELHI** has developed an early agricultural drought monitoring and assessment technique in collaboration with university of Nebraska, Lincoin USA and was successfully validated on two distinct drought prone regions namely, Marathwada and Karnatka.

• **Transgenic Rice with reduced arsenic Accumulation**

Researchers at Lucknow based CSIR National Botanical Research Institute have developed transgenic rice by inserting a novel fungal gene which results in **reduced arsenic accumulation in rice grain.** (18th Nov., 2018)

Early flowering mustard variety: Researchers at Delhi-based TERI School of Advanced Studies have developed an early flowering transgenic variety of mustard. The work is important as Indian mustard is a major oilseed crop in the country and due to changing climate the number of cold days required for full plant cycle has shrunken. Scientists have been working on improving crop varieties by modifying plant genes to make them more adaptable changing climate. Early flowering and maturing varieties can help cope with climate change, without compromising on yields.

Researchers led by Dr. Anandita Singh figured out the role of a regulatory gene, called MIR172 belonging to the micro RNA family, that is present in functionally varied forms in Indian mustard (Brassica juncea) and is responsible for a number of traits including timing of flowering. In plants, this microRNA enacts by negatively regulating the expression of some protein-coding genes involved in flowering time.

By increasing the expression of this microRNA, the researchers have developed an early flowering transgenic plant of Indian mustard cultivar Brassica juncea cv. Varuna and shown that by manipulating flowering time, one can develop a plant variety with shorter life cycle and thus help have better yields through reduced exposure to the harsh climatic conditions in the field. (7th June, 2018)

• **Gene Editing of Banana Genome:** Researchers from National Agri food Biotechnology Institute Mohali have edited the Banana Genome using the gene editing technique CRISPR/CAS9. This is the first such attempt on any fruit in India. This editing now can be deployed for improving nutritional quality argonomically traits as well as pathogen resistance in the Banana crop. CRISPR/Cas9 based genome editing has emerged as most powerful tool for crop improvement due to its capability of creating precise alterations in plant genome trait stacking through multiplexing.

- **Poly-Oxime Gel** (19 October 2018)**:** Scientists from the Institute of Stem Biology and Regnerative Medicine Bangalore have developed a protective gel called Poly-Oxime prepared from a nucleophilic polymer. The gel can be applied on the skin and can breakdown toxic chemicals into safe substances preventing them from going deep into the skin and organs like the brain and the lungs. For Indian farmers it will be a boon as Indian farmers do not wear any protective gear while spraying chemicals in the field and get exposed to pesticides, insecticides and fungicides etc. exposure and toxicity.

Thinnest material in the world: Scientists from IIT Gandhinagar (Gujrat) have developed a material that is 1000 times thinner than a sheet of paper. They have developed a two dimensional material of just one-nanometer thickness using magnesium diboride (as compound of boron). It can be have applications from next generation batteries to ultraviolet absorbing films.

ICAR launches NAHEP (National Agriculture Higher Education Project).

Objective: To support participating Agricultural Universities and ICAR in providing more relevant and higher quality education to Agricultural University students.

The project would contribute to the achievement of four sustainable development goals,

(*i*) Ensure including and equitable quality education and promote lifelong leaving opportunities for all.

(*ii*) Promoting including and sustainable economic growth, employment and decent work for seeks higher level.

- Country's second IARI at Ranchi (Jharkhand) and setting up work already started
- Third IARI inaugurated by Minister of Agriculture in Sept 2020 at Assam at Gaumukh (District Dhemji. Cabinet has approved the sanction of Rs.155 crore in 2017 with and also approved

appointments of 98 positions. IARI-Assam will work on the agrarian challenges and complexities of North- Eastern India in coordination with all existing central and state government research and development institutions and private sector enterprises".

Cabinet approves outlay of 2225.46 crore in the three year action plan (2017-2020) of agricultural education and ICAR Institutes for strengthening and developing higher agricultural education in India.

- Government launches PAISA Portal under Deendayal Antyodaya Yojana (Portal for Affordable Credit and Interest Subvention Access). It is a electronic platform for quick processing of loans. 26 Nov 2018)

- **IMD has developed new technology** called impact based forecasting approach to assess rise of water level in rivers and reservoirs by rain. It will help authorities to minutely monitor impact of rainfall and take real time decisions to avoid any disastrous situation of flood. (24 Nov 2018)

- **Digital Green Revolution** (28 Feb. 2022): Digital Green Revolution: ICAR's plan to digitise breeding programmes of eight crops to double annual yield. The digitisation project will cover 8 crops, Rice, Wheat, Maize, Surghum, Pearl millet, Chickpea, Pigeon pea and Potato.

  The Rs 52 crore project is funded equally by ICAR and the Bill & Melinda Gates Foundation.

  The Indian Council of Agricultural Research (ICAR) is moving its plant breeding management system to a digital platform over the next four years. The objective is to improve the efficiency of plant breeding. It intends to double the average annual yield gain of 0.8% obtained through genetic enhancement since the start of the Green Revolution in the mid-1960s, to 1.5% annually over the next couple of decades.

Newly Released Varieties by IARI PUSA

1. During 2018-19, IARI had released a high yielding, non-basmati, medium slender grain, blast resistant rice variety named "Pusa Sambha 1850" for Chhattisgarh and Odisha.
2. Two new wheat varieties "HI 1612" and "HD 8777" were developed by ICAR-IARI, Regional Station, Indore for North Eastern Plain Zone and Peninsular Zone respectively.

3. Institute has also released Maize hybrid "Pusa Super Sweet Corn 1" with enhanced sweetness with a good grain (9.3 t/ha) and fodder (16.2 t/ha) yield.
4. In Pearl millet, a number of inbreds having high iron (72-113ppm) and zinc content (40-55ppm) have been developed for nutraceuticals importance.
5. In chickpea, high yielding extra-large Kabuli chickpea (>50g/100 seeds) genotypes having high export potential; in pigeonpea, extra early and early varieties of dwarf stature and amenable to mechanical harvesting have been developed
6. Institute has identified germplasm lines of rice and wheat with >20% higher water use efficiency (WUE) as compared to the known checks viz., Nagina 22 and C306, respectively, and QTLs for WUE are also identified.
7. Grape hybrid "Pusa Aditi" was released by the Delhi State Variety Release Committee for its commercial cultivation in NCR region.
8. The varieties of cherry tomato *i.e.* (Pusa Cherry Tomato-1) and onion (Pusa Sobha) have been released and notified by CVRC for cultivation under protected condition in Delhi NCR and major onion growing states of the country, respectively
9. Onion variety 'Pusa Sona' has been released by All India Network Project on Onion and Garlic and recommended for notification by CVRC.
10. Ten new varieties, viz. Longmelon 'Pusa Utkarsh', round melon 'Pusa Raunak', cucumber (parthenocarpic, gynoecious) 'Pusa Seedless Cucumber-6', muskmelon 'Pusa Madhurima' & 'Pusa Sunahari', brinjal 'Pusa Safed Baingan-1' & 'Pusa Hara Baingan-1', okra (resistant to YVMV) 'Pusa Bhindi-5, garden pea 'Pusa Prabal' and chenopodium 'Pusa Green' and two hybrids viz. sponge gourd 'Pusa Shrestha' and bitter gourd 'Pusa Hybrid-4' have been released for cultivation in Delhi NCR region.

- New variety of onion NHO-920 released , this variety plant fall down automatically after maturity which is easy for picking crop by farmers.
- Released and notification of three climate resilient rice varieties "Swarna Purvi Dhan 1, Swarna Purvi Dhan 2 and Swarna Shusk Dhan"

- Climate resilient rice varieties "Swarna Unnat Dhan" and "Swarna Sukha Dhan"
- Climate resilient rice varieties "Swarna Shakti Dhan" and "Swarna Samriddhi Dhan"

New Varieties released by ICAR

- Low erucic acid (<2%) mustard variety first in the world released by IARI- PM 30 (Pusa Mustard 30)
- Swarna Vaideh– first New Varieties released by ICAR Makhana was developed and released by ICAR.
- Virat– Early maturing Summer Mungbean
- New Varieties released by ICAR Pusa Arhar 16– dwarf pigeon pea, Short duration (120 days) redgram variety (Miracle red gram variety)
- Co-0238 (Karan 4) – Wonder variety of sugar cane, released by Sugarcane Breeding Institute, Regional Centre, Karnal.
- Pusa rudhira is an improved variety of-Carrot
- Virat– Early maturing Summer Mungbean
- Pusa Arhar 16– dwarf pigeon pea, Short duration (120 days) redgram variety (Miracle red gram variety)

- **IIT Hyderabad scientists and researchers** has developed smart phone based system to detect adulteration in the milk by Using electrospinning to produce paper like material made of nanosized fibres of nylon, loaded with combination of three dyes. The paper is halochromic, *i.e*, change colour in response to change in acidity. (21 Nov 2018)

- Kerla state has declared **Jackfruit** as the state's official fruit.

- **First World Agriculture Prize** – initiated by Indian council food and agriculture (ICFA) has given to Prof. M.S. Swaminathan and was presented by Vice-President Venkaiah Naidu at Vigyan Bhawan at a special session, named Swaminathan Global Dialogue on Climate Change and Food Security organised by the Indian Council of Food and Agriculture (ICFA).

  2nd World Agriculture Prize was presented on September 18th, 2024 at Mumbai, India. The award presentation was preceded by a special session, named "Norman Borlaug Global Dialogue on Climate Change and Food Security".

Greenhouse gas emission from Indian agriculture

Source	Methane	Nitrous oxide	CO_2 eq.
	Million ton		
Rice cultivation	3.37	—	84.24
Agricultural soil	—	0.22	64.70
Crop residue burning	0.25	0.01	8.21
Total	**3.62**	**0.23**	**157.15**

Scientists from Bangladesh has developed a method to convert jute fibre into low cost Biodegradable cellulose sheets dubbed as SONALI (in Bengali SONALI means GOLDEN) which can be used as wrapping material &carrying bag. Commercial production may start by end of 2019.

GPS Enabled Ksheer tester: (22 Sept 2016). "The Ksheer Tester will empower the common man with a technology to tackle a national level health hazard due to adulteration in milk", Union Minister of Science & Technology and Earth Sciences. On the occasion of the 64th Foundation Day Celebrations of Central Electronics Engineering Research Institute (CSIR-CEERI) in Pilani, inaugurated the product.

A rare variety of tea named **"golden butterfly"** from Dikom tea estate Assam sets new record of auction as it was auctioned for Rs.75,000 per kg at GTAC (Guwahati tea Auction control). (13 August 2019)

IUCN Red List (Created in 1948) H.Q. at Geneva, Switzerland

International Union for the Conservation of Nature (IUCN) has added more than 7,000 animals, fish and plants to its endangered Red List. IUCN's Red List of Threatened Species shows that more and more species assessed are being threatened with extinction. It also warned against mankind's destruction of nature that is driving species to brink of extinction at an unprecedented rate. As on in 2024 more than 45,300 species are threatened with extinction. In 2024 Borneo elephant is declared endangered due to human activity.

INDIA BIGGEST EMITER OF SULPHUR DIOXIDE

According to recently released report by Greenpeace India, India is the largest emitter of sulphur dioxide (SO_2) in the world. The findings of report of report were based on data obtained from NASA's Ozone Monitoring Instrument (OMI) satellite. (Oct., 2020). It is due ot India's high emission output due to expansion of coal based electricity generation.

China is the largest emitter of CO_2 in the world, (10,668 million metric tons) India stands third. (2022).

A team of researchers at **Indian Institute of Technology, Guwahati** (IIT-G) has developed portable diagnostic device to detect bacteria without cell culture or microbiological assays. Researchers team work has been patented as well as published in July 2019 issue of Journal of Materials Chemistry of Royal. Society of Chemistry.

IIT-KGP (Indian Institute of Technology Kharagpur) researchers develop new technology- **Hydro Thermal Carbonization (HTC)** for solid waste management.This Hydro Thermal Carbonization Technology can convert municipal solid waste to biofuel, soil amendment and absorbents.

Khadi Prakartik Paint launched: A special type of paint developed by Khadi and Village Industries commission made out of gobar of cattle. This paint is environment friendly and non-poisonous. (Launched in January 12, 2021)

New Species of Plant Alpine discovered in state of Arunachal Pradesh; A new species of plant alpine in the Tawang district of Arunachal Pradesh. This new species belongs to the family of Himalayan Sunflower. It has been named as *Cremanthodium Indicum*. (February 2021)

National Urban Digital Mission launched (NUDM). This mission was launched on Feb., 2021, in order to create a digital infrastructure for all the cities in the country. Initiatives like **Smart Code, India Urban Data Ind ex (UDX)** Smart cities 2.0 website and Geospatial Management Information System (GMIS) were also launched. Launched by Ministry of Urban Affairs with Ministry of Electronics and Information Technology.

Kerala state to convert conventional Anganwadis into **"Smart anganwadis"**. Smart anganwadis will be built in order to provide early childhood care in a phased manner. Department of Women & Child Development (WCD). (First Smart Anganwadis was constructed in 2022 at Poojappura.)

Meghalaya State of North East will revive its first **ginger processing plant** established in 2004 which remained non-functional, situated at Export Promotion Council Park at Raja Bhagan , Byrnihat Haat. It will likely to be functional in the year 2021. This responsibility has been taken by (NERAMAC) North Eastern Regional Agricultural Marketing Corporation through PPP mode.

CSIR-CMERI unveils First Aqua rejuvenation Plant to facilitate on Organic farming model through treated water. It is a integrated waste water Rejuvenation model which has six stage waste water purification profiled for comprehensive treatment of waste water, based upon diverse purification parameters. Approx. 24,000 litres of that can be rejuvenated will be sufficient for almost 4 acres of agricultural land (barring seasonal variations in water requirements) (Feb., 2021).

Two new species of Rare Genus of Ants discovered.

Species of the ant genus Ooceraea found in Kerala state and TN state which add to diversity of this rare genus. They differ from others of the same genus on the basis of number of antennal segments. One species is found in Periyar Tiger Reserve Kerala has been named **"Ooceraea joshi"** in honor of Amitabh Joshi a distinguished evolutionary biologist from JLN centre for advanced scientific Research (JNCASR) working under DST GOI.

Union Education Mister inaugurates **Agri Food Techathon** at IIT Kharagpur to promote technological innovation and Entrepreneurship on 25th January 2021 virtually. IIT Kharagpur in collaboration with NABARD organized the "Agri-Food Techathon 2021", a first of its kind event to promote participation of India's youth and entrepreneurship in different domains of Agri Food Sector.

TRIFED and State Government Agencies across 21 states of the country have included 14 new minor forest produce items under the scheme **"Mechanism of Minor Forest Products MSP scheme".** This is an initiative to support tribal gatherers of forest produce for fair and remunerative price to them . Ministry of Tribal Affairs has revised the MSP for Minor forest produce and included new items to the list.

India's first CNG tractor launched (Feb. 12, 2021)

Union road transport and highways minister, **Nitin Gadkari** has launched India's first-ever diesel tractor, converted to a much cleaner **(compressed natural gas) CNG variant.** With this, farmers will be able to save more than **Rs. 1 trillion annually** on fuel cost, which will help them to improve their livelihood. Natural gas is utilized in a compressed form for combustion.

- This conversion is jointly done by **Rawmatt Techno Solutions** and **Tomasetto Achille India.** The CNG tractor will help farmers to increase their income, by lowering costs and help to create job opportunities in rural India.
- The farmers can save up to **50 per cent** on the fuel cost as the current Diesel prices are around **89 rupees per litre** whereas CNG is only **43.50 rupees per kilogram.**
- The conversion from diesel to CNG will be beneficial as it is a clean fuel with the lowest carbon and pollutant contents. It is also economical as it has zero lead and is non-corrosive, non-dilutive and non-contaminating which helps in increasing the life of the engine

Tamilnadu launches largest cattle park of Asia in Salem (Feb. 22, 2021

Tamilnadu CM inaugurates the Asia's biggest cattle park - Advanced Institute for Integrated Research on Livestock and Animal Sciences (AIIRLAS) - spread over 1,600 acres and constructed at the cost of Rs 1,023 crore on Koot Road near Thalaivasal in Salem district. CM also inaugurated a veterinary medical college and Thalaivasal revenue taluk office. A frozen semen station to develop high milk yielding indigeneous and cross breed cows will be setup in Karumandurai.

Scientists from **Agharkar Research Institute Pune** with Indian Council of Agricultural Research developed a new variety of **soybean** MACs 1407 high yield variety and gives 39 quintals per hectare and it takes 104 days to mature. Seeds have 41% protein and 19.81% oil content. (April. 2021)

United Nations released the Global Forest Goals report 2021.

According to report around 1.6 billion people on earth rely on forests. This is about 25% of global population.

Of the rural areas in the world more than 40% live in Savanah and forests. This is 20% of global population

According to the report India has to add 200,000 hectares of tree and forest cover per year for sustainable ecology goals.
Team of Zoological Survey of India has discovered a new insect eating mammal species. It has been named as *Crocidura Norcondamica.* This species help to control insect population in the dense forests. (4 May 2021)
Telengana state government bans the use of controversial herbicide "**Glyphosate**". It is used to kill weeds in cotton farms. (May 7, 2021)
United Nations recently launched **Global Methane Assessment Programme.** It was released by Climate and clean Air Collision (CCAC) and United nations Environment programme. Report says that human caused methane can be reduced by 45% by 2030. Annual Global Methane Assessment report released by International Energy Agency (IEA) annually.
IEA (International Energy Agency) released the 2021 **" Renewable Energy Market Update"** IEA increased its forecast for the global growth of solar and wind energy by 25%. Report also revels that 280GW of the renewable energy was installed worldwide in the year 2020. It is 45% increase as compared to 2019.
National bamboo mission recently launched MIS (Management Information system) to strengthen the domestic Aggarbati Industry in the country. (12th May, 2021)
Cyclonic storm YAAS hit Bay of Bengal on 23-28 May, 2021.
India Biodiversity Awards 2021 under category of sustainable use of biological resources won by KRISHI AVAM PARISTHITIKI VIKAS SANTHAN (KRAPAVIS) and individual prize won by Sh. **SHAJI NM who is also called Tuber man of Kerala.** (For conservation of domesticated species). First round of award was launched in 2012. (The National Biodiversity Authority (NBA). The UNDP and MoEFCC collobrated the award)
Bihar famous litchi exported to United Kingdom by Air. First consignment was sent on 24th May 2021.
Ministry of Power launched the National Mission on Use of Biomass in Thermal Power Plants (SAMARTH) on July 12, 2021.
A new species of plant called **African violet** discovered by IISER (Indian Institute of Science Education & Research) (Bhopal). African violet is found in parts of state of Mizoram and adjoining areas of Myanmar. (May 2021)

IFFCO to Launch Nano Urea (Liquid) (August 2021)

It is urea in the form of a nanoparticle. It is a nutrient (liquid) to provide nitrogen to plants as an alternative to the conventional urea.

- Urea is a chemical nitrogen fertiliser, white in colour, which artificially provides nitrogen, a major nutrient required by plants.

It is developed to replace conventional urea and it can curtail the requirement of the same by at least 50%.

It contains 40,000 mg/L of nitrogen in a 500 ml bottle which is equivalent to the impact of nitrogen nutrient provided by one bag of conventional urea.

Nano-fertilizer : Nano-formulations or nano-sized fertilisers are made from ammonium humate, ammonia, urea, peat, plant wastes, and other synthetic fertilisers.

An example of nano-formulation is nano-sized nitrogen (N) fertiliser which is prepared as a result of deposition of urea on calcium cyanamide.

Formulation: Nano-fertilisers are developed using both mechanical and biochemical processes, *i.e.*, materials are grounded to obtain nano-sized particles through mechanical means and biochemical techniques are employed to obtain effective nanoscale formulations.

Recently, the Indian Farmers Fertiliser Cooperative Limited (IFFCO) introduced the world's first Nano Urea Liquid for farmers across the

TRIFOOD Parks in Madhya Pradesh (August 20, 2020)

TRIFED (Tribal Co-operative Marketing Development Federation of India) operating under Ministry of Tribal Affairs signed a MOU with Akhil Bhartiya Vanavasi Kalyan Ashram to set up TRIFOOD parks in state of MP. According to MOU TRIFED and Akhil Bhartiya Vanavasi Kalyan Ashram will together implement Van Dhan Yojna in the state.

TRIFood is a joint initiative of Ministry of Tribal Affairs, Ministry of Food Processing Industry and TRIFED.

Unified Web portal for monitoring programme of **GOBERDHAN** launched. (3rd Feb., 2021)

- **e-Nam (National Agriculture Market).** At present it is integrating 1389 markets across 23 states and 4 UT, During Covid-19, e-Nam platform has been further strengthened by launching FPO trading module in P-Nam whereby FPO can trade their produce from their collection centre without bringing produce to APMC, So far 3366 FPOs have been on board.

- **ICAR-AICRP Setting up a Central Research Institute in Kalahandi for Soyabean**

The Kalahandı district of Odisha produces Soybean on a very limited scale. The ICAR has established a national level research Institute on Soybean in 1987 namely, ICAR-Indian Institute of Soybean Research, Indore (Madhya Pradesh) to conduct basic and strategic research on the crop. The ICAR-AICRP on Soybean has one voluntary center at Regional Research & Technology Transfer Station, Bhawanipatna (Kalahandi) under the Odisha University of Agriculture & Technology, Bhubaneswar. This station is undertaking need-based and location-specific research on Soybean for hot & moist sub-humid climate of Odisha comprising of Kalahandi, Bolangir and Koraput. (2021)

Red Rice Export to USA from India

In a major boost to India's rice exports potential, the first consignment of 'red rice' was flagged off on 5th March (2021) to the United States of America (USA).

Iron-rich 'red rice' is grown in the Brahmaputra valley of Assam, without the use of any chemical fertilizer. The rice variety is referred as 'Bao-dhaan', which is an integral part of Assamese food.

CAZRI Jodhpur develops Agri Voltage technology

Agri-voltaic system of 105 KW capacity was developed by ICAR-Central Arid Zone Research Institute, Jodhpur. This **technology** can increase the income of farmers by generation of electricity and growing of cash crops simultaneously on the same piece of land. The National Solar Energy Federation of India (NSEFI) has prepared a document to develop 13 operational agri-voltaic systems in the country, managing by various solar PVs workers and public institutions

Ladakh launches India 's first solar power lift irrigation system

In **Ladakh**, the **first Solar Lift Irrigation** Plant of Kargil with five KW **Solar** Plant and three HP Submersible Pump at Yak Breeding Farm, Bodh Kharboo has been inaugurated. This prototype will help farmers to solve the shortage of water to irrigate, particularly those lands which are at the banks of the river.

Madhu Kranti Portal Launched (8th April, 2021)

Madhu Kranti Portal and Honey Centers is an initiative of National Bee Board (NBB), Ministry of Agriculture and Farmers Welfare under

National Beekeeping & Honey Mission (NBHM). This portal is being developed for online registration to achieve traceability source of Honey and other beehive products on a digital platform.

Objective: The Madhu kranti Portal has been developed for online registration to achieve the traceability source of Honey as well as other beehive products on the digital platform. It will also help in checking the quality and the source of adulteration of honey.

AAHAR KRANTI MISSION

- The Government has launched 'Aahaar Kranti', a **mission dedicated to spread awareness about nutrition.**

Motto of the Mission is "Uttam Aahaar Uttam Vichar" or "Good Diet-Good Cognition"

Significance:

- The United Nations (UN) has also declared **2021 as the International Year of Fruits and Vegetables**, which vibes very well with Aahaar Kranti.
- Further, it is in line with the **UN sustainable goal 3** that emphasizes human well-being.
- It will **help in applying the rich knowledge of Ayurveda-based nutrition** to practice.

E-SANTA (Electronic Solution for Augmenting) Oct., 2021

Recently the Union Commerce and Industry Minister has inaugurated an **electronic platform** named **e-SANTA**, which will connect **aqua farmers and the buyers.**

About:

The term **e-SANTA** was coined for the web portal, meaning **Electronic Solution for Augmenting NaCSA farmers' Trade in Aquaculture.**

e-SANTA is a **digital bridge** to end the market divide and will act as an alternative marketing tool between farmers & buyers by eliminating middlemen.

NATIONAL NURSERY PORTAL (14th April, 2021)

Agriculture Minister launches a **'National Nursery Portal'** to connect nursery operators with consumers.

The National Nursery Portal, developed by the National Horticulture Board, will help consumers to connect with nurseries and get information about the availability and prices of quality planting material, an official statement said.

UN FOOD SUMMIT 2021

The **United Nations** Secretary General has called for the **first ever UN Food Systems Summit** to be held in September 2021 to strategize the actions for positive change in Agri-**food systems** in the World to realize the vision of the 2030 Agenda for Sustainable Development. 15 Apr. 2021

A consignment of **Dragon Fruit**, was exported to after sourcing it from farmers of Tadasar village in Sangli District, Maharashtra. They are grown mostly in states including, Odisha, Karnataka, Kerala, Tamil Nadu, Maharashtra, West Bengal, Gujarat, Andhra Pradesh, and Andaman & Nicobar Islands

UN Decade on Ecosystem Restoration Report

UN Environment Programme (UNEP) and Food & Agriculture Organisation of UN (FAO) launched 'UN Decade on Ecosystem Restoration' report for 2021-2030 and called on nations to meet commitments of restoring 1 billion hectares of land As per the report, world is facing triple threat of climate change, loss of nature and pollution. So, world must deliver on its commitment to restore at least one billion degraded hectares of land (equal to size of China) in next decade.

- Report highlights, countries also need to meet similar commitments for oceans.
- Humanity is using 1.6 times the number of services that nature can provide sustainably.
- Thus, conservation efforts alone are insufficient to prevent ecosystem collapse and biodiversity loss on larger scal

MIT generates Electricity from Carbon Nanotubes (June 7, 2021)

MIT (Massachusetts Institute of Technology) engineers have discovered a new way of generating electricity using carbon nanotubes. A new material made up of carbon nanotubes have potential to generate electricity by collecting energy from its environment

How electricity is generated?

- These micro carbon particles can create current by interacting with liquid around them.
- Liquid, an organic solvent, attracts electrons out of the nanoparticles, which generates current.

- This current can be used to drive chemical reactions or to power micro or nanoscale robots.
- This electric current can also be used to drive a reaction known as alcohol oxidation

IIT-Kharagpur develops Early Cyclone Detection Technique

June 10, 2021

Indian Institute of Technology (IIT) Kharagpur have developed an early cyclone detection technique which will help in early detection of development or strengthening of tropical cyclones in North Indian Ocean region.

IIT developed this technique using Eddy detection technique. It will investigate formative stages and advance detection time of tropical cyclogenesis. This technique was studied under Climate Change Program with support from Department of Science and Technology. The technique was developed with the aim of identifying initial traces of pre-cyclonic eddy vortices in atmospheric column. It also seeks to track its spatio-temporal evolution. Researchers used a coarser grid resolution of 27 kilometres to identify the characteristics of eddy vortices.

India to add 20 GW of Wind Energy Capacity by 2025: GWEC June 10, 2021

India wind energy market outlook was released by Global Wind Energy Council. As per the report, India has a pipeline of 10.3 GW in Central and State markets. These pipelines will drive installations until 2023. Report highlights India will install 20 GW of wind energy capacity by 2021-25.

India exports GI certified 'Jardalu' Mangoes to UK June 15, 2021

Bihar exported the first commercial consignment of GI certified Jardalu mangoes from Bhagalpur to the United Kingdom. Jardalu mangoes from Bhagalpur district were given GI certification in 2018

GI certified Jalgaon Banana exported to Dubai

Consignment of fibre & mineral rich "Jalgaon Banana" was exported to Dubai on June 16, 2021, marking a major boost to exports. Jalgaon Banana got GI certification in 2016. It was registered with Nisargraja Krishi Vigyan Kendra (KVK) Jalgaon.

Biotech-KISAN programme Mission Programme.

Biotech – Krishi Innovation Science Application Network (Biotech-KISAN)) is a scientist-farmer partnership scheme launched in 2017.

Biotech-Krishi Innovation Science Application Network (Biotech-KISAN) is being implemented in 15 agro-climatic zones of India in phased manner with the following objectives:

- Linking available science and technology to the farm by first understanding the problem of the local farmer and provide solutions to those problems.
- The working together, in close conjunction, of scientists and farmers is the only way to improve the working conditions of small and marginal farmers.
- This programme aims to work with small and marginal farmers especially the woman farmer for better agriculture productivity through scientific intervention and evolving best farming practices in the Indian context

Virtual Water analysis. June 22, 2021.

Researchers from Indian Institute of Technology, Guwahati have found way for better water management policies in India through Virtual Water analysis. This study was led by IIT-Guwahati professor.

Virtual Water

- According to the professor, Virtual Water (VW) is water involved in production & trade of food & non-food commodities & services.
- Virtual water is that "invisible" water that has been consumed in the entire lifecycle of product or service.

Study of IIT Guwahati will help in bridging knowledge governance gap to mitigate water scarcity in Virtual Water flows assessment. As per the study, states with chronic water scarcity, planning & implementation of sustainable agriculture are crucial to achieve water and food security. Study finds that, pressure on freshwater resources in water parched states can be reduced by diversifying production areas by using Virtual Water flows analysis to produce agro-climatically suitable food grains.

Cassava: Feedstock for Bioethanol Production

ICAR-Central Tuber Crops Research Institute (CTCRI) have found CASSAVA as a raw material for bioethanol production to meet Ethanol Blending Petrol (EBP) programme target of 2025 by India

Yellow Gold 48: Yellow Watermelon commercially introduced June 22, 2021

German agrochemicals major Bayer has commercially launched first-ever yellow watermelon variety called "Yellow Gold 48" in India under Seminis brand.

World's first GM Rubber planted in Assam (June 23, 2021)
Rubber Board has started field trial of world's first GM (genetically modified) rubber in the state of Assam

Green Hydrogen: IOC, India's largest oil firm will build the country's first 'green hydrogen' plant at its refinery in Mathura. This decision has been undertaken with the view of preparing for a future to cater to the growing demands for cleaner forms of energy. (July 20, 2021)

The National Green Hydrogen Mission was approved by the Union Cabinet on 4 January 2022, with the intended objectives of: Making India a leading producer and supplier of Green Hydrogen in the world. Creation of export opportunities for Green Hydrogen and its derivatives.

FSSAI set NEW Trans-fat levels
FSSAI recently announced that all edible refined oils, vegetable fat spreads, vanaspati oils, margarines, bakery shortenings and mixed fat spreads can contain only 3% or less trans-fats.
This is to be implemented by January 2021. By January 2022 this is to be further reduced to 2%. This was done after WHO has called for elimination of trans-fats by 2023.

ICAR-Indian Rice Research Institute recently released four new varieties of rice
1. DRR Dhan 53 2. DRR Dhan 54, 3. DRR Dhan 55 and 4. DRR Dhan 56
DRR Dhan 53: Is a novel bacterial blight resistant high yielding fine grain rice variety. It has average yield of 5.5 tone /ha and has seed to seed maturity of 130-135 days. This variety has been released for cultivation in irrigated and bacterial blight endemic areas of Telengana, Karnataka, Andhra Pradesh, Tamilnadu, Chattishgarh, Odisha, Maharastra, Gujrat & Bihar.
DRR Dhan-54 and DRR Dhan-55: These were developed for cultivation under the aerobic system and under dry seed rice. These varieties were developed for cultivation mainly for water limiting areas in the country such as Zone II-Haryana, Zone III- Odisha, Bihar, Jharkhand, Zone IV Gujrat and Zone-V Telengana. These varieties have pest as well as diseases resistance like neck blast, gall midge, leaf blast and rice thrips and moderate to stem borer and plant hopper.

DRR Dhan-56: It is a cross breed developed from Huang-zhan*2/ Phalguna that is built under the collaborative project 'Green Super Rice' for resource poor Africa and Asia phase 3 between the IIRR Hyderabad and IRRI Phillipines. This variety is resistant to false smut and leaf blast. DRR Dhan-56 has been released for cultivation in irrigated conditions of Punjab and Haryana.

ICAR Varieties during 2019-2021: ICAR released 562 new high yielding varieties of filed crops. 89 oilseeds. Pulses 101, forage crops 37, fibre crops 90, sugarcane 14 and potential crops 8. Of 562 varieties 7 were developed through market-assisted selection and 451 varieties are able to withstand different biotic and abiotic stresses. Biofortified varieties to support Kuposhan Mukta Bharat 21 varieties of different crops rich in Fe, Zn, protein, pro-vitamin A etc. in the edible parts or with reduced levels of anti nutritional factors have been developed.

ICAR varieties for enhanced nutritional benefits

Crop	*Variety*	*Distinctive feature*
Carrot	Pusa Rudhira	Lycopene
	Pusa Asita	Anthocyanin
Palak	Pusa Jyoti, Pusa Bharti	Vit A, Vit C., Fe, Ca
Tomato	Pusa Uphar, Pusa Hybrid-2, Pusa Rohini	Vit C, Lycopene
Vegetable mustard	Pusa Sag-1	Vit C, Carotene
Amaranths	Pusa Kiran, Pusa Kirti	Vit A, Vit C, Ca, Fe
Bitter gourd	Pusa Hybrid-1,2	Vit A, Vit C, Fe, Ca
Broad bean	Pusa Sumit	Protein
Pumpkin	Pusa Hybrid-1, Pusa Vikas	Vit A
Cauliflower	Pusa Beta Kesari 1	ß-carotene (8.0-10.0 ppm)
Potato	Bhu-Sona	ß-carotene (14.0/100g)
Sweet Potato	Bhu Krishna	Anthocyanin 90mg/100 gm
Pomegranate	Solapur Lal	Iron, Zinc and Vitamins

SEED PRODUCTION

During 2019-21 (2.30 lakh Quintal) breeder seed of 1330 varieties in 55 field crops were produced and supplied as per indent for downstream multiplication to produce foundation and certified seeds by various public and private agencies. In addition, 248534 and 392232 quintals of quality seeds were produced for pulses and oilseed crops respectively through seed hubs.

India's 14 Tiger Reserves have received the **Global Conservation Assured | Tiger Standards** (CA|TS) accreditation. The 14 tiger reserves are Orang, Manas and Kaziranga in Assam Kanha, Panna and Satpura in MP Pench in Maharashtra Valmiki Tiger Reserve in Bihar Dudhwa in UP Sunderbans in West Bengal Parambikulam in Kerala Bandipur Tiger Reserve of Karnataka ..Annamalai and Mudumalai Tiger Reserve in Tamil Nadu. (July 2021)

RAJA MIRCHA

Raja Mircha, also known as Naga King Chilli, was exported to London from Nagaland. This the first time Raja Mirch is being exported from the state of Nagaland that too during the the Covid-19 pandemic along with a drought-like situation in Nagaland. Key Points In the year 2008 Raja Mircha received Geographical Indication certification...

Biodiesel from the Waste of Slaughtered Chicken

John Abraham, a veterinary-doctor-turned-inventor, has received the patents for inventing biodiesel from the waste of slaughtered chicken. This biodiesel offers mileage of around 38 km a litre. This biodiesel costs around 40% of the current diesel price and cuts down on pollution by half. (July 26, 2021)

"Golden Rice" which is genetically modified has been given approval for commercial production by Philippines. Health experts believe that this decision will save lives and combat childhood blindness in the developing regions of the world. Golden rice a new type of rice that contains beta carotene. (July 2021)

Dairy Investment Accelerator

Government of India's Department of Animal Husbandry & Dairying (DAHD), under its Investment Facilitation Cell, has setup Dairy Investment Accelerator. This has been set up with a view of providing dedicated focus towards the promotion and facilitation of investments in dairy sector of the country.

Agar Trees

The government of Tripura has taken initiatives to promote commercial cultivation of Agar trees and has set a target of achieving Rs 2,000 crore business from this sector in the coming three years. Agar wood is used for incense, perfume, traditional medicines and other products in world market. (19 Jul 2021)

Cultivation of **'monk fruit'** from China has started in Kullu, Himachal Pradesh. The fruit was introduced for field trials in Himachal Pradesh by Palampur-based Council of Scientific Research and Industrial Technology-Institute of Himalayan Bio-resource Technology (CSIR-IHBT). Monk fruit is known for its properties as non-calorie natural sweetener. (July 2021)

Drone Rules, 2021

Ministry of Civil Aviation (MoCA) has released reorganized "Drone Rules, 2021" for public consultation. Draft rules were built on premise of trust, self-certification and non-intrusive monitoring. It will replace UAS Rules 2021 that was released on 12th March 2021. Draft Drone Rules 2021 As per new rules, there will be minimal human interface on digital sky platform and most permissions will be self-generated

Ministry of Civil Aviation released an **"interactive digital airspace map"** for flying drones, that demarcates areas into different zones for **flying drones** across India. Zones for flying drones also include yellow and red zones.

IIT Roorkee- **National Disaster Management Authority developed** an APP to predict natural disasters in a joint venture. (31 Aug 2021)

BRICS Partnership for **Strengthening Agro Biodiversity for Food and Nutrition Security** was theme for virtual meeting of Agricultural Minister of BRICS nations held on August 27, 2021.

SUJALAM Campaign launched by Jal Shakhti Minister for 100 days. This compaign seeks to create More ODF Plus Villages by Undertaking Waste Water Management across Villages of India. (25 Aug 2021)

Labour ministry has launched **e-Shram portal** on August 2021.

Plastic Waste Management Amendment Rules notified by GoI. Rules prohibit specific single use plastic items with low utility and high littering potential by year 2022.

China plan to become **carbon neutral by 2050**. Carbon neutral means a balance between emitting carbon and absorbing carbon from the atmosphere in carbon sinks.

Delhi get its First Animal DNA lab has set-up at Forensic Science Laboratory (Aug., 2021). It will help police for timely disposal of cases related to cow slaughter illegal trade of animals and related cases in Rohini, New Delhi.

World's largest plant that has been designed to **suck carbon dioxide** (CO_2) from air and turn it into rock was started recently in Iceland. (ORCA PLANT) (September 11, 2021)

Ladakh adopted two endangered species, **snow leopard and black-necked crane,** as its State animal and State bird respectively on September 1, 2021.
World Meteorological Organization (WMO) has published a new report titled "Climate Indicators and Sustainable Development: Demonstrating the Interconnections" on September 22,2021.
Intergovernmental Panel on Climate Change (IPCC) recently published the Assessment Report from Working Group I titled **"Climate Change 2021: The Physical Science Basis".** This Sixth Assessment Report addresses the most up-to-date physical understanding of the climate system and climate change, bringing together the latest advances in climate sciences.
The **World Health Organisation (WHO)** has revised **its global air quality guidelines and recommended more stringent standards for key pollutants.** WHO set the new standards in the first update of its air quality guidelines since 2005. PM 2.5 norms for 24 hours average has been changed to 15 micro-g/m^3 against 25 micro-g/m^3
An Indian Project, **Takachar's Innovation,** that recycles agricultural waste into fuel has won the **"Prince William's inaugural Earth shot Prize",** which is also called as "Eco Oscars. (October 18, 2021)
Ministry of Environment, Forest, and Climate Change released the **"Vision Plan: 2021-2031"** for Indian zoos. This plan was released during a two-day National Conference for zoo directors & veterinarians in Gujarat. (October 13, 2021)
International Energy Agency (IEA) has **invited India** to become its full-time member in February 2024. (October 2023)
Multidimensional Poverty Index-2022 released by (UNDP) In 111 countries 1.2 billion people (19.1 %) live in acute multidimensional poverty. In India over the past 15 years the number of Poor people has declined by 415 million. MPI value and incidence of poverty fell from 0.283 in 2005-06 to 0.069 in 2019-21. Incidence of poverty declined from 55.1 % to 16.4 % over the last 15 years. However India still host highest number of poor people in the world and Nigeria has the second highest.
On November 29, 2021, Indian Parliament passed the Farm Laws Repeal Bill, 2021 during first day of the winter session. The three farm laws of centre have been repealed by Lok Sabha as well as

Rajya Sabha in the fastest repeal in recent years, following a year-long protests by farmers.
On November 28, 2021, India's only certified **organic kiwi** from Ziro Valley Farms of Arunachal Pradesh was launched at a mega National Tribal Festival called Aadi Mahotsav.
National Commission for Women (NCW) launched a pan-India capacity building programme called **"She is a Changemaker"** on December 7, 2021.
France-based World Inequality Lab published its report titled **"World Inequality Report 2022.**
The Hornbill Festival, which is called the **'Festivals of Festivals',** is a 10-day annual cultural fest of Nagaland that showcases the rich and diverse Naga ethnicity through folk dances, traditional music, local cuisine, handicraft, art workshops etc. (December, 2021)
The Government of India is to launch **SRESTHA Scheme** for the Scheduled Caste students. The scheme will provide quality residential education to the students. (Launched on June 2022)
The Cabinet has recently approved building 780-km **Green National Highway project** at a cost of Rs 7,660 crore. It includes upgradation of various National Highways in four states – Himachal Pradesh, Rajasthan, Uttar Pradesh and Andhra Pradesh. (Launched in 2020)
The first **geological park** of India will be built in Lamheta, Jabalpur in Madhya Pradesh. Approval for the park was given by Geological Survey of India, under ministry of mining. Park will be constructed with an investment of Rs 35 crores, over five-acre land. (Jan. 2022)
India emerged as the **'Pickle King of The World',** in a recent development. India crossed the exports of worth USD 200 million of agricultural processed product, pickling gherkins, cucumber or cornichons, in the financial year 2020-2021. (January, 2022)
The Digital Government Mission was launched by the Ministry of Electronics and Information Technology. The mission will deliver government services to the citizens. It will use machine learning and artificial intelligence to deliver the services. (January 21, 2022)
Digital Agristack: The Government of India is working on creating a **digital stack** of agricultural data. That is storing data related to farmers. Apart from farmer details, the digital agristack also holds all information about which seeds to buy, weather updates, what manures and fertilizers to use, insurance, best agricultural practices to maximize the yield. (January 16, 2022)

The **tearless onions** are called **"Sunions".** It is to be sold in United Kingdom. These onions were produced by cross breeding onion varieties that are less pungent. (January 15, 2022)

World bank's Global Economic Prospects' Report was released on January 11, 2022. The report notes that, Global economic growth will decrease sharply to 4.1% in 2022 and 3.2% in 2023, from 5.5% in 2021. (January 12, 2022)

Fresh fruits export from India has witnessed a major growth. The major destinations of exports of fresh fruits include the Netherlands, Bangladesh, the UK, the UAE, Iran, Nepal, Saudi Arabia, Russia, Qatar, and Oman. (February 28, 2022)

A 1.7-megawatt **solar photovoltaic plan**t has been commissioned by Bharat Heavy Electricals Ltd. (BHEL) at Bina, Madhya Pradesh for the **Indian Railways.** (26 Feb 2022)

A megadrought is a drought that lasts at least two decades. In the past, megadroughts have been linked to prolonged La Niña conditions. Megadrought is a term that refers to the length of a drought rather than its acute intensity. (February 25, 2022)

The Zoological Survey of India (ZSI), Pune, has discovered a **new frog species** which are growing in the freshwater regions along all the states spanning the Western Ghats. (25 Feb 2022)

The Indian Institute of Technology Roorkee (IIT Roorkee) has launched a mobile app named **'KISAN'** for the farmers. Through this app, agro-meteorological advisory services will be provided to the farmers. (23 Feb 2022)

The National Institute of Animal Biotechnology, Hyderabad, has developed the **"Fluorine Doped Tin Oxide (FTO)** electrode fabricated with reduced Graphene Oxide (rGO)" to detect the Japanese Encephalitis Virus. (February 19, 2022)

Green Hydrogen Policy 2022: The Green Hydrogen/ Green Ammonia Policy was notified by the central government in February 2022. It is a major step, which will help India achieve its climate targets and aid in the smooth implementation of the National Hydrogen Mission. The policy also aims to make India an export Hub for Green Hydrogen and Green Ammonia. The National Hydrogen Mission was launched on August 15, 2021.

The aims of the mission are as follows:

- To help the government in making India a green hydrogen hub.
- To achieve the environment and climate related targets.
- To help in meeting the target of production of 5 million tonnes of Green hydrogen by 2030
- To aid in the development of related renewable energy capacity.

Central government has set the target to **replace diesel** with renewable energy in agricultural sector by **2024.**

On February 9, 2022, Union government put **ban on the import of drones** other than for security, research and defence purposes. Government banned drone import with the aim of promoting **'Made in India' drones.**

A new **Carangid marine fish species** has been identified by the **Central Marine Fisheries Research Institute** (CMFRI) from the Indian coast of Kerala. This newly identified fish belongs to the group named 'queen fish'. The scientific name of this fish is *Scomberoides pelagicus.*(March 26, 2022)

An e-book titled **"Climate Hazards and Vulnerability Atlas of India-State:** Tamil Nadu" has been launched by the India Meteorological Department on 23rd March as part of World Meteorological Day celebrations. (March 24, 2022)

On 23rd March 2022, the **Sujalam 2.0 campaign** was launched by the Union Minister of Jal Shakti. This campaign has been launched for **greywater management.**

The **blending of biomass pellets** has been made mandated by the Ministry of Power. Under the **"Revised Policy for Biomass Utilization for power generation** (on October 8, 2021)

AquaMap

The new water management and policy centre, **AquaMAP** has been inaugurated by the Principal Scientific Adviser to the Government of India. The aim of AquaMAP is to provide solutions regarding the challenges faced due to water problems by implementing smart water . (March 19, 2022)

Acetabularia Jalakanyakae is the new marine algae species discovered in India by Indian scientists from the Central University of Punjab. (March 8, 2022). It is known as "mermaid's wineglass" because of its umbrella shaped cap.

Kanya Shiksha Pravesh Utsav is a campaign to bring out-of-school adolescent Indian girls back into the formal education and skilling system.The campaign was launched on the eve of International Women's Day (8th March 2022) by the Ministry of Women and Child Development in collaboration with the Ministry of Education.
Climateprenurship is a form of social entrepreneurship to fight climate change. (8 Mar 2022)
SVEP: The **Start-up Village Entrepreneurship Program** is a sub-scheme of Deendayal Antyodaya Yojana – National Rural Livelihoods Mission (DAY-NRLM). About SVEP SVEP is being implemented since 2016 by the Ministry of Rural Development. SVEP's objective is to support entrepreneurs in rural areas to set-up enterprises at the village-level in non-agricultural sectors. (March 7, 2022)
Unnat Bharat Abhiyan 2.0 (UBA 2.0) has successfully completed four years on 25th April 2022. **The Unnat Bharat Abhiyan (UBA)** is a flagship programme that falls under the ambit of the Ministry of Education and aims to link the country's Higher Education Institutions with a set minimum of five villages so that these higher institutions can help in contributing to the social and economic betterment of the communities of these villages by using their knowledge base. The National Coordinating Institute of the UBA programme is the Indian Institute of Technology Delhi (IIT Delhi)
Palli village located in the **Samba district of Jammu** has become India's first panchayat to become **carbon neutral**, fully solar energy powered, and all its records have been digitized which will help in receiving all benefits of Central schemes easily.
On 17th April 2022, PM Modi, during his Gujarat tour, inaugurated a **Community Radio Station (CRS)** named **Dudh Vani** that is dedicated to animal husbandry of the state. This radio station will link around 5 lakh dairy producers in 1700 communities in Gujarat.
The United Nations Food and Agriculture Organisation (UN-FAO) along with Arbor Day Foundation has recognized Hyderabad and Mumbai jointly as the **'2021 Tree City of the World'.** Mumbai city awarded the prestigious title for the second consecutive year in 2022
World meterological organization releases State of Global Climate report 2021.
Average global temperature: It was about 1.11 (± 0.13) °C above the pre-industrial level in 2021. The report observed that the past seven years have been the warmest seven years on record.

Greenhouse gas concentrations: It has reached a new global high in 2020. The concentration of carbon dioxide (CO_2) reached 413.2 parts per million (ppm) globally, or 149% of the pre-industrial level. **Ocean heat:** The upper 2000 m depth of the ocean continued to warm in 2021 and it is expected that it will continue to warm in the future. **Ocean acidification:** The oceans absorbs around 23% of the annual emissions of anthropogenic CO_2 to the atmosphere. This reacts with seawater and leads to ocean acidification. As per report, there is a persistent rise in ocean acidification. **Global mean sea level:** It also reached a new record high in 2021, after increasing at an average 4.5 mm per year over the period 2013-2021. This is more than double the rate of between 1993 and 2002 and is mainly due to the accelerated loss of ice mass from the ice sheets. **Ozone Hole:** The ozone hole over the Antarctic was unusually large and deep, reaching its maximum area of 24.8/ million/ km. square (the size of Africa). The size increased as a result of a strong and stable polar vortex and colder than average conditions in the lower stratosphere.
Honey Testing Lab inaugurated at **ICAR-KVK, Hirehalli, Tumakuru** on 20th May 2022.
Bharat Drone Mahotsav inaugurated by PM in New DELHI on May 28th 2022 (Two day event)
China hosted 7th BRICS cultural ministers meeting "Theme" – **Establishing a cultural partnership featuring inclusiveness and mutual learning.** (2022)
MP Ladli Lakshmi Yojana: Daughters are progressing in Madhya Pradesh.The State Government implemented Ladli Laxmi Yojana on 1st April, 2007. Its objective was to create a positive attitude among the public towards the birth of daughters, improve the sex ratio, improve the educational and health status of the girls and lay the foundation for their bright future. Seeing the need, CM Chouhan has recently started **Ladli Laxmi Yojana 2.0 'Atmanirbhar Ladli'**, whose aim is to make every daughter of the state socially and financially empowered. An assurance certificate of Rs 1,43,000/- is issued by the government in the name of the girl child. From the time of registration, Rs 6,000/- will be deposited in the Madhya Pradesh Ladli

Laxmi Yojana. Fund for five consecutive years, the total amount of Rs 30,000/- will be deposited in the name of the girl child.
Severe Cyclonic Storm Asani was a strong tropical cyclone that made landfall in India in May 2022. The third depression and deep depression, and the first named storm of the 2022 North Indian Ocean cyclone season, Asani originated from a depression that the Indian Meteorological Department first monitored on May 7.
The Union Cabinet approved a host of amendments to the National Policy on Biofuels, 2018 and advanced the target of blending 20% **ethanol-blended petrol** by five years, from 2030 to 2025-26. It will now allow for more feedstocks for the production of biofuel.
The Union Government has announced the formation of the **Cotton Council of India** under the Chairmanship of renowned veteran cotton man **Suresh Bhai Kotak. Ahmedabad is known as cotton city of India.** First meeting held on 28th May, 2022)
India **banned wheat exports** on **13 May 2022**, in order to 'manage the overall food security of the country and to support other vulnerable countries'.
Mangroves are critical ecosystems, bridging land, freshwater, and sea. They host tremendous diversity, and protect and provide for countless coastal communities around the world. The 2024 edition of The State of the World's Mangroves highlights the tremendous advances that have been made on multiple fronts to safeguard these ecosystems. It showcases progress in: science and understanding; collaboration and information sharing; practical management interventions; and the many policy, legal, and financial tools.
India's first organic waste-powered Electric Vehicle (EV) charging station has been inaugurated at **Keshavrao Khadye Marg, Mumbai.**
"Drought in Numbers, 2022" report was **presented on 21st May 2022 by the United Nations Convention to Combat Desertification (UNCCD).** The report involves the collection of data on the effects of droughts globally and how droughts can be mitigated.
Green bond : A green bond is a fixed-income debt instrument earmarked to raise money for climate and environmental projects. In 2024, India's government plans to raise Rs. 20,000-25,000 crore through green bonds to fund its clean energy and sustainability initiatives. The government's green bond issuance is a significant step towards integrating India's financial markets with global green finance initiatives.
Scientists have for the first time been able to grow plants in soil from the moon. The soil samples were collected during the Apollo

11, 12 and 17 ventures, according to NPR. The research has significant implications for prolonged space exploration as well agricultural innovation on Earth, says NASA.

PM PRANAM (Promotion of Alternate Nutrients for Agriculture Management): Scheme announced to use lesser amounts of chemical fertilizers in recent budget. Scheme will provide incentive to farmers use lesser amounts of chemical fertilizer by opting natural soil nutrients and improving soil health. (Sept. 2022)

The Global Liveability Index 2022 was published by the "Economist Intelligence Unit (EIU), recently. The Global Liveability index ranks cities across the globe for their "liveability quotient", on the basis of certain criteria.

India's first ever **"Animal Health Summit 2022"** was organised on July 6, 2022 at NASC Complex, New Delhi. The summit was organised by Indian Chamber of Food and Agriculture (ICFA) in collaboration with the Agriculture Today Group.

- **Mission Shakti Scheme** was launched by Government of India.
- It will be implemented for the period of 2021-22 to 2025-26.
- It an integrated women empowerment programme.
- It will act as an umbrella scheme for security, safety, and empowerment of women

Largest floating Solar Power Plant: World's largest floating Solar Power Plant will be constructed in Madhya Pradesh. It is aimed at boosting power generation capacity the state and addressing electricity problems in the central region.

Important Facts to know about the plant:

- The floating solar power plant will be constructed in Khandwa.
- It will generate 600-Megawatt power by 2022-23,
- This project is estimated to be worth Rs. 3000 crores.
- Khandwa district will become the only district in state to house thermal power station, hydel power plant as well as solar power plant.

On August 16, 2022, State Bank of India (SBI) launched the first **"state-of-the-art"** dedicated branch for start-ups to extend support and facilitate them. This branch is located in Koramangala, Bengaluru. It was launched by chairman of SBI Dinesh Khara.

The state-of-the-art facility;

Will provide end-to-end services to start-ups and enhance start-up potential.

World Electric Vehicle Day is observed on September 9 every year in order to mark the celebration of e-mobility.

KRITAGYA 3.0 was launched by ICAR to promote "speed breeding" for crop improvement. The Indian Council of Agricultural Research along with its National Agricultural Higher Education Project (NAHEP) and Crop Science Division is organizing a hackathon KRITAGYA 3.0 aimed at promoting speed breeding for the improvement of crops.

The updated Asian Development Outlook predicts that developing economies will faster than China, For the first time in 30 years, rest of developing Asia will grow at faster pace than China.

A new study published in the journal Nano Today revealed that **nanoplastics** travel up the human food web via plants, insects and fish. Nonplastics are plastic debris particles that are smaller than 1,000 nanometre, which is equal to one billionth of a metre. Because of their miniscule size, nanoplastics are capable of passing through physiological barriers and enter different organisms through the food chain. (September, 2022)

Asian Palm Oil Alliance (APOA) was formed at the Global Summit organized in Agra on September 21, 2022.

The APOA brings together the apex edible oil industry associations from India, Pakistan, Sri Lanka, Bangladesh and Nepal, which are the major importers of palm oil. Next meeting held at Indonesia in 2023.

The Indian government recently launched the JALDOOT App and JALDOOT App e-brochure This app can be used by Gram Rojgar Sahyak to measure the water level in a well twice a year before and after monsoon (September 28, 2022)

5G telephony service was rolled out by Prime Minister Narendra Modi at the 6th Edition of India Mobile Congress 2022 organized at Pragati Maidan in New Delhi. (on October 01, 2022)

The Swachh Survekshan 2024: Housing Ministry launches 9th edition of Swachh Survekshan 2024 to evaluate value chain of waste management.

Swachh Survekshan is an annual cleanliness survey that assesses the progress of India's cities in sanitation and hygiene. The 2024 survey includes several new indicators and is divided into four quarters:

First two quarters: Citizens provide feedback over the phone on various aspects of city cleanliness

Third quarter: Processing facilities are assessed

Fourth quarter: A field assessment is conducted on all indicators

Winners were: Indore (1st), Agra (2nd), Thane (3rd)

Some of the new indicators for 2024 include:

Service Level Progress (SLP)

This indicator now accounts for 60% of the total marks, up from 51% in 2023.

Urban aesthetics and citizen engagement

This indicator assesses the cleanliness of tourist attractions, monuments, and parks.

Participation in campaigns

This indicator assesses the participation of urban local bodies in campaigns driven by the Ministry of Housing and Urban Affairs.

The World Bank's report titled **"Poverty and Shared Prosperity 2022: Correcting Course"** provides the first comprehensive assessment of the COVID-19 pandemic's impact on poverty in developing countries. Consumer Pyramids Household Survey (CPHS). According to the CPHS, 5.6 crore people in India have slipped into poverty in 2020. This is higher than the World Bank's previous estimates of 2.3 crore additional Indians slipping into poverty in 2020.

The World Bank's private sector arm, IFC, launched the **Global Food Security Platform** to address global food crisis. (October 2022)

The report titled **"Global Status of Multi-Hazard Early Warning Systems (2023)**— Target G" was jointly released by United Nations Office for Disaster Risk Reduction (UNDRR) and the World Meteorological Organization (WMO) on the occasion of **International Day for Disaster Reduction.**

Target year set by India to achieve net zero Carbon emission in 2070 at COP 26.

The National Surveillance Programme for Aquatic Animal Diseases (NSPAAD) has launched a mobile app called **'Report Fish Disease'.** This app is designed to help fish farmers across India to track and monitor fish diseases.

The **"One Nation, One Fertilizer"** scheme was launched recently by Prime Minister Narendra Modi.

Key facts

- The "One Nation, One Fertiliser" scheme or **Pradhan Mantri Bhartiya Jan Urvarak Pariyojana** Single brand for fertilizers and logo **Bharat** was launched by the Central Government.

• This is the largest fertilizer initiative implemented by the Indian Government till date.

Union Minister of Petroleum and Natural Gas recently inaugurated **Asia's largest Compressed Bio Gas (CBG)** plant in Lehragaga, Punjab. (22 Oct 2022),

A new research found that climate change could increase the risk of **"viral spillover"** in several parts of the world. This could cause pandemics in the future

Global tree cover has increased by 130.9 million hectares over the past 20 years. Three-quarters of the increase was concentrated in 13 countries. The most significant increase was seen in Russia, Canada, the United States, Brazil and China.

China recorded the largest net increase in green cover of 2.1 million hectares. India recorded 0.87 million hectares increase in tree cover. Around 90 per cent of total tree cover gain in the world is attributed to natural regeneration and assisted natural regeneration that occurred outside plantation.

The increase in tree cover does not cancel out tree loss nor does it eliminate the adverse impact of forest degradation in terms of carbon storage, biodiversity, or ecosystem services.

The **United Nations Educational, Scientific and Cultural Organization (UNESCO)** warned that several world heritage sites may become **glacier-free by 2050.** A survey of 18,600 glaciers at 50 World Heritage Sites found that the glaciers at one-third of these sites will perish because of global warming regardless of the global efforts towards limiting the temperature rise.

UN Adpttion Gap Report: The report titled "The Adaptation Gap Report 2022: Too Little, Too Slow – **Climate adaptation failure puts world at risk"** was released by the United Nations Environment Programme. Report is published each year since 2014 UNEP.

State of the Global Climate in 2022 report was released by the World Meteorological Organization (WMO) on November 6.

• The past eight years were the warmest in record. This was caused because of the ever-increasing greenhouse gas concentration and accumulated heat.

• The global mean temperature in 2022 is around 1.15°C above the 1850-1900 pre-industrial average, ranging between 1.02°C to 1.28°C.

•This is close to 1.5°C – the lower limit set by the 2015 Paris Agreement mandate. The upper limit is 2°C.

• 2022 is slightly cooler when compared with the past few years because of the rare phenomenon of triple-dip cooling La Niña. This is likely to be 6th or 7th warmest year ever recorded.

Rajasthan's First Handicraft Policy: The Handicraft Policy 2022 seeks to harness the full potential of local handicraft industry and promote new jobs. (Launched on September 17, 2022)

• It will boost investments for the promotion of cultural heritage and revival of traditional arts and crafts.

The International Drought Resilience Alliance (IDRA) was launched on November 7 during the 27th Conference of Parties (COP27) Spain and Senegal jointly launched the IDRA to promote tion to increase preparedness against future droughts. The Alliance has 30 countries and 20 organisations. It seeks to create a shift in how the world responds to the growing drought risks by moving away from emergency responses to creating a long-term resilience.onvention on Climate Change (UNFCC)

The Ministry of New & Renewable Energy announced recently that the **National Bio Energy Programme will be implemented until 2025-26**

The National Bioenergy Programme aims to promote bioenergy usage and create an investor-friendly ecosystem based on the circular economy. Bioenergy is derived from once-living organic materials called biomass that are used for producing transportation fuels, heat, electricity and other such products.

The National Bioenergy Programme will be implemented in 2 phases. The programme will have three sub-schemes. They are: 1. Waste to Energy Programme, 2. Biomass Programme, 3. Biogas Programme

Global Carbon Budget 2023 The 2023 Global Carbon Budget projects that global fossil carbon dioxide (CO2) emissions will increase by 1.1% to a record 36.8 billion tonnes. This is due to increases in emissions from all fossil fuels, including coal, oil, and natural gas. The breakdown of emissions by fuel type is:

- Coal: 41% of global CO_2 emissions, up 1.1%
- Oil: 32% of global CO_2 emissions, up 1.5%
- Natural gas: 21% of global CO_2 emissions, up 0.5%
- Cement: 4% of global CO_2 emissions, up 0.08%

Other findings from the 2023 Global Carbon Budget include:

- Atmospheric CO_2 concentration: 419.3 ppm, which is 51% higher than pre-industrial levels
- Land-use change emissions: 4.1 billion tonnes of CO2, which is a net emission
- Ocean and terrestrial vegetation: Absorbed about half of all CO2 emissions, which is the same as it has been for the past six decades

The Global Shield Against Climate Risks initiative was launched on November 14, 2022 by the Vulnerable Twenty (V20) countries and G7 countries. While V20 countries represent 58 countries that are vulnerable to climate change, the G7 represent seven of the world's most industrialized countries. It is a social protection and insurance-based finance mechanism for loss and damage outside the UNFCCC process.

Climate Change Performance Index 2023 was released jointly by Germanwatch, NewClimate Institute and Climate Action Network.

The Climate Change Performance Index (CCPI) is released every year at the UN Climate Change Conference. It is an instrument that ensures transparency in the national and international climate politics. Since 2005, it has provided standardized assessment of countries' climate protection performance. (November 14, 2022)

Carbon Credit Trading: The 2022 Amendment Bill empowers the GoI to specify the carbon credit trading scheme. Carbon credit is a tradeable permit that allows the production of specified amount of carbon dioxide and other GHG emissions. The carbon credit certificates, issued by the government, can be traded. (Dec. 2022.)

India-based **Kalyani Group** has launched India's first **"green"** steel brand called *Kalyani Ferresta.* (Dec 2022)

FSSAI has given millets a full group standards as per FSSAI second amendment Regulations 2023. A govt. statement said that the rules would go in to effect on sep 1, 2022. FSSAI regulations 2011 covers only a few millets, like jowar, ragi, bajra and amaranth crop with their own standards.

Kerala government is to establish **"Green Hydrogen Hubs".** The State Government is allocating Rs 200 crores for this. The hubs are to be set up in **Kochi and Thiruvananthapuram.** The project is to be implemented along with the private partner India Hydrogen Alliance.

The plan is to build green hydrogen plants that can produce 60 tonnes of hydrogen per day.
The plant will be established with a 150 MW Electrolyser and storage infrastructure.

INDIAENERGYWEEK2023: The India Energy Week was inaugurated by PM in Bengaluru, Karnataka. The event is to focus on energy security and India's energy transition goals. This includes net-zero emission target achievement by 2070. The major attraction of the event is the launch of E20 fuel in more than 84 outlets in 11 states and UTs.

INDIA"s FIRST DRONE MANAGEMENT SYSTEM: Skye Air is a startup delivering drones in different parts of India. The startup recently launched a Traffic Management System for drones. It is software and will allow drone operators to choose their routes. Just like flight traffic management, the need for drone traffic management is increasing. With e-commerce businesses looking for options for drone deliveries, drone usage is to increase in the future exponentially.

SEA CAGE FARMING : Sea cage farming is growing fish in the sea using an enclosed net. Even if we use just 1% of inshore water in this, we can produce 3.2 million tonnes of fish. However, India is lagging in the method. To increase sea cage farming in the country, **CMFRI identified 146 spots.**

INDIAN RICE CONGRESS 2023: The second edition of the Indian Rice Congress was recently inaugurated by President Murmu at National Rice Research Institute at Cuttack, Odisha. The 2023 congress was held on the theme **"Transforming Rice Research".** 2024 Rice congress will held from Dec. 5-7, 2024 a Cuttack.

Wolf Prize Winner 2023: Agriculture: Martinus Th Rien from the Federal University of Rio De Janeiro; for his research on **water flow and contaminant transport in soils**

Millet International Initiative for Research and Awareness: The Government of India is to launch Millet International Initiative for Research and Awareness. It is in line with the United Nations' plans for the International Year of Millets, which is 2023. India brings in the concept of "SEED MONEY" in the initiative. Research activities on millet, and increasing their productivity will be focused on the initiative. Indian Institute of Millet Research located in Hyderabad will provide technical support to the initiative.
Key Focus of the mission: Gender Equity, Inclusion, and Equity
Upon India's insistence, the United Nations declared 2023 as the International Millet Year.

Teja Chilli: The Teja Chilli of Telangana is the most popular chilli in many countries. Despite high international demand, the farmers of Teja Chilli are facing heavy losses. They are struggling to recover their investment. The Teja Chilli is the major ingredient of Oleoresin. (Feb 2023)

The Cross Dependency Initiative is a business consultant that specializes in climate risk. It released a report on **"Gross Domestic Climate Risk".** According to the report, the top 50 provinces facing climate risks are China, the US, and India. It says nine Indian states are most vulnerable to climate risks. They are Punjab, UP, Maharashtra, Rajasthan, TN, Gujarat, Kerala, and Assam. Mumbai has the highest risk of all. 26 provinces in China, 9 in India, and 5 in the US are the most vulnerable of all. (Feb 2023)

The Indian Council of Agricultural Research recently introduced a climate-smart wheat variety called the **HD-3385** (Feb 2023)

The Agriculture Innovation Mission for Climate is a joint initiative of the US. It was started by UAE and USA. The main objective of the initiative is to address hunger and climate change. This is to be achieved by inviting countries and other private organizations to invest in innovative and climate-smart agricultural systems.

Following this, India introduced an international brand name for millet **"SHREE ANNA** during the budget. Now, taking a step further, the Food Safety and Standards Authority of India has issued standards for millets. These standards are to be implemented from September 2023

The **recent frequent earthquakes** have helped scientists discover a new core. **This fifth inner core is a solid ball of nickel and iron** and is the innermost layer. (Feb 2023)

Plasticosis in Birds: Scientists from Australia and the UK have discovered a new fibrotic disease called **"Plasticosis"** in birds. This disease is caused by the consumption of plastics, which leads to long-term inflammation and the formation of scar tissue in the digestive system of birds. The study noted that the birds had consume d some 12.5% of their body weight in plastics. (March 2023)

Millets Giveaway: The Small Farmers' Agri-Business Consortium, a society promoted by the Ministry of Agriculture and Farmers Welfare of the Indian government, has launched a special marketing campaign called **Millets Giveaway.** The campaign is aimed at promoting direct purchases from small and marginal farmer producer organisations (FPOs) to support their livelihoods. (Mar 2023)

e-fUELS: E-fuels are a type of fuel that is produced by capturing carbon emissions and combining them with hydrogen made from renewable or CO_2-free electricity. Their examples include e-kerosene, e-methane, and e-methanol. (Mar 2023)

Green National Highway Corridors: India and the World Bank have recently signed a loan agreement to construct 781 kilometers of green highways across four states in India. The project aims to build the Green National Highway Corridors Project in 4 Indian states. The loan assistance of 500 million dollars will be used for the construction of the highways. The Green National Highway Corridors Project will promote the use of green technologies and climate resilience to demonstrate safe and environmentally friendly highways. The project aims to establish a new standard for sustainable infrastructure in India. (Mar 2023)

Global Millets (Shree Anna) Conference: On March 18, 2023, the Prime Minister of India, inaugurated the **Global Millets Shree Anna Conference 2023** in the national capital. The event marks the beginning of a two-day global conference on millets, where PM unveiled a customised postal stamp and commemorative currency coin to mark the International Year of Millets, being celebrated in 2023.

PM MITRA Parks: The Indian government recently announced the establishment of seven PM MITRA Parks in Tamil Nadu, Telangana, Karnataka, Maharashtra, Gujarat, MP, and UP. These parks are inspired by Prime Minister 5F vision, which aims to boost the Indian textile industry from farm to fashion to foreign markets. The PM MITRA scheme was announced in October 2021, and the parks are expected to be set up by 2026-27, with a total outlay of Rs. 4,445 crore. The initial allocation in the 2023-24 Budget is only Rs. 200 crore, but the government expects the parks to attract crores of investment and create lakhs of job (18 Mar 2023)

The study titled "Pathways to Atmanirbhar Bharat" was released recently by Lawrence Berkeley National Laboratory and The India Energy and Climate Center (IECC). According to this report, India has an opportunity to achieve energy independence by 2047 through investments in clean energy technologies. (Mar 2023)

The UN 2023 Water Conference is being held from March 22 to 24 in New York, marking the first summit on water in 50 years. The conference will be a watershed moment to mobilize member states, the UN system, and stakeholders alike to take action and bring successful solutions to a global scale

United Nations World Water Development Report 2023: According to the United Nations World Water Development Report 2023, India is expected to be the most severely affected nation by water scarcity by 2050. This is a worrying trend, given that water scarcity is already a significant issue in India, with millions of people facing severe water shortages. (March 23, 2023)

The National Centre for Sustainable Coastal Management (NCSCM) was established in 2011 as a research institute to support the conservation, protection, restoration, and management of Indian coasts. Its vision is to promote sustainable coasts through increased partnerships, conservation practices, scientific research, and knowledge management for the benefit and well-being of current and future generations. Claude Lorius, a pioneering glaciologist, passed away on Tuesday morning at the age of 91. He was best known for his research into climate change, specifically his work in Antarctica which helped prove that humans were responsible for global warming. (24 Mar 2023)

National Crop Insurance Portal Launched: On March 24, 2023, Union Minister of Agriculture launched the National Crop Insurance Portal's digitized claim settlement module 'DigiClaim' in New Delhi, India. The module falls under the ambit of Pradhan Mantri Fasal Bima Yojana (PMFBY), and will disburse claims electronically, benefitting farmers from six states. The launch marks a revolutionary step towards ensuring that farmers can receive claim amounts digitally in a time-bound and automated manner, making them self-reliant and empowered.

The National Dairy Research Institute (NDRI) in Karnal, Haryana, has produced the country's **first cloned female calf** of the desi breed Gir, named **'Ganga.'** The NDRI had initiated a project to clone indigenous cow breeds such as Gir and Sahiwal to increase milk production. (30 Mar 2023)

Cruelty to Animals New Rules: Prevention of Cruelty to Animals Act was implemented in 1960 to prevent the mistreatment of animals. Recently, the Central Government announced new rules on March 27, 2023, to prevent cruelty to animals during specific procedures, including dehorning of cattle and castration, bra ew rules specify the procedures for dehorning of cattle and castration, branding, or nose-roping of any animal. The rules also require the use of pain-reducing methods during these procedures. Additionally, the law suggests that tagging, branding, and radio-telemetry devices should be placed on dead tissues.

Three New Species of Bush Crickets: The world of insects is vast and diverse, with new species being discovered all the time. Recently, a team of researchers led by Aarini Ghosh and Ranjana Jaiswara discovered three new species of bush crickets in India. These new species are members of the **Hexacentrus cricket family,** and they are unique in their ability to produce ultrasonic sounds of up to 60 kHz. (5 Apr 2023)

World Energy Transitions: Outlook 2023: The International Renewable Energy Agency (IRENA) released its World Energy Transitions: Outlook 2023 report, highlighting the need for a more aggressive approach to transition towards renewable energy sources. The report outlines the current state of renewable energy generation and provides insights into the challenges that need to be overcome to achieve the goals outlined in the Paris Agreement. (8 Apr 2023)

Food conclave 2023: On April 28 and 29, the Telangana government hosted the Food Conclave-2023. The event is an annual brainstorming session that brings together 100 thought leaders from the agri-food industry. The conclave's objective is to identify the main challenges and opportunities for the growth of the Indian agri-food sector in the present decade.

Nagarjunasagar-Srisailam Tiger Reserve: Nagarjunsagar-Srisailam Tiger Reserve is located in the states of Andhra Pradesh and Telangana, making it one of the few tiger reserves in India that spans across two states. It is also the largest tiger reserve in India, covering a total area of 3,728 km2 (1,439 sq mi). The core area of the reserve is 1,200 km2 (460 sq mi), which is an area dedicated solely to the conservation of tigers and other endangered species. (15 Apr 2023)

Global Food Policy Report, 2023: The International Food Policy Research Institute (IFPRI) released the Global Food Policy Report, 2023 on April 13, urging stakeholders to invest in long-term solutions for building resilient and equitable food systems. The report highlights that multiple crises, such as the COVID-19 pandemic, climate change, natural disasters, civil unrest, and political instability, have led to a rise in food insecurity worldwide during 2020-2022.

GLOBAL CLIMATE REPORT: The State of the Global Climate report is an annual publication from the World Meteorological Organization (WMO) that presents the latest findings on the state of climate change worldwide. The report combines observations from global networks of weather and climate stations, ocean buoys, and satellites, as well as modeling studies, to provide a comprehensive assessment of the Earth's climate system. The most recent report, released ahead of Earth Day 2023, reveals several alarming trends that highlight the urgent need for climate action. (April 22, 2023)
The Status of Women in Agrifood Systems Report (April 2023): The Food and Agriculture Organization (FAO) recently released a report titled "The Status of Women in Agrifood Systems" which sheds light on the gender disparities in agriculture and its impact on global economy and food security.
MOREL MUSHROOM: Guchhi, also known as Morel mushroom, is a prized delicacy found in the Himalayan region. It is a wild mushroom belonging to the Morchellaceae family and is known for its unique spongy head. This mushroom is not only delicious but is also packed with essential nutrients. It is expected to receive GI tag in the near future. (April 25, 2023).
CENSUS REPORT OF WATER BODIES : The Ministry of Jal Shakti conducted its first-ever census of water bodies in India between 2018-19. The objective of this census was to collect comprehensive data on all aspects of water bodies in the country, including their condition, encroachment status, and usage. The Water Bodies: First Census Report includes information on various types of water bodies, their distribution across different states and districts, and their current state. (April 25, 2023). According to report, 24,24,540 water bodies have been enumerated in the country, out of which 97.1% (23,55,055) are in rural areas and only 2.9% (69,485) are in urban areas. 59.5% (14,42,993) ponds. Tanks- 15.7%, (3,81,805). Reservoirs-12.1%, (2,92,280). Water conservation schemes- 9.3%, (2,26,217). Lakes 0.9%, (22,361) and others-2.5%, (58,884).
WORLD DEVELOPMENT REPORT 2023: The world population has reached eight billion and is expected to grow for decades, but many countries are facing a sharp drop in working-age adults. The competition for workers and talent will intensify globally, and many countries will be reliant on migration to realise their long-term growth potential. A recent report from the World Bank, titled "World Development Report 2023: Migrants, Refugees, and Societies",

proposes policies for better migration management in destination, transit and origin countries, underscoring its urgency. (April 26, 2023)

The Confederation of Indian Industry (CII) organized the third edition of the Summit on **Farm Machinery Technology,** which was held at the India Habitat Centre in New Delhi from 27-28 April 2023 in physical format. The previous two editions of the summit were conducted virtually, due to the ongoing COVID-19 pandemic. The recent edition was held in physical format.

Navegaon Nagzira Tiger Reserve (NNTR) in Maharastra recently recorded a slight rise in tiger population with a tigress giving birth to four cubs in the protected area. (May 19, 2023)

Biotransformation technology is a **novel approach to ensure plastics that escape refuse streams** are processed efficiently & broken down.

Plastics made using this technology are given a pre-programmed time during which the manufactured material looks and feels like conventional plastics without compromising on quality.

Once the product expires and is exposed to the external environment, **it self-destructs, and bio transforms into bioavailable** wax.

This wax is then consumed by **microorganisms, converting waste into water**, CO_2, and biomass.

This biotransformation technology is the **world's first that ensures polyolefins** fully biodegrade in an open environment causing no microplastics.

DEATH CAP MUSHROOM: The death cap mushroom, scientifically known as **Amanita phalloides,** carries a notorious reputation for its deadly consequences when consumed. With liver failure and even kidney failure as potential outcomes, this mushroom poses a significant threat to those who mistakenly ingest it. Recently, a research has identified a potential antidote for the death cap mushroom, offering a ray of light in the fight against its toxicity. (May 18, 2023)

FAO's **Agriculture and Market Information System (AMIS)** is an inter-agency platform that was set up to boost transparency in the food market and promote policy response for food security. In its latest monitor, the AMIS revealed that India's corn, soyabean and rice production will be affected due to the El Nino weather event. (May 6, 2023)

The **Global Report on Food Crises (GRFC)** for 2023 was produced by the Food Security Information Network (FSIN) and released by the Global Network against Food Crises (GNAFC). The GRFC report serves as a reference document for acute food insecurity at the global, regional, and country levels. The primary objective of the report is to provide independent and consensus-based evidence and analysis that can inform humanitarian and development actions.

Dr P. Indira Devi Conferred Prestigious ISAE Fellow 2023 Title (December 27, 2023)

Dr P. Indira Devi, an eminent agricultural economist from Kerala, has been conferred the distinguished title of ISAE Fellow 2023 by the Indian Society. of Agricultural Economics (ISAE). This recognition highlights her trailblazing contributions spanning over three decades in the domains of agricultural research, policymaking and teaching.

Jute Corporation of India Launches 'Paat-Mitro' App for Jute Farmers (December 22, 2023)

The Jute Corporation of India took a significant step toward supporting jute farmers by launching the 'Paat-Mitro' app. Rachna Shah, Secretary, Ministry of Textiles, inaugurated the app, which aims to provide essential knowledge to farmers for optimal jute cultivation and increased income.

NCRB Data on Farmer Suicides (December 5, 2023)

he year 2022 witnessed a distressing surge in farmer suicides across India, according to the latest data from the National Crime Records Bureau (NCRB), released on December 4, 2023. The grim statistics reveal a 3.7% increase in such cases, with 11,290 reported deaths, compared to 10,281 in 2021 and a 5.7% increase from 2020 figures.

Climate Risks for Women in Agri-Food Systems: Hotspot Map Highlights Vulnerability (November 20, 2023)

A new hotspot map, developed by an international team of researchers, reveals that women engaged in agri-food systems in Africa and Asia, including India, face the highest climate risks. The map, ranking 87 countries based on climate change threats to women in agricultural sectors, identifies areas where climate hazards, exposure faced by women, and vulnerability due to gender inequalities converge.

Tea Mosquito Bug Infestation (April 20, 2023)

India is the world's second-largest tea producer after China. However, the tea industry in India has been facing a major setback due to the infestation of tea mosquito bugs. These pests, which are a type of mirid bug, feed on the tender tea leaves and cause extensive damage to the tea bushes. This has led to a significant decline in tea production, especially in the Valparai region of Tamil Nadu, where the production has fallen by around 50 percent over a decade.

Mizoram to export Bird's eye chilli to the US (March 1, 2023)

The Birds Eye Chilli is widely used in Asian cuisines. The chilli is also called Thai chilli. It is native to Mexico. The Spanish and Portuguese people spread the chilli all over the world. The chilli variety has received a GI tag from GoI. The state of Mizoram is to export the chilli variety.

PM PRANAM (Launched on June 28, 2023)

Features of the scheme:

1. The scheme will not have a separate budget and will be funded by the "savings of existing fertilizer subsidy" under schemes run by the Department of Fertilizers.
2. 50% subsidy savings will be given as a grant to the state that saves the money.
3. 70% of the grant provided under the scheme can be used for asset creation (adoption of alternate fertilisers at the village, block and district levels).
4. 30% grant money can be used for encouraging farmers, panchayats, FPOs and self-help groups that are involved in the reduction of fertiliser use and awareness generation.
5. Data available on a fertiliser Ministry dashboard, iFMS (Integrated Fertilisers Management System), will be used for this purpose.

Organic Aadhaar (June 2021)

Features:

1. Agricultural and Processed Food Products Export Development Authority (APEDA) may introduce "Organic Aadhaar" for the farmers who do organic farming under the National Programme for Organic Production (NPOP).
2. Organic Aadhaar is an identity for the farmer for the organic certification process based on the Aadhar number and Unique Land Parcel Identification Number (ULPIN).
3. ULPIN is a 14-digit identification number given by the government to every plot of land in the country.
4. Linking with the ULPIN could also help in identifying organic land.
5. Organic Aadhaar along with Tracenet will create a strong system for effectively planning and monitoring policy measures.

6. Tracenet system, which started in 2009, gives information on products exported from the country.
7. Organic Aadhaar will give the farmer in a growers group exclusive protection of organic status and income.

Seed Treaty

1. India recently hosted the 9th session of governing body of the Seed treaty.
2. International Accord for Plant Genetic Resources: Nurturing Biodiversity and Collaboration.
3. The Seed Treaty, formally known as the International Treaty of Plant Genetic Resources for Food and Agriculture (ITPGRFA), constitutes a significant internationally legally binding agreement aimed at conserving, utilizing, and overseeing plant genetic resources.
4. Enacted by the FAO in 2001, it gained enforcement in 2004.
5. India is a participating member

Purse Seine Fishing

It is a method of fishing that employs a surrounding net, called a seine that hangs vertically in the water with its bottom edge held down by weights and its top edge buoyed by floats. It is a non-targeted fishing gear and catches all sorts of fishes come in the way of net including juveniles. Hence, are detrimental to marine resources.

1. Some coastal states (such as Tamil Nadu, Odisha, Kerala, etc.) have enforced restrictions on **purse seine fishing.**
2. This move is criticised by the central government, which deems the action as unwarranted.

Regenerative agriculture

Regenerative agriculture is a farming and ranching approach that aims to improve soil health and restore the land for future generations. It's a holistic way of farming that considers the connection between farming systems and the ecological system as a whole

1. Regenerative agriculture focuses on improving the health of soil, which has been degraded by the use of heavy machinery, fertilizers and pesticides in intensive farming.
2. There may not be enough soil left to grow food to feed the world within 50 years.

3. Regenerative agriculture and other farming methods that don't harm the climate can improve farmers' incomes, as well as cutting emissions and boosting soil health.

Lumpy Skin Disease : Lumpy skin disease is a viral disease that affects cattle. It is transmitted by blood-feeding insects, such as certain species of flies and mosquitoes, or ticks.

It causes fever, nodules on the skin and can also lead to death, especially in animals that that have not previously been exposed to the virus. Control options include vaccinations and culling of infected animals.

Lumpy skin disease can lead to significant economic losses

Koronivia Joint Work on Agriculture (KJWA): The Koronivia Joint Work on Agriculture (KJWA) is a landmark decision under the United Nations Framework Convention on Climate Change (UNFCCC) that recognizes the unique potential of agriculture in tackling climate change.

1. India has taken a stand against the KJWA, a measure aimed at decreasing greenhouse gas emissions from agricultural activities.
2. India's opposition is based on the argument that emissions from the agricultural sector are not considered "luxury" emissions but rather essential for the survival of impoverished communities.
3. KJWA is a significant initiative introduced within the framework of UNFCCC with the purpose of curbing greenhouse gas emissions originating from agricultural practices.
4. Despite India having relatively low per capita carbon emissions, it ranks as the third-largest global contributor to carbon emissions overall.
5. Approximately 7 percent of worldwide greenhouse gas emissions are attributed to the agricultural sector.

Pesticide Action Network

The Pesticide Action Network (PAN) has released a report indicating that several states are exceeding the recommended levels of specific pesticides in their usage.

The pesticides in question include:

1. Chlorpyrifos, which is utilized to control soil-borne insects but is notorious for causing brain damage; it has been prohibited in numerous countries.
2. Fipronil, employed to manage ants, beetles, and fleas.

7. This variety is grown using traditional methods and is highly nutritious.
8. It is in great demand in Kerala and Goa

Diammonium Phosphate

1. The Government of India is supporting Indian fertilizer companies to strengthen their backend supply chain, as part of the Atma-Nirbhar Bharat initiative.
2. One of the fertilizers that they are supporting is Diammonium Phosphate (DAP), which is the world's most widely used phosphorus fertilizer.
3. DAP is a preferred fertilizer in India because it contains both Nitrogen (18% N) and Phosphorus [46% P_2O_5 (20% P)], which are primary macronutrients and part of 18 essential plant nutrients.
4. DAP is manufactured by reacting ammonia with phosphoric acid under controlled conditions in fertilizer plants.
5. DAP is an excellent source of P and N for plant nutrition.
6. It is highly soluble and thus dissolves quickly in soil to release plant-available phosphate and ammonium

New Turmeric Cousin Discovered in Nagaland's Ungma Village

September 9, 2024

During a field trip to Nagaland, researchers found a new species of plant closely related to turmeric. This new species, named **Curcuma ungmensis,** was discovered near the village of Ungma

Namo Drone Didi Initiative: Recently, Prime Minister Narendra Modi distributed drones to 1000 Namo drone Didis from across the nation. The scheme aims to provide drones to 15,000 selected Women SHGs for providing rental services to farmers for agriculture purpose. It is a Central Sector Scheme where 80% of the cost of the drone will be provided by the union government. Time period: 2023-24 to 2025-26

Subsidy of 80% of drone cost or a maximum of Rs. 8 lakh will be provided. Nominal Interest Rate of 3% will be payable on the loan. Training to fly drone will also be provided.

International Year of Millets (IYM) 2023.

India has successfully persuaded the UNGA to designate 2023 as the International Year of Millets India is the largest producer of millets in the world, with a 41% share in 2020. . The main millet-growing states are Rajasthan, Maharashtra, Uttar Pradesh and Karnataka. India has also launched a Sub-Mission on NutriCereals under the National Food Security Mission to boost millet cultivation and consumption. Millets were renamed "Nutri Cereals" by the government in 2018, which was also declared the National Year of Millets.

The Union Budget of 2023-24 referred to millets as **"Shree Anna"** - the mother of all grains. The Prime Minister inaugurated the Global Millets (Shree Anna) Conference at IARI Campus, PUSA New Delhi. The ICAR's Indian Institute of Millets Research (Hyderabad) was announced as a Global Centre of Excellence.

AgriSURE Fund Empowers Startups Supporting Indian Farmers (September 4, 2024)

The Indian government has launched the AgriSURE Scheme, an important initiative designed to support agricultural start-ups and rural businesses. This scheme is essential for strengthening India's agricultural sector, which plays a vital role in the country's economy.

AIM Invites Applications for Australia-India Agritech Program

August 20, 2024

The Atal Innovation Mission (AIM) has opened applications for Indian and Australian startups and small businesses to join the climate-smart Agritech group of the India Australia Rapid Innovation and Startup Expansion (RISE) accelerator. This program is designed to help businesses grow internationally and collaborate between India and Australia, especially in the Agritech field.

New Orchid Species Discovered in Sikkim (August 17, 2024)

A newly discovered orchid species, *Gastrodia indica*, was found in **Fambonglho Wildlife Sanctuary in Sikkim,** India, by botanist Madhusudan Khanal. This orchid is notable because it does not open its flowers, a rare trait among orchids, making it the first known cleistogamous orchid in India.

New Plant Species Discovered in Arunachal Pradesh Sanctuary

July 22, 2024

Researchers from the Botanical Survey of India (BSI) have identified a new plant species, *Phlogacanthus sudhansusekharii*, in the Itanagar Wildlife Sanctuary, Arunachal Pradesh. This find brings attention to the area's wide range of plants and animals.

Nagaland Wins Best State in Horticulture Award 2024 (July 14, 2024

Nagaland won the **"Best State in Horticulture Award 2024"** at the Agriculture Leadership Awards, which shows how clever its horticulture projects are. Minister Salhoutuonuo Kruse took the award in New Delhi and said that the government's policies are aimed at promoting prosperity in rural areas and long-term farming.

Krishi-DSS Digital Geospatial Platform (August 19, 2024)

The Indian government introduced the Krishi-Decision Support System (Krishi-DSS), a digital platform designed to manage agricultural data across the country, on August 16 2024. This tool provides a wide range of information, including satellite images, weather data, and soil health ..

Denmark Imposes the World's First Carbon Tax on Livestock Emissions (June 29, 2024)

In a historic move, Denmark's coalition government approved a plan to put in place the world's first carbon emissions tax that targets farms.

Haryana Removes Age Cap on Farmer Aid Scheme (June 7, 2024)

The government of Haryana recently made a major change to the "Mukhyamantri Kisan Evam Khetihar Mazdoor Jeevan Suraksha Yojana," a program that helps farmers and farm workers get money.

Caterpillars Sense Electric Fields to Detect Predators (May 28, 2024)

British scientists have found something very interesting about caterpillars: they have a sixth sense that most land animals don't have. Through special hairs on their bodies called setae, they can pick up on electric fields with this sense, called electroreception. The study, which was published in the Proceedings of the National Academy of Sciences, looked at four types of caterpillars: the cinnabar moth, the rare vapourer moth, the European peacock butterfly, and the common wasp.

7th Session of Codex Committee on Spices and Culinary Herbs (CCSCH) held in Kochi (February 4, 2024)

The 7th Session of Codex Committee on Spices and Culinary Herbs (CCSCH) was recently held in Kochi. During this session, the committee finalised and recommended quality standards for five spices — allspice, juniper berries, star anise, turmeric, and cardamom.

The Punganur Cow – India's Shortest Humped Cattle Breed

January 18, 2024

Prime Minister Narendra Modi recently fed Punganur cows from Andhra Pradesh at his Delhi residence on the Pongal/Makar Sankranti festival. These indigenous cattle have unique characteristics.

Supreme Court Questions Govt on GM Crops (January 15, 2024)

On January 11, the Supreme Court asked why the government's Genetic Engineering Appraisal Committee (GEAC) did not consider expert reports on genetically modified (GM) crops. The GEAC approves the release of GM organisms.

Field Studies Find Nano Urea Failing to Match Conventional Fertilizers in Crop Yields (January 4, 2024)

A two-year Punjab Agricultural University trial discovered paddy and wheat harvests declined substantially from nano urea usage versus normal nitrogen fertilizers, contradicting tall corporate claims.

Performance evaluation of Fremont mandarin on different rootstocks under the hot arid environment of India: Karna khatta rootstock found best for growth and yield traits, whereas **Pectinifera rootstock delivered superior fruit quality with high stionic effects.** Pectinifera rootstock induced dwarfism in **Fremont mandarin** with high rate of scion-rootstock compatibility which enabled high density orcharding and leads to microclimate effect on crop management in arid climate.

Tecomanthe hillii - the Fraser Island Creeper - is a spectacular vine from Queensland, Australia, where it is rare. Each winter & spring it's loaded with rosy, bell shaped blooms in a circular, skirt-like arrangement. When grown over an arbor or trellis, it makes an amazing floral display. The plant is quite rare in cultivation, and seldom seen for sale.

"Fraser Island Creeper" is somewhat of a misnomer, because the plant climbs rather than creeps along the ground. It twines itself around any narrow support, including other plants, however it doesn't smother them. Unlike most vines, the flowers emerge directly from the woody stems, not from the shoots. The big, 3 inch blossoms are a delightful reddish-pink, with a creamy white throat. The vine is attractive even without blooms, thanks to its glossy, hand-shaped leaves.

It comes from a warm climate, where temperatures are usually between 60 and 90°F (16-32°C). It tolerates cooler conditions, and even a degree or two of frost, but it's happiest in warmer temps. It can take anything from full sun down to bright shade, but it seems to do best mid-way, with a bit of afternoon shade.

Govt approves schemes to increase farmers' income, ensure food security: The Pradhan Mantri Rashtriya Krishi Vikas Yojana and Krishonnati Yojana will be implemented with a total proposed expenditure of Rs 1,01,321.61 crore. (3rd Oct. 2024.)

Novel Bacillus thuringiensis isolate based biopesticide for control of White grub **Holotrichia serrata** in sugarcane.

Pear hybrid indentified CITH- Pear hybrid-08: Parent Kashmiri Nakh with good taste, crunchy juicy without grit cells. Good post harvest shelf life.

Nosema disease or nosemosis identified as a significant threat to Agriculture: Nosema caused by microsporidian parasides is a source disease affecting adult honey bees. Domesticated and native bees face a variety of deadly threats causing mortality and reduced fecundity and thus endanger agriculture for bees for pollination.

Cabinet approves ambitious Clean Plant Programme (CPP) to revoluting Horticulture. (9 Aug. 2024) : With a substantial investment of Rs.1,765.67 crore, this pioneering initiative is set to revolutionize the horticulture sector in India and expected to set new standards for excellence and sustainability. Announced earlier in the Budget Speech by the Finance Minister in February 2023, the CPP represents a major leap forward in enhancing the quality and productivity of fruit crops across the nation.

Fijivirus is a genus of viruses that infect monocotyledonous plants, such as those in the Gramineae and Liliaceae families. The viruses are transmitted by **delphacid planthoppers** and cause diseases like Fiji disease and galls in infected plants. It caused dwarfism in the paddy crops in Northen State of India.

Price stabiliation fund (PSF) is a program that aims to prevent sudden price changes of essential commodities in India. The PSF was established in 2014-15 to: Protect consumers by supplying commodities at reasonable prices, Discourage hoarding and speculation, and Incentivize domestic production.

Paddy straw torrefaction is a process that improves the quality of paddy straw as a fuel by increasing its energy density, grindability, and storability. The torrefaction process can also reduce the moisture and oxygen content of the straw. Recently Environment Ministry released guidelines for grant of one time financial support for establishment of torrefaction plants to promote utilization of paddy straw and stubble in thermal power plants.

The fishbone channel plantation method is a conservation technique that uses a fishbone-shaped network of canals to reforest mangroves in areas that don't regularly receive tidal inundation:

Green manure : The manure which is prepared by decomposing of fast growing grasses as sunhemp, cow pea, etc. in field itself is called green manure. For example -Sun hemp is grown in fields, mulched by ploughing and allowed to decompose in field for the preparation of green manure.

A Primary Agricultural Credit Society (PACS) is a grassroots-level cooperative institution that provides short-term and medium-term loans to farmers and other weaker sections of society. PACS are the lowest tier of the short-term cooperative credit structure in India, with District Cooperative Banks (DCCBs) and State Cooperative Banks (SCBs) at the higher tiers

Fodder-centric Farmer Producer Organizations (FPOs) are producer-led organizations that focus on fodder as their primary activity, while animal husbandry is a secondary activity. The goal of these organizations is to address India's fodder deficit. The Ministry of Fisheries, Animal Husbandry and Dairying proposed the idea in 2020, and the National Dairy Development Board (NDDB) was tasked with implementing the scheme.

Indian Council of Agricultural Research (ICAR) (6th August 2024)**:** The Indian Council of Agricultural Research (ICAR) launched its 'One Scientist-One Product' programme on July 16, enhancing research in agriculture and animal husbandry. During its 96th foundation day celebration in Delhi, ICAR announced 323 new crop varieties, including 289 climate-resilient and 27 bio-fortified types.

ADDITIONAL INFORMATION

Ethanol Production in India increased from 40 crore litres to 340 crore litres in just seven years, The **Ethanol Blending Programme (EBP) (2003)** which aims to blend ethanol with petrol, has been a major driver behind this surge. This increase aligns with India's strategic goal to reduce its dependence on imported petroleum and to promote cleaner, **alternative fuels**. Present Target of **ethanol bending** is **10%** and set for 2025-26 is **20%** and 30% by 2029-30 Govt. Policy supports using maize, surplus rice and damaged grains for ethanol. Current capacity in 2023 of ethanol production is about **433 crore litres.**

The **"One District One Product" (ODOP)** initiative in India is a transformative approach to foster balanced regional development across all districts of the country. It aims to select, brand, and promote at least one product from each district, thereby enabling holistic socio-economic growth.

Some of the brands developed under the ODOP initiative include:

Amrit Phal: Developed for Gurugram, Haryana.

Cori Gold: A brand for coriander powder from Kota, Rajasthan.

Kashmiri Mantra: A brand of spices developed under the ODOP concept in J&K.

Madhu Mantra: A brand for honey from Saharanpur, Uttar Pradesh.

Somdana: A brand for millets from Thane, Maharashtra

Whole Wheat Cookies of Dilli Bakes: Developed under the bakery ODOP concept for Delhi.

Scheme adopts the One District One Product (ODOP) approach to reap the benefit of scale in terms of procurement of inputs, availing common services and marketing of products. ODOP for the scheme will provide the framework for value chain development and alignment of support infrastructure. There may be more than one cluster of ODOP products in one district. There may be a cluster of ODOP products consisting of more than one adjacent district in a State.

NECTAR (North East Centre for Technology Application and Reach) established in 2012 at Shillong in Meghalaya set up by Department of Science and technology. It was esablished by merging two missions viz. National Mission for Bamboo Application (NMBA) and mission for geospatial applications (MGA). NECTAR will be ensuring applications of appropriate technologies for development in the areas of biodiversity concerns, watershed management, telemedicine, horticulture, infrastructure planning & development, planning and monitoring, and tele-schooling using cutting-edge MESHNET solutions, employment generation etc. through utilization of local products/resources and associated skill development. *Achievemens:* 1. Training on National Packaging for Agricultural and Horticultural Organic producs. 2. New low cost powerless Heating system activated with water can heat food in remote areas.

India's Stand at COP-26: The Government of India has articulated and put across the concerns of developing countries at the 26 session of the Conference of the Parties (COP26) to the United Nations Framework Convention on Climate Change (UNFCCC) held in Glasgow, United Kingdom. Further, India presented the following:

Five nectar elements (Panchamrit) of India's climate action:

i) Reach 500GW Non-fossil energy capacity by 2030.

ii) 50% of its energy requirements from renewable energy by 2030.

iii) Reduction of total projected carbon emissions by one billion tonnes from now to 2030.

iv) Reduction of the carbon intensity of the economy by 45 per cent by 2030, over 2005 levels.

v) Achieving the target of **net zero emissions** by **2070.**

The **"Mantra of LIFE"** at COP 26 refers to a concept introduced by India's Prime Minister,.Narendra Modi, during the COP26 Summit in Glasgow. **LIFE** stands for **Lifestyle for Environment**, and it emphasizes the need for a collective approach towards a more sustainable and environmentally conscious lifestyle.

COP27 was held from 6 to 20 November 2022 in **Sharm El-Sheikh, Egypt.** One of the significant outcomes of COP27 was the historic decision to establish and operationalize a loss and damage fund, which is a step towards providing financial support to those most affected by **climate change.**

COP 28, the 28th Conference of the Parties to the United Nations Framework Convention on Climate Change (UNFCCC), was held from November 30 to December 13, 2023, in **Dubai, United Arab Emirates.** The conference was the largest of its kind, with some 85,000 participants, including more than 150 Heads of State and Government. The decisions made at COP 28 are expected to guide stronger **climate action plans due by February 2025.**

MICRO IRRIGATION IN INDIA

National Mission On Micro Irrigation (NMMI) and **National Mission for sustainable Agriculture (NMSA)** were launched with a clear mission to promote Micro irrigation systems. The Gujarat Government and **Gujarat Green Revolution Company Ltd**, is a **public private partnership** model for the implementation of micro irrigation. Many such initiatives are being run by various state governments.

The top five states in India for micro-irrigation coverage in 2023 are:

Karnataka - 21.59% of the total area covered under micro-irrigation.

Tamil Nadu - 13.16% of the total area covered.

Gujarat - 13.02% of the total area covered.

Maharashtra - 11.24% of the total area covered.

Andhra Pradesh - 11.02% of the total area covered

The top three states in India by area covered under Drip Irrigation in 2023 are:

1. Maharashtra : 1,414,185 hectares
2. Andhra Pradesh : 1,400,780 hectares
3. Tamil Nadu : 867,898 hectares

Top Three states in Sprinkler Irrigation in India in 2023

1. Rajasthan : 1,777,985 ha 2. Karnatka : 1604,399 ha

3. Gujrat : 788,672 ha

Total area covered under micro Irrigation was **14,491,877 ha**

The average penetration of micro irrigation in India is 19% (as on February 3, 2021), which is much lesser than many countries.

CSIR- Central Mechanical Engineering Research Launched Electric Tiller to Empower Small Farmers (May 2024)

Tiller is designed with the specific needs of farmers owning less than 2 hectares of land in mind

Share of Area under major cops of India

Crop	% Gross Cropped Area	Share of value of output%
Paddy	22.77	13.96
Wheat	15.45	9.68
Coarse Cereals	11.70	3.78
Total Cereals	49.41	27.41
Total Pulses	14.40	5.26
Total foodgrains	63.81	32.67

Total Oilseeds	13.52	8.44
Groundnut	2.62	2.28
Sugarcane	2.43	4.93
Total Fruits & Vegetables	8.64	27.77

World Comparison of Productivity of Major crops with India kg/ha

Crop	World Average	World Highest	All India average	State Highest
Paddy	4764	8640 (USA)	2809	Punjab (6510)
Maize	5879	11111 (USA)	3387	T. Nadu (7066)
Sorghum	1499	4919 (Egypt)	1092	AP (3372)
Sesamum	508	1636 (China)	485	WB (979)
Soybean	2870	3455 (USA)	1669	Telengana (1731)
Groundnut	1648	4635 (USA)	1777	TN (2812)
Sunflower	1970	3032 (China)	923	Bihar (1422)

Pesticide consumption pattern

World = **Herbicide** > Fungicide > Insecticide
Asia = **Insecticide** > Herbicide > Fungicide
America = **Herbicide** > Fungicide > Insecticide
Europe = **Fungicide** > Herbicide > Insecticide
Africa = **Fungicide** > Herbicide > Insecticide
Oceania = **Herbicide** > Insecticide > Fungicide
India = **Insecticide** > Fungicide > Herbicide

❖ **Precipitation scenario in India**

- Total precipitation = 400 m ha m (million hectare meter) or 4000 Cubic km.
- Immediate evaporation = 70 m ha m
- Infiltration = 215 m ha m
- Soil moisture = **165 m ha m**
- Ground water = 50 m ha m
- Surface runoff=115 m ha m
- Rain = 105 m ha m
- Snow 10 m ha m
- Average rainfall in India 118 cm.
- India receives 4% of global precipitation but ranks 133 in the term of waer availability per person/year
- With a rainfall of 1 mm every square meter receives 1 lit. of water or 0.001 m^3. Thus 1 ha. receives 10000 lit. of water.

❖ In India the average annual rainfall is computed based on 30 consecutive year data.

NATURAL FARMING

National Mission on Natural Farming: To motivate farmers to adopt chemical free farming and enhance the reach of natural farming, the Government has formulated **National Mission on Natural Farming (NMNF)** as a separate and independent scheme from 2023-24 by up scaling the Bhartiya Prakritik Krishi Paddati (BPKP).

National Green Hydrogen Mission to Facilitate the Transition Towards A Low Carbon Economy; India Sets a Target of **Annual Production of 5 mmt by 2030**

Green Credit Programme to be Notified for Incentivizing Environmentally Sustainable and Responsive Actions

PM-Pranam to Promote Alternative Fertilizers and Balanced use of Chemical Fertilizers

Proposal to Establish 500 New **'Waste To Wealth'** Plants Under **Gobardhan Scheme** with a Total Investment of Rs. 10,000 Crore

Government now intends to promote natural farming in a mission mode with due focus on farm production of livestock based natural farming inputs like Beejamrut, Jeevamrut, Neemastra etc. In the Budget Speech 2023-24, it has been announced that "over the next 3 years, 1 crore farmers will be facilitated to adopt natural farming. For this, **10,000 Bio-Input Resource Centres (BRC) will be set-up".** These BRCs will be set up for continuous supply of local livestock based natural farming inputs.

- **Knowledge partner** for natural farming extension- National Institute of Agricultural Extension Management (MANAGE)
- **Area under transgenic plants in India (2022-23)**- 13.05 Million Hectares (Mha)
- **India's rank in Cultivation of GM crops-** 4th
- **Bt-Cotton is the only GM crop Cultivated** in India with an area of- 13.05 Million Hectares
- *Central Government recently approved field trials of Indigenous Bt Brinjal Varieties Janak and BSS-793 (Developed by IARI contains Bt Cry1Fa1 gene) (IMP)
- Which state government recently issued a no-objection certificate (NOC) to conduct field trials on Mahyco BG-2 RRF (herbicide tolerant Bt cotton)- Haryana

LATEST INFORMATIONS

A. 109 Varieties of Field and Horticultural Crops Released by PM Narendra Modi, PUSA, New Delhi (11th August 2024)

These 109 varieties of 61 crops include 34 field crops and 27 horticultural crops

Field Crops (69)

1. **Cereals (23):** Rice-9; Wheat-2; Barley 1; Maize-6, Sorghum-1, Pearl millet-1; Finger millet-1; Proso millet-1; Barnyard millet-1;
2. **Pulses (11):** Chickpea -2, Pigeon pea – 2, Lentil -3, Field pea – 1; Fababean -1; Mungbean -2;
3. **Oilseeds (7):** Safflower–2, Soybean–2, Groundnut–2, Sesame -1;
4. **Forage crops (7):** Forage pearl millet-1; Berseem–1, Oats–2; Forage maize-2, Forage sorghum–1,
5. **Sugar crops (4):** Sugarcane - 4;
6. **Fibre crops (6):** Cotton - 5; Jute -1;
7. **Potential crops (11):** Buckwheat-1, Amaranth-4, Winged bean-1, Adzuki bean-1, Pillipesara-1, Kalingda-1, Perilla-2

Horticultural Crops (40)

1. **Fruits (8):** Mango –3, Pomegranate – 1; Guava – 2; Bael – 1; Pummelo – 1;
2. **Vegetable crops (8):** Tomato -2; Bottle gourd -1; Okra – 1; Indian beans – 2; Muskmelan – 1; Water melan – 1;
3. **Tuber crops (3):** Potato – 3;
4. **Spices (6):** Netmug – 1, Small cardamom -2; Fennel – 1; Ajwain – 01; Mango ginger – 1;
5. **Plantation crops (6):** Cocoa -2, Cashew – 2; Coconut – 2;
6. **Flowers (5):** Merigold – 1; Tuberose – 1; Crossandra – 1; Gladiolus – 2;
7. **Medicinal plants (4):** Velvet Bean – 2; Ashwagandha -1; Mandukaparni – 1

RICE

1. **CR Dhan 416 (IET 30201):** Suitable for **coastal saline areas**, yield 48.97 q/ha, maturity 125-130 days, (W.B., Maharashtra, Gujarat)
2. **CR DHAN 810(IET 30409):** Suitable for **rainfed shallow low land**, yield 42.38 q/ha, maturity 150 days, (Odisha, W. B. Assam)

3. **CR Dhan 108 (IET29052):** Suitable for early **direct seeded rainfed condition**, yield 34.46 q/ha, maturity 110-114 days (Odisha, Bihar)
4. **CSR 101 (IET-30827):** Suitable for irrigated **alkaline/saline stress areas,** yield of 35.15 q/ha (alkaline stress); 39.33 q/ha (saline stress) and 55.88 q/ha (normal condition), maturity 125-130 days (U. P., Haryana, T. N., Karnataka)
5. **Swarna PurviDhan 5 IET 29036 (RCPR 68-IR83929-B-B-291-2-1-1-2)**: Suitable for **direct seeded aerobic condition in droughtprone rainfed** as well as water limiting areas during *Kharif*, yield (normal condition - 43.69 q/ha, under moderate drought condition - 29.02 q/ ha), maturity early (110-115 days (Bihar, West Bengal, Jharkhand)
6. **DRR Dhan 73 (IET 30242):** Suitable for irrigated and **rainfed shallow low land areas with low soil P for both *Kharif* and *Rabi*,** yield 60 q/ ha (under normal conditions; 60 kg/ha of P i.e. recommended dose), 40 q/ ha (under low P; 40 kg/ha of P) and 40.0 q/ ha (under low Phosphorus; 0 kg/ha of P), maturity 120-125 days, (Karnataka, Odisha and Telangana)
7. **DRR Dhan 74 (IET 30252):** Suitable for irrigated and **rainfed shallow low land areas with low soil P for both *Kharif* and *Rabi*,** yield 70 q/ ha (under normal conditions; 60 kg/ha of P i.e. recommended dose), 44 q/ ha (under low P; 40 kg/ha of P) and 45.6 q/ ha (under low Phosphorus; 0 kg/ha of P), maturity 130-135 days. (Karnataka, Maharashtra, Telangana, Jharkhand, and regions with P deficit soil of India)
8. **DRR Dhan 78 (IET 30240):** Suitable for irrigated and rainfed shallow low land areas with **low soil P for both***Kharif* and *Rabi*, yield 58 q/ ha (under nornal conditions; 60 kg/ha. (Karnataka and Telangana)
9. **KKL(R) 4 (IET 30697)(KR 19011):** Suitable for **submergence stress conditions**, yield 38q/ha under stress situations and 56 q/ha under normal conditions, maturity mid-early (120-125 days). (Tamil Nadu, Andhra Pradesh, Telangana and Puduchery)

WHEAT

10. **Pusa Gehun Sharbati (HI 1665):** Suitable for timely sown, **restricted irrigated condition**, yield 33.0 q/ha, maturity 110 days (Maharashtra, Karnataka and Plains of Tamil Nadu)
11. **Pusa Gehun Gaurav (HI- 8840):** Suitable for irrigated conditions durum wheat variety, average grain yield 30.2 q/ha, **terminal heat tolerant,**resistance to stem and leaf rusts, biofortified durum wheat with **higher zinc (41.1 ppm) and iron (38.5 ppm) and protein content (~12%)** (ICAR- Indian Agricultural Research Institute Regional Station, Indore- Madhya Pradesh)

BARLEY

12. **DWRB-219:** Irrigated/ limited irrigationcondition of NWPZ, **average yield** 54.49 q/ha, **Maturity** 132 days (ICAR- IIWBR (Indian Institute of Wheat & Barley Research) Karnal- Haryana)

MAIZE

13. **Pusa Popcorn Hybrid – 1 (APCH 2):** Suitable for irrigated *Rabi* ecology, yield: 46.04 q/ha (NWPZ), 47.17 q/ha (PZ), maturity 120.2 days (Punjab, Haryana, Delhi, Uttarakhand (Plain), Uttar Pradesh (Western region), Maharashtra, Karnataka, Andhra Pradesh, Telangana, Tamil Nadu)
14. **Pusa Bio fortified Maize Hybrid – 4 (APH4):** Suitable for *Kharif* season, yield 84.33 q/ha (NWPZ), 71.13 q/ha (PZ), 56.58 q/ha (CWZ), maturity 79.8 days (NWPZ), 93.9 days (PZ), 86.4 days (CWZ) (Punjab, Haryana, Delhi, Uttarakhand (Plain),Uttar Pradesh (Western region), Maharashtra, Karnataka, Andhra Pradesh, Telangana, Tamil Nadu, Gujarat, Madhya Pradesh, Chhattisgarh, Rajasthan)
15. **Pusa HM4 Male Sterile Baby Corn-2 (ABSH4-2) :** Bihar, Jharkhand, Odisha, Uttar Pradesh (Eastern region), West Bengal, Maharashtra, Karnataka, Andhra Pradesh, Telangana, Tamil Nadu, Gujarat, Madhya Pradesh, Chhattisgarh, Rajasthan Suitable for irrigated conditions during *Kharif* season, yield 19.56 q/ha (NEPZ), 14.07 q/ha (PZ) & 16.03 q/ha (CWZ), maturity 53 days
16. **IMH 230 IMHSB 20R-6:** Suitable for irrigated ***Rabi*** **season, high yield 92.36 q/ha,**maturity 145.2 days (NEPZ viz. Eastern Uttar Pradesh, Bihar, Jharkhand, Orissa, West Bengal.)
17. **IMH 231 IMHSB 20K-10:** Suitable for *Kharif* irrigated condition, high yield 70.28 q/ha, maturity 90 days (NEPZ viz. Eastern Uttar Pradesh, Bihar, Jharkhand, Orissa, West Bengal, Assam)
18. **Pusa Popcorn Hybrid – 2 (APCH 3):** Suitable for irrigated ***Rabi*** **season, yield 45.13 q/ha,** maturity 102.5 days (Maharashtra, Karnataka, Andhra Pradesh, Telangana and Tamil Nadu)

SORGHUM

19. **DSH-6 (CSH-49) (SPH-1943) :** Suitable for **rainfed ecology *Kharif* season**, grain yield 38.77 q/ha, fodder yield 116 q/ha, maturity early (100-105 days (Tamil Nadu, Karnataka, Telangana,Gujarat, Rajasthan)

PEARL MILLET

20. **MH 2417 (Pusa- 1801):** ICAR - Indian Agricultural Research Institute, New Delhi It is a **dual purpose** hybrid suitable for *kharif* cultivation

under irrigated and **rainfed conditions**; high grain yield 33.34 q/ha and dry fodder yield 175 q/ha..

FINGER MILLET

21. **VL Mandua-402:** Suitable for **rainfed situations**, average seed yield 2261 kg/ha, matures in 111 days (Uttarakhand)

PROSO MILLET

22. **CPRMV-1 (DHPM- 60-4/PMV 466):** Suitable for **rainfed *Kharif* season**, yield 24-26 q/ha, maturity 70-74 days (Karnataka and Tamil Nadu)

BARNYARD MILLET

23. **VL Madira-254:** Suitable for **rainfed situations**, average yield 1,719 kg/ha, matures in 101 days (Uttarakhand)

PULSES

CHICKPEA

24. **Pant Gram 10 (PG 265):** Suitable for *desi* timely sown **rainfed / irrigated** conditions of *Rabi* season in NEPZ, yield 17.79 q/ha, maturity 130 days (Uttar Pradesh, Bihar, Jharkhand, West Bengal and Assam)
25. **Nandyal Gram 1267 (NBeG 1267):** Suitable for **mechanical harvesting in rainfed** as *desi* variety during *Rabi*, could also be grown with 1-2 protected irrigations in normal fertility condition, yield of 20.95 q/ha, maturity early (90-95 days), (South zone (Andhra Pradesh, Telangana, Karnataka and Tamil Nadu)

PIGEON PEA

26. **Phule Pallavi (Phule Tur-12-19-2):** Suitable for normal sown **rainfed /irrigated areas** of Cenrtal Zone in *Kharif* season, yield 21.45 q/ha, mid-early 157-159 days (Maharashtra, Gujarat, Madhya Pradesh and Chhattisgarh during Kharif)

PIGEONPEA

27. **NAM-88:** Suitable for **rainfed/irrigated areas in *Kharif* season**, yield 14.90 q/ha, maturity early (142 days) Karnataka, Telangana, Andhra Pradesh, Tamil Nadu). Developed by ICRISAT by technique Nested association mapping (NAM)

LENTIL

28. **Pant Lentil 14 (PL 320):** Suitable for timely sown **rainfed / irrigated** conditions of *Rabi* season in NWPZ, yield 15.55 q/ha, maturity 128 days, (Punjab, Haryana, Delhi, North-West & Central Rajasthan,

Western Uttar Pradesh, Plains of Uttarakhand and Jammu and Kashmir)

29. **RKL 20-26(D) Kota Masoor 6:** Suitable for **rainfed normal sown conditions** in *Rabi* season, yield 17.37 q/ha in NWPZ, 16.0 q/ha in CZ, maturity 125 days (NWPZ), 111 days (North Western Plain Zone and Central Zone of India)

30. **PSL-17:** Suitable for salinity conditions, seed yield 12.95 q/ha, maturity 125 days, (Division of Genetics, ICAR-Indian Agricultural Research Institute, New Delhi)

FIELD PEA

31. **Pant Pea 484:** Suitable for timely sown **rainfed / irrigated** conditions of *Rabi* season in NWPZ, yield 23.33 q/ha, maturity 120 days (Punjab, Haryana, Delhi, North-West & Central Rajasthan, Western Uttar Pradesh, Plains of Uttarakhand and Jammu and Kashmir)

FABA BEAN

32. **HFB-3 (HB 14-21) :** Suitable for irrigated, timely sown *Rabi* season in high fertility conditions, yield 23.65 q/ha, medium maturity (129 – 137 days) (Northern plain zone of the country (Haryana, Punjab, Delhi) & Central Zone (Uttar Pradesh and Chhattisgarh)

MUNGBEAN

33. **Lam Pesara 610 (LGG 610):** Suitable for **rice fallows and upland situations** during *Rabi* season for mechanical harvesting, yield 11.17 q/ha, maturity 74 days (South zone states, Andhra Pradesh, Telangana, Tamilnadu, Karnataka, Kerala and Odisha)

34. **PMS-8:** Suitable for **Salt affected conditions** [ECe 5.8-6.5 dS/m], Average seed yield 494.5 q/ha, Matures in 70 days (Division of Genetics, ICAR-Indian Agricultural Research Institute, New Delhi)

OILSEEDS

SAFFLOWER

35. **ISF-123-sel-15 :** Suitable for **late sown rainfed condition**, yield 16.31 q/ha, maturity 127 days (Karnataka, Maharashtra, Andhra Pradesh, and Telangana)

36. **ISF-300:** Suitable **for timely sown rainfed/irrigated** condition, yield17.96 q/ha, maturity134 days (Maharashtra, Karnataka, Andhra Pradesh, Telangana, Madhya Pradesh, Chhattisgarh)

SOYBEAN

37. **NRC 197:** Suitable for **rainfed *Kharif* season**, yield 16.24 q/ha, maturity 112.67 days (Himachal Pradesh and Uttarakhand)

38. **NRC 149:** Suitable for **rainfed *Kharif* season,** yield 24.0 q/ha, maturity 127 days (Punjab, Haryana, Delhi, North Eastern Plains of Uttar Pradesh, Plains of Uttarakhand and Eastern Bihar)

GROUNDNUT

39. **Girnar 6 (NRCGCS 637):** Suitable for timely sown *Kharif* season, yield 30.30 q/ha, maturity123 days (Recommended for Zone I (Rajasthan, Uttar Pradesh, Punjab states and Haryana) of India)
40. **TCGS 1707 (ICAR KONARK) Spanish Bunch**: Suitable for timely **sown rainfed/ irrigated *Kharif***, yield 24.76 q/ha, maturity 110-115 days (Odisha and West Bengal)

SESAME

41. **Tanjila (CUMS-09A):** Suitable for irrigated, **summer crop with early or late sown condition,** seed yield 963 kg/ha - 1147.7 kg/ha, oil yield 438.5 kg/ha - 558.0 kg/ha, oil content 46.17%, maturity 91 days (West Bengal, Bihar, Odisha, Maharashtra, Chhattisgarh, Telangana, Karnataka, Tamil Nadu and Kerala)

FORAGE CROPS

FORAGE PEARL MILLET

42. **JPM 18-7 (Jawahar Pearl Millet 18-7):** Suitable for **rainfed/ irrigated** under normal fertility condition during rainy season, yield 440-480 q/ha (green fodder), maturity 120-130 days (Rajasthan, Punjab, Haryana, Gujarat, Madhya Pradesh,Maharashtra, Uttar Pradesh, Chhattisgarh, Telangana, Andhra Pradesh, Tamil Nadu and Karnataka

BERSEEM

43. **Jawahar berseem 08-17 (JB 08-17):** Suitable for irrigated **multi-cut** during winter season, yield 620-650 q/ha (green fodder), maturity 190-200 days (Madhya Pradesh, Maharashtra, Gujarat, Chhattisgarh and Uttar Pradesh)

FORAGE MULTICUT OAT

44. **Him Palam Forage Oat-1 (PLP-24):** Suitable for timely sown, normal fertility and irrigated condition, yield 260-300 q/ha (green fodder yield), maturity 180-185 days (Himachal Pradesh, Jammu and Kashmir and Uttarakhand)

OAT

45. **Jawahar Oat 13-513 (JO- 13-513):** Suitable for Oat growing areas of eastern and north-western zone, yield 225-250 q/ha (green fodder yield), maturity 135-145 days (Punjab, Haryana, Uttar Pradesh, Uttarakhand, Odisha, Bihar, Jharkhand, Assam and Uttar Pradesh)

FORAGE MAIZE

46. **Pusa Forage Maize Hybrid-1 (AFH-7)** Suitable for irrigated, Kharif season, green fodder yield 413.1 q/ha, maturity 95-105 days ((Tarai region of Uttarakhand, Punjab, Haryana and Rajasthan)
47. **HQPM 28:** Suitable for Central Zone during K*harif*season, yield 427.6 q/ha (green fodder), 79.06 q/ha (dry matter), 20.9 q/ha. (All fodder growing areas of Central Zone (Uttar Pradesh (Bundelkhand region), Maharashtra, Madhya Pradesh, Chhattisgarh)

FORAGE SORGHUM

48. **CSV 57F (SPV 2801) (UTFS 111):** Suitable for **rainfed *Kharif*,** yield - 435q/ha (green fodder); 139q/ha (dry fodder) (Haryana, Punjab, Uttarakhand, Gujarat, Uttar Pradesh, Rajasthan and Delhi)

SUGARCANE

49. **Karan 17(Co 17018):** Suitable for irrigated timely to late sown season,yield914.8 q/ha, maturity 330-360 days (North West Zone (Haryana, Punjab, western and central Uttar Pradesh, Rajasthan, Uttarakhand)
50. **IKHSU-16 (CoLk 16202):** Suitable for irrigated condition, yield 932 q/ha, **sucrose (%) 17.74, CCS 114.3 q/ha, maturity early** (10 months) (Punjab, Haryana, Uttarakhand, Rajasthan, Central and Western Parts of Uttar Pradesh)
51. **IKHSU-17 (CoLk 16470):** Suitable for irrigated condition, yield 825.0 q/ha, CCS yield 95. 9 q/ha, sucrose 17.37%, maturity 360 days (Eastern Part of Uttar Pradesh, Bihar, Jharkhand, West Bengal and Assam)
52. **CoPb 99 (CoPb 17215):** Suitable for growing in medium and high fertile soil under irrigated subtropical climatic conditions during spring season, yield - CCS 112.7 q/ha, mean **cane yield 901.4 q/ha, sucrose 18.01%**, maturity – mid-late (12 months) (Punjab, Haryana, Uttarakhand, Rajasthan, Central and Western Parts of Uttar Pradesh)

FIBRE CROPS

COTTON

53. **CICR-H Bt Cotton 65 (ICAR -CICR 18 Bt) :** Suitable for **rainfed condition**, yield 15.47 q/ha, maturity 140-150 days (Central Zone)
54. **CICR- H Bt Cotton 40 (ICAR-CICR- PKV 081 Bt):** Suitable for **rainfed condition**, yield 17.30 q/ha, maturity 140 to 150 days (South Zone)
55. **Shalini (CNH 17395) (CICR-H Cotton 58) :** Suitable for rainfed*Kharif* condition, yield 14.41 q/ha, maturity 160 to 165days (Madhya Pradesh,Maharashtra & Gujarat)

56. **CNH-18529 (CICR-H NC Cotton 64):** Suitable for **rainfed and irrigated conditions** of Central Zone, yield of 10.11 q/ha, maturity 160-165 days (Chhattisgarh,Gujarat, Madhya Pradesh and Maharashtra)
57. **PDKV Dhawal (AKA-2013-8):** Suitable for timely sown *Kharif* under **rainfed situation**, yield 12.84 q/ha, maturity 160-180 days (Madhya Pradesh, Maharashtra and Gujrat)

WHITE JUTE

58. **JRC 9:** Suitable for timely **sown rainfed/ irrigated** condition, yield 31.97 q/ha, maturity110 – 120 days (fibre), 120-130 days (seed) (West Bengal, Odisha, Bihar, Uttar Pradesh, Tripura and Assam)

POTENTIAL CROPS

BUCKWHEAT

59. **Him Tara (EC125940):** Suitable **for rainfed *Kharif* hilly** areas, seed yield 14.04 q/ha, maturity 92.06 days (Northern Hill zone (Himachal Pradesh and Uttarakhand)

AMARANTH

60. **Him Gauri (IC037156):** Suitable for grain purposes cultivation during rainfed*Kharif* season in **mid to high hilly areas**, seed yield 24.06 q/ha, maturity 128.46 days (Northern Hills Zone (NHZ) Himachal Pradesh and Uttarakhand)

GRAIN AMARANTH

61. **RMA 120 (Jodhpur Rajgira 2):** Suitable for plains areas of the country in *Rabi* irrigated high fertility condition, 14.05 q/ha seed yield, maturity early (119-128 days (WZ (Rajasthan, Gujarat, Maharashtra); NPZ (U.P.); EZ (Odisha, Jharkhand) and CZ (Chhattisgarh)
62. **Gujarat Amaranth 8 (GA 8) SKNA 1407:** Suitable for cultivation in plain areas of the country in Rabi season, yield 14.55 q/ha, maturity 97 – 168 days (WZ (Gujarat, Rajasthan, Maharashtra), NPZ (Utter Pradesh, CZ (Chhattisgarh), EZ Jharkhand and Odisha)
63. **VL Chua 140:** Suitable for **rainfed *Kharif* ecology**, yield 16.86 q/ha, maturity 125-127 days (Uttarakhand, Himachal Pradesh)

WINGED BEAN

64. **PWB 17-18 (Phule Shrawani):** Suitable for irrigated well drained soil having medium fertility in *Kharif* season, seed yield 13.81 q/ha, green pod yield 142.96 q/ha, maturity late - 160 – 168 days. (WZ (Maharashtra), CZ (Chhattisgarh), NPZ (Uttar Pradesh) and EZ (Jharkhand state)

ADZUKI BEAN

65. Him Jwala (IC341939): Suitable for Northern hilly areas during rainfed*Kharif* season, yield 19.54 q/ha, maturity 108 days (Northern Hills zone (Himachal Pradesh, Uttarakhand) and part of Northeastern States)

PILLIPESARA

66. Prathama (OUAT Kalinga Pillipesara-1) (IC 524667): Suitable for early sowing irrigated uplands / medium lands in *Kharif*, yield – seed 1.82 q/ha, green biomass - 157.46 q/ha maturity medium (83 days) (Odisha (Eastern Zone) & Tamil Nadu (Southern Zone)

KALINGADA(WATERMELAN)

67. SKNK 1407 (Gujarat Kalingada 3): Suitable **for rainfed *Kharif* season,** yield 217 kg/ha (seed), 89.81q/ha (fruit), maturity 77 - 87 days (Gujarat and Rajasthan)

PERILLA

68. Poorvottar Perilla-2IC-615382: Suitable for rainfed*Kharif* hilly ecosystem for Northern Hill Zone (NHZ) and North Eastern Hill Zone (NEHZ), yield 10.33 q/ha, maturity medium-late (163.63 days (NHZ and NEHZ (Uttarakhand, Manipur, Meghalaya, Mizoram, Sikkim, Arunachal Pradesh and Nagaland)

69. Poorvottar Perilla-1(IC-615369): Suitable for rainfed*Kharif* hilly ecosystem for Northern Hill Zone (NHZ) and North Eastern Hill Zone (NEHZ), yield 11.15 q/ha, maturity medium-late (160.63 days). (NHZ and NEHZ (Uttarakhand, Manipur, Meghalaya, Mizoram, Sikkim, Arunachal Pradesh and Nagaland))

HORTICULTURAL CROPS

FRUITS

MANGO

70. Arka Udaya: It is a hybrid between Amrapali x Arka Anmol. Trees are semi-vigorous and having the bunch bearing habit.The fruits have pink blush on peel and good quality. (All mango growing states)

71. Ambika: It is a climate resilient hybrid (Amrapali x Janardan Pasand) recommended for subtropical and tropical regions. It is a regular bearer, high yielding and late in maturity type. (All mango growing states)

72. Arunika: It is a dwarf hybrid developed by crossing between Amrapali x Vanraj. It is a regular bearer and late in maturity. (All mango growing states)

POMEGRANATE

73. **Solapur Anardana:** It is a processing variety developed from three-way cross of Bhagawa x [(Ganesh x Nana) x Daru]. Fruits big (280 g), maturity period is 148-150 days (All pomegranate growing states)

GUAVA

74. **Lalit:** Fruits are round, medium sized with thick pericarp (>1.5 cm). The variety is heavy having saffron yellow-colored fruits (185-200 g) (Uttar Pradesh)
75. **Arka Kiran:** Climate resilient variety having semi-vigorous growth habit. Fruits are round, medium sized (180 to 190 g), (Andhra Pradesh, Karnataka and Maharashtra Tamil Nadu)

BAEL

76. **Swarna Vasudha:** It is a medium vigorous, regular bearing, big fruits (1100 to 1800 g), high yieid under rainfed condition (43.8 kg/plant), high pulp recovery (>80%), TSS(>40°Brix), acidity (0.54%). (Jharkhand)

PUMMELO

77. **Arka Chandra:** Arka Chandra (clone 18-5)is a clonal selection from accession 18 that is being maintained in the field gene bank at IIHR, Bengaluru. The tree is medium-sized (2.0-3.0 m) and spreading. (All Citrus growing states)

VEGETABLES

TOMATO

78. **Pusa Shakti:** This tomato variety is suitable for open-field cultivation during May-October. Ripe fruits have a thick pericarp (7.00 mm), moderate TSS (4.8°Brix), and lycopene (6mg/100 gq/ha) (Chhattisgarh, Odisha,Andhra Pradesh and Telangana)
79. **Pusa Tomato Hybrid 6: This tomato hybrid is resistant to four diseases including, ToLCD (tomato leaf curl disease, late blight, Fusarium-wilt and bacterialwilt.** (Chhattisgarh, Orissa, Telangana and A. P.)

BOTTLE GOURD

80. Kashi Shubhara: This variety is suitable for *kharif, zaid* and off-season production under low tunnel/ protected structure. (Uttar Pradesh, Bihar, Jharkhand and Punjab)

OKRA

81. **Arka Nikita:** Genic Male sterility based F_1 hybrid resistant to YVMV disease. (Karnataka)

INDIAN BEAN

82. **Kashi Bouni Sem- 207: Bushy in growth habit** (plant height 65-70 cm. (Punjab, Uttar Pradesh, Bihar, Jharkhand, Rajasthan, Gujarat, Haryana and Delhi)
83. **Arka Vistar: It a climate resilient, pole type and photo-insensitive variety.** Pods are long, thick, very broad and dark green coloured. It has pod yield of 37.0 t/ha after 120 days. (Karnataka)

MUSKMELON

84. **Thar Mahima:** Fruits are round (780-900 g) having TSS of 11.58-11.80^{0}Brix. Flesh is salmon orange and 2.8-3.2 cm thick. (Rajasthan)

WATERMELON

85. **Thar Tripti:** Fruits are attractive and characterized by green stripes and round shape. The fruit weight 2.3-3.4 kg with light red flesh colour. **TSS ranges from 11.5-12.17^{0}Brix** and rind thickness can range from 5.0-1.8 cm. (Madhya Pradesh, Maharashtra and Goa)

TUBER CROPS

POTATO

86. **Kufri Chipsona-5:** It produces white cream, ovoid tubers with shallow medium eyes and creamy flesh and has good storability under ambient conditions. Medium maturing (90-100 days) (Haryana, Uttarakhand, Uttar Pradesh, Madhya Pradesh, Gujarat Rajasthan and Chhattisgarh)
87. **Kufri Jamunia:** It produces dark purple oblong tubers with shallow eyes and purple flesh. Medium maturing (90-100 days) (Haryana, Punjab, Uttar Pradesh, Uttarakhand Plains, Madhya Pradesh, Rajasthan, Chhattisgarh, Gujarat, Odisha, Assam, W. B. and Bihar)
88. **Kufri Bhaskar: It is a heat tolerant variety having mite and hopper burn tolerance.**Early-medium maturing (85-90 days) (Haryana, Punjab, plains of Uttarakhand, Uttar Pradesh-early season planting) and Central plains (Rajasthan, Gujarat, Chhattisgarh for main planting)

SPICES

NUTMEG

89. **Kerala Shree: First farmers participatory variety with bold nuts and fully covered mace**. Dry nut weight of 9.1-11.2 g (Kerala, Tamil Nadu, and other nutmeg growing areas of India)

SMALL CARDAMOM

90. **IISR Manushree: Drought tolerant variety** with average yield of 550 kg dry capsules/ ha under irrigated conditions and 360 kg dry capsules/ ha under moisture stress conditions (Karnataka and Kerala)

91. **IISR Kaveri: Drought tolerant variety** with average yield of 482 kg dry capsules/ha under irrigated conditions and 308 kg dry capsules/ha under moisture stress conditions. (Karnataka)

FENNEL

92. **RF-290 :** The seed yield of this variety is 2065 kg/ha. **This variety is resistant to ramularia blight.** (Raj., Gujarat, U. P, Bihar, and Haryana)

AJWAIN

93. **Gujarat Ajwàin 3:** High yielding with an average seed yield of 1035 kg/ha. (Gujarat, Rajasthan, Haryana, A. P., U. P. and Chhattisgarh)

MANGO GINGER

94. **IISR Amrit:** Bold rhizomes, high yield (45.75 t/ha), light yellow core, desirable flavour (myrcene and â pinene), and contains 0.32%. (Kerala, Bihar, Odisha, West Bengal, Chhattisgarh and Gujarat)

PLANTATION CROPS

COCOA

95. **Vittal Cocoa Hybrid-1:** Early bearing, stable, high yielder with medium canopy both under arecanut & coconut shades and suitable for high density planting. High dry beans yield 1.5-2.5 kg/tree/year in 15-18 m^2 canopy. (Kerala, Karnataka and Tamil Nadu)
96. **Vittal Cocoa Hybrid-2:** Early bearing, high yielder with **black pod rot disease resistance** both under arecanut & coconut gardens. Dry beans yield 1.5-2.5 kg/tree/year in 14-20 m^2 canopy. (Kerala, Karnataka, Tamil Nadu, Andhra Pradesh and Gujarat)

CASHEW

97. **Nethra Jumbo-1:** It is a variety suitable for **rainfed** conditions jumbo nuts (12 g), with premium grade kernel (W130). (Karnataka)
98. **Nethra Ganga (H-130):** It has bold nuts (12-13 g) and high yielding hybrid with cluster bearing habit, high precocity with 6-8 nuts per panicles. It has high shelling value (29.5%). (Karnataka)

COCONUT

99. **Kalpa Suvarna:** It is a dwarf, high yielding, dual purpose coconut variety with green coloured, oblong fruits having **sweet tender coconut water and good quality copra.** (Karnataka and Kerala)
100. **Kalpa Shatabdi:** This is a dual purpose coconut variety, with large fruits, suitable for copra and tender nut production. (Kerala, Karnataka and Tamil Nadu)

FLOWER CROPS

MARIGOLD

101. **Pusa Bahar:** This variety is vigorous having plant height of 75-85 cm. It produces compact, flattened, attractive and large size (8-9 cm) flowers of yellow colour. (Punjab, U. P., Bihar and Jharkhand, Chhattisgarh, Odisha, A.P. and Telangana, Raj., Guj., Haryana and Delhi, M. P., Maharashtra and Goa.)

TUBEROSE

102. **Arka Vaibhav:** It produces double-type flowers on medium tall spikes. The flower buds are greenish in colour, florets are white in colour and ideal as a cut flower. (West Bengal, Assam, Punjab, Uttar Pradesh, Bihar, Rajasthan, Delhi, Maharashtra, Karnataka and Tamil Nadu)

CROSSANDRA

103. **Arka Shreeya:** Novel flower colour orange-red florets (Orange Red group 32A), high loose flower yield (2269.2 kg/ac/year), (Andhra Pradesh, Maharashtra and Karnataka)

GLADIOLUS

104. **Arka Amar:** The flowers are red colour having red margin and white line on tepals with yellow blotch. It has a long spike (101 cm) with good rachis length (50 cm) (Himachal Pradesh, Punjab, Jharkhand, Andhra Pradesh and Maharashtra)
105. **Arka Aayush:** Its florets are with a unique colour: Blotch Red with Yellow border. The florets are open-faced, thick, slightly ruffled and arranged in double rows. (W. B., Punjab, Rajasthan and Maharashtra)

MEDICINAL PLANTS

VELVET BEAN

106. **Arka Dhanvantari:** Long duration variety (180-190 days) with non-itchy trichomes on pods. High seed yield of 2- 2.5 t/ha with high L-Dopa content of 4.5 to 5.5%. (Karnataka)

VELVET BEAN

107. **Arka Daksha:** Medium duration variety comes to maturity 150-160 days after sowing. It has non-itchy trichomes on pods. (Karnataka)

ASHWAGANDHA

108. **Arka Ashwagandha:** It is pure line selection with high dry root yield of 11.95 q/ha and total withanolide content of 0.580%. (Karnataka)

MANDUKAPARNI

109. **Arka Prabhavi :** It produces 13 tonnes of fresh herb and 2.3 tonnes of dry herb/ ha/ year with **higher asiaticoside content (3.85%) and higher tri-terpenoid content (7.27%).** (Karnataka)

B. Agricultural Budget 2024-25

Transforming Agriculture Research

- A comprehensive review of the agriculture research setup.
- Focus on raising productivity and developing climate-resilient varieties.
- Funding is provided in challenge mode, including to the private sector.
- Oversight by domain experts from both the government and private sector.

Release of New Varieties

- Release 109 high-yielding and climate-resilient varieties of 32 field and horticulture crops for farmers.

Natural Farming

- Initiate one crore farmers into natural farming over the next two years.
- Support through certification and branding.
- Implementation by scientific institutions and willing gram panchayats.
- Establish 10,000 need-based bio-input resource centres.

Missions for Pulses and Oilseeds

- Strengthen production, storage, and marketing of pulses and oilseeds.
- Strategy for achieving self-sufficiency ('atmanirbharta') in oilseeds such as mustard, groundnut, sesame, soybean, and sunflower.

Vegetable Production & Supply Chains

- Develop large-scale clusters for vegetable production near major consumption centres.
- Promote Farmer-Producer Organisations, cooperatives, and start-ups for vegetable supply chains, including collection, storage, and marketing.

Digital Public Infrastructure for Agriculture

- Implement Digital Public Infrastructure (DPI) in agriculture in partnership with states.
- Cover farmers and their lands within three years.
- Conduct a digital crop survey for Kharif in 400 districts this year.
- Include details of 6 crore farmers and their lands in farmer and land registries.
- Enable issuance of Jan Samarth-based Kisan Credit Cards in five states.

Shrimp Production & Export

- Provide financial support for setting up Nucleus Breeding Centres for Shrimp Broodstocks.
- Facilitate financing for shrimp farming, processing, and export through NABARD.

National Cooperation Policy

- Introduce a National Cooperation Policy for systematic and all-round development of the cooperative sector.
- Aim to fast-track rural economic growth and generate large-scale employment opportunities.

Union Budget Highlights 2024-25

- In 2024-25, the Ministry of Rural Development has been allocated Rs 1,80,233 crore. The Department of Rural Development has been allocated Rs 1,77,566 crore, 4% higher than the revised estimates of 2023-24. The Department of Land Resources has been allocated Rs 2,667 crore, which is a 41% increase over the revised estimates of 2023-24.

Table 1: Budgetary Allocation to the Ministry of Rural Development (in Rs crore)

Department	22-23 Actuals	23-24 RE	24-25 BE	% Change*
Rural Development	1,76,837	1,71,069	1,77,566	4%
Land Resources	1,002	1,898	2,667	41%
Total	**1,77,839**	**1,72,967**	**1,80,233**	**4 %**

Note: BE is budget estimate and RE is revised estimate.

**% change is change in 2024-25 BE over 2023-24 RE.*

Sources: Demands for Grants of the Ministry of Rural Development 2024-25; PRS

Budget Highlights

- Three crore houses will be provided under the PM Awas Yojana in rural and urban areas.
- Phase-IV of the Pradhan Mantri Gram Sadak Yojana will be launched to provide all-weather road connectivity to 25.000 rural habitations.
- Unique land parcel identification number (ULPIN) will be assigned for all lands in rural areas. Initiatives will be taken for establishment of land registry and linking it to farmers registry

C. LIVESTOCK CENSUS — 2019

The Livestock Census has been conducted in the country periodically since 2019-20. Since then it has been conducted once every 5 years. It covers all domesticated animals and their headcounts.

Key Points

- The total Livestock population is 53.6 crore in the country showing an increase of 4.6% over Livestock Census 2012.
  - o **West Bengal** observed the **highest increase of 23%,** followed by **Telangana (22%).**
  - o Cross-bred animals contributed around **28% to India's total milk production in 2018-19.**
- Total Bovine population (Cattle, Buffalo, Mithun and Yak) is 30.3 crore in 2019 which shows an increase of 1.0% over the previous census.
- The Female Cattle (Cows population) is 14.5 crore, increased by 18.0% over the previous census (2012).
  - o **Uttar Pradesh** has observed a **maximum decrease in cattle population** though the state has taken several steps to save cattle.
  - o **West Bengal** has seen the **highest rise of 15%** in cattle population.
- The total sheep in the country is 7.4 crore in 2019, increased by 14.1% over previous Census.
- The Goat population in the country in 2019 is 15 crore showing an increase of 10.1% over the previous census.
- The total Poultry in the country is 85.2 crore in 2019, increased by 16.8% over previous Census.

Distribution of livestock population

Categories	Livestock population (%)
Cattle	35.94
Buffaloes	20.45
Sheep	13.87
Goats	27.80
Pig	1.69
Others	0.23

NUMBER OF LIVESTOCK AND POULTRY DURING 20th LIVESTOCK CENSUS — 2019

Species	Top Indian state		
Cattle	West bengal	UP	MP
Buffalo	UP	Rajasthan	Gujrat
Sheep	Telangana	AP	Karnataka
Goat	Rajasthan	WB	UP
Pig	ASSAM	Jharkhand	Meghalaya
Poultry	TN	AP	Telangana

D. Minimum Support Price (MSP) for Kharif Crops 2024-25

Commodity	Variety	MSP for 2023-2024 (Rs per quintal)	MSP for 2024-2025 (Rs per quintal)	Increase over previous year (Rs per quintal)
Paddy	Common	2183	2300	117
	Grade 'A'	2203	2320	117
Jowar	Hybrid	3180	3371	191
	Maldandi	3225	3421	196
Bajra		2500	2625	125
Maize		2090	2225	135
Ragi		3846	4290	444
Arhar (Tur)		7000	7550	550
Moong		8558	8682	124
Urad		6950	7400	450
Cotton				
	Medium Staple*	6620	7121	501
	Long Staple **	7020	7521	501
Groundnut in shell		6377	6783	406
Sunflower seed		6760	7280	520
Soyabeen				
	Yellow	4600	4892	292
Sesamum	-	8635	9267	632
Nigerseed	-	7734	8717	983

RABI CROPS (Rabi Marketing Season (RMS) 2024-25)

Commodity	MSP for 2023-2024 (Rs per quintal)	MSP for 2024-2025 (Rs per quintal)	Increase over previous year (Rs per quintal)
Wheat	2125	2275	150
Barley	1735	1850	115
Gram	5335	5440	105
Masur (Lentil)	6000	6425	425
Rapeseed & Mustard	5450	5650	200
Safflower	5650	5800	150
Toria	5050	5450	400

D. Various Soil Colours

There are various factors or soil constituents that influence in the soil colour which are as follows:

(*i*) **Organic matter.** Soils containing high amount of organic matter show the colour variation from black to dark brown

(*ii*) **Iron compounds:** Soils containing higher amount of iron compounds generally impart red, brown and yellow tinge colour

(*iii*) **Silica, lime and other salts:** Sometimes soils contain either large amounts of silica and lime or both. Due to presence of such materials in the soil the colour of the soil appears like white or light coloured

(*iv*) **Mixture of organic matter and iron oxides**. Very often soil contain a certain amount of organic matter and iron oxides. As a result of their existence in soil, the most common soil colour is found and known as brown.

(*v*) **Alternate wetting and drying condition:** During monsoon period due to heavy rain the reduction of soil occurs and during dry period the oxidation of soil also takes place. Due to development of such altermating oxidation and reduction condition, the colour of soil in different horizons of the soil profile is variegated or mottled. This mottled colour is due to residual products of this process especially iron and manganesc compounds

(*vi*) **Oxidation-reduction conditions:** When soils are waterlogged for a longer period the permanent reduced condition will develop. The presence of ferrous compounds resulting from the reducing condition in waterlogged soils bluish and greenish colour.

E. Inflorescences of Different Crops

Crops	Inflorescence
Rice, Litchi, Mango, Grape, Palak, Moringa, Sorghum	Panicle
Wheat,Barley Oat,Bajra, Jowar and other grasses	Spikelet/Head/Ear
Onion, Carrot, Coriander (Umbeliferae/ Apiacae)	Umbel
Mulberry, Poplar	Amentum (Catkin)
Euphorbiacae	Cyathium
Compositae/Asteracae Family like-sunflower, safflower, Marigold	Capitulum (advance type of Inflorescence)
Ficus spp., Fig, Pomgranate	Hypanthodium
Radish, Mustard (Brassicaceae family)	Raceme
Amaranthus, latjeera, Beet root	Spike
Banana, Colocasia, Palmae family like-palm, Arecanut, Coconut, Datepalm	Spadix
Sugarcane	Arrow/Open panicle
Maize –Male	Tassel
Female	Silk
Guava, Papaya, Sapota, Citrus, Strawberry	Solitary
Ber, Sweet orange	Fasicle
Sweet potato, Broccoli, Brinjal, Tomato, Potato	Cyme

F. Test Weight of Field Crops

Crops	Test Weight (gm)	Crops	Test Weight (gm)
Rice	25	Mung,Urd	34-36
Basmati rice	21	Rajma	38-44
Wheat, Oat, Barley	30-40	Bajra	5-7
Cotton, Arhar	70-72	Mustard	3-5
Pea	100	Lucern	2-4
Maize	278	Tobacco	2.5-3
Soybean	55	Linseed, safflower	10
Cowpea	80	**Phalaris Minor**	**2.0**
Sunflower	40-50	**Echinochloa spp**	**1.40**
Sorghum	25-30	Chenopodium	0.70
Gram desi	17-26	Cyperus spp	0.19
Kabuli	>26 gm		

G. GST on Agricultural products July, 2024

- **Zero per cen GST rate:** Fresh fruits, Vegetables, Cereals, and Pulses are exempt from GST.
- **5 per cent GST rate:** Items like milk, butter, and other dairy roducts such as frozen vegetables and sugar.
- **12 per cent GST rate:** Packaged and branded agricultural products such as nuts and dry fruits.
- **18 per cent GST rate:** Processed food items like sauces, ketchup and packaged foods.
- **12 per cent GST rate:** Luxury food items and processed conatining preservatives.

As of June 2024, the GST Council is considering exempting the fertilizer sector from the current 5% GST rate. The decision was made after discussions on recommendations from the Standing Committee on Chemicals and Fertilisers. The committee proposed reducing GST rates on raw materials and nutrients to support farmers and fertilizer manufacturers.

Currently, the GST rate on fertilizers is 5%, while the GST rate on raw materials like ammonia and sulfuric acid is 18%. The higher GST rate on inputs results in a large amount of unutilized input tax credit. Recently the council has made changes cerain renewable energy deviced new GST rate 12% instead of old 5%. All types of milk cans new rates is 12%. Solar cookers new rate is 12% instead of 5%.

H. The Pesticide Management Bill, 2020

(Bill seeks to replace the Insecticides Act, 1968)

- **The Pesticide Management Bill, 2020** was introduced in Rajya Sabha by the Minister of Agriculture and Farmers Welfare, on March 23, 2020. It seeks to regulate the manufacture, import, sale, storage, distribution, use, and disposal of pesticides, in order to ensure the availability of safe pesticides and minimise the risk to humans, animals, and environment. The Bill seeks to replace the Insecticides Act, 1968.
- **Pest and pesticide:** The Bill defines a pest as any species of animal, plant, or pathogenic agent that is unwanted, or injurious to plants, humans, animals, and the environment. A pesticide is any substance of chemical or biological origin intended for preventing or destroying any pest in agriculture, industry, public health, pest control operations, or for ordinary use.

- **Central Pesticides Board:** The central government will constitute the Central Pesticides Board to advise the central and state governments on scientific and technical matters arising under the Act. It will also advise the central government in formulating standards and best practices for: (*i*) pesticide manufacturers, laboratories, and pest control operators, (*ii*) working conditions and training of workers, and (*iii*) recall and disposal of pesticides.
- **Registration of pesticides:** Persons seeking to import or manufacture a pesticide for ordinary use, agriculture, industry, pest control, or public health, are required to obtain a certificate of registration for the pesticide from the Registration Committee.
- **Registration criteria:** The Committee will evaluate the information submitted in the application about the pesticide on factors such as safety, efficacy, necessity, end-use, risks, and availability of safer alternatives. It will not register a pesticide if the applicant submits false or misleading information, or if the maximum limits for the residue of the pesticide on crops and commodities are not specified under the Food Safety and Standards Act, 2006. It may also not register the pesticide if there is scientific uncertainty regarding its risks and benefits, and threats of serious and irreversible damage to human health, other living organisms, or the environment.
- **Licences:** A person seeking to manufacture, distribute, exhibit for sale, sell, or stock pesticides, or undertake pest control operations is required to obtain a licence from the Licensing Officer, who may be appointed by the state government.
- A person's licence will be revoked if he is convicted of an offence under the Act. It can also be revoked if he: (*i*) violates the conditions under which the licence was granted, (*ii*) contravenes the provisions of the Act or the rules made under the Act, or (iii) had submitted false or misleading information to obtain the licence.
- **Offences:** Under the Bill, manufacturing, importing, distributing, selling, exhibiting for sale, transporting, storing, or undertaking pest control operations, without a licence or certificate is punishable with imprisonment of up to three years, or a fine of up to Rs 40 lakh, or both.

I. Awards

Environment Protection Award 2024 : M.S. Swaminathan Award for Environment Protection 2024: Anant Darshan Shankar, the Chairman of the Indian Institute of Ayurveda and Integrative Medicine, was awarded this award in 2024. Shankar is known for his work in establishing a conservation network for wild medicinal plants.

National Water Awards: First national water awards started in 2018 by the Ministry of Jal Shakti. Second edition of the award. **Among the states category,** Tamil Nadu bagged the Best State Award, followed by Maharashtra and Rajasthan.

5th National Water Awards (NWA) for 2023 are announced by Ministry of Jal Shakti.

Objective: To encourage the stakeholders to adopt a holistic approach toward water resources management.

WORLD FOOD PRIZE

First awarded : 1987.   Last awarded : 2024

World Food Prize laureates (2010 onwards)

2010	David M. Beckmann Jo Luck	United States	For their leadership in two grassroots organisations tackling hunger and nutrition
2011	John Kufuor	Ghana	For creating and implementing government policies to alleviate hunger and poverty in their countries
	Luiz Inácio Lula da Silva	Brazil	
2012	Daniel Hillel	Israel	For conceiving and implementing micro-irrigation in arid and dry land regions
2013	Marc Van Montagu	Belgium	For their individual achievements in modern agricultural biotechnology supporting sustainability and global food security (see controversy[c])
	Mary-Dell Chilton Robert Fraley	United States	

2014	Sanjaya Rajaram	India	For developing 480 varieties of disease resistant wheat and increasing global production by 200 million tons
2015	Fazle Hasan Abed	Bangladesh	For building an organization (BRAC) that is effective in reducing poverty in Bangladesh and 10 other countries
2016	Maria Andrade	Cape Verde	For the "single most successful example of biofortification" in the form of biofortified orange-fleshed sweet potato that is resistant, tolerant and high yielding
	Robert Mwanga	Uganda	
	Jan Low	United States	
	Howarth Bouis	United States	For "the implementation of a multi-institutional approach to biofortification as a global plant breeding strategy"
2017	Akinwumi Adesina	Nigeria	For leadership and innovation in building political will to transform African agriculture at all levels
2018	Lawrence Haddad	U.K. South Africa	For elevating maternal and child malnutrition to a central issue at national and international levels
	David Nabarro	U.K. United Nations	
2019	Simon N. Groot	Netherlands	For empowering smallholder farmers in more than 60 countries through enhanced vegetable production
2020	Rattan Lal	India United States	For a soil-centric approach to sustainably increasing food production
2021	Shakuntala H. Thilsted	Trinidad and Tobago Denmark	For achievements in pioneering fish-based nutrition-sensitive approaches to food systems

2022	Cynthia Rosenzweig	United States	For her pioneering work in modeling the impact of climate change on food production worldwide
2023	Heidi Kühn	United States	For her farmer-focused development model that revitalizes farmland, food security, livelihoods and resilience after devastating conflict.
2024	Dr. Geoffrey Hawtin	UK	Recognised for their work in preserving the worlds crop diversity and seed heritage.
	Dr. Cary Fowler	US	

Other Indians who won Food Prize award

1987	M.S. Swaminathan	India's Green Revolution and leadership at IRRI
1998	B. R. Barwale	High Quality seeds throughout India
1989	Verghese Kurien	Operation Flood, largest milk production revolution
1996	Gurdev Khush:	For advances in rice breeding leading to substantially
2000	Surinder Vasal	For research and leadership for quality protein maize
2005	Modadugu Vijay Gupta:	Low-cost techniques for freshwater fish farming.

J. 11TH AGRICULTURE CENSUS

- Agriculture Census in India is conductedevery five years. The field work of Agriculture Census is carried out in three phases, each phase of one-year duration.Next Agriculture Census (11th in the series) was postponed by a year due to outbreak of COVID-19 pandemic. Consequently, the Agriculture Census 2021-22 was launched on 28th July, 2022. At present, Phase-I of Agriculture Census 2021-22 is in progress which is targeted to be completed by December, 2023.
- The 11th Agriculture Census of India, which began in August 2022, is the most recent agricultural census in the country. The census was launched by the Union Minister of Agriculture and Farmers Welfare on July 28, 2022.
- **The 11th Agriculture Census includes the following features:**
- **Data collection**: The census collects data on a variety of agricultural parameters, including the number and area of operational holdings, size of holdings, land use, tenancy, and cropping pattern.
- **Digital land records:** The census uses existing land records, including records of land ownership and season-wise area enumeration for crops.
- **Apps and software:** The census collects data through apps and software.

- **Real-time monitoring:** The census monitors progress in real-time.
- **Smartphones and tablets:** The census was the first to collect data on smartphones and tablets.
- The census is conducted every five years, but the 2021-22 census was delayed by a year due to the COVID-19 pandemic. Phase-I of the 2021-22 census is currently in progress and is expected to be completed by December 2023. T
- For Agriculture Census 2021-22, an amount of Rs. 7685 Lakh has been allocated to States during Financial Year 2023-24. State-wise budget allocation during FY 2023-24 is given below.

State wise budget allocation during FY 2023-24

S. No	States/UTs	Amount (in Lakh)
1	A & N Islands	39
2	Andhra Pradesh	375
3	Arunachal Pradesh	240
4	Assam	35
5	Bihar	755
6	Chandigarh	48
7	Chhattisgarh	401
8	Daman & Diu and D & N Haveli	17
9	Delhi	43
10	Goa	35
11	Gujarat	421
12	Haryana	66
13	Himachal Pradesh	40
14	Jammu & Kashmir	104
15	Jharkhand	482
16	Karnataka	378
17	Kerala	543
18	Ladakh	11
19	Lakshadweep	45
20	Madhya Pradesh	108
21	Maharashtra	561
22	Manipur	162
23	Meghalaya	291
24	Mizoram	58

25	Nagaland	150
26	Odisha	521
27	Puducherry	14
28	Punjab	17
29	Rajasthan	112
30	Sikkim	76
31	Tamil Nadu	238
32	Telangana	95
33	Tripura	59
34	Uttar Pradesh	572
35	Uttarakhand	175
36	West Bengal	398
	Total	7685

World Census of Agriculture

The **International Institute of Agriculture** conducted the **first Programme for the World Census of Agriculture (WCA)** in 1930.

·Beginning with 1950, the **Food and Agriculture Organization (FAO)** of the United Nations has been preparing and advocating the decennial World Programme for the Census of Agriculture (WCA).

The concept of **agricultural operational holdings adopted in India** differs to some extent with that of FAO, as it does not include those holdings which are not operating any agricultural land and are engaged exclusively in the poultry, livestock & fishing etc.

K. MODERN TECHNOLOGIES FOR AGRICULTURAL DEVELOPMENT

PRECISION FARMING

Precision agriculture is the practice of applying inputs (fertilizers, seeds, chemicals, etc.) across a farm at variable rates based on soil conditions, weed pressure and yield map. It refers to the ***farm management concept in which inputs are precisely used as per site condition and crop needs in order to achieve maximum crop yield***. This avoids wastage of input and aids cost minimization. It is also called as ***precision farming, GPS farming, prescription farming, farming by satellite, spatially variable agriculture, farming by the foot, site-specific management*** and ***variable rate application***.

Where precision agriculture is applicable?

- It is applicable where land holding is larger. In case of small land holdings, implementation of precision farming is difficult as operation of advanced instrumentation (variable rate applicator, drones, etc.) needs bigger size land.
- It is not suitable for small farmers as it warrants more money for site condition assessment, and purchase of costly instruments and devices.
- It is applicable if the farmer is literate. Farmers should have sufficient knowledge about precision farming, advanced technologies (RS, GIS, VRT, etc.)
- It can be implemented wherever hi-tech machineries and technologies available.

Components of precision farming

• Remote Sensing • Geographical information system • Global Positioning System • Soil testing • Yield monitors

Steps involved in precision farming

1) **Assessing variation:** In this process, site condition is assessed by quantifying various site factors *viz.,* soil fertility, slope, erosion, organic matter, soil moisture and others.

2) **Managing variation** : Farmer should **match economic inputs as per site variability** in the farm. **Poor site condition warrants more inputs and *vice versa*.** In case of soil fertility management, more fertilizer should be applied whenever soil fertility is poor.

Higher spatial variability has **higher potential for precision farming** and *vice versa*. It is difficult to execute precision informing in case of increasing temporal variability during the crop period.

3) **Evaluation**: The yield obtained, input used and cost involved is worked.

VARIABLE RATE TECHNOLOGY

Variable rate Technology

It refers to the **technology that applies inputs at varying level based on site condition.** Inputs such as fertilizers, lime, seeds and pesticides can be applied at different rate depending upon site condition. It aids precise input application in time.

Components of variable rate technology

1. **Differential GPS receiver**: This provides location information of the farm/site.

2. **Computer**: It aids **processing of information** received from sensor, soil analysis and other variability assessment made towards precision farming.
3. **Variable rate software:** This is the computer program which processes the data and communicates to variable rate applicator and controller.
4. **Controller** is the device which controls the rate of flow of inputs (fertilizer, seed, pesticides, lime, etc.).
5. **Variable rate applicator** is the machine which applies input to farm / site depending upon the farm variability.

Methods of variable rate application

1. Map based variable rate application
2. Sensor based variable rate application

CROP MAPPING AND YIELD PREDICTION

Crop mapping

It refers to the **process and preparation of maps showing crops in a region.**

Remote sensing and GIS in crop mapping

Remote sensing approach utilizes multi temporal satellite data (*i.e.,* satellite image obtained from different seasons) for the defined area. Crops are identified and differentiated based on spectral reflectance value. Spectral reflectance of crops varies with type of crop and growth stages of crop. Leaf colour, chlorophyll content, pigments, size and shape of leaves, arrangement of leaves, etc. determines the spectral reflectance of particular crop. All these parameters changes with growth stages of plant. Accordingly, spectral reflectance value of crop varies with type of crop and stages of crop. In addition to this, ancillary data such as rainfall and temperature is also useful in crop identification and mapping.

SOIL MAPPING AND FERTILISER PRESCRIPTION

Soil information is one of the vital input for agriculture scientists, planners, teachers, engineers, land use planners and many others. This vital input is provided to users in the form of soil map.

Soil map represents the ***special kind of maps which shows the spatial distribution of soils in an area***. The pictorial depiction of soil in a landform is showed in soil map ***with the aid of defined symbols, legends, direction arrow in a specified scale.***

Soil Mapping

Soil mapping refers to the **process of producing soil map for a region by adopting standard soil survey protocols.** It is **laborious and time consuming work.** Soil map is the final output from a soil survey.

Methods of Soil mapping

1. Soil mapping through conventional soil surveys
2) Soil mapping through remote sensing and GIS approach

Soil mapping through RS&GIS involves the following two methods:

a) Soil mapping through visual interpretation of satellite image

This process involves following essential steps;

- **Image procurement and base map preparation**
- **Image interpretation**
- **Field survey**
- **Laboratory testing and soil classification**
- **Cartography and printing**

b) Computer aided approach for soil mapping

In order to overcome this, computers are used to classify the satellite image. In this approach, the **entire spectral reflectance in the satellite image is grouped at specified interval** automatically.

CROP SIMULATION MODELS

Crop simulation model is a simulation model that describes the processes of growth and development as a function of weather, soil and crop management (De Wit, 1993). Parameters such as soil, climate, crop, pest, pathogen, management measures, etc. can be used as input for crop simulation model depending upon the simulation.

Simulation model processes the input information as per logic defined by user and produce results. Outputs of crop simulation model are expected growth, yield, fertilizer requirement, water requirement, quantity of inputs (pesticide, amendments, etc.) needed for management and others.

Crop simulation models can be used both at individual field and at regional level. It can also be utilized for different climatic regions, different soil conditions, varied cropping systems, and different varieties of same crop.

DRONES AND ITS APPLICATIONS

Drone refers to the **flying machinc without an on-board human pilot.** Simply **it is a pilotless aircraft.** It is also called as **unmanned aerial vehicle, quadcopter, etc.**

Quadcopter refers to the **drone containing four blades.** Four blades ensure more stability. **Anything that is carried by the drone** during flight is called as **payload.** ***Attitude*** concerns with the **orientation of UAV** during flying which includes pitch, roll and yaw. **Pitch** concerns with **orientation of UAV, whether UAV is tilted up or down** whereas **roll** concerns with **orientation of UAV, whether UAV is twist** around its control axis. **Yaw** is concerned with the **left and right movement of UAV.**

Gyroscope informs whether the **flying of UAV is at level or at certain angle of inclination to the earth surface. Gimbal** is the assembly or **mount on which camera of UAV is fixed.** It ensures stability of camera while flying of UAV.

Components of drones

1. Unmanned aerial vehicle (UAV)

a) GPS b) Motors, gears and blade system c) Sensors

2. Ground-based controller

3. System of Communications

Advantages of drones

- Image can be recorded at any time or season or date
- Drones are smaller in size
- Possible to access inaccessible and narrow region
- It is easier to operate and more efficient
- Cheaper in cost
- Provides data more precise than the satellite image
- Multiple utility

Applications of drones in agriculture

- **Plant health monitoring**

✓ Water stress reduction - helps in scheduling irrigation

✓ Pathogen detection - helps in disease identification and management

✓ Detection of insect damage – aids pest management

- Aero-biological sampling
- Estimation of nitrogen level by comparing leaf greenness
- Mapping of invasive weeds
- Monitoring applications
- Forest fire monitoring
- Monitoring biodiversity
- Assessing erosion in agricultural fields

NANO TECHNOLOGY

Nanotechnology is the science which aims to create molecule based computer chips and other devices that are thousand of times smaller than the existing technologies.

A nanometer is about the width of six bonded atoms, 1/40000 width of hair. DNA, our genetic material is in 2.5 nanometer range, while red blood cells are approximately 2.5 micrometers (2,500 nanometers).

Institute of Nano Science and Technology (GoI) is situated at Mohali (Punjab).

GoI has launched a mission, Nanotechnology mission in 2007. It focuses on fundamental aspects of nanoscience development of product and processes for national development, especially in areas of national relevances such as safe drinking water, materials development sensors development and drug discovery etc.

Second phase of mission is extended till 2017.

HYDROPONICS

Hydroponic developed in Latin America and can be defined as **"the cultivation of plants in water without soil.** In 1978 hydroponic pioneer Dr. Howard Resh published the book Hydroponics Food Production. Prof William Ferderick Gericke of California University, Berkley defined hydroponics as crop growth in mineral nutrition solutions with no solid medium for the roots. Now Growers all over the world are using hydroponics technique due to the lack of a large water supply or fertile farmland. Home gardeners use hydroponics on smaller scale to grow fresh vegetables year round at the apartment or balconies of the homes. Green houses and nurseries also grow plants in a soil less peat or bark based growing mix, then nutrients are applied through water supply.

Advantages of hydroponics

1. There are no weeds to remove
2. Soil borne pests and diseases are minimized.
3. Hydroponic plants are healthier and more vigrous.
4. Plants can mature faster and yield an earlier harvest.
5. Hydroponic gardens use less space since the roots need not to spread out.
6. Hydroponic can offer cultivating plants in desert sands and they can be transformed into productive lands.
7. Water can be reused in hydroponics and less is wasted as runoff and evaporation.

Disadvantages:

1. Hydroponic conditions create an environment that stimulates *salmonella* growth.
2. Pathogens attack including damp off due to *verticillium wilt* due to high moisture levels and over watering of soil based plants.
3. Different hydroponic plants require different fertilizers.

Two main types of hydroponics are;

1. Solution culture; 2. Medium culture

Solution culture does not use solid medium for the roots just the nutrient solution. Three main types of solution culture are; (*i*) state (*ii*) Continuous flow and (*iii*) aeroponics

Medium culture method have a solid medium for the roots *e.g.* sand culture, gravel, or rockwood.

For all techniques most hydroponic reservoirs are now built of plastic but other materials can be used including concrete, glass, metal, vegetable solids and woods.

Hydroponic averages compared with ordinary soil yields as under:

Crop	*Hydroponic equivalent per acre*	*Agricultural average per acre*
Wheat	3624 kg	2536 kg
Rice	5436 kg	407 kg
Potatoes	70666 kg	8150 kg
Cabbage	8150 kg	5890 kg
Tomatoes	181,200 kg	9000 kg

AQUAPONICS

It is a food production system that couples aquaculture with hydroponics whereby the nutrient-rich aquaculture water is fed to hydroponically grown plants.

Aquaponics is seen as a highly efficient farming solution that addresses these multiple farming challenges. By using less water compared to conventional farming, reducing the need for synthetic fertilizers, and minimizing waste, aquaponics presents a promise for producing food without damaging our environment.

AEROPONICS

It is a system where roots are continuously or discontinuously kept in an environment saturated with fine drops of nutrient solution (a mist or a aerosol). The method require no substrate and entails growing plants with their roots suspended in a deep air or growth chamber with the roots periodically wetted with a fine mist of atomized nutrients. This technique have proved to be commercial successful for propagation, seed germination, seed potato production, tomato production, leaf crop and micro greens etc.

Aeroponics is an innovative method of plant cultivation that offers several advantages. It allows for the efficient and precise delivery of nutrients and oxygen to plant roots, promoting rapid growth and higher yields while conserving water and space. Additionally, aeroponics reduces the risk of soil-borne diseases and pests, making it a clean and sustainable agricultural technique.

One of the standout advantages of aeroponics is its ability to promote rapid plant growth and higher yields. Traditional soil-based farming often relies on the plant's ability to access nutrients through the soil, which can be an inefficient process.

National Mission for Sustainable Agriculture

Govt. to promote organic farming methods, combining modern technology with traditional farming practices.

Sustainable Sugar Initiative

It is a practical approach to sugarcane production and in this approach the cost of the seed is very low as compared to conventional method of cultivation. Drip Irrigation practices can be effectively used. Scope of intercropping is more in this approach compared o the conventional methods of cultivation.

L. GI Tag (Geographical Indications Tags)

A Geographical Indication (GI) tag is a sign or name given to products that have a specific geographical origin and qualities that are due to that origin. GI tags are a type of intellectual property that can be used to protect products and brand equity.

Inforced with 15th Sep. 2003. Darjeeling Tea was the first Indian product to get a GI tag. As of April 2024: 643 GI Tag registered in india.

Indian State	GT Tags
Andhra Pradesh	Srikalahasthi Kalamkari, Kondapalli Bommallu, Machilipatnam Kalamkari, Budithi Bell & Brass Metal Craft. Andhra Pradesh Leather Puppetry, Uppada Jamdani Sarees, Venkatagiri Sarees, Bobbili Veena, Mangalagiri Sarees and Fabrics, Dharmavaram Handloom Pattu Sarees And Paavadas, Udayagiri Wooden Cutlery, Durgi Stone Carvings, Etikoppaka Toys, Allagadda Stone Carving, Tirupati Laddu, Bandar Laddu, Guntur Sannam Chilli, Araku Valley Arabica Coffee
Arunachal Pradesh	Idu Mishmi Textiles, Arunachal Orange
Assam	Muga Silk of Assam, Muga Silk of Assam (Logo), Assam Karbi Anglong Ginger, Joha Rice of Assam, Tezpur Litchi, Boka Chaul, Kaji Nemu
Bihar	Madhubani Paintings, Applique (Khatwa) Work of Bihar, Sujini Embroidery Work of Bihar, Sikki Grass Products of Bihar, Bhagalpur Silk, Applique (Khatwa) Work of Bihar (Logo), Sikki Grass Products of Bihar (Logo), Sujini Embroidery Work of Bihar (Logo), Bhagalpuri Zardalu, Katarni Rice, Maghai Paan, Shahi Litchi of Bihar, Silao Khaja
Chhattisgarh	Bastar Dhokra, Bastar Wooden Craft, Bastar Iron Craft
Goa	Khola Chilli, Feni
Gujarat	Sankheda Furniture, Agates of Cambay, Kutch Embroidery, Tangaliya Shawl, Surat Zari Craft, Kachchh Shawls, Patan Patola, Sankheda Furniture (Logo), Kutch Embroidery (Logo), Agates of Cambay (Logo), Jamnagari Bandhani, RajKot Patola, Pethapur Printing Blocks, Gir Kesar Mango, Bhalia Wheat

Himachal Pradesh	Kullu Shawl, Kangra Tea, Chamba Rumal, Kinnauri Shawl, Kullu ShawL (Logo), Kangra Paintings, Himachali Kala Zeera, Himachali Chulli Oil
J& K	Kani Shawl, Kashmir Pashmina, Kashmir Paper Machie, Kashmir Walnut Wood Carving, Khatamband, Kashmiri Hand Knotted Carpet, Saffron (Mongra, Lachha, Guchhi)
Jharkhand	Sohrai – Khovar Painting
Karnataka	Mysore Silk, Bidriware, Channapatna Toys & Dolls, Mysore Rosewood Inlay, Kasuti Embroidery, Mysore Traditional Paintings, Ilkal Sarees, Ganjifa Cards of Mysore, Navalgund Durries, Karnataka Bronzeware, Sandur Lambani Embroidery, Kinhal Toys, Udupi Sarees, Mysore Silk (Logo), Coorg Orange, Mysore Betel leaf, Nanjangud Banana, Udupi Mallige, Mysore Mallige, Hadagali Mallige, Coorg Green Cardamom, Devanahalli Pomelo, Appemidi Mango, Kamalapur Red Banana, Bangalore Blue Grapes, Coorg Arabica Coffee, Chikmagalur Arabica Coffee, Bababudangiri Arabica Coffee, Sirsi Supari, Gulbarga Tur Dal, Dharwad Pedha, Mysore Agarbathi, Mysore Sandalwood Oil, Mysore Sandal soap
Kerala	Tirur Betel Leaf (Tirur Vettila), Alleppey Coir, Navara Rice, Palakkadan Matta Rice, Alleppey Green Cardamom, Maddalam of Palakkad, Screw Pine Craft of Kerala, Brass Broidered Coconut Shell Crafts of Kerala, Pokkali Rice , Vazhakulam Pineapple, Cannanore Home Furnishings, Balaramapuram Sarees and Fine Cotton Fabrics, Kasaragod Sarees, Kuthampully Sarees, Central Travancore Jaggery, Wayanad Jeerakasala Rice A, Wayanad Gandhakasala Rice, Payyannur Pavithra Ring, Chendamangalam Dhoties & Set Mundu, Kaipad Rice , Chengalikodan Nendran Banana, Kuthampully Dhoties & Set Mundu, Maddalam of Palakkad (Logo), Brass Broidered Coconut Shell Craft of Kerala (Logo), Screw Pine Craft of Kerala (Logo, Nilambur Teak, Wayanad Robusta Coffee, Marayoor Jaggery (Marayoor Sharkara),
Madhya Pradesh	Chanderi Sarees, Leather Toys of Indore, Bagh Prints of Madhya Pradesh Bell Metal Ware of Datia and Tikamgarh Maheshwar Sarees & Fabrics, Bell Metal Ware of Datia and Tikamgarh (Logo), Leather Toys of Indore (Logo), Ratlami Sev, Bagh Prints of Madhya Pradesh (Logo), Jhabua Kadaknath Black Chicken Meat

Maharashtra	Solapur Chaddar, Solapur Terry Towel, Puneri Pagadi, Nashik Valley Wine, Paithani Sarees and Fabrics, Mahabaleshwar Strawberry, Nashik Grapes, Kolhapur Jaggery, Ajara Ghansal Rice, Mangalwedha Jowar, Sindhudurg & Ratnagiri Kokum, Waghya Ghevada, Navapur Tur Dal, Vengurla Cashew, Lasalgaon Onion, Sangli Raisins, Beed Custard Apple, Jalna Sweet Orange, Waigaon Turmeric, Purandar Fig, Jalgaon Bharit Brinjal, Solapur Pomegranate, Bhiwapur Chil, Ambemohar Rice, Dahanu Gholvad Chikoo, Jalgaon Banana, Marathwada Kesar Mango, Karvath Kati Sarees & Fabrics, Alphonso, Sangli Turmeric
Manipur	Shaphee Lanphee, Wangkhei Phee, Moirang Phee, Kachai Lemon, Chak-Hao
Meghalaya	Khasi Mandarin, Memong Narang
Mizoram	Mizo Chilli, Pawndum, Ngotekherh, Hmaram, Tawlhlohpuan, Mizo Puanchei
Nagaland	Naga Mircha, Naga Tree Tomato, Chakhesang Shawl
Odisha	Kotpad Handloom fabric, Orissa Ikat, Konark Stone carving, Orissa Pattachitra, Pipli Applique Work, Khandua Saree and Fabrics, Gopalpur Tussar Fabrics, Ganjam Kewda Rooh, Ganjam Kewda Flower, Dhalapathar Parda & Fabrics, Sambalpuri Bandha Saree & Fabrics, Bomkai Saree & Fabrics, Habaspuri Saree & Fabrics , Berhampur Patta (Phoda Kumbha) Saree & Joda, Orissa Pattachitra (Logo), Kandhamal Haladi, Odisha Rasagola
Rajasthan	Kota Doria, Blue Pottery of Jaipur, Molela Clay Work, Kathputlis of Rajasthan, Bikaneri Bhujia, Kota Doria (Logo), Bagru Hand Block Print, Thewa Art Work, Makrana Marble, Molela Clay Work of Rajasthan (Logo). Blue Pottery of Jaipur (Logo), Kathputlis of Rajasthan (Logo), Pokaran Pottery
Tamil Nadu	Salem Fabric, Kancheepuram Silk , Kancheepuram Silk, Madurai Sungudi, Coimbatore Wet Grinder, Thanjavur Paintings, Temple Jewellery of Nagercoil, Thanjavur Art Plate, East India Leather, Salem Silk known as Salem Venpattu, Kovai Kora Cotton Sarees, Arani Silk, Swamimalai Bronze Icons, Eathamozhi Tall Coconut, Thanjavur Doll, Nilgiri (Orthodox), Virupakshi Hill Banana, Sirumalai Hill Banana. Madurai Malli, Pattamadai Pai ("Pattamadai Mat"), Nachiarkoil Kuthuvilakku

	("Nachiarkoil Lamp"), Chettinad Kottan, Toda Embroidery, Thanjavur Veenai, Thanjavur Art Plate (Logo), Swamimalai Bronze Icons (Logo), Temple Jewellery of Nagercoil (Logo), Mahabalipuram Stone Sculpture, Erode Manjal (Erode Turmeric), Thirubuvanam Silk Sarees, Kodaikanal Malai Poondu, Palani Panchamirtham, Dindigul Locks, Kandangi Saree, Srivilliputtur Palkova, Kovilpatti Kadalai Mittai, Thanjavur Pith Works, Arumbavur Wood Carvings
Tripura	Tripura Queen Pineapple
Telangana	Pochampally Ikat, Silver Filigree of Karimnagar, Nirmal Toys and Craft, Nirmal Furniture, Nirmal Paintings, Gadwal Sarees, Hyderabad Haleem, Cheriyal Paintings, Siddipet Gollabhama, Narayanpet Handloom Sarees, Pochampally Ikat (Logo), Adilabad Dokra, Warangal Durries, Telia Rumal
Uttar Pradesh	Allahabad Surkha Guava, Lucknow Chikan Craft, Malihabadi Dusseheri mango, Banaras Brocades and Sarees, Hand Made Carpet of Bhadohi. Agra Durrie, Farrukhabad Prints, Lucknow Zardozi. Banaras Brocades and Sarees (Logo), Kalanamak Rice, Firozabad Glass, Kannauj Perfume , Kanpur Saddlery, Moradabad Metal Craft, Saharanpur Wood Craft, Meerut Scissors, Khurja Pottery, Banaras Gulabi Meenakari Craft, Varanasi Wooden Lacquerware & Toys, Mirzapur Handmade Dari, Nizamabad Black Pottery, Banaras Métal Repoussé Craft, Varanasi Glass beads, Ghazipur Wall Hanging, Varanasi Soft Stone Jali Work, Chunar Balua Patthar. Gorakhpur Terracotta, Kalpi Handmade Paper, Barabanki Handloom Product, Baghpat Home Furnishings, Sambhal Horn Craft, Mainpuri Tarkashi, Mahoba Gaura Patthar Hastashilp, Amroha Dholak
West Bengal	Darjeeling Tea (word & logo), Nakshi Kantha, Santiniketan Leather Goods, Malda Laxman Bhog Mango, Khirsapati (Himsagar) mangoes, Malda Fazli Mango, Santipur Saree, Baluchari Saree, Dhaniakhali Saree, Joynagar Moa, Bardhaman Sitabhog, Bardhaman Mihidana, Gobindobhog Rice, Tulaipanji Rice, Banglar Rasogolla, Bankura Panchmura Terracotta Craft, Bengal Patachitra, Purulia Chhau Mask, Wooden Mask of Kushmandi, Madur kathi

List of GI Tags 2024

Products	Categories	States/UTs
Kashmir Saffron	Agriculture	Jammu & Kashmir
Manipuri Black Rice	Food Stuff	Manipur
Kandhamal Haladi	Agricultural	Odisha
Rasagola	Food Stuff	Odisha
Kodaikanal Malai Poondu	Agricultural	Tamil Nadu
Pawndum	Handicraft	Mizoram
Ngotekherh	Handicraft	Mizoram
Hmaram	Handicraft	Mizoram
Palani Panchamirtham	Food Stuff	Tamil Nadu
Tawlhlohpuan	Handicraft	Mizoram
Mizo Puanchei	Handicraft	Mizoram
Gulbarga Tur Dal	Agricultural	Karnataka
Tirur Betel Leaf (Tirur Vettila)	Agricultural	Kerala
Khola Chilli	Agricultural	Goa
Idu Mishmi Textiles	Handicraft	A. Pradesh
Dindigul Locks	Manufactured	Tamil Nadu
Kandangi Saree	Handicraft	Tamil Nadu
Srivilliputtur Palkova	Food Stuff	Tamil Nadu
Amroha Dholak	Handicraft	Uttar Pradesh
Mahoba Gaura Patthar Hastashilp	Handicraft	Uttar Pradesh
Mainpuri Tarkashi	Handicraft	Uttar Pradesh
Sambhal Horn Craft	Handicraft	Uttar Pradesh
Baghpat Home Furnishings	Handicraft	Uttar Pradesh
Barabanki Handloom Product	Handicraft	Uttar Pradesh
Kalpi Handmade Paper	Handicraft	Uttar Pradesh
Cuttack Rupa Tarakasi, Banglar Muslin,	Handicraft	Odisha
Similipal kai chutney	Food	Odisha

IMPORATNT DAYS AND THEMES

Celebrated as	Day	Themes 2024
National Youth Day	Jan, 12	"Arise, Awake, and Realize the Power You Hold".
National Girl Child Day	Jan, 24	"Mere Viksit Bharat Ka Sapna"
Republic Day/	Jan 26	Viksit Bharat' and 'Bharat – Loktantra ki Matruka',
International Customs Day	Jan, 26	"Customs Engaging Traditional and New Partners with Purpose".
World Wetland Day	Feb., 2	"Wetlands and Human Wellbeing".
World Pulses Day	Feb., 10	"Pulses: nourishing soils and people".
National Science Day	Feb, 28	"Indigenous Technologies for Viksit Bharat".
Zero Discrimination Day	Mar, 1	"To protect everyone's health, protect everyone's rights"
National Safety Day	Mar, 4	"Safety Leadership for ESG Excellence".
International Women's Day	Mar, 8	"Invest in women: Accelerate progress". Central Industrial, Security Froce (CISF)
Commonwealth Day	Mar, 11	"One Resilient Common Future: Transforming our Common Wealth".
International Day of Mathematics	Mar, 14	"Playing with Math".
World Consumer Day	Mar, 15	"Fair and Responsible Artificial Intelligence for Consumers".
World Forestry Day	Mar, 21	"Forests and Innovation: New Solutions for a Better World".
World Water Day	Mar, 22	"Water for Peace".
World Meteorological Day	Mar, 23	"At the Frontline of Climate Action".
World Theatre Day	Mar, 27	Theatre and a Culture of Peace".
World Autism Awareness Day	April, 2	"Empowering Autistic Voices".
International Carrot Day Founded in year 2003	April, 4	
National Maritime Day,	April, 5	Navigating the future: safety first
World Health Day	April, 7	"My health, my right".

Celebrated as	Day	Themes 2024
National Safe Motherhood Day	Apri, 11	"Empowering Families, Strengthening Communities for Optimal Maternal Health".
Jallianwala bagh Massacre Day (1919)	April 13	
National Civil Service Day,	April, 21	"Celebrating the Contribution of Civil Servants".
Earth Day	April, 22	"Planet vs. Plastics"
National Civil Service Day	April, 23	"Celebrating the Contribution of Civil Servants"
World Malaria Day	April, 25	"Accelerating the fight against malaria for a more equitable world".
World Intellectual Property Day	April, 26	"IP and the SDGs: Building our common future with innovation and creativity".
International Seed Day		"Preserving Indigenous seed varieties"
Labour day (Kamgar Din)	May, 1	"Ensuring workplace safety and health amidst climate change".
World Tuna Day	May, 2	"Our Tuna, Our Legacy".
World Red Cross Day	May, 8	"I give with joy, and the joy I give is a reward"
Mother's Day	May, 10	"Celebrating Motherhood: A Timeless Bond"
National Technology Day	May, 11	"From Schools to Startups: Igniting Young Minds to Innovate"
International Nurses Day	May, 12	"Our Nurses. Our Future. The economic power of care".
International Day of Plant Health		"Plant health, safe trade, and digital technology".
International Day of the Family	May, 16	"Families and Climate Change: International Year of the Family + 30"
National Endangered Species Day	May, 17	"Celebrate Saving Species".
World Telecommunication Day		"Digital Innovation for Sustainable Development".
World Bee Day	May, 20	"Bee engaged with Youth".
Anti Terrorism Day	May, 21	"Voices for Peace: Victims of Terrorism as Peace Advocates and Educators"

Celebrated as	Day	Themes 2024
International Biodiversity Day	May, 22	"Be part of the Plan".
World No Tobacco Day	May, 31	"Protecting children from tobacco industry interference".
World Milk Day	June, 1	"Moving towards a healthy tomorrow: Ensuring the nutritional benefit of milk for our overall wellbeing".
World Environment Day	June, 5	"Land restoration, desertification, and drought resilience".
World Food Safety Day	June, 7	"Food safety: prepare for the unexpected".
World Ocean Day	June, 8	Awaken New Depths".
World Blood Donor Day	June, 14	"20 years of celebrating giving: Thank you, blood donors!".
World Refugee Day	June, 20	"For a World Where Refugees Are Welcomed".
International Yoga Day	June, 21	"Yoga for Self and Society".
United Nations Day of Public Service	June, 23	"Fostering Innovation amid Global Challenges: a Public Sector Perspective".
World Microbiome Day	June, 24	"Feed Your Microbes".
World Drug Day,	June, 26	"The evidence is clear: invest in prevention".
National Statistics Day	June, 29	Use of Data for Decision Making"
International Day of the Tropics	June, 29	"The future belongs to the Tropics".
International Asteroid Day	June, 30	"Deflection Strategies".
International Day of Parliamentarism	June, 30	"Parliamentary diplomacy: Building bridges for peace and understanding".
National Doctor's Day (Birthday of B.C. Roy)	July, 1	"Healing Hands, Caring Hearts".
World Zoonoses Day	July, 6	"One World, One Health: Prevent Zoonoses".
World Population Day	July, 11	"Leave No One Behind, Count Everyone".
NABARD foundation day,	July, 12	"Climate Risk Management and banking".
World Youth Skill Day	July, 15	"Youth Skills for Peace and Development".

Celebrated as	Day	Themes 2024
Nelson Mandela International Day	July, 16	"Combating poverty and inequality is in our hands".
World Chess Day, International Moon Day,	July, 20	"Chess for Sustainable Development". "Illuminating the shadows".
Kargil Vijay Diwas	July, 26	"Kargil Victory Day".
World Hepatitis Day	July, 28	"It's time for action".
International Tiger Day	July, 29	"Call for Action".
National Handloom Day	Aug., 7	"Weaving Sustainable Futures".
World Senior Citizen Day	Aug, 8	"Ageing with Dignity: The Importance of Strengthening Care and Support Systems for Older Persons Worldwide".
World Biofuel Day	Aug., 10	"Sustainable Biofuels: Fueling a Greener Future".
International Youth Day	Aug., 12	"From Clicks to Progress: Youth Digital Pathways for Sustainable Development".
Indian Independence Day	Aug, 15	"Viksit Bharat"
World Photography Day	Aug., 19	"An Entire Day".
Sadbhavna Diwas	Aug, 20	"Unity in Diversity: Embracing Our Differences".
National Sports Day	Aug , 29	"Sport for the Promotion of Peaceful and Inclusive Societies".
World Coconut Day	Sep., 2	"Coconut for a Circular Economy: Building Partnership for Maximum Value".
International Day of Clean Air for Blue Skies	Sep., 7	"Invest in #CleanAirNow".
International Literacy Day (UNESCO)	Sep, 8	"Promoting multilingual education: Literacy for mutual understanding and peace".
World Suicide Prevention Day	Sep., 10	"Changing the Narrative on Suicide"
Hindi Diwas	Sep, 14	"Hindi – the bridge between traditional knowledge and artificial intelligence".
World Ozone Day	Sep, 16	"Montreal Protocol: Advancing Climate Action".

Celebrated as	Day	Themes 2024
World Bamboo Day,	Sep., 18	"Next Generation Bamboo: Solution, Innovation, and Design"
World Rhino Day	Sep, 22	"Keep the Five Alive".
International Day for the Total Elimination of Nuclear Weapons	Sep., 26	"Forging a path towards human security, justice, and peace".
World Wildlife Day	Oct, 4	"Connecting People and Planet: Exploring Digital Innovation in Wildlife Conservation".
World Teachers Day,	Oct, 5	"Valuing Teacher Voices: Towards a New Social Contract for Education".
International Girl Child Day	Oct., 11	"Girls' Vision for the Future".
International Day of Rural Women	Oct, 15	"Rural Women Cultivating Good Food for All".
World Food Day	Oct ,16	"Right to foods for a better life and a better future".
Int. Poverty Eradication Day	Oct, 17	"Ending Social and Institutional Maltreatment Acting together for just, peaceful and inclusive societies".
Children's Day	Nov, 14	"Investing in our future means investing in our children"
Universal Children's Day	Nov, 20	Theme for 2023 'For Every Child, Every Right.
World Television Day	Nov, 21	"Television – Then and Now"
World Olive Tree Day	Nov., 26	"Celebrating the olive tree as a symbol of peace".
World AIDS Day	Dec, 1	"Remember and Commit".
International Day of People with Disabilities	Dec., 3	"Transformative Solutions for Inclu sive Development: the role of Innovation in Fuelling an Accessible and Equitable World".
World Soil Day	Dec, 5	"Caring for Soils: Measure, Monitor, Manage".
Human Rights Day	Dec,10	"Three Decades of Respect for and Promotion of Human Rights".
National Farmers Day	Dec, 23	"Delivering Smart Solutions for Sustainable Food Security and Resilience".

The World Cooperative Economic Forum (WCopEF) has been established by cooperative sector experts to represent the interests of over three million cooperatives worldwide. The forum aims to promote cooperative thinking and movement, as well as raise concerns with governments and other stakeholders.

Peer-to-peer social investing platform - Rang De - has recently announced that it has disbursed over Rs 85 lakhs worth of credit to small, marginal, and landless farmers in Assam through the Xamahar initiative. This initiative aims to assist more than 10,000 potato farmers with access to credit over the next two years, up to 50 percent being first-time borrowers.

The government has announced a scheme to computerize state cooperative registrars and Agriculture and Rural Development Banks (ARDBs) offices with a budget of Rs 225.09 crore. This comes after the government started computerizing all Primary Agriculture Credit Societies (PACS) in the country. The scheme aims to bring transparency and efficiency to these offices and save time for users. The government has also decided to computerize the offices of Registrar of Cooperatives of all states and Union Territories. The central government has been taking steps to strengthen all cooperative societies in the country.

The Indian government aims to bring 60% of the total wheat area under climate-resistant varieties for the upcoming rabi sowing season, due to strong El Nino conditions. Despite the challenges of climate change, the Ministry of Agriculture has set a target of achieving a record wheat output of 114 million tonnes in the 2023-24 season.

The Indian government has announced plans to **integrate 28 new wholesale mandis,** or market yards, with the online trading platform eNAM. This will bring the total number of integrated mandis to 1,389. The Agriculture Secretary, highlighted the importance of providing key information to stakeholders in order to remove information asymmetry. He also emphasized the need to make the entire value chain more efficient and reduce wastage.

The Indian government has launched the **Unified Portal for Agri-Statistics (UPAg)**, a digital platform that aims to provide stakeholders with credible and standardized data for informed decision-making in the agriculture sector. The portal, developed by the Ministry of Agriculture, will address challenges related to data governance and offer features such as data standardization, analysis, granular production estimates, and commodity profile reports.

The United Nations Development Programme (UNDP) and the **National Bank for Agriculture and Rural Development (NABARD)** have signed an MoU to collaborate on data-driven innovations in agriculture and food systems. The partnership aims to support smallholder farmers by sharing open-source data for product development, technology transfer, and policy framing.

The G20 Technical Workshop on "Climate Resilient Agriculture" has begun in Hyderabad, India, with the aim of discussing climate change challenges and promoting international cooperation to address climate-related agricultural issues. The workshop focuses on research needs and innovations for climate resilient agriculture and includes participants from G20 member states, guest-invited countries, and international organizations. The workshop will also involve visits to research institutes and cultural experiences. (4-6 Sept. 2023)

Union Cooperation Minister Amit Shah launched the **'Bharat Organics'** brand of the newly created National Cooperative Organics Ltd (NCOL) and announced that it will become a trusted brand in India and abroad. The NCOL will initially sell six organic products in India, with plans to expand internationally.

India's **higher educational institutes (HEIs)** have come together to create a first-of-its-kind HEI-led **'Agri-Consortium'** – a move initiated by IIM Kashipur. An in-principal framework in this regard was adopted by participating institutions (HEIs) during the 'Agri-Entrepreneurship in South Asia' Consortium-2023 organized by the Foundation for Innovation & Entrepreneurship Development (FIED), an incubation center of IIM Kashipur, in collaboration with the Ministry of Agriculture and Farmers Welfare

The **Haryana government** is giving special incentives to farmers for adopting crop diversification, said an official statement here on Sunday. This will not only enhance their income but will also ensure the protection of the environment.

ROLES OF VITAMINS

Vitamin A : Important for vision, skin, bones and reproduction. Found in: carrots, sweet, potatoes, spinach

Vitamin B_1 : Converts carbs into energy & is vital for glucose metabolism. Found in : Sunflower seeds, whole grains, pork

Vitamin B_2 : Plays a role in energy production, skin and eye health. Found in: Dariy, organ, meat, almonds

Vitamin B_6: Amino Acid. Metabolism and red blood cell production. Found in : Bananas, Chickpea, Salmon

Vitamin B_5 : Necessary for fatty acid synthesis. Found in Avocado, chicken, broccoli

Vitamin B_3 : DNA Repair, and metabolic functions as well as energy. Found in Salt-water Fish, Popultry, Avacado.

Vitamin B7: It enables the body to metabolize proteins, fats, and carbohydrates. It also contributes to keratin, a structural protein in the skin, hair, and nails.

Vitamin B_9 : It is essential for making DNA and RNA. Deficiency: During pregnancy, this can affect the fetus's nervous system. Doctors recommend folic acid supplements before and during pregnancy.

Vitamin B_{12}: It is essential for a healthy nervous system. Deficiency: Low levels may lead to neurological problems and some types of anemia. Good sources: Examples include fish, shellfish, meat, poultry, eggs, milk and other dairy products, fortified cereals, fortified soy products, and fortified nutritional yeast.

Vitamin C: It contributes to collagen production, wound healing, and bone formation. It also strengthens blood vessels, supports the immune system, helps the body absorb iron, and acts as an antioxidant.

Vitamin D: It is necessary for the healthy mineralization of bone. Good sources: Exposure to UVB rays from the sun or other sources causes the body to produce vitamin D. Fatty fish, eggs, beef liver, and mushrooms also contain the vitamin.

Vitamin E: Its antioxidant activity helps prevent oxidative stress, an issue that increases the risk of widespread inflammation and various diseases. Good sources: These include wheat germ, kiwis, almonds, eggs, nuts, leafy greens, and vegetable oils.

List of UN International Years

Sr. No	Year	Designated as
1.	2020	International Year of Plant Health
2.	2019	International Year of Indigenous Languages
3.	2017	International Year of Sustainable Tourism for Development
4.	2015	International Year of Light and Light-based Technologies
5.	2015	International Year of Soils
6.	2014	International Year of Family Farming
7.	2013	International Year of Water Cooperation
8.	2013	International Year of Quinoa
9.	2012	International Year of Sustainable Energy for All
10.	2011	International Year of Forests
11.	2010	International Year of the Seafarer
12.	2010	International Year of Biodiversity
13.	2009	International Year of Natural Fibres
14.	2009	International Year of Human Rights Learning
15.	2009	International Year of Astronomy
16.	2008	International Year of the Potato/Planet Earth
17.	2021	International Year of the Fruits and Vegetables
18.	2022	International Year of the Artisanal and Aquaculture
19.	2023	International Year of the Millets
20.	2024	International Year of the Camelids (Camels, llamas, alpacas, vicuńas and guanacos).

- **Viticulture:** Science of grapes.
- **Olericulture:** Science of vegetables.
- Cloud cover is measured in unit OKTA.
- Convection heat transfer is faster than conduction.
- Lapse rate is vertical decrease or gradient. Normal lapse rate is 6.5 °C per km or 3.5 F per 1000 feet.
- A **cyclone** is a more or less circular area of low atmospheric pressure in which winds blow counter clockwise in Northern hemisphere.
- An **anticyclone** is an area of relatively high pressure in which the winds tends to blow spirally outwards in a clockwise direction in Northern hemisphere.
- The **degree day** is a departure of one degree for one day in the mean daily temperature from a specified base temperature.
- A **degree day** is a measure of heating or cooling.

HIGHEST PRODUCING STATES FOR DIFFERENT LIVESTOCK PRODUCTS (2021-22)

Livestock Product	*Max. Producing State*
Indigenous Cow Milk	**Rajasthan**
Non-descript cow milk	**Madhya Pradesh**
Exotic Cow Milk	**Punjab**
Crossbred cow milk	**Tamilnadu**
Buffalo Milk	**Uttar Pradesh**
Goat Milk	**Rajasthan**
Total Milk	Uttar Pradesh
Buffalo Meat	**Uttar Pradesh**
Pig Meat	**Uttar Pradesh**
Sheep Meat	**Andhra Pradesh**
Poultry Meat	**Andhra Pradesh**
Goat Meat	**West Bengal**
Total Meat	Maharashtra
Fowl Egg and Total egg	**Andhra Pradesh**
Duck Egg	**West Bengal**
Ram wool	**Rajasthan**
Ewe wool	**Andhra Pradesh**
Lamb wool	**Gujarat**
Total wool	Rajasthan

INDIA'S GLOBAL RANKING WITH RESPECT TO DIFFERENT LIVESTOCK STATISTICS

Rank	*Particular*
First	Livestock population Buffalo population Goat population Milk production Goat milk Cara beef
Second	Cattle population Sheep population

Third	Egg production
Fourth	Chicken production
Fifth	Poultry production
Eighth	Duck production Meat Production
Ninth	Camel population
Tenth	Pig population

N percentage in Different important nitrogenous fertilizers

Ammoniacal fertilizers :-

1) Ammonium sulphate (20.5%N)
2) Ammonium chloride (26.5%N)
3) Anhydrous ammonia (82.0%N)

Nitrate fertilizers

1) Sodium nitrate (16.5%N)
2) Calcium Nitrate (15.5%N)
3) Pottasium Nitrate (13%N)

Ammonium And Nitrate fertilizers

1) Ammonium nitrate (33%N)
2) Calcium ammonium nitrate (26%N)

Amide fertilizers

1) Urea (46%N)
2) Calcium cynamide (21%N)

P percentage in Phosphorus Fertilizers

Water soluble

1) SSP(16%P) 2) DSP(32%P
3) TSP(48%P) 4) DAP(46%P)
5) MAP(48%P) 6) Ammonium phosphate (20%P)

Citrate Soluble

1) DCP(33-40%P) 2) Basic slag (14-18%P)
3) Rhemenea (23-26%P)

Citrate & water insoluble

1) Rock Phosphate (20-40%P)
2) Row bone meal (20-25%P)
3) Steamed Bone meal (20-30%P)

Ammonium polyphosphate (62%P)

PLACES AND FRUITS

S. No.	*Places*	*Fruit*
1.	Kotgarh (Himachal Pradesh)	Apple bowl of India
2.	Goa	Cashew
3.	Nashik	Grape capital of India
4.	Muzaffarpur district of Bihar	Litchi (Lychee Kingdom of India)
5.	Solapur	Pomegranate
6.	Nagpur	Orange City of India
7.	Mahabaleshwar (Maharashtra)	Strawberry
8.	Ratnagiri (Maharashtra)	Alphonso mango
9.	Siliguri	Pineapples
10.	Kanthalloor	Peach
11.	Jalgaon	Bananas
12.	Allahabad	Guava
13.	Amalsad, Navsari (Gujarat)	Chiku or Sapota
14.	Pune (Maharashtra)	Anjeer or Fig
15.	Anantapur (A.P.)	Papaya
16.	Himachal Pradesh	Plum and Peach fruit
17.	Punjab	Pear
18.	Andhra Pradesh	Custard Apple, Banganapalle Mango & Papaya
19.	Madhya Pradesh	Ber
20.	Arunachal Pradesh	Kiwi
21.	Beed (Maharashtra)	Custard Apple
22.	Manipur	Kachai Lemon
23.	Ganganagar (Rajasthan)	Kinnow
24.	Ladakh	Apricot
25.	Uttarkhand	Chestnut
26.	Andaman and Nicobar	Water Apple

LIST OF RAMSAR SITES IN INDIA 2024

Sl.	Ramsar Site	State/UT	Designated Year	Area (Hectares)
1	Kolleru Lake	Andhra Pradesh	2002	90100
2	Deepor Beel	Assam	2002	4000
3	Kabartal Wetland	Bihar	2020	2620
4	Nagi Bird Sanctuary	Bihar	2024	205.8
5	Nakti Bird Sanctuary	Bihar	2024	332.6
6	Nalsarovar	Gujarat	2012	12000
7	Wadhvana Wetland	Gujarat	2021	630
8	Thol Lake Wildlife Sanctuary	Gujarat	2021	699
9	Khijadia Wildlife Sanctuary	Gujarat	2021	512
10	Nanda Lake	Goa	2022	42
11	Sultanpur National Park	Haryana	2021	143
12	Bhindawas Wildlife Sanctuary	Haryana	2021	412
13	Pong Dam Lake	Himachal Pradesh	2002	15662
14	Chandertal Wetland	Himachal Pradesh	2005	49
15	Renuka Wetland	Himachal Pradesh	2005	20
16	Wular Lake	J&K	1990	18900
17	Surinsar-Mansar Lakes	J&K	2005	350
18	Hokera Wetland	J&K	2005	1375
19	Hygam Wetland Conservation Reserve	J&K	2022	802
20	Shallbugh Wetland	J&K	2022	1675
21	Ranganathittu Bird Sanctuary	Karnataka	2022	518
22	Ankasamudra Bird Conservation Reserve	Karnataka	2023	98.76
23	Aghanashini Estuary	Karnataka	2023	4801
24	Magadi Kere Conservation Reserve	Karnataka	2023	54.38
25	Asthamudi Wetland	Kerala	2002	6140
26	Sasthamkotta Lake	Kerala	2002	373
27	VembanadKol Wetland	Kerala	2002	151250
28	Tso Kar Wetland Complex	Ladakh	2020	9577

29	Tsomoriri Lake	Ladakh	2002	12000
30	Bhoj Wetlands	Madhya Pradesh	2002	3201
31	Sirpur Wetland	Madhya Pradesh	2022	161
32	Sakhya Sagar	Madhya Pradesh	2022	248
33	Yashwant Sagar	Madhya Pradesh	2022	823
34	Tawa Reservoir	Madhya Pradesh	2024	20050
35	Nandur Madhameshwar	Maharashtra	2019	1437
36	Lonar Lake	Maharashtra	2020	427
37	Thane Creek	Maharashtra	2022	6521
38	Loktak Lake	Manipur	1990	26600
39	Pala Wetland	Mizoram	2021	1850
40	Chilka Lake	Odisha	1981	116500
41	Bhitarkanika Mangroves	Odisha	2002	65000
42	Satkosia Gorge	Odisha	2021	98197
43	Tampara Lake	Odisha	2021	300
44	Hirakud Reservoir	Odisha	2021	65400
45	Ansupa Lake	Odisha	2021	231
46	Harike Lake	Punjab	1990	4100
47	Kanjli Lake	Punjab	2002	183
48	Ropar Lake	Punjab	2002	1365
49	Beas Conservation Reserve	Punjab	2019	6429
50	Keshopur – Miani Community Reserve	Punjab	2019	344
51	Nangal Wildlife Sanctuary	Punjab	2019	116
52	Keoladeo Ghana National Park	Rajasthan	1981	2873
53	Sambhar Lake	Rajasthan	1990	24000
54	Point Calimere Wildlife and Bird Sanctuary	Tamil Nadu	2002	38500
55	Koonthankulam Bird Sanctuary	Tamil Nadu	2021	72
56	Chitrangudi Bird Sanctuary	Tamil Nadu	2021	260
57	Karikili Bird Sanctuary	Tamil Nadu	2022	58
58	Pichavaram Mangrove	Tamil Nadu	2022	1479
59	Pallikaranai Marsh Reserve Forest	Tamil Nadu	2022	1248

60	Gulf of Mannar Marine Biosphere Reserve	Tamil Nadu	2022	52672
61	Vembannur Wetland Complex	Tamil Nadu	2022	20
62	Vellode Bird Sanctuary	Tamil Nadu	2022	77
63	Udhayamarthandapuram Bird Sanctuary	Tamil Nadu	2022	44
64	Vedanthangal Bird Sanctuary	Tamil Nadu	2022	40
65	Suchindram Theroor Wetland Complex	Tamil Nadu	2022	94
66	Vaduvur Bird Sanctuary	Tamil Nadu	2022	113
67	Kanjirankulam Bird Sanctuary	Tamil Nadu	2022	97
68	Karaivetti Bird Sanctuary	Tamil Nadu	2023	453.72
69	Longwood Shola Reserve Forest	Tamil Nadu	2023	116.007
70	Nanjarayan Bird Sanctuary	Tamil Nadu	2024	125.865
71	Kazhuveli Bird Sanctuary	Tamil Nadu	2024	5151.6
72	Rudrasagar Lake	Tripura	2005	240
73	Upper Ganga River	Uttar Pradesh	2005	26590
74	Nawabganj Bird Sanctuary	Uttar Pradesh	2019	225
75	Parvati Arga Bird Sanctuary	Uttar Pradesh	2019	722
76	Saman Bird Sanctuary	Uttar Pradesh	2019	526
77	Samaspur Bird Sanctuary	Uttar Pradesh	2019	799
78	Sandi Bird Sanctuary	Uttar Pradesh	2019	309
79	Sarsai NawarJheel	Uttar Pradesh	2019	161
80	Sur Sarovar	Uttar Pradesh	2020	431
81	Haiderpur Wetland	Uttar Pradesh	2021	6908
82	Bakhira Wildlife Sanctuary	Uttar Pradesh	2021	2894
83	Asan Conservation Reserve	Uttarakhand	2020	444
84	East Calcutta Wetlands	West Bengal	2002	12500
85	Sunderbans Wetland	West Bengal	2019	423000

What is Dead Zone?

It is also known as hypoxia, which means reduced level of oxygen in the water. There are 146 dead zones in the world's oceans where marine life could not be supported due to depleted oxygen levels, according to the Global Environment Outlook Year Book, 2003 which is now becoming more than 400 zones. It can occur naturally, but contemporary it is increasing day by day due to the nutrient pollution created by humans.

Reason of the development of Dead Zone

It is caused by an interaction between biological, chemical and physical factors.

1. Increase in chemical nutrients in the water, leading to excessive blooms of algae that deplete underwater oxygen levels
2. Agricultural runoff of Nitrogen and Phosphorous
3. Sewage, vehicular and industrial emissions and even natural factors

What is Carbon Sequestration?

It is the process involved in carbon capture and the long-term storage of atmospheric carbon dioxide. In other words, we can say it is a process of removing carbon dioxide from the Earth's atmosphere and then stored in liquid or solid form. It can be done through the forest and soil conservation practices that enhance the storage of carbon (such as restoring and establishing new forests, wetlands, and grasslands) or reduce CO_2 emissions (such as reducing agricultural tillage and suppressing wildfires).

How Organic Farming helps in Carbon Sequestration?

Organic Farming is complimented the Carbon Sequestration because it focus on sustainability without using synthetic fertilizers, pesticides etc. In this farming, organic mulch play an important role in carbon sequestration. **Organic mulching** refers to covering the soil with any organic matter such as applying compost or farm yard manure over the soil surface followed by adding a layer of dry organic matter over it. This compost contains aligned beneficial microbes, where the dry matter is rich in carbon and the green matter is rich in nitrogenous substances. When decomposition of these components takes place the carbon nitrogen ratio in the soil becomes 10:1, ideal for the proliferation of microbes. This type of farming practices not only improve the soil fertility but also increase farmers' income.

Carbon Footprint.

Carbon footprint can be defined as the total amount of greenhouse gases produced to directly and indirectly support human activities, usually expressed in equivalent tons of carbon dioxide (CO_2). There are two types of carbon foot printing which are :

1. Organisational, 2. Product

It is a very powerful tool to understand the impact of personal behaviour on global warming. **Kyoto Protocol** recognizes six **Green House Gasses (GHGs)** which are given below:

1. Carbon dioxide (CO_2) 2. Methane (CH_4)
3. Nitrous Oxide (N_2O) 4. Hydrofluorocarbons (HFCs)
5. Perfluorocarbon (PFCs) 6. Sulphur hexafluoride (SF_6)

So the release of each of the above gasses would be counted in the Carbon Footprint.

Clean Development Mechanism (CDM) developed by Kyoto Protocol

Steps to reduce carbon footprint

1. By replacing a regular light bulb with a compact fluorescent lamp (CFL) which can save 150 pounds of carbon dioxide each year.
2. Use of public transport and carpool: We can reduce carbon emissions through walking, ridding bike, carpool and using public transport.
3. We can reduce carbon emissions by using recycled products and recycling materials.
4. Planting a tree is always a good idea. A single tree can absorb one ton of carbon dioxide over its lifetime.
5. By using certified energy efficient appliances and equipment will also reduce carbon footprint.
6. Encourage and support greenhouse-reducing practices in community.

India ranks 2nd in Carbon Footprint - 2 516.967 Mt CO_2.

Carbon Offsetting

Carbon Offsetting can be defined as the mitigation of carbon footprints through the development of alternative projects such as solar, wind, tidal energy or reforestation. The emission of greenhouse gases is a global problem and carbon offsets work on the idea that any reduction in any area is worthwhile. It basically measures how much carbon dioxide (CO_2) we have produced just by going through our daily lives. Offsets are measured in tonnes of carbon dioxide equivalent (CO_2e). One tonne of carbon offset represents the reduction of one tonne of CO_2 or its equivalent in other green house gases.

The Global Carbon Budget

The sum of all exchanges (inflows and outflows) of carbon compounds between the earth's carbon reservoirs (such as land mass and atmosphere) in the carbon cycle. It is used to determine, whether a particular pool/reservoir/circle is working as a source of the Carbon Dioxide. It is important to be in line what is scientifically required to keep global warming and climate change tolerable.

Global-warming potential (GWP), is a term used to describe the relative potency, molecule for molecule, of a greenhouse gas, taking account of how long it remains active in the atmosphere.

Global warming potential is the heat absorbed by any greenhouse gas in the atmosphere, as a multiple of the heat that would be absorbed by the same mass of carbon dioxide. GWP is 1 for CO_2. For other gases it depends on the gas and the time frame. Carbon dioxide equivalent is calculated from GWP

Chlorofluorocarbons (CFCs), hydrofluorocarbons (HFCs), hydrochlorofluorocarbons (HCFCs), perfluorocarbons (PFCs), and sulfur hexafluoride (SF_6) are sometimes called high-GWP gases because, for a given amount of mass, they trap substantially more heat than CO_2. (The GWPs for these gases can be in the thousands or tens of thousands).

Species (Time	**Chemical formula**	**Lifetime (years)**	**Global Warming Potential Horizon)**		
			20 years	**100 years**	**500 years**
Carbon dioxide	CO_2	variable	1	1	1
Methane	CH_4	12 ± 3	56	21	6.5
Nitrous oxide	N_2O	120	280	310	170
HFC-23	CHF_3	264	9100	11700	9800
HFC-32	CH_2F_2	5.6	2100	650	200
HFC-41	CH_3F	3.7	490	150	45
HFC-43-10mee	$C_5H_2F_{10}$	17.1	3000	1300	400

HFC-125	C_2HF_5	32.6	4600	2800	920
HFC-134	$C_2H_2F_4$	10.6	2900	1000	310
HFC-134a	CH_2FCF_3	14.6	3400	1300	420
HFC-152a	$C_2H_4F_2$	1.5	460	140	42
HFC-143	$C_2H_3F_3$	3.8	1000	300	94
HFC-143a	$C_2H_3F_3$	48.3	5000	3800	1400
HFC-227ea	C_3HF_7	36.5	4300	2900	950
HFC-236fa	$C_3H_2F_6$	209	5100	6300	4700
HFC-245ca	$C_3H_3F_5$	6.6	1800	560	170
Sulphur hexafluoride	SF_6	3200	16300	23900	34900
Perfluoromethane	CF_4	50000	4400	6500	10000
Perfluoroethane	C_2F_6	10000	6200	9200	14000
Perfluoropropane	C_3F_8	2600	4800	7000	10100
Perfluorobutane	C_4F_{10}	2600	4800	7000	10100
Perfluorocyclobutane	c-C_4F_8	3200	6000	8700	12700
Perfluoropentane	C_5F_{12}	4100	5100	7500	11000
Perfluorohexane	C_6F_{14}	3200	5000	7400	10700

Ozone depletion potential: Ozone depleting potential is **a measure of how much damage a chemical can cause to the ozone layer compared with a similar mass of trichlorofluoromethane (CFC-11).** CFC-11, with an ozone depleting potential of 1.0, is used as the base figure for measuring ozone depleting potential

- According to IPCC, by the year 2100 the global average surface warming will increase by- **1.1-6.4 °C**
- According to IPCC, by the year 2100 the sea level and occurence of natural disaster will increase by- **10 %**
- National Mission for Sustaining Himalayan Ecosystem was launched in- **2010**
- Term Carbon Trading introduced in the year 2002 by **-Koyoto Protocol**
- Regenerative agriculture term coined by - **Robert Rodale**

(1) **Live stock and rice production** are found to be the main source of GHG emissions in Indian Agriculture. Enteric fermentation-54.6% -Rice cultivation–17.5% and 19.1% from fertilizer use.

(2) **Global greenhouse gas emission** for 2022 is estimated to be 58 Gigatons; the largest annual level ever recorded.

(3) If current trend of growth and demography continues it may touch to 62 GT in **2030.**

(4) One gigatonne is equal to 1,000,000,000 tons.

(5) **Biochar** is a high-carbon, fine grained residue that is produced in pyrolysis. Burning biomass in low oxygen conditions create biochar. Main purpose of creation of biochar is for **carbon sequestration**.

(6) **Global fertilizer** use is projected to increase from around 90 million tonnes in 2000 to more than 150 million tonnes by 2050.

(7) **Global chemical** used for agriculture currently amount to 2 million tons per year with herbicides-47.5% insecticides-29.5%, fungicides, 17.5 % and other chemicals 5.5 %.

(8) **Carbon price** is tool to lower emissions of CO_2 & other greenhouse gases. In theory, carbon price is equal to the social cost of carbon.

Water footprint

Water footprint is an environmental indicator that measures the volume of fresh water (in litres or cubic metres) used throughout the entire production chain of a consumer item or service It can be used to measure the consumption of water for practically anything - from the manufacture of a pair of trousers to the entire consumption of a country, including for example a harvest, or a company's annual activities. The concept of the "water footprint" was conceived in 2002 by Arjen Hoekstra

According to the Water Footprint Network, a water footprint is made up of three types of water depending on where the water comes from:

- **Green water footprint:** is water from precipitation (rain or snow) that is stored in the root zone of the soil and evaporated, transpired or incorporated by plants. It is particularly relevant for agricultural, horticultural and forestry products.
- **Blue water footprint:** is water that has been sourced from surface or groundwater resources and is either evaporated,

incorporated into a product or tipped into the sea. Irrigated agriculture, industry and domestic water use can each have a blue water footprint.

- **Grey water footprint:** is the amount of fresh water required to assimilate pollutants in the production process to meet water quality standards.
- The water footprint, for *e.g,* of a country is measured through four factors as follows:
- **Total consumption volume** There is, a correlation between the Gross Domestic Product (GDP) of a country and its water footprint. The greater the GDP, the bigger the water footprint.

  Consumption patterns: Consumption habits in a country condition its water footprint, especially the consumption of items that require a large amount of water for their production.
- **Climate**: Countries with higher mean temperatures, and therefore, greater evaporation, crops need more water.
- **Agricultural practices:** Countries with low agricultural performance because of inefficiencies have high water footprints.
- India's water footprint — 980 m^3 rank below the global average of 1.243 m^3.
- **India ranks:** Among the various countries of the world, India has the largest total WF at 1047 billion cubic meters per year (BCM/year), followed by China (967 BCM/year) and the USA (826 BCM/year).
- **Water Pollution:** Around 70% of India's water is contaminated, with nearly half of the country's rivers unsafe for drinking or irrigation. This makes India rank 120th out of 122 countries on a Global Water Quality Index 2024.

BASIC INFORMATION RELATED TO AGRICULTURE AND ALLIED SCIENCES IN INDIA (Updated upto 2023)

Average Annual Rainfall	: 1,099 mm
Water Resources of India	
Total geographical area	: 328.762 M ha (~329 M ha) which is 2.45% of global land area
Total precipitation	: 3880 km^3 or 390 ~ 400 BCM
Natural runoff (surface water and ground water)	: 1999.2 km^3(BCM)
Natural recharge	: 432 km^3
Available water	: 1869 + 432 = 2301 km^3 (60%)
Evapotranspiration (ET)	: 3838 – 2301 = 1537 km^3 (40%)
Estimated Utilisable surface potential	: 690 km^3 (37% of the surface water resources)
Utilisable groundwater potential	: **432 km^3**
Total utilizable water flow Irrigation	: **690 + 432 = 1122 km^3**
Available Ground water resources for Irrigation	: **361 km^3**
(National Institute of Hydrology)	
The Generated Rainfall Volume (2022-23)	: 1099.2 mm
Maximum area under Irrigation is in	: Ganga basin
Maximum ave. annual runoff occurs in	: Brahmaputra basin
Gross Irrigated area (2022-23)	: 73 Mha (2022-23) and 109.02 (2022 Provisional)
Net Irrigated area (2022-23)	: 71.6 Mha (74.841 Mha 2022)
Total Irrigation potential	: 108.2 Mha
Total foodgrain production (2022-23)	: 3296.87 Lakh/t.
Total Cropped Area (2022-23)	: 204.76 Mha
Net Cropped area	: 139.90 Mha
Cropping Intensity	: 141.60%
Average number of rainy days	: 57 days
Total degraded land	: 130 Mha
Water resources potential of India	: 186.9 Mha-m

Global current water withdrawal	:	**4600 km³** (availability – 4200 km³)
By **2030**, expected withdrawal	:	**6900 km³** (expected availability will be– 4100 km³)
Expected global water deficit by 2030	:	**40%**
By **2030**, expected withdrawal in India	:	**1500 km³** (expected availability will be – 744 km³)
Expected water deficit by 2030 in India	:	**50%**
India has tength largest forest area in the world	:	7,13,789 sq km
Total forest cover in India (2023)	:	7,13,789 square km (21.71%)
Optimum forest area required	:	At least 33 % of Geographical area
Total % area under forage crops	:	9.13 Mha
Food grain produced during (2022-23)		
Rice	:	2838 kg/hac
Wheat	:	2531 kg/hac
Pulses	:	902 kg/ha
Total area under horticultural crops (2023)	:	28.12 Mha
Total Horticultural crop production (2023-24)	:	355.25 MT
Total fruit production in India (2023-24)	:	113.39 MT
Total Vegetable production in India (2023-24)	:	209.39 MT
Total Milk production in India (2023-24)	:	236.35 MT
Minimum requirement of Milk	:	300 g / person / day
Availability of animal protein + Meat in India diet (2022)	:	18 g / person / day
Maximum area under drip irrigation is in	:	Maharastra>AP>TN
Maximum area under sprinkler irrigation	:	RJ >Karnataka>Gujarat
Fertilizer consumption per unit of cropped area	:	146.7 kg/ha (2022-23)
All-India NPK ratio widened from	:	7.7:3.1:1 during 2021-22 to 11.8:4.6:1 during 2022-23.
Highest fertilizer consumption rate (2022-23)	:	Punjab (191.1 kg/ha) among States
Consumption of total fertilizer nutrients at	:	0.542 million MT during 2022-23 witnessed a decline of 7.8% over 2021-22.
Total fertilizer consumption is maximum in	:	Punjab (36.61 MT in 2022-23)

Annual pesticides use in India (2022-23)	:	52466 MT
Irrigated area in India (2022-23)	:	52% of the total cultivated area.
Maximum area under irrigation is in	:	Punjab (98·7% of cultivated area)
Major source of irrigation in India is	:	Underground water
Number of Electric pumps in India	:	30 million.
Per capita availability of arable land (2019)	:	0·12 ha.
Average size of operational holdings (2015-16)	:	1·08 ha
Maximum operational holding size is in	:	Nagaland (5 ha)
Minimun operational holding size is in	:	Kerala (0·18 ha)
Number of farm operational holdings (2015-16)	:	146.5 million
Per capita water availability (2021)	:	1486 cu.m / person
Percentage of fresh water on the earth	:	2.5%.
Agricultural & Livestock Census conducted	:	Every 5 years.
Highest Population below poverty line (2021)	:	Chhattisgarh (40% of total population)
Maximum no. of persons below poverty line	:	Uttar Pradesh (598·19 per lakh)
Population below poverty line is estimated	:	Every 5 years
Percentage of Agril. exports to total exports	:	11.9 % (in 2021-22)
Percentage of Agril. imports to total imports	:	5.23 % (in 2022-23)
Share of Agril. and Allied sectors	:	21·6 % of GDP at current prices (2022-23)
Share of LIvestock & fisheries	:	5 % of GDP (2022)
Record highest food grains yield production	:	3296.87 Lakh tonnes (2022-23
Lowest production of food grains in last ten years	:	251.54 MT (in 2016-17)
Total area under organic certification in India	:	10.17 Mha (2022-23)

Additional information :

- ❋ India is the Second country after China to have commercialized the hybrid rice technology.
- ❋ Highest number of hybrids in Rice are in use in China followed by India.
- ❋ Rice is approximately 42% of total food grains produced in the country.
- ❋ Incorporation of 'Norin 10' dwarfing gene from CIMMYT improved wheat yields during green revolution time.

- Indian variety HD 2172 has paved way for self sufficiency in food grain production in Sudan.
- HD 2329 wheat variety contributed maximum production during Green revolution in India.
- India is the first country to develop and use hybrids in pearlmillet
- The tobacco contributes the highest excise revenue among agricultural commodities in India.
- India is the only country to have developed and grown hybrid cotton commercially.
- Cotton contributes approximately 33% of the total forex earned by country.
- India contributes 15 per cent of the World's cotton production.
- India is the second largest country after China in production of vegetables.
- India ranks sixth largest in world coffee production.
- Karnataka is largest producer of coffee in the country.
- China is the largest producing and consuming country of tobacco in the world, India is the third highest tobacco producer of tobacco
- India ranks first in production and consumption of Sugar in the world.
- India ranks first in the production of tea.
- India holds fourth place in World cotton production.
- Maharastra is the largest cotton as well as cereals producing area in India.
- A grain, vegetable and fodder crop, the amaranth has high protein (83%) with high lysine content, a good balance of essential amino acids.
- India stands first in Grapes productivity (40 t/ha).
- Crop having maximum area under irrigation : Wheat
- Most extensively grown crop in the world : Mango
- Indias share in world mango production is nearly 65%
- Banana plant gives fruits only once in life.
- Sugar content in Sugarcane available in India is 10-11%.
- Sugar recovery in sugarcane in India is 6-8%.
- In sugarcane red rot is caused by 'fungus'
- White bud of maize is due to deficiency of Zinc.
- Black heart of potato is due to deficiency of Oxygen
- Fruit cracking of tomato is due to deficiency of Boron.

* Potato is unfit for consumption if solanine is greater than 20 mg/ 100 gm
* Protato (nickname) is India's high protein GM potato (NBPGR)
* Catface is important disorder in Tomato
* Powdery mildew disease in mango is caused due to "Odium mangiferae"
* Nepthalene Acetic Acid (NAA) is used to control malformation in mango.
* Damping off in papaya is caused due to Fungus
* Mango is propogated through 'veener grafting'.
* Browning of caluliflower is caused by Boron deficiency.
* Canker disease in cirtus is caused due to "Xanthomònal citrii".
* Redness of apple is due to 'Anthocynin'.
* Red colour in tomato is due to "Lycopin".
* Red colour in Chilli is due to "Capcyanthin".
* Orange colour in Carrots is due to "Carotin".
* Bitterness in Bittergourd (karela) is due to "Memordicocide".
* Pungency in Chilli is due to the presence of "Capsaicin" alkaloid.
* Pungency is Onion is due to the presence of "Allyle Propyl Di-Sulphide".
* Growth hormone helps in enlargement of grape fruit is 'Abscic acid'.
* Bordeaux mixture a fungicide spray is a combination of copper sulphate, Lime and water (in 4:4:50 or 5:5:50 ratio).
* World's first genetically modified non-human primate, monkey, named ANDi.
* Recently Scientists seperated germs in Maize corn for manufacturing contraceptives.
* Minimum Support Prices of various crops are recommended by Agricultural Cost & Price Commission.
* National Agricultural Insurance Scheme (NAIS) or Rashtriya Krishi Bima Yojana is sponsored by General Insurance Corporation.
* Recently Germany has stopped import of textiles from India due to use of Azo Dyes in textile making.
* 'sui generis' used in GATT proposals is a Latin word which means 'a system of its own'.
* Ricetec a company of America has got a marketing patent of 'Basmati Rice'

FREQUENTLY USED TERMS/SUBJECTS, DERIVED FROM AND THEIR MEANING

S.No.	Term/Subject	Derived from	Meaning
1.	Agriculture	Latin	Ager or Agri : Soil or field Cultra: Cultivation
2.	Agronomy	Greek	Agros : Field Nomos: to manage
3.	Atmosphere	Greek	Atmos: Vapour Sphaira: Sphere
4.	Genetics	Greek	Gen: to become or to grow into
5.	Entomology	Greek	Entomon: Cut up or divided into segments
6.	Horticulture	Latin	Hortus: Garden Culture: Cultivation
7.	Insect	Latin	Insectum: Cut up or divided into segments
8.	Pathology	Greek	Pathos: Disease Logos: a treatise
9.	Meteorology	Greek	Meteoro: Air Logy : to study
10.	Monsoon	Arabic	Mausium: Season
11.	Entomology	Greek	Entomon: Insect, Logia: Stydy
12.	Mycology	Greek	Mykos: Fungus Logos: Study of
13.	Cytology	Greek	Cyto : Cell Logos : Study of
14.	Psychology	Greek	Psyche : Soul (life) Logos : Study of
15.	Ecology	Greek	Oikos : House, Dwelling Place Logos : Study of
16.	Soil	Latin	Solum : Ground or earth
17.	Extension	Latin	Ex : Out Tensio : Stretching
18.	Economics	Greek	Eco : Home Nomos : Account

TERMINOLOGY RELATED TO CROP CULTURAL PRACTICES IN INDIA

- *Mixed farming* is a system of farming on a particular farm which includes crop production, raising livestock, poultry, fisheries, bee keeping etc., to sustain and satisfy as many needs of the farmer as possible.
- Crop rotation refers to the sequence of crops grown on the same land in a recurring succession.
- *Mono cropping* or Mono culture refers to growing of only one crop on a piece of land year after year.
- Growing two or more crops on the same piece of land in one calender year is known as *multiple cropping*. It includes inter cropping, mixed cropping and sequence cropping.
- *Inter cropping* is growing two or more crops simultaneously on the same piece of land with a define row pattern.
- *Mixed cropping* is growing of two or more crops simultaneously intermingled without any row pattern.
- Relay cropping refers to growing a second crop between the rows of the main crop which is ready to harvest shortly.
- *Sequence cropping* can be defined as growing of two or more crops in sequence on the same piece of land in a farming year.
- *Strip cropping:* refers to system of growing of few rows of erosion resisting crops and erosion permitting crops in alternate strips on contour (across the slope) with the objective of breaking long slopes to prevent soil loss and runoff.
- *Contour strip cropping*: The erosion permitting crops and erosion resisting crops are grown in alternate strips along the contour.
- *Field strip cropping*: The erosion permitting crops and erosion resisting crops are grown across the general slope irrespective of contour.
- *Wind strip cropping*: The erosion permitting crops and erosion resisting crops are grown across the prevailing wind direction irrespective of contour.

- *Buffer strip cropping*: is practiced in areas having steep slopes and badly eroded soils where strips of permanent cover crops or perennial legumes or grasses or shrubs are alternated with field crops
- *Catch crops:* Crop which is grown as substitute for the main crop that has failed on account of unfavourable condition.
- *Cover crops:* Cover crops are grown primarily to cover the soil and to reduce loss of moisture and erosion by wind and water.
- *Trap crops:* Crop which is grown on boundary of the field (Lady finger in cotton) or as inter crop (Marigold in tomato) for protection against insects, diseases, nematodes etc.
- *Ratooning:* The practice of taking the second crop from previous one is known as ratooning.
- *Nipping:* Act of removal of ear heads from plant stalk.
- *Ley farming* : Growing of grasses on agriculturally usable land in a agri-system and thereafter keeping it as pasture for few years.
- *Ideotype*: refers to plant type in which physiological and morphological characteristics are ideally suited to high production with reliability.
- Cultivation of crops in areas where rainfall is less than 750 mm per annum is called *dry farming.*
- Cultivation of crops in areas receiving rainfall above 750 mm per annum is known as *dry land* farming.
- Cultivation of crops in areas receiving rainfall above 1150 mm per annum is known as *rainfed farming*
- In low rainfall regions (<750 mm/annum) mono cropping is practiced, where as in high rainfall regions (>750 mm/annum) inter cropping is practiced.
- The term "arid" is derived from a Latin word, "arere" which means 'dry'.
- Aridity and drought are due to insufficient water. Aridity is a permanent climatic feature and is theculmination of a number of long term processes, whereas drought is a temporary condition that occurs for a short period due to deficient precipitation.

- *Agro-forestry* is a production technique that combines agriculture and forestry, horticulture or animal husbandry on the same piece of land in order to fully utilize the natural resources.
- *Bonsai* is a Japanese art, where in excess dwarfing of plants done to prepare plants with very less height as compared to free growth of plant.
- High yield varieties of wheat and rice are dwarf in nature due to presence of "Norin" and "Dee-Gee-Woo-Gen" genes respectively.
- The word "cereal" has been derived from Roman Godess "ceres" which means "giver of grains".
- *Parboiling*: hydrothermal treatment of rice grain immediately followed by drying prior to miling to prepare parboiled grains with gelatinized starch with minimum grain swelling followed by slow drying is called parboiling.
- Wheat crop was supposed to be introduced in India by Aryans
- *Triticum Aestivum* is the common bread wheat grown in India
- *Triticum Dicoccum* is the common wheat species used in granular preparations such as 'Upma/daliaya'.
- **Harvest index** is a measure of crop yield, the weight of a harvested product as a percentage of total plant weight of a crop.

$$\textbf{Harvest index} = \frac{\text{Grain yield}}{\text{Biological yield}} \times 100$$

- The energy content of edible oil is much higher (39.8 MJ/kg) than protein (23.9 MJ/kg) or carbohydrate (16.8 MJ/kg)
- **Saturated Fatty Acids** (palmitic and stearic acid) have direct relation with blood cholesterol and coronary heart diseases as it increases low density lipoprotein which is harmful.
- In unsaturated fatty Acids (Oleic, Linoleic, Linolenic and Erucic acids) Linoleic & Linolenic acids are essential fatty acids (not synthesized by human body and are to supplied from outside) and their absence in body would lead to physiological disorders.
- The ratio of oleic to linoleic acid affects the storage ability of edible oil and it should be greater than 1.6 for longer shelf life.
- Consumption of one g of oil is equivalent to 1.66 g of protein and 2.37 g of carbohydrates.

- King of vegetable oil seed crop "Ground Nut"
- Queen of vegetable oil seed crop "Sesame"
- Safflower oil is rich in poly unsaturated fatty acids as linoleic acid up to 78% which play an important role in reducing cholesterol content and is also recommended for heart patients.
- **Methane** is a major end product of anaerobic breakdown of organic matter in wetland rice soils. It is released from these soils to the atmosphere by diffusion and ebullition and through roots and stems of rice plants.
- Methane production is negatively correlated with a soil redox potential and positively correlated with soil temperature, soil carbon content, and rice growth.
- Methane is exclusively produced by methanogenic bacteria that can metabolize only in the absence of free oxygen and at redox potentials of < 150 mV.
- *Delta* represents total depth of water required by a crop.
- *Duty* is the volume of irrigation water required to mature a crop. It is the ratio of irrigated area to the quantity of water used (ha/l).
- *Base period* is the period for which water is supplied to a crop
- Number of days between irrigations during the crop period without rainfall is *irrigation interval.*
- The common water distribution pattern in canal command areas in Northern India is *Warabandhi.*
- Ratio of total cropped area in different seasons to the total land area is known as *'Cropping Intensity '.*
- **Bolting:** When a plant produces flowers or seed prematurely instead of crop. This is usually due to excessive heat and sun exposure.
- **Caryopsis:** The grain or fruit of grasses including cereals.
- **Crop year:** Generally 12 month period from the end of one year's harvest to the next.
- **Pistil:** The female sexual reproductive seed bearing organ of a flower, consisting of an ovary, style and stigma.
- **Stamen:** Male reproductive organ of a flower, consisting of a filament and pollen containing anther

CRITERIA FOR HARVESTING AND STORAGE OF IMPORTANT CROPS

Crop	Criteria for harvesting	Moisture Content % for safe storage of grains
Rice	Moisture Content of grains < 20 %	14
Wheat	Moisture Content of grains 15%	12
Sorghum	Moisture Content of grains < 28%	12
Maize	Moisture Content of grains 22-25 %	12
Sugarcane	Brix 18-20 %	
Cotton	Bolls fully opened	
Groundnut		6

RECOMMENDED SEED RATE/PLANTING MATERIAL OF FIELD AND HORTICULTURE CROPS

Crops	Seed rate (kg/ha)	Crops	Seed rate (kg/ha)
Rice:-		**Gram:-**	
SRI	6-8 kg/ha	Desi	60-80 kg
Hybrid	15-18 kg/ha	Kabuli	80-100
Transplanting	40-50 kg/ha	**Arhar/pigeon pea:-**	
Line sowing	60-80 kg/ha	Normal	12-15 kg
Broadcasting	100-110 kg/ha	For bund	4 kg/ha
Dapog m[illegible]d	**2.5-3.0 kg/m²** **(ADS,CGPSC 2016)**	**Green gram/Mung:-**	
Direct seeded rice	**20-25 kg/ha**	Kharif	12-15
Aerobic rice	**40-50 kg/ha**	**Summer**	**20-25 kg**
Wheat:-		**Black gram/Urd/mash:-**	
Normal	100-125	Kharif	20-25 kg
Late sown/Rained condition	25 % more than normal condition	**Summer**	30-35 kg
Dibbler	25-30	**Pea:-**	75-80
FIRB	70-80	**Lathyrus:-**	

SWI (System of Wheat Intensification)	**20-25**	Line sowing	40-50
Maize:-		**Utera**	**70-80**
Normal grain/QPM	20 kg	**Lentil:-**	
Sweet corn:	**8.0 kg**	Sole crop	30-40 kg
Pop corn	**12 kg**	**Utera**	60 kg
Baby corn	**25 kg**	**Rajma/frenchbean**	100-120 kg
Forage crop	40-50 kg	**Horse bean/Kulthi**	25 kg
Hybrid	20-25 kg	**Mothbean**	10-15 kg
Composite	15-18 kg	**Groundnut:-**	
Sorghum:-		**Bunch variety**	100-120 kg
Irrigated Transplanted :	7.5 kg/ha	**Spreading variety**	80-100 kg
Rainfed (Direct sown	15 kg/ha	**Soybean**	75-80 kg
Average	12-15 kg/ha	**Rapeseed &Mustard**	
Bajra/Pearl millet	4-5 kg	Sole crop	4-6 kg
Shoof fly endemic area	**12.5 kg**	**Intercropping**	**2-3 kg**
Transplanting	**3.75 kg**	**Sunflower Normal**	**8-10 kg**
Barley:-		**For Hybrid**	**5-6 kg (IMP)**
Irrigated	80-100 kg	**Safflower**	15 kg
Rain fed	100 kg	**Linseed/Flax**	30-40 kg
Oat:-		**For Dual purpose (Oil + Flax)**	**35-45 kg**
Grain	80 kg	**Castor**	10 kg
Fodder	100 kg	**Til**	3-4 kg
Rye (Secle cereal)		For intercropping Til	1.0kg
Grain	60 kg	**Niger**	5-8 kg
Fodder	80 kg	**Berseem:-**	
Triticale	80-110 kg	Diploid	25-30 kg
Finger millet		Tetraploid	35 kg

Transplanting (3-4 week old seedling-SADO cg vyapam 2016)	**5 kg**	**Lucern**	**20-25 kg**
Line sowing	8-10 kg	**Napier grass:-**	
Broadcasting	10-12 kg	Rooted slip	35000-40000 (slip)
Kodo:-	6-8 kg		
Cotton:-		**Jute:-**	
Desi	10-12kg/ha	White jute	6-8
American	18-20	Tosa jute	4-6
Hybrid	2-3 kg	**Sunhemp**	15-20 kg
Bt cotton	1.5 kg	For green manuring	60 kg
Mesta:-	10-12 kg		
Tobacco:-	2.5-3.0 kg		
Sugarcane:-			
3 budded	35000-40000 setts/ha		
2 budded	80000 setts/ha		
1 budded	120000 setts/ha		
For SSI (Sustainable Sugarcane Initiative)	**5000 setts/ha**		

INFORMATION FROM AGRICULTURAL AND ALLIED SCIENCES

Genetics & Plant Breeding:

- The term Genetics was coined by : ***Bateson, W.***
- Life was first originated in water
- The theory of genes in the chromosomes was introduced by : ***Morgan***
- The theory of acquired characteristics was proposed by : ***Lamark***
- The theory of inheritance was proposed by : ***G. Mendel***
- Mendel's laws of inheritance were rediscovered in 1900 by : ***de Varies & Tschermak***
- The theory of natural selection was proposed by : ***Charles Darwin***
- The term heterosis was first used by : ***Mendel***
- Hugo de Vries defined the term ***Mutation.***
- ***Carolus Linnaeus*** recognised the importance of sex in plants
- The physical basis of life is ***'protoplasm'.***
- Scientific name of bread wheat is " Triticum durum'.
- Cytoplasmic genetic male sterility is exploited ***in hybrid seed production.***
- A single gene affecting more than one character is known as ***'Pleiotorpy'.***
- Ginger & Turmeric are examples of modified ***stem rhizome.***
- Onion is a modified ***form of leaf.***
- Onion & Garlic are examples of ***bulb***
- Edible part of apple : ***Thalamus***
- Edible part of mango : ***Mesocarp.***
- Edible part of coconut : ***Endosperm***
- The inflorescence of Banana is known as ***'Spadix'.***
- Development of Individual from an unfertilized egg is known as ***'Parthenogenesis'.***
  or it is a natural form of asexual production
- A single gene affecting more than one character is known as ***'Pleiotropy'.***
- ***Meiosis:*** Cell division in which one cell with two set of chromosomes gives rise to four cells with one set of chromosomes each.
- ***Mitosis:*** Cell division in which one cell gives rise to two cells of genetically identical to the cell which produced them.

- ***Heterosis:*** Occurrence of offspring more vigorous than either parent.
- ***Heterobeltosis:*** A form of heterosis where hybrid is superior to its best parents.
- ***Allogamy or Cross Pollination:*** Pollen grains from flowers of one plant transferred to receptive stigmas of flowers of another plant.
- ***Autogamy or Self Pollination:*** When pollen grains from an anther falls on receptive stigma of the same flower.
- ***Chasmogamy:*** The flower opens but only after the pollination has taken place. *Ex.* Mung.
- ***Cleistogamy:*** Flower does not open at all, ensuring complete self pollination. *Ex.* Wheat
- ***Geitonogamy:*** Pollen grains from flowers of one plant fall on the receptive stigams of other flower of the same plant. Ex. Maize.
- ***Breeder's seed:*** Seed which is directly controlled by breeding institution or individual and which has the maximum genetic purity.
- ***Pedigree record:*** Record that is used for maintaing relationship between parents and offspring.
- ***Monoecious:*** Male and Female reproductive structures are found on the same plant but in separate flowers.
- ***Dioecious:*** Male and Female reproductive structures are found on separate plants.
- Robert Brown discovered ***nucleus of cell.***
- During metaphase the structure of chromosome can be best observed.
- ***Examples of C_3 type Plants :*** Rice, Wheat, Cotton, Soybean, Potato etc.
- ***Examples of C_4 type Plants :*** Maize, Sorghum, Pearl millet, Sugarcane etc.
- **Examples of CAM** (*Crassulaceae Acid Metabolism*) type plants: Pineapple Sisal, Cactus etc.
- C_3 plants have lower water use efficiency
- C_4 plants possess **Kranz anatomy**
- CAM plants perform double ***CO_2 fixation.***
- In CAM plants ***transpiration is negligible*** since stomata are closed during the day.
- C_4 plants uses ***low concentration*** of CO_2.
- ***Light saturation*** is high in C_4 plants as compared to C_3 plants.

Latest India Ranking in Different Indexes 2024

Index	Published by	Topped by	India's Rank in World
Global Innovation Index 2024	World Intellectual Property Organization (WIPO)	Switzerland Sweeden, U.K.	39th
Climate Change Performance Index 2024	Bonn-based Environmental think tank Germanwatch	Rank 4: Denmark (The top 3 positions are unoccupied)	7th
World Happiness Report 2023	United Nations Sustainable Development Solution Network	Finland, Denmark, Iceland and Isreal	126th
Global Terrorism Index 2024	Institute for Economics and Peace (IEP)	Afghanistan	14th
Global Environment Performance Index 2024	Yale and Columbia University Researchers	Estonia Luxembourg	176 th
Human Development Index 2023-24	United Nations Development Program's (UNDP)	Switzerland	134th
Global Hunger Index 2023	Irish aid agency Concern Worldwide and German organisation Welt Hunger Hilfe	17 nations in all, with Belarus, Chile, China, and Croatia sharing first place	111th
World Press Freedom Index 2024	Reporters Without Borders	Norway, Denmark Sweeden	159
Global Energy Transition Index	World Economic Forum in collaboration with Accenture	Sweden Denmark	63rd

Global Crypto Adoption Index 2024	Block chain Analytic	India, Nigeria	1st
Global Gender Gap Report 2024	World Economic Forum	Iceland	129th
International Intellectual Property Index 2024	US Chamber of Commerce Global Innovation Policy Centre	United States, United Kingdom, and Germany	42rd
Global Manufacturing Risk Index 2024	US-based property consultant Cushman Wakefield	China	42rd
Cryptocurrency Gains by Country Index 2024	Crypto Analytics Firm Chainalysis	USA UK	21
Global Skills Index 2024	Salesforce	Switzerland Japan, Germany	87th
The Global Liveability Index 2024	Economic Intelligence Unit's (EIU)	Vienna Austria	141st
Commitment to Reducing Inequality Index (CRII) 2024	Oxfam International and Development Finance International (DFI)	Norway Germany Australia	123
Corruption Perception Index 2023	Transparency International	Denmark	93

Indian Institute of Natural Resins and Gums, Ranchi (Renamed as National Institute of Secondary Agriculture) Sept. 2024. Originally established as the Indian Lac Research Institute established in 1924 an is one of the oldest institutions within the ICAR

SOIL SCIENCE

Soil types:

Alifsols	: Relatively Young soils, acid soils
Andisols	: Volcanic soils
Ardisols	: Desert soils
Entisols	: Alluvial soils (Recently formed soils)
Histosols	: Organic soils
Inceptiosols	: Young soils (covers maximum % of total area of India)
Mollisols	: Prairie soils
Oxysols	: Highly weathered soils
Ultisols	: Low nutrient soils
Vertisols	: Black soils, Swelling clay soils

- **Igneous Rocks** : Granite, Basalt and Syenite (consists 95% on earth crust)
- **Sedimentary rocks** : Lime stone, Sand stone, Dolomite, Peat, Lignite
- **Metamorphic rocks** : Gneiss Marble, Quartzite and Slate
- **Igneous rocks** are the source of parent material for other rocks.
- **Sedimentary rocks** are formed from the breaking down of pre-existing rocks, they are also known as stratified rocks or aqueous rocks
- Four stages of sedimentary rocks formation is **i) weathering; ii) transportation; iii) deposition; iv) digenesis.**
- Gneiss from Granite; Marble from Lime stone; Quartzite from Sand stone; Slate from Shales.
- Mineral is a naturally occurring, homogenous element or inorganic compound that has a definite chemical composition and a characteristic geometric form.
- The primary minerals are those which are formed owing to the crystallization of the molten magma
- The secondary minerals are formed at the earth's surface by the weathering of the pre-existing primary minerals under variable conditions of temperature and pressure.
- Parent material is the unconsolidated and more or less chemically weathered mineral or organic matter, from which the solum of soils is developed by pedogenic processes.
- Relief or topography relates to the configuration of the land surface and is described in terms of difference in elevation/slope
- Empty space between soil particles is called : Pore space

- Bulk density is the mass of a unit volume of dry soil in its natural structure.
- Soil texture refers to the relative proportions of various types of soil particles.
- Most preferable soil structure : Granular & Crumb
- Average particle density : 2·67 g/cc(2·60 to 2·75)
- Optimum bulk density: 1·2 to 1·6 g/cc
- Relationship between particle density & bulk density: Porosity = (1 – Bulk density / Particle density) × 100
- Pore space : Clay soils higher than sandy soils
- Bulk density of sub soils is higher as compared to surface soils
- On addition of organic matter bulk density decreases
- According to Stokes' Law the terminal velocity of a spherical particle settling under the influence of gravity in a fluid of a given density and viscosity is proportional to the square of the particle's radius
- Dominant soil group in India : **Alluvial**; (The only Azonal soil)
- Alluvial soils are most recently formed soils
- In red soils the dominant clay mineral is Kaolinite
- The Red soils are red in colour due to presence of **Iron Oxide**
- Best Agreculture soils in the world are - **Mollisol**
- Most important soil for Agreculture - **Alluvial soil**
- **Red soils** are best suitable for Irrigated agriculture
- Organic matter is responsible for the dark brown to very dark grey colour of soil.
- Black soils are best suitable for dry land agriculture
- In Black soils the dominant clay mineral is **Montmorllonite**
- Black cotton soils are deficient in Nitrogen
- The particles of size greater than 0.84 mm in diameter are generally not eroded by wind but they are easily eroded by water.
- **Eutrophication** refers to the process of enrichment of watercourses or surface water bodies by inorganic plant nutrients.
- The first experiment aimed at elucidating the increase in the weight of plant during its growth was reported by Nicholas (1401-1446)

- Jean Baptiste Boussingault (1802-1882) was called as 'father of field plot technique'
- Justus von Liebig (1803-1873) put forth the 'law of minimum' which states that the yield is governed by the limiting nutrient and is directly proportionalto the factor which is minimum in the soil. He observed that the Crop yields were increased by adding minerals or inorganic elements to the soil.
- *Pedology* is the study of soil as a natural body and deals with its genesis, survey and classification.
- *Edaphology* is the science deals with the influence of soil on the living things such as human beings, plants etc.
- *Atterberg's limits* are used to relate soil water content limits and various states of soil consistency
- Soil consistence is a dynamic physical property of soils, which varies with the variation of soil moisture and applied stress.
- *Liquid limit* is the water content at which soils is practically liquid but possesses a small shearing strength. This is also termed as upper plastic limit.
- *Plastic limit* is the minimum water content at which a soil is plastic and the soil threads start to crumble. This is also known as lower plastic limit. In other words at which a fine-grained soil can no longer be remolded without cracking.
- *Plastic limit* is the lowest moisture content at which a soil can be deformed without cracking.
- *Plastic limit* is the upper limit of moisture content for tillage operation for most crops except rice.
- *Shrinkage limit* is the water content of soil at which soil transforms from the semi-solid state to the solid state.
- *Soil plasticity* is defined as the property which enables a clay / soil to take-up water, to form a mass that can be deformed in to any desirable shape and to maint ain the shape after the deformation pressure is removed.
- Soils with bulk density less than 1.0 are considerd as **problematic soils.**
- Waterlogging is serious when hydraulic conductivity is less than **2.5 cm/hr.**

SOIL AS A MEDIUM OF PLANT GROWTH

Soil is a crucial medium for plant growth, providing physical support, nutrients, water, and aeration to plants.

Composition of Soil: 1. Mineral Constituents:

- **Sand:** Large particles, providing good aeration and drainage.
- **Silt:** Medium-sized particles, contributing to soil fertility.
- **Clay:** Small particles, helping to retain water and nutrients.

2. Inorganic Components

- **Minerals:** Essential elements like nitrogen, phosphorus, potassium, sulphur, calcium, magnesium, etc.

Cations and Anions: Charged particles that play a role in nutrient availability.

Mineral Particles: • Quartz • Feldspar • Mica:

Organic Constituents:

1. Humus:

• Decomposed Organic Matter • Rich in Nutrients • Improves Soil Structure

2. Microorganisms:

• Bacteria • Fungi • Actinomycetes:

3. Plant and Animal Residues:

• Roots • Leaves and Stems • Animal Residues

4. Soil Fauna:

• Earthworms • Insects and Arthropods

5. Acids:

• Humic Acids • Fulvic Acids

Nutrient Elements:

1. Macroelements: • Nitrogen (N), Phosphorus (P), Potassium (K): Essential for plant growth.

2. Micronutrients: • **Iron (Fe), Zinc (Zn), Copper (Cu), Manganese (Mn), Boron (B), Molybdenum (Mo), Chlorine (Cl), Nickel (Ni):**

Required in smaller quantities but crucial for plant development.

Water:

- **Soil Moisture:** Water held in the soil pores.
- **Groundwater:** Water present below the soil surface.
- **Capillary Water:** Water held against gravity in the soil.

Air:

- **Pores:** Spaces between soil particles that contain air.
- **Oxygen and Carbon Dioxide:** Gases essential for root respiration and microbial activity.

- **Soil Aeration:** The exchange of gases between soil and the atmosphere.

Microorganisms:

- **Bacteria, Fungi, and Protozoa:** Microorganisms that decompose organic matter and contribute to nutrient cycling.
- **Nematodes and Earthworms:** Soil-dwelling organisms that influence soil structure and nutrient availability.
- **Plant Roots:** Extract nutrients and water from the soil.
- **Soil Moisture:** Water held in the soil pores.
- **Groundwater:** Water present below the soil surface.
- **Capillary Water:** Water held against gravity in the soil

Mineral And Organic Constituents (Role and Impact)

The mineral and organic constituents of soil play crucial roles in supporting crop production by influencing soil fertility, structure, and overall health.

Mineral Constituents:

1. Sand, Silt, and Clay:

- **Role:** Determine soil texture, influencing water retention, drainage, and aeration.
- **Impact on Crop Production:** Proper texture ensures an optimal balance of water and air for root development.

2. Mineral Particles (Quartz, Feldspar, Mica):

- **Role:** Contribute to the overall mineral composition of the soil.
- **Impact on Crop Production:** Influence nutrient availability, especially potassium from feldspar.

Organic Constituents:

1. Humus:

- **Role:** Improves soil structure, water retention, and nutrient availability.
- **Impact on Crop Production:** Enhances soil fertility, promotes microbial activity, and provides a stable reservoir of nutrients for plant uptake.

2. Microorganisms (Bacteria, Fungi, Actinomycetes):

- **Role:** Decompose organic matter, fix nitrogen, and participate in nutrient cycling.
- **Impact on Crop Production:** Improve soil fertility by breaking down organic material and making nutrients available to plants.

3. Plant and Animal Residues:

- **Role:** Contribute to organic matter content.

- **Impact on Crop Production:** Decomposed residues release nutrients, enhance soil structure, and support microbial activity.

4. Soil Fauna (Earthworms, Insects, Arthropods):

- **Role:** Aerate the soil, break down organic matter, and contribute to nutrient cycling.
- **Impact on Crop Production:** Improve soil structure, nutrient availability, and overall soil health.

5. Humic and Fulvic Acids:

- **Role:** Enhance nutrient availability, transport, and soil structure.
- **Impact on Crop Production:** Improve nutrient uptake by plants, promote root development, and enhance overall soil fertility.

Overall Impact of Nutrients on Crop Production:

1. Soil Fertility:

- Proper levels of nutrients, both macro and micronutrients, contribute to soil fertility, supporting healthy crop growth.

2. Water Retention and Drainage:

- Soil texture influences water retention and drainage, affecting the availability of water to crops.

3. Soil Structure:

- Organic constituents, such as humus and soil fauna, improve soil structure, promoting root penetration and aeration.

4. Microbial Activity:

- Microorganisms contribute to nutrient cycling, organic matter decomposition, and disease suppression, supporting plant health.

CHEMICAL, PHYSICAL AND BIOLOGIAL PROPERTIES OF SOIL

Chemical Properties of Soil:

1. pH (Acidity/Alkalinity):

- pH levels influence nutrient availability to plants.
- Most plants prefer slightly acidic to neutral soils (pH 6-7).

2. Nutrient Content:

- Essential nutrients include nitrogen, phosphorus, potassium, calcium, magnesium, sulphur, and micronutrients.
- Nutrient availability is influenced by soil Ph and organic matter.

3. Cation Exchange Capacity (CEC):

- The soil's ability to hold and exchange essential cations (positively charged ions) with plant roots.
- CEC is influenced by clay and organic matter content.

4. Soil Organic Matter (SOM):
- Decomposed plant and animal residues in various stages.
- Enhances soil structure, water retention, and nutrient availability.

5. Soil Texture:
- Relative proportions of sand, silt, and clay particles.
- Influences water retention, drainage, and nutrient availability.

Physical Properties of Soil:

1. Texture:
- Determines the soil's ability to retain water and nutrients.
- Sandy soils have larger particles and drain quickly, while clayey soils retain water.

2. Structure:
- Arrangement of soil particles into aggregates or clumps.
- Affects porosity, aeration, and water movement.

3. Porosity:
- Percentage of open spaces (pores) in the soil.
- Influences water retention and drainage.

4. Bulk Density:
- Mass of soil per unit volume.
- High bulk density can indicate soil compaction, limiting root growth.

5. Water Holding Capacity:
- Amount of water a soil can retain against gravitational forces.
- Influenced by soil texture and organic matter content.

Biological Properties of Soil:

1. Microorganisms:
- Bacteria, fungi, protozoa, and other microbes play crucial roles in nutrient cycling and organic matter decomposition.
- Earthworms, insects, and other larger organisms contribute to soil structure and nutrient cycling.

2. Roots:
- Plant roots help stabilize the soil and contribute to organic matter through root exudates.

3. Soil Fauna:
- Nematodes, mites, and other organisms contribute to nutrient cycling and decomposition.

4. Biodiversity:
- Diverse soil ecosystems contribute to overall soil health and resilience.

Table-1: Production of Important Crops in three Largest Producing States in 2023-24*

(Production in Million Tonnes)

Crops/Groups of Crops	States	Production	Per cent Share of Production to All India	Cumulative per cent Share of Production
I. Foodgrains				
Rice	Telangana	16.63	12.17	12.17
	Uttar Pradesh	15.72	11.50	23.67
	West Bengal	15.12	11.06	34.73
Wheat	Uttar Pradesh	35.43	31.38	31.38
	Madhya Pradesh	21.28	18.84	50.22
	Punjab	17.78	15.75	65.97
Maize	Karnataka	5.49	15.39	15.39
	Bihar	4.61	12.93	28.32
	Madhya Pradesh	4.33	12.14	40.46
Total Nutri/ Coarse Cereals	Rajasthan	8.03	14.66	14.66
	Karnataka	7.61	13.90	28.56
	Madhya Pradesh	5.49	10.02	38.58
Tur	Karnataka	1.02	30.13	30.13
	Maharashtra	0.86	25.41	55.54
	Uttar Pradesh	0.38	11.23	66.77
Gram	Madhya Pradesh	3.19	27.52	27.52
	Maharashtra	2.86	24.72	52.24
	Rajasthan	2.23	19.31	71.55
Total Pulses	Madhya Pradesh	6.18	25.23	25.23
	Maharashtra	4.00	16.33	41.56
	Rajasthan	3.63	14.83	56.39
Total Foodgrains	Uttar Pradesh	59.29	18.03	18.03
	Madhya Pradesh	39.84	12.12	30.15
	Punjab	32.59	9.91	40.06

Crops/Groups of Crops	States	Production	Per cent Share of Production to All India	Cumulative per cent Share of Production
II. Oilseeds				
Groundnut	Gujarat	4.64	45.12	45.12
	Rajasthan	2.02	19.59	64.71
	Madhya Pradesh	0.99	9.63	74.34
Rapeseed & Mustard	Rajasthan	5.98	45.40	45.40
	Uttar Pradesh	1.87	14.24	59.64
	Madhya Pradesh	1.75	13.28	72.92
Soyabean	Madhya Pradesh	5.47	41.92	41.92
	Maharashtra	5.23	40.09	82.01
	Rajasthan	1.17	8.96	90.97
Sunflower	Karnataka	0.07	40.72	40.72
	Haryana	0.03	17.37	58.09
	Odisha	0.02	13.77	71.86
Total Oilseeds	Rajasthan	9.57	24.17	24.17
	Madhya Pradesh	8.37	21.15	45.32
	Gujarat	7.19	18.15	63.47
Sugarcane	Uttar Pradesh	205.56	46.45	46.45
	Maharashtra	112.09	25.33	71.78
	Karnataka	41.81	9.45	81.23
Cotton@	Gujarat	9.06	27.86	27.86
	Maharashtra	8.05	24.74	52.60
	Telangana	5.08	15.62	68.22
Jute & Mesta$	West Bengal	7.87	80.97	80.97
	Bihar	0.99	10.15	91.12
	Assam	0.68	7.05	98.17

Source: ES&E Division, Department of Agriculture and Farmers Welfare.

@ Production in million bales of 170 kg. each. $ Production in million bales of 180 kg. each.

* As per 3rd Advance Estimates

* From the year 2022-23, for agricular and Commercial Crops, the rabi season has been seggregatged into Rabi & Sum